Embedded Software

Newnes Know It All *Series*

PIC Microcontrollers: Know It All

Lucio Di Jasio, Tim Wilmshurst, Dogan Ibrahim, John Morton, Martin Bates, Jack Smith, D.W. Smith, and Chuck Hellebuyck

ISBN: 978-0-7506-8615-0

Embedded Software: Know It All

Jean Labrosse, Jack Ganssle, Tammy Noergaard, Robert Oshana, Colin Walls, Keith Curtis, Jason Andrews, David J. Katz, Rick Gentile, Kamal Hyder, and Bob Perrin

ISBN: 978-0-7506-8583-2

Embedded Hardware: Know It All

Jack Ganssle, Tammy Noergaard, Fred Eady, Lewin Edwards, David J. Katz, Rick Gentile, Ken Arnold, Kamal Hyder, and Bob Perrin

ISBN: 978-0-7506-8584-9

Wireless Networking: Know It All

Praphul Chandra, Daniel M. Dobkin, Alan Bensky, Ron Olexa, David Lide, and Farid Dowla

ISBN: 978-0-7506-8582-5

RF & Wireless Technologies: Know It All

Bruce Fette, Roberto Aiello, Praphul Chandra, Daniel Dobkin, Alan Bensky, Douglas Miron, David Lide, Farid Dowla, and Ron Olexa

ISBN: 978-0-7506-8581-8

For more information on these and other Newnes titles visit: **www.newnespress.com**

Embedded Software

Jean Labrosse
Jack Ganssle
Tammy Noergaard
Robert Oshana
Colin Walls
Keith Curtis
Jason Andrews
David J. Katz
Rick Gentile
Kamal Hyder
Bob Perrin

ELSEVIER

AMSTERDAM • BOSTON • HEIDELBERG • LONDON
NEW YORK • OXFORD • PARIS • SAN DIEGO
SAN FRANCISCO • SINGAPORE • SYDNEY • TOKYO
Newnes is an imprint of Elsevier

Newnes

Newnes is an imprint of Elsevier
30 Corporate Drive, Suite 400, Burlington, MA 01803, USA
Linacre House, Jordan Hill, Oxford OX2 8DP, UK

Recognizing the importance of preserving what has been written, Elsevier
prints its books on acid-free paper whenever possible.

Library of Congress Cataloging-in-Publication Data
Embedded software/Jean Labrosse . . . [et al.]. – 1st ed.
　　p. cm. – (Newnes know it all series)
　Includes bibliographical references and index.
　　ISBN-13: 978-0-7506-8583-2 (pbk. : alk. paper)　　1.　Embedded computer systems–
Programming.　　2.　Computer software–Development.　　I.　Labrosse, Jean J.
　TK7895.E42E588 2008
　005.26–dc22

　　　　　　　　　　　　　　　　　　　　　　　　　　　　2007023369

British Library Cataloguing-in-Publication Data
A catalogue record for this book is available from the British Library.

ISBN: 978-0-7506-8583-2

For information on all Newnes publications
visit our Web site at www.books.elsevier.com

07　08　09　10　　9　8　7　6　5　4　3　2　1

Printed in the United States of America

Contents

About the Authors .... *x*

Introduction .... *xiii*

Chapter 1: Basic Embedded Programming Concepts *1*

  1.1 Numbering Systems 2
  1.2 Signed Binary Numbers 5
  1.3 Data Structures 13
  1.4 Communications Protocols 29
  1.5 Mathematics 37
  1.6 Numeric Comparison 46
  1.7 State Machines 59
  1.8 Multitasking 74

Chapter 2: Device Drivers *85*

  2.1 In This Chapter 85
  2.2 Example 1: Device Drivers for Interrupt-Handling 89
  2.3 Example 2: Memory Device Drivers 110
  2.4 Example 3: Onboard Bus Device Drivers 134
  2.5 Board I/O Driver Examples 143
  2.6 Summary 168

Chapter 3: Embedded Operating Systems *169*

  3.1 In This Chapter 169
  3.2 What Is a Process? 175
  3.3 Multitasking and Process Management 177
  3.4 Memory Management 213
  3.5 I/O and File System Management 230
  3.6 OS Standards Example: POSIX (Portable Operating System Interface) 232
  3.7 OS Performance Guidelines 235
  3.8 OSes and Board Support Packages (BSPs) 237
  3.9 Summary 239

Chapter 4: Networking **241**
  4.1   Introduction to the RCM3200 Rabbit Core 243
  4.2   Introduction to the Dynamic C Development Environment 244
  4.3   Brief Introduction to Dynamic C Libraries 246
  4.4   Memory Spaces in Dynamic C 247
  4.5   How Code Is Compiled and Run 256
  4.6   Setting Up a PC as an RCM3200 Development System 259
  4.7   Time to Start Writing Code! 259
  4.8   Embedded Networks 274
  4.9   Dynamic C Support for Networking Protocols 275
  4.10 Typical Network Setup 279
  4.11 Setting Up a Core Module's Network Configuration 282
  4.12 Project 1: Bringing Up a Rabbit Core Module for Networking 288
  4.13 The Client Server Paradigm 293
  4.14 The Berkeley Sockets Interface 294
  4.15 Using TCP versus UDP in an Embedded Application 298
  4.16 Important Dynamic C Library Functions for Socket Programming 300
  4.17 Project 2: Implementing a Rabbit TCP/IP Server 303
  4.18 Project 3: Implementing a Rabbit TCP/IP Client 311
  4.19 Project 4: Implementing a Rabbit UDP Server 322
  4.20 Some Useful (and Free!) Networking Utilities 328
  4.21 Final Thought 331

Chapter 5: Error Handling and Debugging **333**
  5.1   The Zen of Embedded Systems Development and Troubleshooting 333
  5.2   Avoid Debugging Altogether—Code Smart 340
  5.3   Proactive Debugging 341
  5.4   Stacks and Heaps 342
  5.5   Seeding Memory 344
  5.6   Wandering Code 346
  5.7   Special Decoders 347
  5.8   MMUs 348
  5.9   Conclusion 349
  5.10 Implementing Downloadable Firmware with Flash Memory 350
  5.11 The Microprogrammer 350
  5.12 Advantages of Microprogrammers 351
  5.13 Disadvantages of Microprogrammers 351
  5.14 Receiving a Microprogrammer 352
  5.15 A Basic Microprogrammer 354
  5.16 Common Problems and Their Solutions 355
  5.17 Hardware Alternatives 362
  5.18 Memory Diagnostics 364
  5.19 ROM Tests 365

5.20 RAM Tests.. 367
5.21 Nonvolatile Memory 372
5.22 Supervisory Circuits.. 372
5.23 Multibyte Writes.. 374
5.24 Testing 378
5.25 Conclusion 378
5.26 Building a Great Watchdog................................ 379
5.27 Internal WDTs 382
5.28 External WDTs 384
5.29 Characteristics of Great WDTs 386
5.30 Using an Internal WDT 389
5.31 An External WDT 391
5.32 WDTs for Multitasking 393
5.33 Summary and Other Thoughts............................ 395

Chapter 6: Hardware/Software Co-Verification **399**
  6.1 Embedded System Design Process 399
  6.2 Verification and Validation 401
  6.3 Human Interaction... 403
  6.4 Co-Verification 405

Chapter 7: Techniques for Embedded Media Processing **443**
  7.1 A Simplified Look at a Media Processing System 445
  7.2 System Resource Partitioning and Code Optimization 451
  7.3 Event Generation and Handling 452
  7.4 Programming Methodology................................. 455
  7.5 Architectural Features for Efficient Programming 456
  7.6 Compiler Considerations for Efficient Programming 465
  7.7 System and Core Synchronization 472
  7.8 Memory Architecture—the Need for Management 476
  7.9 Physics of Data Movement 488
  7.10 Media Processing Frameworks 495
  7.11 Defining Your Framework................................. 497
  7.12 Asymmetric and Symmetric Dual-Core Processors 505
  7.13 Programming Models 507
  7.14 Strategies for Architecting a Framework 510
  7.15 Other Topics in Media Frameworks 523

Chapter 8: DSP in Embedded Systems **529**
  8.1 Overview of Embedded Systems and Real-Time Systems 536
  8.2 Real-Time Systems... 536
  8.3 Hard Real-Time and Soft Real-Time Systems 537
  8.4 Efficient Execution and the Execution Environment 541

8.5   Challenges in Real-Time System Design 542
8.6   Summary 553
8.7   Overview of Embedded Systems Development Life Cycle
      Using DSP 554
8.8   The Embedded System Life Cycle Using DSP 554
8.9   Optimizing DSP Software 580
8.10 What Is Optimization? 580
8.11 The Process 581
8.12 Make the Common Case Fast 584
8.13 Make the Common Case Fast—DSP Architectures 584
8.14 Make the Common Case Fast—DSP Algorithms 587
8.15 Make the Common Case Fast—DSP Compilers 588
8.16 An In-Depth Discussion of DSP Optimization 595
8.17 Direct Memory Access 595
8.18 Using DMA 596
8.19 Loop Unrolling 604
8.20 Software Pipelining 610
8.21 More on DSP Compilers and Optimization 620
8.22 Programmer Helping Out the Compiler 633
8.23 Profile-Based Compilation 646
8.24 References 653

Chapter 9: Practical Embedded Coding Techniques 655
9.1   Reentrancy 655
9.2   Atomic Variables 656
9.3   Two More Rules 658
9.4   Keeping Code Reentrant 659
9.5   Recursion 661
9.6   Asynchronous Hardware/Firmware 661
9.7   Race Conditions 662
9.8   Options 664
9.9   Other RTOSes 665
9.10 Metastable States 666
9.11 Firmware, Not Hardware 668
9.12 Interrupt Latency 671
9.13 Taking Data 674
9.14 Understanding Your C Compiler: How to Minimize
      Code Size 677
9.15 Modern C Compilers 677
9.16 Tips on Programming 687
9.17 Final Notes 695
9.18 Acknowledgments 696

Chapter 10: Development Technologies and Trends *697*
  10.1 How to Choose a CPU for Your System on Chip Design.......................697
  10.2 Emerging Technology for Embedded Systems Software Development 700
  10.3 Making Development Tool Choices 707
  10.4 Eclipse—Bringing Embedded Tools Together 721
  10.5 Embedded Software and UML 725
  10.6 Model-based Systems Development with xtUML 739
  10.7 The Future 743
Index *745*

About the Authors

Jason Andrews (Chapter 6) is the author of *Co-verification of Hardware and Software for ARM SoC Design.* He is currently working in the areas of hardware/software co-verification and testbench methodology for SoC design at Verisity. He has implemented multiple commercial co-verification tools as well as many custom co-verification solutions. His experience in the EDA and embedded marketplace includes software development and product management at Verisity, Axis Systems, Simpod, Summit Design, and Simulation Technologies. He has presented technical papers and tutorials at the Embedded Systems Conference, Communication Design Conference, and IP/SoC and written numerous articles related to HW/SW co-verification and design verification. He has a B.S. in electrical engineering from The Citadel, Charleston, S.C., and an M.S. in electrical engineering from the University of Minnesota. He currently lives in the Minneapolis area with his wife, Deborah, and their four children.

Keith Curtis (Chapter 1) is the author of *Embedded Multitasking.* He is currently a Technical Staff Engineer at Microchip, and is also the author of Embedded Multitasking. Prior to and during college, Keith worked as a technician/programmer for Summit Engineering. He then graduated with a BSEE from Montana State University in 1986. Following graduation, he was employed by Tele-Tech Corporation as a design and project engineer until 1992. He also began consulting, part time, as a design engineer in 1990. Leaving Montana in 1992, he was employed by Bally Gaming in Las Vegas as an engineer and later as the EE manager. He worked for various Nevada gaming companies in both design and management until 2000. He then moved to Arizona and began work as a Principal Application Engineer for Microchip.

Jack Ganssle (Chapters 5 and 9) is the author of *The Firmware Handbook.* He has written over 500 articles and six books about embedded systems, as well as a book about his sailing fiascos. He started developing embedded systems in the early 70s using the 8008. He's started and sold three electronics companies, including one of the bigger embedded tool businesses. He's developed or managed over 100 embedded products, from deep-sea navigation gear to the White House security system . . . and one instrument that analyzed cow poop! He's currently a member of NASA's Super Problem Resolution Team, a group of outside experts formed to advise NASA in the wake of Columbia's demise, and serves on the boards of several high-tech companies. Jack gives seminars to companies world-wide about better ways to develop embedded systems.

Rick Gentile (Chapter 7) is the author of Embedded Media Processing. Rick joined ADI in 2000 as a Senior DSP Applications Engineer and he currently leads the Processor Applications Group, which is responsible for Blackfin, SHARC, and TigerSHARC processors. Prior to joining ADI, Rick was a member of the Technical Staff at MIT Lincoln Laboratory, where he designed several signal processors used in a wide range of radar sensors. He has authored dozens of articles and presented at multiple technical conferences. He received a B.S. in 1987 from the University of Massachusetts at Amherst and an M.S. in 1994 from Northeastern University, both in Electrical and Computer Engineering.

Kamal Hyder (Chapters 4 and 5) is the author of *Embedded Systems Design Using the Rabbit 3000 Microprocessor*. He started his career with an embedded microcontroller manufacturer. He then wrote CPU microcode for Tandem Computers for a number of years and was a Product Manager at Cisco Systems, working on next-generation switching platforms. He is currently with Brocade Communications as Senior Group Product Manager. Kamal's BS is in EE/CS from the University of Massachusetts, Amherst, and he has an MBA in finance/marketing from Santa Clara University.

David J. Katz (Chapter 7) is the author of Embedded Media Processing. He has over 15 years of experience in circuit and system design. Currently, he is the Blackfin Applications Manager at Analog Devices, Inc., where he focuses on specifying new convergent processors. He has published over 100 embedded processing articles domestically and internationally and has presented several conference papers in the field. Previously, he worked at Motorola, Inc., as a senior design engineer in cable modem and automation groups. David holds both a B.S. and M. Eng. in Electrical Engineering from Cornell University.

Jean Labrosse (Chapter 3) is the author of *MicroC/OS-II* and *Embedded Systems Building Blocks*. Dr. Labrosse is President of Micrium whose flagship product is the Micrium μC/OS-II. He has an MSEE and has been designing embedded systems for many years.

Tammy Noergaard (Chapters 2 and 3) is the author of *Embedded Systems Architecture*. Since beginning her embedded systems career in 1995, she has had wide experience in product development, system design and integration, operations, sales, marketing, and training. Noergaard worked for Sony as a lead software engineer developing and testing embedded software for analog TVs. At Wind River she was the liaison engineer between developmental engineers and customers to provide design expertise, systems configuration, systems integration, and training for Wind River embedded software (OS, Java, device drivers, etc.) and all associated hardware for a variety of embedded systems in the Consumer Electronics market. Most recently she was a Field Engineering Specialist and Consultant with Esmertec North America, providing project management, system design, system integration, system configuration, support, and expertise for various embedded Java systems using Jbed in everything from control systems to medical devices to digital TVs. Noergaard has lectured to engineering classes at the University of California at Berkeley and Stanford, the Embedded Internet Conference, and the Java User's Group in San Jose, among others.

Robert Oshana (Chapter 8) is the author of *DSP Software Development Techniques*. He has over 25 years of experience in the real-time embedded industry, in both embedded application development as well as embedded tools development. He is currently director of engineering for the Development Technology group at Freescale Semiconductor. Rob is also a Senior Member of IEEE and an adjunct at Southern Methodist University. He can be contacted at: robert.oshana@freescale.com

Bob Perrin (Chapters 4 and 5) is the author of *Embedded Systems Design Using the Rabbit 3000 Microprocessor*. He got his start in electronics at the age of nine when his mother gave him a "150-in-one Projects" kit from Radio Shack for Christmas. He grew up programming a Commodore PET. In 1990, Bob graduated with a BSEE from Washington State University. Since then, Bob has been working as an engineer designing digital and analog electronics. He has published about twenty technical articles, most with *Circuit Cellar*.

Colin Walls (Introduction and Chapter 10) is the author of *Embedded Software: The Works*. He has over twenty-five years experience in the electronics industry, largely dedicated to embedded software. A frequent presenter at conferences and seminars and author of numerous technical articles and two books on embedded software, Colin is a member of the marketing team of the Mentor Graphics Embedded Systems Division. Colin is based in the UK.

Introduction

Colin Walls

Embedded systems are everywhere. You cannot get away from them. In the average American household, there are around 40 microprocessors, not counting PCs (which contribute another 5–10 each) or cars (which typically contain a few dozen). In addition, these numbers are predicted to rise by a couple of orders of magnitude over the next decade or two. It is rather ironic that most people outside of the electronics business have no idea what "embedded" actually means. Marketing people are fond of segmenting markets. The theory is that such segmentation analysis will yield better products by fulfilling the requirements of each segment in a specific way. For embedded, we end up with segments like telecom, mil/aero, process control, consumer, and automotive. Increasingly though, devices come along that do not fit this model. For example, is a cell phone with a camera a telecom or consumer product? Who cares? An interesting area of consideration is the commonality of such applications. The major comment that we can make about them all is the amount of software in each device is growing out of all recognition. In this book, we will look at the inner workings of such software. The application we will use as an example is from the consumer segment—a digital camera—that is a good choice because whether or not you work on consumer devices, you will have some familiarity with their function and operation.

I.1 Development Challenges

Consumer applications are characterized by tight time-to-market constraints and extreme cost sensitivity. This leads to some interesting challenges in software development.

I.1.1 Multiple Processors

Embedded system designs that include more than one processor are increasingly common. For example, a digital camera typically has two: one deals with image processing and the other looks after the general operation of the camera. The biggest challenge with multiple processors is debugging. The code on each individual device may be debugged—the tools and techniques are well understood. The challenge arises with interactions between the two processors. There is a clear need for debugging technology that addresses the issue of debugging the system—that is, multicore debugging.

I.1.2 Limited Memory

Embedded systems usually have limited memory. Although the amount of memory may not be small, it typically cannot be added on demand. For a consumer application, a combination of cost and power consumption considerations may result in the quantity of memory also being restricted. Traditionally, embedded software engineers have developed skills in programming in an environment with limited memory availability. Nowadays, resorting to assembly language is rarely a convenient option. A thorough understanding of the efficient use of C and the effects and limitations of optimization are crucial.

If C++ is used (which may be an excellent language choice), the developers need to fully appreciate how the language is implemented. Otherwise, memory and real-time overheads can build up and not really become apparent until too late in the project, when a redesign of the software is not an option. Careful selection of C++ tools, with an emphasis on embedded support, is essential.

I.1.3 User Interface

The user interface (UI) on any device is critically important. Its quality can have a very direct influence on the success of a product. With a consumer product, the influence is overwhelming. If users find that the interface is "clunky" and awkward, their perception of not just the particular device, but also the entire brand will be affected. When it is time to upgrade, the consumer will look elsewhere.

Therefore, getting it right is not optional. However, getting it right is easier to say than do. For the most part, the UI is not implemented in hardware. The functions of the various controls on a digital camera, for example, are defined by the software. In addition, there may be many controls, even on a basic model. Therefore, in an ideal world, the development sequence would be:

1. Design the hardware.

2. Make the prototypes.

3. Implement the software (UI).

4. Try the device with the UI and refine and/or reimplement as necessary.

However, we do not live in an ideal world.

In the real world, the complexity of the software and the time-to-market constraints demand that software is largely completed long before hardware is available. Indeed, much of the work typically needs to be done even before the hardware design is finished. An approach to this dilemma is to use prototyping technology. With modern simulation technology, you can run your code, together with any real-time operating system (for example, RTOS) on your development computer (Windows, Linux, or UNIX), and link it to a graphical representation of the UI. This enables developers to interact with the software as if they were holding the device in their hand. This capability makes checking out all the subtle UI interactions a breeze.

I.2 Reusable Software

Ask long-serving embedded software engineers what initially attracted them to this field of work and you will get various answers. Commonly though, the idea of being able to *create* something was the appeal. Compared with programming a conventional computer, constrained by the operating system and lots of other software, programming an embedded system seemed like working in an environment where the developer could be in total control. (The author, for one, admits to a megalomaniac streak.)

However, things have changed. Applications are now sufficiently large and complex that it is usual for a team of software engineers to be involved. The size of the application means that an individual could never complete the work in time; the complexity means that few engineers would have the broad skill set. With increasingly short times to market, there is a great incentive to reuse existing code, whether from within the company or licensed from outside.

The reuse of designs—of intellectual property in general—is common and well accepted in the hardware design world. For desktop software, it is now the common implementation strategy. Embedded software engineers tend to be conservative and are not early adopters of new ideas, but this tendency needs to change.

I.2.1 Software Components

It is increasingly understood that code reuse is essential. The arguments for licensing software components are compelling, but a review of the possibilities is worthwhile.

We will now take a look at some of the key components that may be licensed and consider the key issues.

1.3   Real-Time Operating System

The treatment of a real-time operating system (RTOS) as a software component is not new; there are around 200 such products on the market. The differentiation is sometimes clear, but in other cases, it is more subtle. Much may be learned from the selection criteria for an RTOS.

1.3.1   RTOS Selection Factors

Detailed market research has revealed some clear trends in the factors that drive purchasing decisions for RTOS products.

Hard real time: "Real time" does not necessarily mean "fast"; it means "fast enough." A real-time system is, above all, predictable and deterministic.

Royalty free: The idea of licensing some software, and then paying each time you ship something, may be unattractive. For larger volumes, in particular, a royalty-free model is ideal. A flexible business model, recognizing that all embedded systems are different, is the requirement.

Support: A modern RTOS is a highly sophisticated product. The availability of high-quality technical support is not optional.

Tools: An RTOS vendor may refer you elsewhere for tools or may simply resell some other company's products. This practice will not yield the level of tool/RTOS integration required for efficient system development. A choice of tools is, on the other hand, very attractive.

Ease of use: As a selection factor, ease of use makes an RTOS attractive. In reality, programming a real-time system is not easy; it is a highly skilled endeavor. The RTOS vendor can help by supplying readable, commented source code, carefully integrating system components together, and paying close attention to the "out-of-box" experience.

Networking: With approximately one third of all embedded systems being "connected," networking is a common requirement. More on this topic later.

Broad computer processing unit (CPU) support: The support, by a given RTOS architecture, of a wide range of microprocessors is a compelling benefit. Not only does this support yield more portable code, but also the engineers' skills may be readily leveraged. Reduced learning curves are attractive when time to market is tight.

1.3.2 RTOS Standards

There is increasing interest in industry-wide RTOS standards, such as OSEK, POSIX (Portable Operating System Interface), and μiTRON. This subject is wide ranging, rather beyond the scope of this book and worthy of a chapter devoted exclusively to it.

OSEK: The short name for the increasingly popular OSEK/VDX standard, OSEK is widely applied in automotive and similar applications.

miTRON: The majority of embedded designs in Japan use the μiTRON architecture. This API may be implemented as a wrapper on top of a proprietary RTOS, thus deriving benefit from the range of middleware and CPU support.

POSIX: This standard UNIX application-programming interface (API) is understood by many programmers worldwide. The API may be implemented as a wrapper on top of a proprietary RTOS.

1.4 File System

A digital camera will, of course, include some kind of storage medium to retain the photographs. Many embedded systems include some persistent storage, which may be magnetic, or optical disk media or nonvolatile memory (such as flash). In any case, the best approach is standards based, such as an MS-DOS-compatible file system, which would maximize the interoperability possibilities with a variety of computer systems.

1.5 Universal Serial Port (USB)

There is a seemingly inviolate rule in the high-tech world: the easier something is to use, the more complex it is "under the hood."

Take PCs for example. MS-DOS was very simple to understand; read a few hundred pages of documentation and you could figure out everything the operating system (OS) was up to. Whatever its critics may say, Windows is easier to use, but you will find it hard (no, impossible) to locate anyone who understands everything about its internals; it is incredibly complex.

USB fits this model. Only a recollection of a few years' experience in the pre-USB world can make you appreciate how good USB really is. Adding a new peripheral device to a PC could not be simpler. The electronics behind USB are not particularly complex; the really smart part is the software. Developing a USB stack, either for the host computer or for a peripheral device, is a major undertaking. The work has been done for host computers—USB is fully supported on Windows and other operating systems. It makes little sense developing a stack yourself for a USB-enabled device.

USB has one limitation, which restricts its potential flexibility: a strict master/slave architecture. This situation will change, as a new standard, USB On-the-Go (OTG), has been agreed upon and will start showing up in new products. This standard allows devices to change their master/slave status, as required. Therefore, USB gets easier to use and—guess what—the underlying software becomes even more complex. Many off-the-shelf USB packages are available.

1.6   Graphics

The liquid crystal display (LCD) panel on the back of a camera has two functions: it is a graphical output device and part of the user interface (UI). Each of these functions needs to be considered separately.

As a graphical output device, an LCD is quite straightforward to program. Just setting red, green, and blue (RGB) values in memory locations results in pixels being lit in appropriate colors. However, on top of this underlying simplicity, the higher-level functionality of drawing lines and shapes, creating fills, and displaying text and images can increase complexity very rapidly. A graphic functions library is required.

To develop a graphic user interface (GUI), facilities are required to draw screen elements (buttons, icons, menus, and so forth) and handle input from pointing devices. An additional library, on top of the basic graphics functions, is required.

1.7   Networking

An increasing number of embedded systems are connected either to the Internet or to other devices or networks. This may not sound applicable to our example of a digital camera, but

Bluetooth connectivity is quite common and even wireless local area networks (more commonly referred to as Wi-Fi)-enabled cameras have been demonstrated.

A basic Transmission Control Protocol/Internet Protocol (TCP/IP) stack may be straightforward to implement, but adding all the additional applications and protocols is quite another matter. Some key issues are worthy of further consideration.

I.8   IPv6

Internet Protocol (IP) is the fundamental protocol of the Internet, and the currently used variant is v4. The latest version is v6. (Nobody seems to know what happened to v5.) To utilize IPv6 requires new software because the protocol is different in quite fundamental ways. IPv6 addresses a number of issues with IPv4. The two most noteworthy are security (which is an add-on to IPv4 but is specified in IPv6) and address space. IPv6 addresses are much longer and are designed to cover requirements far into the future (see Figure I.1).

```
Standard Format:
        3ffe:2900:0102:0001:0000:0000:0000:0002
Leading Zeros Removed:
        3ffe:2900:102:1:0:0:0:2
Double Colon Notation:
        3ffe:2900:102:1::2
```

Figure I.1: IPv6 Addresses

If you are making Internet-connected devices, do you need to worry about IPv6 yet?

If your market is restricted to nonmilitary/nongovernment customers in North America, IPv6 will not be a requirement for a few years. If your market extends to Asia or Europe or encompasses military applications, you need to consider IPv6 support now. You also need to consider support for dual stacks and IPv6/IPv4 tunneling.

I.8.1   Who Needs a Web Server?

The obvious answer to this question is "someone who runs a web site," but, in the embedded context, there is another angle.

Imagine that you have an embedded system and you would like to connect to it from a PC to view the application status and/or adjust the system parameters. This PC may be local, or it could be remotely located, connected by a modem, or even over the Internet; the PC may also be permanently linked or just attached when required.

What work would you need to do to achieve this? The following tasks are above and beyond the implementation of the application code:

- Define/select a communications protocol between the system and the PC.

- Write data access code, which interfaces to the application code in the system and drives the communications protocol.

- Write a Windows program to display/accept data and communicate using the specified protocol.

Additionally, there is the longer-term burden of needing to distribute the Windows software along with the embedded system and update this code every time the application is changed.

An alternative approach is to install web server software in the target system. The result is:

- The protocol is defined: HyperText Transfer Protocol (HTTP).

- You still need to write data access code, but it is simpler; of course, some web pages (HyperText Markup Language; HTML) are also needed, but this is straightforward.

- On the PC you just need a standard web browser.

The additional benefits are that there are no distribution/maintenance issues with the Windows software (everything is on the target), and the host computer can be anything (it need not be a Windows PC). A handheld is an obvious possibility.

The obvious counter to this suggestion is size: web servers are large pieces of software. An embedded web server may have a memory footprint as small as 20 K, which is very modest, even when storage space for the HTML files is added.

I.8.2   SNMP

SNMP (Simple Network Management Protocol) is a popular remote access protocol, which is employed in many types of embedded devices. The current version of the specification is v3.

SNMP offers a very similar functionality to a web server in many ways. How might you select between them?

If you are in an industry that uses SNMP routinely, the choice is made for you. If you need secure communications (because, for example, you are communicating with your system over the Internet), SNMP has this capability intrinsically, whereas a web server requires a secure sockets layer (SSL). On the other hand, if you do not need the security, you will have an unwanted memory and protocol overhead with SNMP.

A web server has the advantage that the host computer software is essentially free, whereas SNMP browsers cost money. The display on an SNMP browser is also somewhat fixed; with a web server, you design the HTML pages and control the format entirely.

1.9 Conclusion

As we have seen, the development of a modern embedded system, such as a digital camera, presents many daunting challenges. With a combination of the right skills and tools and a software-component-based development strategy, success is attainable. The challenges, tools, and some solutions will be considered in the pages of this book.

CHAPTER 1

Basic Embedded Programming Concepts

Keith Curtis

The purpose of this first chapter is to provide the software designer with some basic concepts and terminology that will be used later in the book. It covers binary numbering systems, data storage, basic communications protocols, mathematics, conditional statements, state machines, and basic multitasking. These concepts are covered here not only to refresh the designer's understanding of their operations but also to provide sufficient insight so that designers will be able to "roll their own" functions if needed. While this chapter is not strictly required to understand the balance of the book, it is recommended.

It is understandable why state machines and multitasking needs review, but why are all the other subjects included? And why would a designer ever want to "roll my own" routines? That is what a high-level language is for, isn't it? Well, often in embedded design, execution speed, memory size, or both will become an issue. Knowing how a command works allows a designer to create optimized functions that are smaller and/or faster than the stock functions built into the language. It also gives the designer a reference for judging how efficient a particular implementation of a command may be. Therefore, while understanding how a command works may not be required in order to write multitasking code, it is very valuable when writing in an embedded environment.

For example, a routine is required to multiply two values together, a 16-bit integer, and an 8-bit integer. A high-level language compiler will automatically type-convert the 8-bit value into a 16-bit value and then performs the multiplication using its standard 16×16 multiply. This is the most efficient format from the compiler's point of view, because it only requires an 8×8 multiply and 16×16 multiply in its library. However, this does creates two inefficiencies; one, it wastes two data memory locations holding values that will always be zero and, two, it wastes execution cycles on 8 additional bits of multiply that will always result in a zero.

The more efficient solution is to create a custom 8×16 multiply routine. This saves the 2 data bytes and eliminates the wasted execution time spent multiplying the always-zero MSB of the 8-bit value. Also, because the routine can be optimized now to use an 8-bit multiplicand, the routine will actually use less program memory as it will not have the overhead of handling the MSB of the multiplicand. So, being able to "roll your own" routine allows the designer to correct small inefficiencies in the compiler strategy, particularly where data and speed limitations are concerned.

While "rolling your own" multiply can make sense in the example, it is *not* the message of this chapter that designers should replace all of the built-in functions of a high-level language. However, knowing how the commands in a language work does give designers the knowledge of what is possible for evaluating a suspect function and, more importantly, how to write a more efficient function if it is needed.

1.1 Numbering Systems

A logical place to start is a quick refresher on the base-ten number system and the conventions that we use with it. As the name implies, base ten uses ten digits, probably because human beings have ten fingers and ten toes so working in units or groups of ten is comfortable and familiar to us. For convenience in writing, we represent the ten values with the symbols "0123456789."

To represent numbers larger than 9, we resort to a position-based system that is tied to powers of ten. The position just to the left of the decimal point is considered the ones position, or 10 raised to the zeroth power. As the positions of the digits move to the left of the decimal point, the powers of ten increase, giving us the ability to represent ever-larger large numbers, as needed. So, using the following example, the number 234 actually represents 2 groups of a hundred, 3 groups of ten plus 4 more. The left-most value, 2, represents 10^2. The 3 in the middle represents 10^1, and the right-most 4 represents 1 or 10^0.

Example 1.1

```
234
2    *10^2=   200
  3  *10^1=    30
    4 *10^0= +   4
              234
```

By using a digit-position-based system based on powers of 10, we have a simple and compact method for representing numbers.

To represent negative numbers, we use the convention of the minus sign "−". Placing the minus sign in front of a number changes its meaning, from a group of items that we have to a group of items that are either missing or desired. Therefore, when we say the quantity of a component in the stock room is −3, which means that for the current requirements, we are three components short of what is needed. The minus sign is simply indicating that three more are required to achieve a zero balance.

To represent numbers between the whole numbers, we also resort to a position-based system that is tied to powers of ten. The only difference is that this time, the powers are negative, and the positions are to the right of the decimal point. The position just to the left of the decimal point is considered 10^0 or 1, as before, and the position just to the right of the decimal point is considered 10^−1 or 1/10. The powers of ten continue to increase negatively as the position of the digits moves to the right of the decimal point. So, the number 2.34, actually presents 2 and 3 tenths, plus 4 hundredths.

Example 1.2

```
2.34
2    *10^0  =   2.
  3  *10^-1 =    .30
  4 *10^-2 = +  .04
               2.34
```

For most everyday applications, the simple notation of numbers and a decimal point is perfectly adequate. However, for the significantly larger and smaller numbers used in science and engineering, the use of a fixed decimal point can become cumbersome. For these applications, a shorthand notation referred to as *scientific notation* was developed. In scientific notation, the decimal point is moved just to the right of the left-most digit and the shift is noted by the multiplication of ten raised to the power of the new decimal point location. For example:

Example 1.3

Standard notation	Scientific notation
2,648.00	2.648x10^3
1,343,000.00	1.343x10^6
0.000001685	1.685x10^-6

As you can see, the use of scientific notation allows the representation of large and small values in a much more compact, and often clearer, format, giving the reader not only a feel for the value, but also an easy grasp of the number's overall magnitude.

Note: When scientific notation is used in a computer setting, the notation 10^ is often replaced with just the capital letter E. This notation is easier to present on a computer screen and often easier to recognize because the value following the ^ is not raised as it would be in printed notation. So, 2.45x10^3 becomes 2.45E3. Be careful not to use a small "e" as that can be confusing with logarithms.

1.1.1   Binary Numbers

For computers, which do not have fingers and toes, the most convenient system is *binary* or base two. The main reason for this choice is the complexity required in generating and recognizing more than two electrically distinct voltage levels. So, for simplicity, and cost savings, base two is the more convenient system to design with. For our convenience in writing binary, we represent these two values in the number system with the symbols "0" and "1." Note: Other representations are also used in Boolean logic, but for the description here, 0 and 1 are adequate.

To represent numbers larger than one, we resort to a position-based system tied to powers of two, just as base 10 used powers of ten. The position just to the left of the decimal point is considered 2^0 or 1. The power of two corresponding to each digit increases as the position of the digits move to the left. So, the base two value 101, represents 1 groups of four, 0 groups of two, plus 1. The left-most digit, referred to as the most significant bit or MSB, represents 2^2 (or 4 in base ten). The position of 0 denotes 2^1 (or 2 in base ten). The right-most digit, referred to as the least significant bit or LSB (1), represents 1 or 2^0.

Example 1.4

```
101
1    *2^2=  100 (1*4 in base ten)
 0   *2^1=   00 (0*2 in base ten)
  1  *2^0= +  1 (1*1 in base ten)
            101 (5 in base ten)
```

Therefore, binary numbers behave pretty much the same as they do for base 10 numbers. They only use two distinct digits, but they follow the same system of digit position to indicate the power of the base.

1.2   Signed Binary Numbers

To represent negative numbers in binary, two different conventions can be used, *sign and magnitude*, or *two's complement*. Both are valid representations of signed numbers and both have their place in embedded programming. Unfortunately, only two's complement is typically supported in high-level language compilers. Sign and magnitude can also be implemented in a high-level language, but it requires additional programming for any math and comparisons functions required. Choosing which format to use depends on the application and the amount of additional support needed. In either case, a good description of both, with their advantages and disadvantages, is presented here.

The sign and magnitude format uses the same binary representation as the unsigned binary numbers in the previous section. And, just as base-ten numbers used the minus sign to indicate negative numbers, so too do sign and magnitude format binary numbers, with the addition of a single bit variable to hold the sign of the value. The sign bit can be either a separate variable, or inserted into the binary value of the magnitude as the most significant bit. Because most high-level language compilers do not support the notation, there is little in the way of convention to dictate how the sign bit is stored, so it is left up to the designer to decide.

While compilers do not commonly support the format, it is convenient for human beings in that it is a very familiar system. The sign and magnitude format is also a convenient format if the system being controlled by the variable is vector-based—i.e., it utilizes a magnitude and direction format for control.

For example, a motor speed control with an H-bridge output driver would typically use a vector-based format for its control of motor speed and direction. The magnitude controls the speed of the motor, through the duty cycle drive of the transistors in the H-bridge. The sign determines the motor's direction of rotation by selecting which pair of transistors in the H-bridge are driven by the PWM signal. Therefore, a sign and magnitude format is convenient for representing the control of the motor.

The main drawback with a sign and magnitude format is the overhead required to make the mathematics work properly. For example:

1.   Addition can become subtraction if one value is negative.

2.   The sign of the result will depend on whether the negative or positive value is larger.

3.   Subtraction can become addition if the one value is negative.

4. The sign of the result will depend on whether the negative or positive value is larger and whether the positive or negative value is the subtracted value.

5. Comparison will also have to include logic to determine the sign of both values to properly determine the result of the comparison.

As human beings, we deal with the complications of a sign and magnitude format almost without thinking and it is second nature to us. However, microcontrollers do not deal well with exceptions to the rules, so the overhead required to handle all the special cases in math and comparison routines makes the use of sign and magnitude cumbersome for any function involving complex math manipulation. This means that, even though the sign and magnitude format may be familiar to us, and some systems may require it, the better solution is a format more convenient for the math. Fortunately, for those systems and user interfaces that require sign and magnitude, the alternate system is relatively easy to convert to and from.

The second format for representing negative binary numbers is two's complement. Two's complement significantly simplifies the mathematics from a hardware point of view, though the format is less humanly intuitive than sign and magnitude. Positive values are represented in the same format as unsigned binary values, with the exception that they are limited to values that do not set the MSB of the number. Negative numbers are represented as the binary complement of the corresponding positive value, plus one. Specifically, each bit becomes its opposite, ones become zeros and zeros become ones. Then the value 1 is added to the result. The result is a value that, when added to another value using binary math, generates the same value as a binary subtraction. As an example, take the subtraction of 2 from 4, since this is the same as adding -2 and $+4$. First, we need the two's complement of 2 to represent -2

Example 1.5

```
0010    Binary representation of 2
1101    Binary complement of 2 (1s become 0s, and
        0s become 1s)
1110    Binary complement of 2 + 1, or -2 in two's
        complement

Then adding 4 to -2
  1110    -2
+0100    +4
  0010    2 with the msb clear indicating a positive
          result
```

Representing numbers in two's complement means that no additional support routines are needed to determine the sign and magnitude of the variables in the equation; the numbers are just added together and the sign takes care of itself in the math. This represents a significant simplification of the math and comparison functions and is the main reason why compilers use two's complement over sign and magnitude in representing signed numbers.

1.2.1 Fixed-Point Binary Numbers

To represent numbers between the whole numbers in signed and unsigned binary values we once again resort to a position-based system, this time tied to decreasing negative powers of two for digit positions to the right of the decimal point. The position just to the left of the decimal point is considered 2^0 or 1, with the first digit to the right of the decimal point representing 2^{-1}. Each succeeding position represents an increasing negative power of two as the positions of the digits move to the right. This is the same format used with base-ten numbers and it works equally well for binary values. For example, the number 1.01 in binary is actually 1, plus 0 halves and 1 quarter.

Example 1.6

```
1.01
1    *2^0  =  1        (1*1 in base ten)
  0  *2^-1 =    .0     (0*½ in base ten)
  1  *2^-2 = +  .01    (1*¼ in base ten)
                1.01   (1¾  in base ten)
```

While any base-ten number can be represented in binary, a problem is encountered when representing base-ten values to the right of the decimal point. Representing a base-ten 10 in binary is a simple 1010; however, converting 0.1 in base ten to binary is somewhat more difficult. In fact, to represent 0.1 in binary (.0000110011) requires 10 bits to get a value accurate to within 1%. This can cause intermittent inaccuracies when dealing with real-world control applications.

For example, assume a system that measures temperature to .1°C. The value from the analog-to-digital converter will be an integer binary value, and an internal calibration routine will then offset and divide the integer to get a binary representation of the temperature. Some decimal values, such as .5°C will come out correctly, but others will have some degree of round-off error in the final value. Then, converting values with round-off error back into decimal values for the user interface will further increase the problem, resulting in a display with a variable accuracy.

For all their utility in representing real numbers, fixed-point binary numbers have little support in commercial compilers. This is due to three primary reasons:

1. Determining a position for the decimal point is often application specific, so finding a location that is universally acceptable is problematic.

2. Multiply, and specifically divide, routines can radically shift the location of the decimal point depending on the values being used.

3. It has difficulty in representing small fractional base-ten values.

One alternative to the fixed-point format that does not require a floating-point format is to simply scale up all the values in a system until they are integers. Using this format, the temperature data from the previous example would be retained in integer increments of .1°C, alleviating the problem of trying to represent .1°C as a fixed-point value. Both the offset and divider values would have to be adjusted to accommodate the new location of the decimal point, as would any limits or test values. In addition, any routines that format the data for a user interface would have to correctly place the decimal point to properly represent the data. While this may seem like a lot of overhead, it does eliminate the problem with round off error, and once the constants are scaled, only minimal changes are required in the user interface routines.

1.2.2 Floating-Point Binary Numbers

Another alternative is to go with a more flexible system that has an application-determined placement of the decimal point. Just as with base-ten numbers, a fixed decimal point representation of real numbers can be an inefficient use of data memory for very large or very small numbers. Therefore, binary numbers have an equivalent format to scientific notation, referred to as *floating-point*.

In the scientific notation of base-ten numbers, the decimal point was moved to the right of the leftmost digit in the number, and an exponent notation was added to the right-hand side. Floating-point numbers use a similar format, moving the decimal point to the right of the MSB in the value, or *mantissa*, and adding a separate *exponent* to the number. The exponent represents the power of two associated with the MSB of the mantissa and can be either positive or negative using a two's complement format. This allows for extremely large and small values to be stored in floating-point numbers.

For storage of the value, typically both the exponent and the mantissa are combined into a single binary number. For signed floating-point values, the same format is used, except the

MSB of the value is reserved for the sign, and the decimal point is placed to the right of the MSB of the matissa.

In embedded applications, floating-point numbers are generally reserved for highly variable, very large or small numbers, and "rolling your own" floating-point math routines are usually not required. It is also beyond the scope of this book, so the exact number of bits reserved for the mantissa and exponent and how they are formatted will not be covered here. Any reader desiring more information concerning the implementation of floating-point numbers and mathematics is encouraged to research the appropriate industry standards for additional information. One of the more common floating-point standards is IEEE 754.

1.2.3   Alternate Numbering Systems

In our discussion of binary numbers, we used a representation of 1s and 0s to specify the values. While this is an accurate binary representation, it becomes cumbersome when we move into larger numbers of bits. Therefore, as you might expect, a couple of shorthand formats have been developed, to alleviate the writer's cramp of writing binary numbers. One format is *octal* and the other is *hexadecimal*. The octal system groups bits together into blocks of 3 and represents the values using the digits 0–7. Hexadecimal notation groups bits together into blocks of 4 bits and represents the values using the digits 0–9, and the letters A–F (see Table 1.1).

Table 1.1

Decimal	Binary	Octal	Hexadecimal
0	0000	00	0
1	0001	01	1
2	0010	02	2
3	0011	03	3
4	0100	04	4
5	0101	05	5
6	0110	06	6
7	0111	07	2
8	1000	10	8
9	1001	11	8
10	1010	12	A
11	1011	13	B
12	1100	14	C
13	1101	15	D
14	1110	16	E
15	1111	17	F

Octal was originally popular because all 8 digits of its format could be easily displayed on a 7-segment LED display, and the 3-bit combinations were easy to recognize on the binary front panel switches and displays of the older mainframe computers. However, as time and technology advanced, problems with displaying hexadecimal values were eliminated and the binary switches and LEDs of the mainframe computer front panels were eventually phased out. Finally, due to its easy fit into 8-, 16-, and 32-bit data formats, hexadecimal eventually edged out octal as a standard notation for binary numbers. Today, in almost every text and manual, values are listed in either binary, decimal (base ten), or hexadecimal.

1.2.4 Binary-Coded Decimal

Another binary numeric format is binary-coded decimal or *BCD*. BCD uses a similar format to hexadecimal in that it groups together 4 bits to represent data. The difference is that the top 6 combinations, represented by A–F in hexadecimal, are undefined and unused. Only the first 10 combinations represented by 0–9 are used.

The BCD format was originally developed for use in logic blocks, such as decade counters and display decoders in equipment, to provide a base-ten display, and control format. The subsequent development of small 8-bit microcontrollers carried the format forward in the form of either a BCD addition/subtraction mode in the math instructions of the processor, or as a BCD adjust instruction that corrects BCD data handled by a binary addition/subtraction.

One of the main advantages of BCD is its ability to accurately represent base-ten values, such as decimal dollars and cents. This made BCD a valuable format for software handling financial and inventory information because it can accurately store fractional base-ten decimal values without incurring round-off errors. The one downside to BCD is its inefficiency in storing numbers. Sixteen bits of BCD can only store a value between 0 and 9999, while 16-bit binary can represent up to 65,535 values, a number over 60 times larger.

From this discussion, you may think that BCD seems like a waste of data storage, and it can be, but it is also a format that has several specific uses. And even though most high-level languages don't offer BCD as a storage option, some peripherals and most user interfaces need to convert binary numbers to and from BCD as a normal part of their operation. So, BCD is a necessary intermediate format for numbers being converted from binary to decimal for display on a user interface, or communication with other systems. Having an understanding of the format and being able to write routines that convert binary to BCD and back are, therefore, valuable skills for embedded designers.

1.2.5 ASCII

The last format to be discussed is ASCII. ASCII is an acronym for the American Standard Code for Information Interchange. It is a 7-bit code that represents letters, numbers, punctuation, and common control codes.

A holdover data format from the time of mainframe computers, ASCII was one of two common formats for sending commands and data serially to terminals and printers. The alternate code, an 8-bit code known as EBIDIC, has since disappeared, leaving ASCII as the de facto standard with numerous file formats and command codes based on it. Table 1.2 is a chart of all 128 ASCII codes, referenced by hexadecimal:

Table 1.2

Hex	ASCII	Hex	ASCII	Hex	ASCII	Hex	ASCII	Hex	ASCII	Hex	ASCII	Hex	ASCII	Hex	ASCII
00	NUL	10	DLE	20	SP	30	0	40	@	50	P	60	`	70	p
01	SOH	11	DC1	21	!	31	1	41	A	51	Q	61	a	71	q
02	STX	12	DC2	22	"	32	2	42	B	52	R	62	b	72	r
03	ETX	13	DC3	23	#	33	3	43	C	53	S	63	c	73	s
04	EOT	14	DC4	24	$	34	4	44	D	54	T	64	d	74	t
05	ENQ	15	NAK	25	%	35	5	45	E	55	U	65	e	75	u
06	ACK	16	SYN	26	&	36	6	46	F	56	V	66	f	76	v
07	BEL	17	ETB	27	'	37	7	47	G	57	W	67	g	77	w
08	BS	18	CAN	28	(	38	8	48	H	58	X	68	h	78	x
09	HT	19	EM	29	)	39	9	49	I	59	Y	69	I	79	y
0A	LF	1A	SUB	2A	*	3A	:	4A	J	5A	Z	6A	j	7A	z
0B	VT	1B	ESC	2B	+	3B	;	4B	K	5B	[	6B	k	7B	{
0C	FP	1C	FS	2C	,	3C	<	4C	L	5C	\	6C	l	7C	|
0D	CR	1D	GS	2D	-	3D	=	4D	M	5D	]	6D	m	7D	}
0E	SO	1E	RS	2E	.	3E	>	4E	N	5E	^	6E	n	7E	~
0F	SI	1F	US	2F	/	3F	?	4F	O	5F	_	6F	o	7F	DEL

Among the more convenient features of the code is the placement of the codes for the numbers 0–9. They are placed such that conversion between BCD and ASCII is accomplished by simply OR-ing on the top 3 bits, or AND-ing them off. In addition, translation between upper and lowercase just involves adding or subtracting hexadecimal 20.

The code also includes all of the more common control codes such as BS (backspace), LF (line feed), CR (carriage return), and ESC (escape).

Although ASCII was among the first computer codes generated, it has stood the test of time and most, if not all, computers use it in one form or another. It is also used extensively in small LCD and video controller chips, thermal printers, and keyboard encoder chips. It has even left its mark on serial communications, in that most serial ports offer the option of 7-bit serial transmission.

1.2.6 Error Detection

One of the things that most engineers ask when first exposed to ASCII is what to do with the eighth bit in an 8-bit system. It seems a waste of data memory to just leave it empty, and it doesn't make sense that older computer systems wouldn't use the bit in some way. It turns out that the eighth bit did have a use. It started out in serial communications where corruption of data in transit was not uncommon. When serially transmitted, the eighth bit was often used for error detection as a *parity* bit.

The method involved including the parity bit, which, when exclusive OR-ed with the other bits, would produce either a one or a zero. Even parity was designed to produce a zero result, and odd parity produced a one. By checking each byte as it came in, the receiver could detect single-bit errors, and when an error occurred, request a retransmission of the data. This is the same parity bit that is still used in serial ports today. Users are given the option to use *even* or *odd*, and can even choose *no* parity, which turns off the error checking.

Parity works fine for 7-bit ASCII data in an 8-bit system, but what about 8-, 16-, and 32-bit data? When computer systems began passing larger and larger blocks of data, a better system was needed—specifically, one that didn't use up 12.5% of the bandwidth—so several other error-checking systems were developed. Some are able to determine multibit errors in a group of data bytes, while other simpler systems can only detect single-bit errors. Other, more complex, methods are even able to detect and correct bit errors in one or more bytes of data. While this area of design is indeed fascinating, it is also well beyond the scope of this book.

For our use here, we will concentrate on two of the simpler systems, the *check sum*, and the *cyclic redundancy check* or *CRC*.

The check sum is the simpler of the two systems and, just as it sounds, it is simply a one- or two-byte value that holds the binary sum of all the data. It can detect single-bit errors, and even some multibit errors, but it is by no means a 100% check on the data.

A CRC, on the other hand, uses a combination of shifts and Boolean functions to combine the data into a check value. Typically a CRC shifts each byte of data in the data block into the CRC value one bit at a time. Each bit, before it is shifted into the CRC value, is combined with feedback bits taken from the current value of the CRC. When all of the bits in the data block have been shifted into the CRC value, a unique CRC value has been generated that should detect single and more of the multibit errors. The number, type, and combination of bit errors that can be detected is determined by several factors. These include both the number of bits in the CRC and the specific combination of bits fed back from the CRC value during the calculation. As mentioned previously, an in-depth description of CRC systems, and even a critique of the relative merits of the different types of CRC algorithms is a subject sufficient to fill a book, and as such is beyond the scope of this text. Only this cursory explanation will be presented here. For more information on CRC systems, the reader is encouraged to research the subject further.

One final note on CRCs and check sums. Because embedded designs must operate in the real world, and because they will be subject to electromagnetic interference (EMI), radio frequency interference (RFI), and a host of other disruptive forces, CRCs and check sums are also typically used to validate the contents of both program and data memory. Periodically running a check sum on the program memory, or a CRC check of the data in data memory is a convenient "sanity check" on the system. So, designers working in noisy environments with high functional and data integrity requirements should continue their research into these valuable tools of the trade.

1.3   Data Structures

In a typical high-level application, once the format for the data in a program has been determined, the next step is to define a data structure to hold the information. The structure will determine what modifying functions, such as assignment and math functions, are available. It will determine what other formats the data can be converted into, and what user interface possibilities exist for the data.

In an embedded design, a data structure not only defines storage for data, it also provides a control conduit for accessing control and data registers of the system peripherals. Some peripheral functions may only need byte-wide access, while others may require single bit control. Still others may be a combination of both. In any case, it is essential that the right type of data structure be defined for the type of data to be stored or the type of control to be exercised over the peripheral.

Therefore, a good understanding of the data structure's inner workings is important both for efficiency in data storage and for efficient connections to the system's peripherals. Of specific interest is:

1. What type of information can be stored in the data structure?

2. What other functions are compatible with the data structure?

3. Can the data structures be used to access peripheral control and data registers?

4. How does the date structure actually store the information in memory?

5. How do existing and new functions access the data?

A good understanding of the data structures is important both for efficiency in data storage and for an efficient conduit to the system's peripherals. Knowing how a language stores information can also be proactive in the optimization process, in that it gives the designer insight into the consequences of using a particular data type as it applies to storage and access overhead. This information may allow the designer to choose wisely enough to avoid the need for custom routines altogether.

The following sections covering data structures will try to answer all five of these questions as they apply to each of the different data structures.

1.3.1   Simple Data Types

The term "simple data type" refers to variables that store only one instance of data and store only one format of data at a time. More complex data types, which hold more than one instance of data or hold more than one type of data, will be covered in the next section titled *Complex Data Types*.

Declaration 1.1

```
BIT        variable_name
```

The simplest data type is the Boolean or BIT (binary digit). This data type has only two possible states, 1 or 0. Alternately, TRUE or FALSE, and YES or NO can also be used with some compilers. It is typically used to carry the result of a Boolean logical expression or the binary status of a peripheral or comparison. It can even be used as part of another data type to hold the sign of a value. In each case, the variable provides a simple on/off or yes/no functionality or status.

When BIT is used as a variable, it is assigned a value just like any other variable. The only difference with the BIT data structure is that it can also be assigned the result of a comparison using combinations of Boolean logic and the standard comparison operators, $<>$ and $=$.

> **Note:** A helpful debugging trick is to assign the result of a comparison to a BIT variable and then use the variable in the conditional statement. This allows the designer to monitor the status of the BIT variable and determine the path of execution without having to step through the entire code block step-by-step.

Code Snippet 1.1

```
Flag = (Var_A > Var_B) & (Var_A < Var_C);
if Flag then printf(Var_A);
```

To use the BIT data structure as a conduit to a peripheral control register, the bit must be defined to reside at the corresponding address and bit of the peripheral function to be controlled. As this is not universally supported in C compilers, compilers that do support the feature may have different syntax. So, this is yet another point that must be researched in the user's manual for the compiler. If the compiler does not allow the user to specify both the address and bit location, there is an alternate method using the STRUCTURE statement and that will be covered in the *Complex Data Types* section of this chapter.

Due to the Boolean's simple data requirements, BIT is almost always stored as a single bit within a larger data word. The compiler may choose to store the binary value alone within a larger data word, or it may combine multiple bits and other small data structures for storage that is more efficient. The designer also has the option to force the combination of BITs and other small data structures within a larger data word for convenience, or for more efficient access to control bits within a peripheral's control register. Additional information on this process is presented in the STRUCTURE data structure following.

To access a BIT, the compiler may copy the specific data bit to be accessed into a holding location and then shift it to a specific location. This allows the high-level language to optimize its math and comparison routines for a single bit location within a data word, making the math and comparison routines more efficient. However, this does place some overhead on the access routines for the BIT's data structure.

Other compilers, designed for target microcontrollers with instructions capable of setting, clearing, manipulating, and testing individual bits within a data word, avoid this overhead by simply designing their Boolean and comparison routines to take advantage of the BIT instructions.

To access the BIT directly in memory, the designer needs two pieces of information, the address of the data word containing the BIT, and the location of the BIT within the data word. The address of the byte containing the BIT is typically available through the variable's label. The specific BIT within the byte may not be readily available, and may change as new variables are added to the design. For these reasons, it is generally best to only use manual access of a BIT defined using either a compiler function that allows the designer to specify the bit location, or a STRUCTURE.

Using a STRUCTURE to define the location of a BIT is also useful in that it can be used to force the compiler to group specific variables together. It can also be used to force a group of commonly used BITs into common bit locations for faster access. Finally, defining a BIT within a STRUCTURE and a UNION, gives the designer the option to access the BITs as either individual values or as a group for loading default states at start-up.

One point that should be noted concerning this data type is that not all high-level language compilers recognize it. And, many compilers that do recognize the data type may not agree on its name or syntax, so the designers should review the user's guide for any compiler they intend to use, as there may be differences in the syntax used or restrictions on the definition of this data type.

Declaration 1.2

```
SIGNED CHAR        variable_name
UNSIGNED CHAR      variable_name
```

The CHAR data type was originally designed to hold a single ASCII character, thus the name CHAR, which is short for character. CHARs are still commonly used for holding single ASCII characters, either for individual testing or as part of an output routine, or even grouped with other CHARs to form an array of characters called a STRING. However, over time, it has also come to be a generic variable type for 8-bit data. In fact, most if not all modern high-level languages allow the use of CHAR variables in math operations, conditional statements, and even allow the definition of a CHAR variable as either signed or unsigned.

In embedded programming, the CHAR is equally as important as the Boolean/BIT data type because most peripheral control registers will be one or more bytes in length and the CHAR variable type is a convenient way to access these registers. Typically, a control register for a peripheral will be defined as a CHAR for byte-wide access, allowing the entire register to be set with one assignment. The CHAR may also be tied to a STRUCTURE of BITs using a UNION definition to allow both bit-wise control of the functions, as well as byte-wise access for initialization. More information on both the UNION and the STRUCTURE will be presented in later sections of this chapter.

An important point to note is that this variable may be assumed to be signed or unsigned by the C compiler if the words SIGNED or UNSIGNED are not included in the definition of the variable. The only American National Standards Institute (ANSI) requirement is that the compiler be consistent in its definitions. Therefore, it is best to specify the form in the definition of the variable to avoid problems migrating between compilers.

Manually accessing a CHAR variable at the language level is very simple, as most compilers recognize the data structure as both a character variable, and a signed or unsigned binary value. Access at the assembly language level is also simple as the name given to the variable can be used as an address label to access the data memory. Because the CHAR represents the smallest data structure short of a BIT, the format used to store the data in memory is also simple. The 8-bit value is simply stored in the lower 8 bits of the data memory word. Because the data is stored as a single byte, no additional information, beyond the address, is required.

Declaration 1.3

```
INT              variable_name
UNSIGNED INT     variable_name
```

INT, short for integer, is the next larger data type. It is typically used to hold larger signed and unsigned binary values, and while the BITs and CHARs have consistent and predefined data lengths, the length of an INT is largely dependent on the specific implementation of the high-level compiler. As a result, the actual number of bits in an INT can vary from as few as 16 bits, to whatever the upper limit of the compiler is. The only limitation on the size of an INT is that it must be larger than a CHAR and less than or equal to the size of a LONG. Therefore, to determine the actual size of an INT in a specific compiler, it is necessary to consult the user's manual for the compiler being used.

Because of an INT's somewhat indeterminate length, it can present a problem for efficiently storing larger data. Some compilers may not allocate sufficient bits to hold an application's data, while others may allocate too many bits, resulting in wasted data storage. This can be a very serious problem if the application using the data is to be shared across several different compilers and processors. To alleviate this problem, the designer has three basic options:

1. The large groups of data can be broken into individual bytes and stored as an array of unsigned CHARs, and then recreated in an INT when needed. This minimizes the storage requirements to the minimum number of required bytes, but it also complicates any math or comparison operation that may be required.

2. The INT can be defined as LONGs within a STRUCTURE, allowing the designer to specify the number of bits to be used for the variable. This eliminates the math problem, but the compiler will incur additional overhead, when it automatically converts the data into a standard-length LONG prior to performing the math, and will then incur additional overhead converting it back when the math is complete.

3. The best solution is to simply get to know the compilers to be used and define the variables appropriately for each implementation. The variable type casting will then force the compiler to use the appropriate math and comparison functions, resulting in a much simpler design, while incurring only a minimal processing overhead.

As with the CHAR variable type, the name given to the variable acts as a label and can be used as a pointer to the data in assembly language. However, the number of bytes reserved for the variable and the order in which the bytes are stored in data memory may differ from compiler to compiler. Therefore, once again, it is up to the designers to do their homework and research the exact storage format used.

One of the important statements in the previous paragraph is often missed: "the order in which the bytes are stored in data memory may differ." Specifically, does the compiler store the MSB in the first or last data memory location allocated to the variable? There are two formats that can be used: *big endian* and *little endian*. In the big endian format, the MSB is stored in the first data memory address (lowest memory address) and the LSB is stored in the last data memory address (highest memory address). In little endian, the reverse is true; the LSB is in the first memory address and the MSB in the last. So, to correctly access an INT in assembly, it is necessary not only to determine the number of bytes stored but also which storage format is used. This information is also typically found in the manual. However, if it is not explicitly stated, a simple test routine can answer the question. The test routine defines an INT variable and loads the value 4660 into the variable. Then, by examining data memory, the format can be determined. If the data in the lower memory address is the hexadecimal value 12 followed by the hex value 34, then the format is big endian; if the first byte is 0x34, then the format is little endian.

Due to the generally variable length and format of the INT, it is not a good choice for accessing peripheral registers containing control bits or data. INTs can be, and often are, used for this purpose, but the practice can cause portability problems, including unexpectedly truncated data, the inclusion of data bits from adjacent peripherals, and even scrambled data. The practice is only recommended if the portability of the resulting routines is not a goal of the project.

Declaration 1.4

```
LONG            variable_name
UNSIGNED LONG   variable_name
```

LONG, short for long integer, is the next larger data type. It is typically used to hold very large signed and unsigned binary values, and while the BITs and CHARs have consistent and predefined data lengths, the length of a LONG is again, dependent on the specific implementation of the high-level compiler. As a result, the actual number of bits in a LONG can vary from as few as 16 bits, up to whatever the upper limit of the compiler defines for data types. The only limitation on the size of a LONG variable is that it must be at least as large, or larger, than an INT. Typically, a LONG is twice the size of an INT, but this is not specified by the ANSI2 standard. Therefore, to determine the actual size of an INT in a specific compiler, it is necessary to consult the user's manual for the compiler being used.

Because the LONG is somewhat nonstandard in length, it can also present problems for portability and efficiently storing larger data. As a result, the storage options that applied to the INT serve equally well for the LONG.

Storage problems for larger groups of data can be handled by breaking the larger data blocks into individual bytes and storing as an array of unsigned CHARs, and then recreating in a LONG when needed. This minimizes the storage requirements to the minimum number of required bytes, but it also complicates any math or comparison operation that may be required.

The portability problems can be alleviated by simply getting to know the compilers being used, and defining the variables appropriately for each implementation. The variable type casting will then force the compiler to use the appropriate math and comparison functions, resulting in a much simpler design, while incurring only a minimal processing overhead.

The actual length of the variable will also affect manual access to a LONG variable. As with the CHAR, the name given to the variable acts as a label when accessing the data in assembly language. However, the number of bytes stored for the variable and the order in which the bytes are stored in data memory may differ from compiler to compiler. Therefore, once again, it is up to the designers to do their homework and research the exact storage format used.

Due to the generally variable length and format of the LONG, and its excess length, it is almost never used for accessing peripheral registers containing control bits or data. In fact, due to their length, LONG data types will generally only be useful for very specialized data within the program, although a variable requiring the number of bits included in a LONG is generally rare.

One place that LONG variables do find use is for intermediate results in calculations involving INTs, or as accumulation variables that hold the summation of a large number of data samples. While the LONG may seem attractive for this use, it is can have some unforeseen consequences. Remember that the compiler will typically convert all data in a math function to the largest data type prior to performing the operation. This can result in a shortage of temporary data storage during math operations on the LONG variables. As an example, performing a multiply on a 24-bit LONG variable can use up 12 bytes of data storage just for the temporary storage of the upgraded term variables. So, it is generally advisable to resort to either an array of CHARs or, in extreme cases, an array of INTs to store large data values. This allows the designer to more tightly regulate the amount of data storage required. It also limits the amount of temporary data storage required for math, even though it will require a custom, and somewhat complicated, math routine.

Manually accessing a LONG variable uses the same process as accessing an INT; there are just more bytes to access. As with other data types, the variable name will act as a label for the starting data memory address of the data, and the appropriate big/little endian format must be used to access the data in the proper sequence.

Declaration 1.5

```
FLOAT    variable_name
DOUBLE   variable_name
```

FLOAT, short for floating-point, and **DOUBLE**, short for double precision floating-point, are another simple data structure common to embedded C programming. Typically the FLOAT and DOUBLE are used to hold very large or very small signed binary values. They accomplish this by using a system similar to scientific notation in base-ten numbers. The data structure maintains a base value, or mantissa, and an exponent that holds the power of two associated with the MSB of the mantissa. Together, the exponent and mantissa are concatenated into a single data structure.

Most implementations assign 32 bits of storage for the exponent and mantissa of a FLOAT, and 64 bits for the DOUBLE. However, it is important to note that, like the INT and LONG, the exact size of the FLOAT is determined by the compiler implementation and, potentially, configuration options for the compiler. Therefore, to determine the actual size of a FLOAT or DOUBLE in a specific compiler, it is necessary to consult the user's manual for the compiler being used.

Because the actual implementation of both FLOATs and DOUBLEs is dependent on the standard used by the compiler, and their size and complex nature tends to limit their application in embedded designs, they will not be discussed in any great detail here. Any reader interested in the specifics of FLOAT or DOUBLE data structures can find additional information in either an advanced computer science text or the IEEE specification IEEE 754.

Code Snippet 1.2

```
pointer_name = *variable_name;
pointer_name = &variable_name;
```

Pointers are the last data structure to be covered in this chapter. A pointer, simply stated, is a variable that holds the address of another variable. With it, designers can access data memory independently of a specifically defined variable name. In fact, one of the primary uses of data pointers is to create dynamically allocated data storage, which is essentially an unnamed variable created "on-the-fly" as the program is running. This ability to create storage is quite powerful, although the responsibility of monitoring the amount of available data storage shifts from the compiler to the designer.

Pointers are somewhat unique in that they are typically associated with another data type. The reason for this is because the pointer needs to know the storage format of the data so it can correctly interpret the data. It also needs this information if it is to be used to dynamically allocate variables, so it can reserve the right amount of memory. This is not to say that a pointer can't be used to access one type of data with another type's definition. In fact, this is one of the more powerful capabilities of the pointer type.

The syntax of the pointer data structure is also somewhat unique. The "*" sign is used as a prefix for the variable being accessed, to indicate that the data held in the variable is to be loaded into the pointer. The "&" sign is used as a prefix for the variable being accessed, to indicate that the address of the variable is to be loaded into the pointer. What this means is that both the data and the address of a variable can be loaded into the pointer data structure. Having the ability to access both gives pointers the ability to not only pass addresses around, but also to perform math on the addresses.

Accessing pointers by machine language is typically not needed as most microcontrollers already have the ability to access data through index registers. This, plus the ability to use variable labels as constant values in assembly language provides all the functionality of a pointer. In addition, the number of bits used for a pointer will be dependent upon the addressing modes used by the compiler and the architectural specifics of the microcontroller. So, an explanation of how to access pointers through assembly language will be highly specific to both the microcontroller and the language, and of little additional value, so no attempt will be made to explain access here.

1.3.2   Complex Data Types

Complex data types refer to those variable types that either hold more than one type of data, STRUCTUREs and UNIONs, or more than one instance of a simple data type, ARRAYs. These data types allow the designer to group blocks of data together, either for programming convenience or to allow simplified access to the individual data elements.

One complex data type that will not be covered is POINTERs, mainly because their ability to dynamically allocate data is, in general, not particularly applicable to small-embedded applications, where the data storage requirements tend to be static. In addition, the amount of memory available in small microcontrollers is insufficient to implement a heap of any reasonable size, so using pointers would be inefficient at best.

Declaration 1.6

```
STRUCT structure_name {
   variable_type variable_name;
   variable_type variable_name;
   } variable_name;
```

The **STRUCTURE** data type is a composite data structure that can combine multiple variables and multiple variable types into a single variable structure. Any simple variable structure available in the language can typically be included within a structure, and included more than once. The specific number of bits allocated to each variable can also be specified, allowing the designer to tailor the storage capacity of each variable.

Each instance of the various data structures within the STRUCTURE is given a specific name and, when combined with the STRUCTURE's name, can be accessed like any other variable in the system. Names for individual fields within a structure can even be repeated in different STRUCTUREs because the name of the different STRUCTUREs allows the high-level language to differentiate the two variables.

Using this capability, related variables can even be grouped together under a single name and stored in a common location. While the improved organization of storage is elegant and using a common group name improves readability, the biggest advantage of common storage for related variables is the ability to store and retrieve groups of data in a faster, more efficient manner. The importance of this capability will become clearer when context storage and switching are discussed later in the chapter.

The STRUCTURE is also very useful for creating control and data variables linked to the system peripherals, because it can be used to label and access individual flags and groups of bits, within an 8- or 16-bit peripheral control register. The labeling, order, and grouping of the bits is specified when the STRUCTURE is defined, allowing the designer to match up names and bits in the variables to the names and bits specified in the peripheral's control and data registers. In short, the designer can redefine peripheral control and data bits and registers and unique variables accessible by the program.

For example, the following is a map of the control bits for an analog-to-digital converter peripheral. In its control register are bits that specify the clock used by the ADC (ADCS1 & ADCS0), bits that specify the input channel, (CHS3–CHS0), a bit that starts the conversion and signals the completion of the conversion (GO/DONE), and a bit that enables the ADC (ADON).

Definition 1.1

ADCON0 (Analog to Digital Control Register)

Bit 7	Bit 6	Bit 5	Bit 4	Bit 3	Bit 2	Bit 1	Bit0
ADCS1	ADCS0	CHS2	CHS1	CHS0	GO/DONE	CHS3	ADON

To control the peripheral, some of these bits have to be set for each conversion, and others are set only at the initial configuration of the peripheral. Defining the individual bit groups with a STRUCTURE allows the designer to modify the fields individually, changing some, while still keeping others at their initialized values. A common prefix also helps in identifying the bits as belonging to a common register.

Declaration 1.7

```
STRUCT REGDEF{
        UNSIGNED INT    ADON:1;
        UNSIGNED INT    CHS3:1;
        UNSIGNED INT    GODONE:1;
        UNSIGNED INT    CHS:3;
        UNSIGNED INT    ADCS:2;
        } ADCON0;
```

In the example, UNSIGNED INT data structures, of a specified 1-bit length are defined for bits 0 through 2, allowing the designer to access them individually to turn the ADC on and off, set the most significant channel select bit, and initiate and monitor the conversion process. A 3-bit UNSIGNED INT is used to specify the lower 3 bits of the channel selection, and a 2-bit UNSIGNED INT is tied to clock selection. Using these definitions, the controlling program for the analog-to-digital converter can now control each field individually as if they were separate variables, simplifying the code and improving its readability.

Access to the individual segments of the STRUCTURE is accomplished by using the STRUCTURE's name, followed by a dot and the name of the specific field. For example, ADCON0.GODONE = 1, will set the GODONE bit within the ADCON0 register, initiating a conversion. As an added bonus, the names for individual groups of bits can be repeated within other STRUCTUREs. This means descriptive names can be reused in the STRUCTURE definitions for similar variables, although care should be taken to not repeat names within the same STRUCTURE.

Another thing to note about the STRUCTURE definition is that the data memory address of the variable is not specified in the definition. Typically, a compiler-specific language extension specifies the address of the group of variables labeled ADCON0. This is particularly important when building a STRUCTURE to access a peripheral control register, as the address is fixed in the hardware design and the appropriate definition must be included to fix the label to the correct address. Some compilers combine the definition of the structure and the declaration of its address into a single syntax, while others rely on a secondary definition to fix the address of a previously defined variable to a specific location. Therefore, it is up to the designer to research the question and determine the exact syntax required.

Finally, this definition also includes a type label "REGDEF" as part of the variable definition. This is to allow other variables to reuse the format of this STRUCTURE if needed. Typically, the format of peripheral control registers is unique to each peripheral, so only microcontrollers with more than one of a given peripheral would be able to use this feature. In fact, due to its somewhat dubious need, some compilers have dropped the requirement for this part of the definition, as it is not widely used. Other compilers may support the convention to only limited degrees, so consulting the documentation on the compiler is best if the feature is to be used.

Access to a STRUCTURE from assembly language is simply a matter of using the name of the structure as a label within the assembly. However, access to the individual bits must be accomplished through the appropriate assembly language bit manipulation instructions.

Declaration 1.8

```
UNION union_name {
    variable_type variable_name;
    variable_type variable_name;
    } variable_name;
```

In some applications, it can be useful to be able to access a given piece of data not only by different names, but also using different data structures. To handle this task, the complex data type UNION is used. What a UNION does is create two definitions for a common word, or group of words, in data memory. This allows the program to change its handling of a variable based on its needs at any one time.

For example, the individual groups of bits within the ADCON0 peripheral control register in the previous section were defined to give the program access to the control bits individually. However, in the initial configuration of the peripheral, it would be rather cumbersome and inefficient to set each variable one at a time. Defining the STRUCTRUE from the previous example in a UNION allows the designer to not only individually access the groups of bits within the peripheral control register, but it also allows the designer to set all of the bits at once via a single 8-bit CHAR.

Declaration 1.9

```
UNION   UNDEF{
        STRUCT REGDEF{
            SHORT         ADON;
            SHORT         CHS3;
            SHORT         GODONE;
            UNSIGNED CHAR     CHS:3;
            UNSIGNED CHAR     ADCS:2;
            } BYBIT;
        UNSIGNED CHAR  BYBYTE;
            } ADCON0 @ 0x1F;
```

In the example, the original STRUCTURE definition is now included within the definition of the UNION as one of two possible definitions for the common data memory. The STRUCTURE portion of the definition has been given the subname "BYBIT" and any access to this side of the definition will require its inclusion in the variable name. The second definition for the same words of data memory is an unsigned CHAR data structure, labeled by the subname "BYBYTE."

To access the control register's individual fields, the variable name becomes a combination of the UNION and STRUCTURE's naming convention: `ADCON0.BYBIT.GODONE = 1`. Byte-wide access is similarly accessed through the UNION's name combined with the name of the unsigned CHAR: `ADCON0.BYBYTE = 0x38`.

Declaration 1.10

data_type variable_name[max_array_size]

The ARRAY data structure is nothing more than a multielement collection of the data type specified in the definition. Accessing individual elements in the array is accomplished through the index value supplied within the square brackets. Other than its ability to store multiple copies of the specified data structure, the variables that are defined in an array are indistinguishable from any other single element instance of the same data structure. It is basically a collection of identical data elements, organized into an addressable configuration.

To access the individual data elements in an array, it is necessary to provide an index value that specifies the required element. The index value can be thought of as the address of the element within the group of data, much as a house address specifies a home within a neighborhood. One unique feature of the index value is that it can either be a single value for a one-dimensional array, or multiple values for a multidimensional array. While the storage of the data is not any different for a one-dimensional array versus a multidimensional array, having more than one index variable can be convenient for separating subgroups of data within the whole, or representing relationships between individual elements.

By definition, the type of data within an array is the same throughout and can be of any type, including complex data types such as STRUCTUREs and UNIONs. The ARRAY just specifies the organization and access of the data within the block of memory. The declaration of the ARRAY also specifies the size of the data block, as well as the maximum value of all dimensions within the ARRAY.

One exception to this statement that should be noted: Not all compilers support ARRAYs of BOOLEANs or BITs. Even if the compiler supports the data type, ARRAYs of BOOLEANs or BITs may still not be supported. The user's manual should be consulted to determine the specific options available for arrays.

Accessing an array is just a matter of specifying the index of the data to be accessed as part of the variables; note:

Code Snippet 1.3

```
ADC_DATA[current_value] = 34;
```

In this statement, the element corresponding to the index value in `current_value` is assigned the value 34. `current_value` is the index value, 34 is the data, and `ADC_DATA` is the name of the array. For more dimensions in an ARRAY, more indexes are added, surrounded by square brackets. For instance:

Code Snippet 1.4

```
ADC_DATA[current_value][date,time];
```

This creates a two-dimensional array with two index values required to access each data value stored in the array.

Accessing an array via assembly language becomes a little more complex, as the size of the data type in the array will affect the absolute address of each element. To convert the index value into a physical data memory address, it is necessary to multiply the index by the number of bytes in each element's data type, and then add in the first address of the array. So, to find a specific element in an array of 16-bit integers, assuming 8-bit data memory, the physical memory address is equal to:

Equation 1.1

$$(\text{Starting address of the ARRAY}) + (2 * (\text{index value}))$$

The factor of 2, multiplied by the index value, accounts for the 2-byte size of the integer, and the starting address of the ARRAY is available through the ARRAY's label. Also note that the index value can include 0, and its maximum value must be 1 less than the size of the array when it was declared.

Accessing multidimensional ARRAYs is even more complex, as the dimensions of the array play a factor in determining the address of each element. In the following ARRAY the address for a specific element is found using this equation:

Equation 1.2

$$(\text{Starting address of the ARRAY}) + (2 * \text{index1}) + (2 * \text{index}$$
$$2 * (\text{max\_index1} + 1))$$

The starting address of the array and index1 are the same as the previous example, but now both the maximum size of index1 and the value in index2 must be taken into account. By multiplying the maximum value of index1, plus 1, by the second index, we push the address up into the appropriate block of data. To demonstrate, take a 3×4 array of 16-bit integers defined by the following declaration:

Declaration 1.11

```
Int K_vals[3][4] = {  0x0A01, 0x0A02, 0x0A03, 0x0B01, 0x0B02, 0x0B03,
                      0x0C01, 0x0C02, 0x0C03, 0x0D01, 0x0D02, 0x0D03}
```

This will load all 12, 16-bit, locations with data, incrementing through the first index variable. And then incrementing the second index variable each time the first variable rolls over. So, if you examine the array using X as the first index value, and Y as the second, you will see the data arrayed as follows:

Table 1.3

X→	0	1	2
Y			
0	0x0A01	0x0A02	0x0A03
1	0x0B01	0x0B02	0x0B03
2	0x0C01	0x0C02	0x0C03
3	0x0D01	0x0D02	0x0D03

There are a couple of things to note about the arrangement of the data: One, the data loaded when the array was declared was loaded by incrementing through the first index and then the second. Two, the index runs from 0 to the declared size -1. This is because zero is a legitimate index value, so declaring an array as K_val[3] actually creates 3 locations within the array indexed by 0, 1, and 2.

Now, how was the data in the array actually stored in data memory? If we do a memory dump of the data memory starting at the beginning address of the array, and assume a big endian format, the data should appear in memory as follows:

Memory 1.1

```
0x0100:   0x0A 0x01 0x0A 0x02 0x0A 0x03 0x0B 0x01
0x0108:   0x0B 0x02 0x0B 0x03 0x0C 0x01 0x0C 0x02
0x0110:   0x0C 0x03 0x0D 0x01 0x0D 0x02 0x0D 0x03
```

So, using the previous equation to generate an address for the element stored at [1][3], we get:

```
Address = 0x0100 + (byte_per_var*1) + (byte_per_
  var*3*3)
Address = 0x0100 + (2*1) + (2*3*3)
Address = 0x0114
```

From the dump of data memory, the data at 0×0114 and 0×0115 is $0 \times 0D$ and 0×02, resulting in a 16-bit value of $0 \times 0D02$, which matches the value that should be in K_vals[1][3].

1.4 Communications Protocols

When two tasks in a multitasking system want to communicate, there are three potential problems that can interfere with the reliable communication of the data. The receiving task may not be ready to accept data when the sending task wants to send. The sending task may not be ready when the receiving task needs the data. Alternatively, the two tasks may be operating at significantly different rates, which means one of the two tasks can be overwhelmed in the transfer. To deal with these timing related problems, three different communications protocols are presented to manage the communication process.

The simple definition of a protocol is "a sequence of instructions designed to perform a specific task." There are diplomatic protocols, communications protocols, even medical protocols, and each one defines the steps taken to achieve a desired result, whether the result is a treaty, transfer of a date, or treating an illness. The power of a protocol is that it plans out all the steps to be taken, the order in which they are performed, and the way in which any exceptions are to be handled.

The communications protocols presented here are designed to handle the three different communications timing problems discussed previously. Broadcast is designed to handle transfers in which the sender is not ready when the receiver wants data. Semaphore is designed to handle transfers in which the receiver is not ready when the sender wants to send data. Buffer is designed to handle transfers in which the rates of the two tasks are significantly different.

1.4.1 Simple Data Broadcast

A simple broadcast data transfer is the most basic form of communications protocol. The transmitter places its information, and any updates, in a common globally accessible variable. The receiver, or receivers, of the data then retrieve the information when they need it. Because the receiver is not required to acknowledge its reception of the data, and the transmitter provides no indication of changes in the data, the transfer is completely asynchronous. A side effect of this form of transfer is that no event timing is transferred with the data; it is purely a data transfer.

This protocol is designed to handle data that doesn't need to include event information as part of the transfer. This could be due to the nature of the data, or because the data only takes on significance when combined with other events. For example, a system that time stamps the reception of serial communications into a system. The current time would be posted by the real time clock, and updated as each second increments. However, the receiver of the current time information is not interested in each tick of the clock, it only needs to know the current time, when a new serial communication has been received. So, the information contained in the variables holding the current time are important, but only when tied to secondary event of a serial communication. While a handshaking protocol could be used for this transfer, it would involve placing an unreasonable overhead on the receiving task in that it would have to acknowledge event tick of the clock.

Because this transfer does not convey event timing, there are some limitations associated with its use:

1. The receiving tasks must be able to tolerate missing intermediate updates to the data. As we saw in the example, the receiver not only can tolerate the missing updates, it is more efficient to completely ignore the data until it needs it.

2. The sending task must be able to complete all updates to the data, before the information becomes accessible to the receiver. Specifically, all updates must be completed before the next time the receiving task executes; otherwise, the receiving task could retrieve corrupted data.

3. If the sending task cannot complete its updates to the date before a receiving task gains access to the data, then:

   • The protocol must be expanded with a flag indicating that the data is invalid, a condition that would require the receiver to wait for completion of the update.

   • Alternatively, the receiver must be able to tolerate invalid data without harm.

As the name implies, a broadcast data transfer is very much like a radio station broadcast. The sender regularly posts the most current information in a globally accessible location, where the receiver may retrieve the data when it needs it. The receiver then retrieves the data when its internal logic dictates. The advantage of this system is that the receiver only retrieves the data when it needs it and incurs no overhead to ignore the data when it does not need the data. The down side to this protocol is simple: the sender has no indication of when the receiver will retrieve the data, so it must continually post updates whether they are ultimately needed or not. This effectively shifts the overhead burden to the transmitter. And, because there is no handshaking between the sender and receiver, the sender has no idea whether the receiver is even listening. Therefore, the transfer is continuous and indefinite.

If we formalize the transfer into a protocol:

- The transmitter posts the most recent current data to a global variable accessible by the receiver.

- The receiver then retrieves the current data, or not, whenever it requires the information.

- The transmitter posts updates to the data, as new information become available.

Because neither party requires any kind of handshaking from the other and the timing is completely open and the *broadcast protocol* is limited to only transferring data, no event timing is included. A receiver that polls the variable quickly enough may catch all the updates, but there is nothing in the protocol to guarantee it. Therefore, the receiving task only really knows the current value of the data and either does not know or care about its age or previous values.

The first question is probably, "Why all this window dressing for a transfer using a simple variable?" One task stores data in the holding variable and another retrieves the data, so what's so complicated? That is correct—the mechanism is simple—but remember the limitations that went along with the protocol. They are important, and they more than justify a little window dressing:

1. The transmitting task must complete any updates to a broadcast variable before the receiver is allowed to view the data.

2. If the transmitting task cannot complete an update, it must provide an indication that the current data is not valid, and the receiving task must be able to tolerate this wait condition.

3. Alternatively, the receiver must be tolerant of partially updated data.

These restrictions are the important aspect of the Broadcast Transfer and have to be taken into account when choosing a transfer protocol, or the system could leak data.

1.4.2 Event-Driven Single Transfer

Data transfer in an event-driven single transfer involves not only the transfer of data but also creates a temporary synchronization between the transmitter and the receiver. Both information and timing cross between the transmitter and receiver.

For example, a keyboard-scanning task detects a button press on the keyboard. It uses an event-driven single transfer to pass the code associated with the key onto a command-decoding task. While the code associated with the key is important, the fact that it is a change in the status of the keyboard is also important. If the event timing were not also passed as part of the transfer, the command decoding task would not be able to differentiate between the initial press of the key and a later repeat of the key press. This would be a major problem if the key being pressed is normally repeated as part of the system's operations. Therefore, event-driven single transfers of data require an indication of new data from the transmitter.

A less obvious requirement of an event-driven single transfer is the acknowledgment from the receiver indicating that the data has been retrieved. Now, why does the transmitter need to know the receiver has the data? Well, if the transmitting routine sends one piece of data and then immediately generates another to send, it will need to either wait a sufficiently long period of time to guarantee the receiver has retrieved the first piece of data, or have some indication from the receiver that it is safe to send the second piece of data. Otherwise, the transmitter runs the risk of overrunning the receiver and losing data in the transfer. Of the two choices, an acknowledge from the receiver is the more efficient use of processor time, so an acknowledge is required as part of any protocol to handle event-driven single transfers.

What about data—is it a required part of the transfer? Actually, no, a specific transfer of data is not necessary because the information can be implied in the transfer. For example, when an external limit switch is closed, a monitoring task may set a flag indicating the closure. A receiving task acknowledges the flag by clearing it, indicating it acknowledges the event. No format data value crossed between the monitoring and receiving tasks because the act of setting the flag implied the data by indicating that the limit switch had closed.

Therefore, the protocol will require some form of two-way handshaking to indicate the successful transfer of data, but it does not actually have to transfer data. For that reason, the protocol is typically referred to as a *semaphore protocol*, because signals for both transfer and acknowledgment are required.

The protocol for handling event-driven single transfers should look something like the following for a single transfer:

- The transmitter checks the last transfer and waits if not complete.

- The transmitter posts the current data to a global variable, accessible by the receiver (optional).

- The transmitter sets a flag indicating new data is available.

- The transmitter can either wait for a response or continue with other activities.

- The receiver periodically polls the new data flag from the transmitter.

- If the flag is set, it retrieves the data (optional), and clears the flag to acknowledge the transfer.

There are a few limitations to the protocol that should be discussed so the designer can accurately predict how the system will operate during the transfer.

1. If the transmitter chooses to wait for an acknowledgment from the receiver, before continuing on with other activities:

    a. Then the transmitter can skip the initial step of testing for an acknowledge prior to posting new data.

    b. However, the transmitter will be held idle until the receiver notices the flag and accepts the data.

2. If, on the other hand, the transmitter chooses to continue on with other activities before receiving the acknowledgment:

    a. The transmitter will not be held idle waiting for the receiver to acknowledge the transmitter.

    b. However, the transmitter may be held idle at the initial step of testing for an acknowledge prior to most new data.

It is an interesting choice that must be made by the designer. Avoid holding the transmitter idle and risk a potential delay of the next byte to be transferred, or accept the delay knowing that the next transfer will be immediate. The choice is a trade-off of transmitter overhead versus a variable delay in the delivery of some data.

Other potential problems associated with the semaphore protocol can also appear at the system level and an in-depth discussion will be included in the appropriate chapters. For now, the important aspect to remember is that a semaphore protocol transfers both data and events.

1.4.3 Event-Driven Multielement Transfers

In an event-driven multielement transfer, the requirement for reliable transfer is the same as it is for the event-driven single transfer. However, due to radically different rates of execution, the transmitter and receiver cannot tolerate the synchronization required by the semaphore protocol. What is needed is a way to slow down the data from the transmitter, so the receiver can process it, all without losing the reliability of a handshaking style of transfer.

As an example, consider a control task sending a string of text to a serial output task. Because the serial output task is tied to the slower transfer rate of the serial port, its execution will be significantly slower than the control task. So, either the control task must slow down its execution to accommodate the serial task, or some kind of temporary storage is needed to hold the message until the serial task is ready to send it. Given the control task's work is important and it can't slow down to the serial task's rate, then the storage option is the only one that makes sense in the application.

Therefore, the protocol will require at a minimum; some form of data storage, a method for storing the data, and a method for retrieving it. It is also assumed that the storage and retrieval methods will have to communicate the number of elements to be transferred as well.

A protocol could be set up that just sets aside a block of data memory and a byte counter. The transmitting task would load the data into the memory block and set the byte counter to the number of data elements. The receiving task can then retrieve data until its count equals the byte counter. That would allow the transmitting task to run at its rate loading the data, and allow the receiver to take that data at a rate it can handle. However, what happens if the transmitting task has another block of data to transfer, before the receiving task has retrieved all the data?

A better protocol is to create what is referred to as a *circular buffer*, or just buffer protocol. A buffer protocol uses a block of data memory for storage, just as the last protocol did. The difference is that a buffer also uses two address pointers to mark the locations of the last store and retrieve of data in the data block. When a new data element is added to the data memory block, it is added in the location pointed to by the storage pointer and the pointer is incremented. When a data element is retrieved, the retrieval pointer is used to access the data and then it is incremented. By comparing the pointers, the transmitting and receiving tasks can determine:

1. Is the buffer empty?

2. Is there data present to be retrieved?

3. Is the buffer is full?

So, as the transmitter places data in the buffer, the storage pointer moves forward through the block of data memory. And as the receiver retrieves data from the buffer, the retrieval pointer chases the storage pointer. To prevent the system from running out of storage, both pointers are designed to "wraparound" to the start of the data block when they pass the end. When the protocol is operating normally, the storage pointer will jump ahead of the retrieval pointer, and then the retrieval pointer will chase after it. Because the circular buffer is essentially infinite in length, because the pointers always wraparound, the storage space will be never run out. In addition, the two pointers will chase each other indefinitely, provided the transmitter doesn't stack up so much data that the storage pointer "laps" the retrieval pointer.

Therefore, how does the buffer protocol look from the pointer of view of the transmitting task and the receiving task. Let's start with the transmit side of the protocol:

*   The transmitter checks to see if the buffer is full, by comparing the storage pointer to the retrieval pointer.

*   If the buffer is not full, it places the data into the buffer using the storage pointer and increments the pointer.

*   If the transmitter wishes to check on the receiver's progress, it simply compares the storage and retrieval pointers.

From the receiver's point of view:

*   The receiver checks the buffer to see if data is present by comparing the storage and retrieval pointers.

*   If the pointers indicate data is present, the receiver retrieves the data using the retrieval pointer and increments the pointer.

Therefore, the two tasks have handshaking through the two pointers, to guarantee the reliable transfer of data. However, using the data space and the pointers allows the receiving task to receive the data at a rate it can handle, without holding up the transmitter.

Implementing a buffer protocol can be challenging though, due to the wraparound nature of the pointers. Any increment of the pointers must include a test for the end of the buffer, so the routine can wrap the pointer back around to the start of the buffer. In addition, the comparisons for buffer full, buffer empty, and data present can also become complicated due to the wraparound.

In an effort to alleviate some of this complexity, the designer may choose to vary the definition of the storage and retrieval pointers to simplify the various comparisons. Unfortunately, no one definition will simplify all the comparisons, so it is up to the designer to choose which definition works best for their design. Table 1.4 shows all four possible definitions for the storage and retrieval pointers, plus the comparisons required to determine the three buffer conditions.

Table 1.4

Pointer Definitions	Comparisons	Meaning

Storage > last element stored
Retrieval > last element retrieved
	IF (Storage == Retrieval)
	IF (Storage+1 == Retrieval)
	IF (Storage <> Retrieval)
		then buffer is empty
		then buffer is full
		then data present

Storage > last element stored
Retrieval > next element retrieved
	IF (Storage+1 == Retrieval)
	IF (Storage == Retrieval)
	IF (Storage+1 <> Retrieval)
		then buffer is empty
		then buffer is full
		then data present

Storage > next element stored
Retrieval > last element retrieved
	IF (Storage == Retrieval +1)
	IF (Storage == Retrieval)
	IF (Storage <> Retrieval+1)
		then buffer is empty
		then buffer is full
		then data present

Storage > next element stored
Retrieval > next element retrieved
	IF (Storage == Retrieval)
	IF (Storage+1 == Retrieval)
	IF (Storage <> Retrieval)
		then buffer is empty
		then buffer is full
		then data present

It is interesting that the comparisons required to test each condition don't change with the definition of the pointers. All that does change is that one or the other pointer has to be incremented before the comparison can be made. The only real choice is which tests will have to temporarily increment a pointer to perform its test, the test for buffer full, or the test for buffer empty/data available. What this means for the designer is that the quicker compare can be delegated to either the transmitter (checking for buffer full) or the receiver (checking for data present). Since the transmitter is typically running faster, then options one or four are typically used.

Also note that the choices are somewhat symmetrical; options one and four are identical, and options two and three are very similar. This makes sense, since one and four use the same sense for their storage and retrieval pointers, while the pointer sense in two and three are opposite and mirrored.

One point to note about buffers, because they use pointers to store and retrieve data and the only way to determine the status of the buffer is to compare the pointers, the buffer-full test always returns a full status when the buffer is one location short of being full. The reason for this is because the comparisons for buffer empty and buffer full turn out to be identical, unless the buffer-full test assumes one empty location.

If a buffer protocol solves the problem of transferring data between a fast and slow task, then what is the catch? Well, there is one and it is a bear. The basic problem is determining how big to make the storage space. If it is too small, then the transmitter will be hung up waiting for the receiver again because it will start running into buffer-full conditions. If it is too large, then data memory is wasted because the buffer is underutilized.

One final question concerning the buffer protocol is how is the size of the data storage block determined? The size can be calculated based on the rates of data storage, data retrieval, and the frequency of use. Or the buffer can be sized experimentally by starting with an oversized buffer and then repeatedly testing the system while decreasing the size. When the transmitting tasks starts hitting buffer-full conditions, the buffer is optimized. For now, just assume that the buffer size is sufficient for the designs need.

1.5   Mathematics

In embedded programming, mathematics is the means by which a program models and predicts the operation of the system it is controlling. The math may take the form of thermodynamic models for predicting the best timing and mixture in an engine, or it may be a simple time delay calculation for the best toasting of bread. Either way, the math is how a microcontroller takes its view of the world and transforms that data into a prediction of how to best control it.

For most applications, the math libraries supplied with the compiler will be sufficient for the calculations required by our models and equations. However, on occasion, there will be applications where it may be necessary to "roll our own" routines, either for a specialized math function, or just to avoid some speed or data storage inefficiencies associated with the supplied routines. Therefore, a good understanding of the math underlying the libraries is important, not only to be able to replace the routines, but also to evaluate the performance of the supplied functions.

1.5.1   Binary Addition and Subtraction

Earlier in the chapter, it was established that both base ten and binary numbering system use a digit position system based on powers of the base. The position of the digit also plays a part in the operation of the math as well. Just as base-ten numbers handle mathematics one digit at a time, moving from smallest power to largest, so do binary numbers in a computer. In addition, just like base-ten numbers, carry and borrow operations are required to roll up over- or underflows from lower digits to higher digits. The only difference is that binary numbers carry up at the value 2 instead of ten.

So, using this basic system, binary addition has to follow the rules in Tables 1.5 and 1.6:

Table 1.5

If the carry_in from the next lower digit $= 0$	
▶ $0 + 0 + $ carry_in	results in 0 & carry_out $= 0$
▶ $1 + 0 + $ carry_in	results in 1 & carry_out $= 0$
▶ $0 + 1 + $ carry_in	results in 1 & carry_out $= 0$
▶ $1 + 1 + $ carry_in	results in 0 & carry_out $= 1$

Table 1.6

If the carry_in from the next lower digit $= 1$	
▶ $0 + 0 + $ carry_in	results in 1 & carry_out $= 0$
▶ $1 + 0 + $ carry_in	results in 0 & carry_out $= 1$
▶ $0 + 1 + $ carry_in	results in 0 & carry_out $= 1$
▶ $1 + 1 + $ carry_in	results in 1 & carry_out $= 1$

Using these rules in the following example of binary addition produces a result of 10101100. Note the carry_in values are in bold:

Example 1.7

```
111 111 <--carry bits
 00110101
+01110111
         0   1 + 1           =  0 with carry_out
        0    1 + 0 + carry_in = 0 with carry_out
       1     1 + 1 + carry_in = 1 with carry_out
      1      0 + 0 + carry_in = 1
     0       1 + 1           =  0 with carry_out
    1        1 + 1 + carry_in = 1 with carry_out
   0         1 + 0 + carry_in = 0 with carry_out
  1          0 + 0 + carry_in = 1
 10101100
```

Converting the two values to decimal, we get $53+119$, for a total of 172. 172 in binary is 1010110, so the math checks.

Binary subtraction operates in a similar manner, using the borrow instead of carry. Building a similar table of rules for subtraction yields Tables 1.7 and 1.8:

Table 1.7

If the borrow_in from the next lower digit $= 0$
▶ 0 − 0 − borrow_in   results in 0 & borrow_out $= 0$
▶ 1 − 0 − borrow_in   results in 1 & borrow_out $= 0$
▶ 0 − 1 − borrow_in   results in 1 & borrow_out $= 1$
▶ 1 − 1 − borrow_in   results in 0 & borrow_out $= 0$

Table 1.8

If the borrow_in from the next lower digit $= 1$
▶ 0 − 0 − borrow_in   results in 1 & borrow_out $= 1$
▶ 1 − 0 − borrow_in   results in 0 & borrow_out $= 0$
▶ 0 − 1 − borrow_in   results in 0 & borrow_out $= 1$
▶ 1 − 1 − borrow_in   results in 1 & borrow_out $= 1$

Using these rules for subtraction in the following example produces a result of 00111110. Note the borrow values are in bold:

Example 1.8

```
111111   <--borrow
10110101
-01110111
        0  1 - 1                    = 0
       1   0 - 1                    = 1 with borrow_out
      1    1 - 1 - borrow_in        = 1 with borrow_out
     1     0 - 0 - borrow_in        = 1 with borrow_out
    1      1 - 1 - borrow_in        = 1 with borrow_out
   1       1 - 1 - borrow_in        = 1 with borrow_out
  0        0 - 1 - borrow_in        = 0 with borrow_out
 0         1 - 0 - borrow_in        = 0
00111110
```

Converting the two values to decimal, we get 181 – 119, for a difference of 62. 62 in binary is 00111110, so, again, the math checks.

In addition, as expected, binary addition and subtraction are not any different than addition and subtraction in base ten. The carry_out carries up a value of 1 to the next digit, and a borrow_out carries up a value of −1 to the next digit. This makes sense—addition and subtraction are universal concepts, and should be independent of the base of the number system.

1.5.2 Binary Multiplication

In addition, we added each digit together, one at a time, and carried the overflow up to the next digit as a carry. In multiplication, we multiply each digit together, one at a time, and carry the overflow up to the next digit as a carry as well. The only difference is that the carry may be greater than 1.

For multipliers with more than one digit, we again handle each one separately, multiplying the digit through all the digits of the multiplicand, and then add the results from each digit together to get a result, making sure to align the digits with the digit in the multiplier. For example:

Example 1.9

```
    123   Multiplicand
   x321   Multiplier

    123   (1 x 123 x 1)      the x1 is due to the position of the
                             1 in the multiplier

   2460   (2 x 123 x 10)     the x10 is due to the position of the
                             2 in the multiplier

 +36900   (3 x 123 x 100)    the x100 is due to the position of the
                             3 in the multiplier

  37483   Result
```

Thus is the essence of long multiplication—straightforward and simple, if somewhat tedious. So, it should come as no surprise that the process is no different for binary multiplication. Each bit in the multiplier, 1 or 0, is multiplied by each of the bits in the multiplicand. And when all the bits have been multiplied, we add together the result, making sure that we keep each interim result lined up with its multiplier bit. Just as straightforward and simple, although a little less tedious as we only have to multiply by 1 or 0.

Therefore, if we convert 6 and 11 into binary and multiply them together, we should get the binary equivalent of 66.

Example 1.10

```
   1011  (11 in decimal)
  x0110  (6 in decimal)
00000000  (0 x 1011, the original value x 1)
00010110  (1 x 10110, the original value x 2)
00101100  (1 x 101100, the original value x 4)
+00000000  (0 x 1011000, the original value x 8)
01000010
```

In addition, 01000010 in decimal is 66, so once again the math checks out.

Before we move on to division, let's take a minute and check out some interesting points in binary multiplication.

1. The digit by digit multiply is only multiplying by 1 or 0, so the process of multiplying each bit of the multiplier with each bit of the multiplicand is very simple. In fact, algorithms for binary multiply typically don't bother with the bit-by-bit multiply; they just check the multiplier bit and if it is set, they shift over the multiplicand and add it into the result.

2. The act of shifting the multiplicand left to align with the multiplier, for each bit in the multiplier, would be a waste of time. It is simpler to just create a temporary variable to hold the shifted form of the multiplier from the last bit, and then shift it once for the next. That way the temporary variable only has to be shifted once for each bit of the multiplier.

3. The bits in the multiplier will have to be tested one at a time, from the LSB to the MSB, to perform the multiplication. If we can use a temporary variable to hold a shift copy of the multiplicand, why not use a temporary variable to hold a shifted copy of the multiplier that shifts to the right? That way the bit tested in the multiplier is always the LSB.

4. The result of the multiply was nearly twice the size of the multiplier and multiplicand. In fact, if the multiplier and multiplicand were both 1111, it would have been twice the size. So, to prevent losing any bits in the result to roll over, the multiply algorithm will have to have a result at least twice the size of input variables, or have a number of bits equal to the total bits in both the input variables, if they are different sizes.

Using these insights, an efficient binary multiply routine can be created that is fairly simple. It is just a matter of shifting and adding inside a loop:

Algorithm 1.1

```
char A       ; multiplicand
char B       ; multiplier
int  C       ; 16-bit result
int  temp_a  ; 16-bit temp holding variable for
               multiplicand
char temp_b  ; 8-bit temp holding variable for
               multiplier

C = 0
Temp_a = A
Temp_b = B
FOR I = 0 to 7                  ; multiplier is 8 bits
     IF (LSB of B = 1) THEN C = C + temp_a
     SHIFT temp_a LEFT 1    ;multiplicand * 2
     SHIFT temp_b RIGHT 1   ; multiplier / 2
NEXT I
```

For each pass through the loop, the LSB of the multiplier is tested. If the bit is set, then the multiplicand is added to the result. If the bit is clear, the multiplicand is not added to the result. In either case, the temporary copy of the multiplicand is multiplied by 2 and the temporary copy of the multiplier is divided by 2 for the next pass through the loop. The loop repeats until all of the multiplicand bits have been tested.

1.5.3   Binary Division

Binary division is also a simplified version of base 10 long division. Remember the techniques for base 10 division from school? Take the divisor and see if it divides into the first digit of the dividend and if it does, put the number of times it does above the line. Then multiply that result by the divisor and subtract it from the dividend. Pass the remainder down to the next line and repeat the process until the remainder is less than the divisor. At that point, you have a result and any left-over remainder.

Example 1.11

```
                0128  ← result
Divisor →  12) 1546  ← dividend
                0000  (0 x 12 x 1000)
                1546
                1200  (1 x 12 x 100)
                 346
                 240  (2 x 12 x 10)
                 106
                  96  (8 x 12 x 1)
                  10  ← remainder
```

Binary division operates in the same way; the divisor is left shifted until its LSB is in the same digit position as the MSB of the dividend, and the divisor is subtracted from the dividend. If the result is positive, the corresponding bit in the result is set, the divisor is right shifted one position, and the process is repeated until the result is less than the remainder. As an example, let's take 15 and divide it by 5. 15 is 1111 in binary, and 5 is 0101. Performing the divide:

Example 1.12

```
                    0011       Result
Divisor     0101 ) 0001111     Dividend
                   0101000     (0 x 0101 x 1000)
                  -0010100
                   0001111
                   0010100     (0 x 0101 x 0100)
                  -0010100
                   0001111
                  -0001010     (1 x 0101 x 0010)
                   0000101
                  -0000101     (1 x 0101 x 0001)
                         0     Remainder
```

We end up with a result of 0011, or 3 in decimal, with a 0 remainder. Since 15 divided by 5 is equal to 3 with no remainder, the math checks.

If we examine the division process, we find some of the same interesting points that we found in examining the multiply process:

1. Prior to beginning the divide, the divisor had to be left-shifted until its LSB was in the same digit position as the dividend's MSB. This means the algorithm will require a temporary variable for the divisor.

2. The difference between the dividend and the shifted devisor will also have to be held in a temporary variable.

3. To accommodate the initial subtractions of the divisor, the dividend had to be padded with additional zeros. So, the minimum length of the temporary variable used for the dividend must be at least equal to the total number of bits in the dividend and the divisor, -1. Remember that the first subtraction is with the LSB of the divisor in the same position as the MSB or the dividend.

4. The temporary variable used to hold the dividend will hold the remainder at the end of the operation.

5. The bit set in the result for each successful subtraction of the divisor is the same digit position as the LSB of the divisor.

Using these insights, an efficient binary multiply routine can be created that is fairly simple. It is just a matter of shifting, testing, and subtracting with a bit set. The resulting algorithm is similar to the multiplication algorithm:

Algorithm 1.2

```
char A          // divisor
char B          // dividend
char C          // result
char R          // remainder
int  temp_a     // 16-bit temp holding variable for divisor
int  temp_b     // 16-bit temp holding variable for dividend

C = 0
temp_a = SHIFT A LEFT 7                 // left shift divisor 7x
temp_b = B                              // dividend
FOR I = 0 to 7                          // loop repeats 8x
      SHIFT C LEFT 1                    // shift to next bit in R
      temp_b = temp_b - temp_a
      if (borrow = 0)
            then
                  C = C + 1;
                  ; set the bit
            else
                  temp_b = temp_b + temp_a // undo subtract
      endif
      SHIFT temp_a RIGHT 1              // shift the divisor 1
NEXT I
R = temp_b
```

At the start of the routine, the divisor and dividend are copied into their temporary variables, and the divisor is left-shifted 7 times. This leaves the divisor LSB in the same digit position as the MSB of the dividend. The divisor is subtracted from the dividend and the result is checked for a borrow; remember that the borrow indicated that the divisor is larger than the dividend, resulting in a negative difference. If the borrow is set, then the divisor is added back into the dividend to undo the subtraction. If the borrow is clear, then the dividend can be subtracted, and the corresponding bit in the result is set. The divisor is right-shifted, and the loop repeats for all 8 bits in the dividend.

Note that the bit set in the result is always the LSB, and the result is shifted one position to the left at the start of each loop. But, from the section above, we expected to set the result bits from the MSB down to the LSB, corresponding with the LSB of the divisor. Why the change? The algorithm could be done as it is described previously, but it would require another temporary variable to hold a single bit corresponding to the LSB of the divisor. The bit would be shifted with the divisor, and if the divisor was subtracted, we would OR the bit into the result. However, by setting the LSB and shifting left each time the divisor is shifted right, we accomplish the same result, and it doesn't require an additional temporary variable for a single bit.

1.6   Numeric Comparison

In the previous example of division, we compared the divisor to the dividend on each pass through the loop to determine if the divisor was less than or equal to the dividend. We did this with a subtraction and a test of the borrow flag. If the result of the subtraction was negative, then the divisor was greater than the dividend. But what about greater than, equal to, less than or equal to, or just plain not equal—how are those comparisons performed? The answer is that we still do a subtraction, but we just have to test for the right combination of positive, negative, or zero.

Fortunately, microcontrollers are well equipped to perform these tests because whenever a microcontroller performs a subtraction, status flags in the microcontroller's status register record information about the result. Typically, this information includes both a borrow and zero flags. The borrow flag indicates whether the result of the operation was positive or negative and the zero flag tells us if the result of the operation was zero.

Therefore, if we look at the results of a subtraction, by testing the flags we should be able to determine every combination of relationships between the two values:

- If the result of the subtraction is zero, then the two values are equal.

- If the borrow flag is set, then the larger value was subtracted from a smaller value.

- If the borrow is clear, then the smaller was subtracted from the larger, unless the zero flag is also set; in which case the values are equal.

Fairly simple, but unfortunately, there is a little more to it than just less than, greater than, and equal. There is also greater than or equal, less than or equal, and just not equal. The microcontroller could just perform the subtraction and test for all of the possible combinations of flags, but if both flags have to be tested for every condition that could be inefficient. Some conditions will require that both flags be tested, and others will require only one test. Assuming that the tests exhibit some symmetry, it should be possible to swap the order of the variables in the subtraction to give us a set of operations that can determine the relationship with only one test. So, let's build a table showing both possible ways the subtraction can be performed for each of the tests and see if we can find a single test for each condition (see Table 1.9).

Table 1.9: Subtraction-based Comparisons

Relationship	Subtraction	Result	Tests Required
A > B	B – A	Negative	Borrow = true
	A – B	Positive & nonzero	Borrow = false and Zero = false
A => B	B – A	Negative or zero	Borrow = true or Zero = true
	A – B	Positive	Borrow = false
A = B	B – A	Zero	Zero = true
	A – B	Zero	Zero = true
A <= B	B – A	Positive	Borrow = false
	A – B	Negative or zero	Borrow = true or Zero = true
A < B	B – A	Positive & nonzero	Borrow = false and Zero = false
	A – B	Negative	Borrow = true
A != B	B – A	Nonzero	Zero = false
	A – B	Nonzero	Zero = false

From the table, we can determine that:

- For A > B, subtract A from B and test for Borrow = true.

- For A => B, subtract B from A and test for Borrow = false.

- For A = B, subtract either variable from the other and test for Zero = true.

- For A <= B, subtract A from B and test for Borrow = false.

- For A < B, subtract B from A and test for Borrow = true.

- For A != B, subtract either variable from the other and test for Zero = false.

As predicted, by swapping the order of the variables in some of the subtractions, we can simplify the tests down to a single bit test for each of the possible comparisons.

1.6.1 Conditional Statements

Now that we have the math of the comparison figured out, what about the conditionals statements that use the comparison? If we assume a C-like programming language, the conditional statements include IF/THEN/ELSE, SWITCH/CASE, DO/WHILE, and FOR/NEXT. While some of the statements are related, each has its own unique function and requirements.

The IF/THEN/ELSE, or IF statement, is the most basic conditional statement. It makes a comparison and, based on the result, changes the flow of execution in the program. The change can be to include an additional section of the program, or to select between two different sections. The comparison can be simple or compound. The statements can even be nested to produce a complex decision tree. In fact, the IF statement is the basis for all of the conditional statements, including the SWITCH/CASE, DO/WHILE, and FOR/NEXT.

For now, let's start with just the basic IF conditional statement. In a typical IF, a comparison is made using the techniques described in the last section. If the result of the comparison is true, the block of instructions associated with the THEN part of the statement is executed. If the result of the comparison is false, the block of instructions associated with the ELSE part of the statement is executed. Note, the ELSE portion of the statement is optional in most high-level languages. If the ELSE is omitted, then a false result will cause the program to fall through the instruction with no action taken. The implementation of the statement typically takes the following form:

Code Snippet 1.5

```
IF (comparison)
    THEN
        {Section_a}
    ELSE
        {Section_b}
ENDIF              ; note some languages use {}
                   around the two
                   ; sections in place of ENDIF
```

A common variation of the basic IF is to combine two or more statements into a more complex comparison. This is commonly referred to as Nested IF statements, and may involve new IF statements in either, or both, the THEN or ELSE side of the statement. By nesting the IF statements, several different comparisons can be obtained:

- Complex combinations, involving multiple variable, can be tested for a single combination.

- A single variable can be compared to multiple values.

- Alternatively, multiple variables can be compared against multiple values.

Let's start with the simpler comparison, comparing multiple variables for a single combination. This comparison can be implemented by nesting multiple IF statements, the first IF comparing the first variable against its value and the THEN portion of the statement, another IF comparing the second variable against its value, and so on for all the variables and values.

Code Snippet 1.6

```
IF (Var_A > 5)
     THEN IF (Var_B < 3)
            THEN IF (Var_C <> 6)
                THEN
                      {Section_a}
ENDIF
IF (Var_A <= 5) THEN {Section_b}
IF (Var_B >= 3) THEN {Section_b}
IF (Var_C == 6) THEN {Section_b}
```

However, this is an inefficient use of program memory because each statement includes the overhead of each IF statement. The ELSE condition must be handled separately with multiple copies of the Section b code.

The better solution is to put all the variables and the values in a single compounded IF statement. All of the variables, compared against their values, are combined using Boolean operators to form a single yes or no comparison. The available Boolean operators are AND (&&), OR (||), and NOT (!). For example:

Code Snippet 1.7

```
IF (Var_A > 5) && (Var_B < 3) && (Var_C <> 6)
   THEN
        {Section_a}
   ELSE
        {Section_b}
   ENDIF
```

This conditional statement will execute Section_a if; Var_A > 5 and Var_B < 3, and Var_C is not equal to 6. Any other combination will result in the execution of Section_b. Therefore, this is a smaller, more compact, implementation that is much easier to read and understand in the program listing.

The next IF statement combination to examine involves comparing a single variable against multiple values. One of the most common examples of this type of comparison is a WINDOW COMPARISON. In a window comparison, a single variable is compared against two values that form a window, or range, of acceptable or unacceptable values. For instance, if the temperature of a cup of coffee is greater than 40°C, but less than 50°C, it is considered to have the right temperature. Warmer or colder, it either is too cold or too hot to drink. Implementing this in a IF statement would result in the following:

Code Snippet 1.8

```
IF (Temperature > 40) && (Temperature < 50)
    THEN
        {Drink}
    ELSE
        {Don't_Drink}
    ENDIF
```

The compound IF statement checks for both a "too hot" and "too cool" condition, verifying that the temperature is within a comfortable drinking temperature range. The statement also clearly documents what range is acceptable and what is not.

Another implementation of comparing a single value against multiple values is the ELSE IF combination. In this configuration, a nested IF is placed in the ELSE portion of the statement, creating a string of comparisons with branches out of the string for each valid comparison. For instance, if different routines are to be executed for each of several different values in a variable, an ELSE IF combination can be used to find the special values and branch off to each one's routine. The nested IF statement would look like the following:

Code Snippet 1.9

```
IF (Var_A = 5)
    THEN
        {Routine_5}
    ELSE IF (Var_A = 6)
        THEN
            {Routine_6}
        ELSE IF (Var_A = 7)
            THEN
                {Routine_7}
            ELSE
                {Other_Routine}
```

And,

- If Var_A is 5, then only Routine_5 is executed.

- If Var_A is 6, then only Routine_6 is executed.

- If Var_A is 7, then only Routine_7 is executed.

- If Var_A is not 5, 6, or 7, then only the Other_Routine is executed.

Now, if each statement checks for its value, why not just have a list of IF statements? What value does nesting the statements have? There are three reasons to nest the IF statements:

1. Nesting the IF statements saves one IF statement. If the comparison was implemented as a list of IF statements, a window comparison would be required to determine when to run the Other_Routine. It is only run if the value is not 5, 6, or 7.

2. Nesting the statements speeds up the execution of the program. In the nested format, if Routine_5 is executed, then when it is done, it will automatically be routed around the rest of the IF statements and start execution after the last ELSE. In a list of IF statements, the other three comparisons would have to be performed to get past the list of IF statements.

3. If any of the routines modify Var_A, there is the possibility that one of the later comparisons in the last might also be true, resulting in two routines being executed instead of just the one intended routine.

Therefore, nesting the ELSE IF statements has value in reduced program size, faster execution speed, and less ambiguity in the flow of the program's execution.

For more complex comparisons involving multiple variables and values, IF/THEN/ELSE statements can be nested to create a decision tree. The decision tree quickly and efficiently compares the various conditions by dividing up the comparison into a series of branches. Starting at the root of the tree, a decision is made to determine which half of the group of results is valid. The branches of the first decision then hold conditional statements that again determine which 1/4 set of solutions are valid. The next branch of the second decision then determines which 1/8 set of solutions is valid, and so on, until there is only one possible solution left that meets the criteria. The various branches resemble a tree, hence the name "decision tree."

To demonstrate the process, assume that the letters of a name—Samuel, Sally, Thomas, Theodore, or Samantha—are stored in an array of chars labeled NAME[]. Using a decision tree, the characters in the array can then be tested to see which name is present in the array. The following is an example of how a decision tree would be coded to test for the three names:

Code Snippet 1.10

```
IF (NAME[0] == 'S')
        THEN IF (NAME[2] == 'm')
                THEN IF (NAME[3] == 'a')
                        THEN Samantha_routine();
                        ELSE Samuel_routine();
                ELSE Sally_routine
        ELSE IF (NAME[2] == 'o')
                THEN Thomas_routine();
                ELSE Theodore_routine();
```

The first IF statement uses the letter in location 0 to differentiate between S and T to separate out Thomas and Theodore from the list of possible solutions. The next IF in both branches uses the letter in location 2 to differentiate between M and L to separate out Sally from the list of possible solutions, and to differentiate between Thomas and Theodore. The deepest IF uses the letter in location 3 to differentiate between Samantha and Samuel. So, it only takes two comparisons to find Thomas, Theodore, or Sally, and only three comparisons to find either Samantha or Samuel.

If, on the other hand, the comparison used a list of IF statements rather than a decision tree, then each IF statement would have been more complex, and the number of comparisons would have increased. With each statement trying to find a distinct name, all of the differentiating letters must be compared in each IF statement. The number of comparisons required to find a name jumps from a worst case of three (for Samantha and Samuel), to four and five for the last two names in the IF statement list. To provide a contrast, the list of IF statements to implement the name search is shown below:

Code Snippet 1.11

```
IF (NAME[0] == 'S') && (NAME[2] == 'm') && (NAME[3]
    == 'a')
    THEN Samantha_routine;
IF (NAME[0] == 'S') && (NAME[2] == 'm') && (NAME[3]
    == 'u')
    THEN Samuel_routine;
IF (NAME[0] == 'S') && (NAME[2] == 'a')
    THEN Sally_routine;
IF (NAME[0] == 'T') && (NAME[2] == 'o')
    THEN Thomas_routine;
IF (NAME[0] == 'T') && (NAME[2] == 'e')
    THEN Theodore_routine;
```

As predicted, it will take four comparisons to find Thomas, and five to find Theodore, and the number of comparisons will grow for each name added to the list. The number of differentiating characters that will require testing will also increase and names that are similar to those in the list increase. A decision tree configuration of nested IF statements is both smaller and faster.

Another conditional statement based on the IF statement is the SWITCH/CASE statement, or CASE statement as it is typically called. The CASE statement allows the designer to compare multiple values against a single variable in the same way that a list of IF statements can be used to find a specific value. While a CASE statement can use a complex expression, we will use it with only a single variable to determine equality to a specific set of values, or range of values.

In its single variable form, the CASE statement specifies a controlling variable, which is then compared to multiple values. The code associated with the matching value is then executed. For example, assume a variable (Var_A) with five different values, and for each of the values a different block of code must be executed. Using a CASE statement to implement this control results in the following:

Code Snippet 1.12

```
SWITCH (Var_A)
{
        Case 0:    Code_block_0();
            Break;
        Case 1:    Code_block_1();
            Break;
        Case 2:    Code_block_2();
            Break;
        Case 3:    Code_block_3();
            Break;
        Case 4:    Code_block_4();
            Break;
        Default:  Break;
}
```

Note that each block of code has a break statement following it. The break causes the program to break out of the CASE statement when it has completed. If the break were not present, then a value of zero would have resulted in the execution of Code block 0, followed by Code block 1, then Code block 2, and so on through all the blocks in order. For this example, we only wanted a single block to execute, but if the blocks were a sequence of instructions and the variable was only supplying the starting point in the sequence, the case statement could be used to start the sequence, with Var_A supplying the starting point.

Also note the inclusion of a Default case for the statement; this is a catchall condition that will execute if no other condition is determined true. It is also a good error recovery mechanism when the variable in the SWITCH portion of the statement becomes corrupted. When we get to state machines, we will discuss further the advantages of the Default case.

1.6.2 Loops

Often it is not enough to simply change the flow of execution in a program. Sometimes what is needed is the ability to repeat a section until a desired condition is true, or while it is true. This ability to repeat until a desired result or do while a condition is true is referred to as an iteration statement, and it is very valuable in embedded programming. It allows designers to write programs that can wait for desired conditions, poll for a specific event, or even fine-tune a calculation until a desired result occurs. Building these conditional statements requires a combination of the comparison capabilities of the IF statement with a simple GOTO to form a loop.

Typically there are three main types of iterating instructions, the FOR/NEXT, the WHILE/DO, and the DO/WHILE. The three statements are surprisingly similar; all use a comparison function to determine when to loop and when not to, and all use an implied GOTO command to form the loop. In fact, the WHILE/DO and the DO/WHILE are really variations of each other, with the only difference being when the comparison is performed. The FOR/NEXT is unique due to its ability to automatically increment/decrement its controlling variable.

The important characteristic of the WHILE/DO statement, is that it performs its comparison first. Basically, WHILE a condition is true, DO the enclosed loop. Its logic is such that if the condition is true, then the code inside the loop is executed. When the condition is false, the statement terminates and begins execution following the DO. This has an interesting consequence: if the condition is false prior to the start of the instruction, the instruction will terminate without ever executing the routine within the loop. However, if the condition is true, then the statement will execute the routine within the loop until the condition evaluates as false. The general syntax of a DO/WHILE loop is shown below:

Code Snippet 1.13

```
WHILE (comparison)
    Routine();
DO
```

DO is a marker signifying the end of the routine to be looped, and the WHILE marks the beginning, as well as containing the comparison to be evaluated. Because the comparison appears at the beginning of the routine to be looped, it should be remembered that the condition is evaluated before the first execution of the routine and the routine is only executed if the condition evaluates to a true.

The mirror of the WHILE/DO is the DO/WHILE statement. It is essentially identical to the WHILE/DO, with the exception that it performs its comparison at the end. Basically, DO the enclosed loop, WHILE a condition is true. Its logic is such that, if the condition is true, then the code inside the loop is executed. When the condition is false, the statement terminates and begins execution following the WHILE. This has the alternate consequence that, even if the condition is false prior to the start of the instruction, the instruction will execute the routine within the loop at least once before terminating. If the condition is true, then the statement will execute the routine within the loop until the condition evaluates as false. The general syntax of a DO/WHILE loop is shown below:

Code Snippet 1.14

```
DO
    Routine();
WHILE (comparison)
```

DO is a marker signifying the beginning of the routine to be looped, and the WHILE marks the end, as well as containing the comparison to be evaluated. Because the comparison appears at the end of the routine to be looped, it should be remembered that the condition is evaluated after the first execution of the routine.

So, why have two different versions of the same statement? Why a DO/WHILE *and* a WHILE/DO? Well, the DO/WHILE could more accurately be described as a REPEAT/UNTIL. The ability to execute the routine at least once is desirable because it may not be possible to perform the comparison until the routine has executed. Some value that is calculated, or retrieved by, the routine may be needed to perform the comparison in the WHILE section of the command. The WHILE/DO is desirable for exactly the opposite reason—it may be catastrophic to make a change unless it is determined that a change is actually needed. Therefore, having the option to test before or test after is important, and is the reason that both variations of the commands exist.

The third type of iteration statement is the FOR/NEXT, or FOR statement. The FOR statement is unique in that it not only evaluates a condition to determine if the enclosed routine is executed, but it also sets the initial condition for the variable used in the conditions, and specifies how the variable is indexed on each iteration of the loop. This forms essentially a fully automatic loop structure, repeating any number of iterations of the loop until the termination condition is reached. For example, a FOR loop could look like the following:

Code Snippet 1.15

```
FOR (Var_A=0; Var_A<100; Var_A=Var_A+5)
    routine();
```

In the example, a variable Var_A is initially set to zero at the beginning of the loop. The value in Var_A is compared to 100, and if it is less than 100, then the routine is executed. After execution of the routine is complete, the variable is incremented by 5 and the comparison is repeated. The result is that the routine is executed, and the variable incremented by 5, over and over until the comparison is false.

Within the general format of the FOR statements are a couple of options:

1. The initial value of the variable doesn't have to be zero. The value can be initialized to any convenient value for a specific calculation in the routine within the loop.

2. The increment value is similarly flexible. In fact, the increment value can be negative, resulting in a decrement of the value, or the increment value can be dynamic, changing on each pass through the loop.

3. The termination condition may also be dynamic, changing for each pass through the loop.

4. The variable used to control the loop is also accessible within the loop, allowing the routine to length, shorten, or even stop the loop by incrementing, decrementing or assigning a value to the variable.

5. If all three terms are left out of the FOR statement, then an infinite loop is generated that will never terminate.

1.6.3 Other Flow Control Statements

Three other program flow control statements are important in later discussions, GOTO, CALL, and RETURN. The GOTO statement is just as the name suggests. It is an unconditional jump from one place in the program to another. The CALL is similar, except it retains the address of the next instruction, following the CALL instruction, in a temporary location. This return address is then used when the RETURN statement is reached to specify the jump-back location.

The use of the GOTO statement is often criticized as an example of poor programming. If the program were properly designed, then looping and conditional statements are sufficient for proper programming. Unfortunately, in embedded programming there are conditions and events beyond the designer's control. As a result, it is sometimes required to break out of the program flow and either restart the program or rearrange its execution to correct a fault.

Therefore, while the GOTO is not a statement that should be used lightly, it is a statement that will be needed for certain fault recovery programming.

The CALL and RETURN are more acceptable to mainstream programming, as they are the means of creating subroutines. When a section of programming is used in multiple places in the program, it is a more efficient use of program memory to build a small separate routine and access it through CALL and RETURN statements.

Although the CALL and RETURN statements are useful, their use should be tempered with the knowledge that each CALL will place two or more bytes of data onto a data structure called the STACK. The purpose of the STACK is to store temporary values that don't have a specific storage location, such as the return address of a CALL. The issue with using the STACK is that:

1. Data memory is often limited with small microcontrollers, and any function that increases data memory usage runs the risk of overwriting an existing variable.

2. The number of locations within the STACK is sometimes limited in small microcontrollers, and unnecessary calls may result in the loss of the oldest return address stored there.

3. Interrupt functions also use the STACK to store return addresses, making it difficult to gauge the exact number of locations in use at any given time.

Therefore, limiting the number of subroutines built into a program is only prudent.

One of the reasons often given for including a large number of subroutines in a program is the ability of subroutines to compress functionality, making the program more readable to anyone following the designer. If the purpose of a subroutine is to alleviate complexity in the listing, then subroutines can still be used, they just have to include the INLINE statement in front of the CALL. What the INLINE statement does is force the language compiler to disregard the CALL/RETURN statements and compile the routines from the subroutine in line with the routines calling the subroutine. In this way, the readability enhancement of the subroutine is still achieved, while eliminating the impact on the amount of data memory available in the STACK. However, it should be noted that the use of the INLINE instruction is not a common practice. Typically, a macro performs the same function and is a more commonly used construct. So, for compatibility and general form, the INLINE statement should only be used if the designer is comfortable with its use and is aware of any impact its use might have on the resulting code.

1.7   State Machines

Control systems that manage electrical or mechanical systems must often be able to
generate, or respond to, sequential events in the system. This ability to use time as part of
the driver equation is in fact one of the important abilities of a microcontroller that makes
it such a good control for electrical and mechanical systems. However, implementing
multiple sequences can become long and involved if a linear coding style is used.

A simple construct, called a *state machine*, simplifies the task of generating a sequence by
breaking the sequence into a series of steps and then executing them sequentially.
While this sounds like an arbitrary definition of a linear piece of code, the difference
is that the individual sections, or steps in the sequence, are encoded within a SWITCH/CASE
statement. This breaks the sequence into logical units that can be easily recognized in the
software listing and, more importantly, it allows other functions to be performed between
the individual steps. It does this by only executing one step each time it is called.
Repeatedly calling the state machine results in the execution of each step in the sequence.
To retain the state machine's place in the sequence, a storage variable is defined that
determines which step in the sequence is to be executed next. This variable is referred to as
the *state variable*, and it is used in the SWITCH/CASE statement to determine
which step, or state, in the state machine is to be executed when the state machine is called.

For this system to work, the state variable must be incremented at the completion of each
state. However, it is also true that the sequence of states may need to change due to changes
in the condition of the system. Given that the state variable determines which state is
executed, it follows that to change the sequence of states, one must simply load the state
variable with a new value corresponding with the new direction the sequence must go. As
we will see in this book, this simple construct is very powerful, and is in fact the basis for
multitasking.

Therefore, the short definition of a state machine is a collection of steps (states) selected for
execution based on the value in a state variable. Further, manipulation of the value in the
state variable allows the state machine to emulate all the conditional statements previously
presented in this chapter.

One of the advantages of the state machine-based design is that it allows the easy
generation of a sequence of events. Another advantage of state machine-based design is its
ability to recognize a sequence of events. It does this by utilizing the conditional change
of the state variable, much as described in the previous paragraph. The only difference is
that the state variable does not normally change its value, unless a specific event is

detected. As an analogy, consider a combination lock: to open the lock, the numbers have to be entered in a specific sequence such as 5, 8, 3, 2. If the numbers were entered 2, 3, 5, 8, the lock would not open, so the combination is not only the numbers but their order.

If we were to create a state machine to recognize this sequence, it would look something like the following:

Code Snippet 1.16

```
State = 0;
SWITCH (State)
{
    CASE 0: IF (in_key()==5)   THEN state = 1;
            Break;
    CASE 1: IF (in_key()==8)   THEN State = 2;
                               Else State = 0;
            Break;
    CASE 2: IF (in_key()==3)   THEN State = 3;
                               Else State = 0;
            Break;
    CASE 3: IF (in_key()==2)   THEN UNLOCK();
                               Else State = 0;
            Break;
}
```

Provided that the values returned by in_key() are in the order of 8, 5, 3, 2, the state variable will step from 0 to 3 and the function UNLOCK() will be called. The state variable is only loaded with the value of the next state when the right value is received in the right state. If any of the values are out of sequence, even though they may be valid for another state, the state variable will reset to 0, and the state machine will start over. In this way, the state machine will step through its sequence only if the values are received in the same sequence as the states in the state machine are designed to accept.

Therefore, state machines can be programmed to recognize a sequence of events, and they can be programmed to generate a sequence of events. Both rely on the history of the previous states and the programmable nature of the state-to-state transitions.

Implementing a state machine is just a matter of:

1. Creating a state variable.

2. Defining a series of states.

3. Decoding the state variable to access the states.

4. Tying actions to the states.

5. Defining the sequence of the states, and any conditions that change the sequence.

For example, consider a state machine designed to make peanut and jelly sandwiches. The sequence of events is:

1. Get two slices of bread.

2. Open peanut butter jar.

3. Scoop out peanut butter.

4. Smear on first slice of bread.

5. Open jelly jar.

6. Scoop out jelly.

7. Smear on second slice of bread.

8. Invert second slice of bread.

9. Put second slice on first slice of bread.

10. Eat.

OK, the first thing to do is create a state variable; let's call it PBJ. It has a range of values from 1 to 10, and it probably defines as a CHAR. Next, we have to define the sequence of steps in the process, and create a means to decode the state variable.

If we take each of these instructions and build them into a CASE statement to handle decoding the state variable, then all it needs is the appropriate updates to the state variable and the state machine is complete.

Algorithm 1.3

```
SWITCH(PBJ)
{
    case 1:   Get two slices.
              PBJ = 2
              break
    case 2:   Open peanut butter jar.
              PBJ = 3
              break
    case 3:   Scoop out peanut butter.
              PBJ = 4
              break
    case 4:   Smear on first slice of bread.
              PBJ = 5
              break
    case 5:   Open jelly jar.
              PBJ = 6
              break
    case 6:   Scoop out jelly.
              PBJ = 7
              break
    case 7:   Smear on second slice of bread.
              PBJ = 8
              break
    case 8:   Invert second slice of bread.
              PBJ = 9
              break
    case 9:   Put second slice on first slice of bread.
              PBJ = 10
              break
    case 10:  Eat
              break
    Default:  break
}
```

The calling routine then simply calls the subroutine 10 times and the result is an eaten peanut butter and jelly sandwich.

Why go to all this trouble? Wouldn't it be simpler and easier to just write it as one long function? Well, yes, the routine could be done as one long sequence with the appropriate delays and timing. But this format has a couple of limitations. One, making a PB and J sandwich would be all the microcontroller could do during the process. And, two, making one kind of a PB and J sandwich would be all the routine would be capable of doing. There is an important distinction in those two sentences; the first states that the microcontroller would only be able to perform one task, no multitasking, and the second states that all the program would be capable of would be one specific kind of PB and J sandwich, no variations.

Breaking the sequence up into a state machine means we can put other functions between the calls to the state machine. The other calls could cover housekeeping details, such as monitoring a serial port, checking a timer, or polling a keyboard. Breaking the sequence up into a state machine also means we can use the same routine to make a peanut butter only sandwich simply by loading the state variable with state 8, instead of state 5 at the end of state 4. In fact, if we include other steps such as pouring milk and getting a cookie, and include some additional conditional state variable changes, we now have a routine that can make several different varieties of snacks, not just a PB and J sandwich.

The power of the state machine construct is not limited to just variations of a sequence. By controlling its own state variable, the state machine can become a form of specialized virtual microcontroller—basically a small, software-based controller with a programmable instruction set. In fact, the power and flexibility of the state machine will be the basis for the multitasking system described later in the book.

Before we dive into some of the more advanced concepts, it is important to understand some of the basics of state machine operation. The best place to start is with the three basic types of state machines: *execution-indexed*, *data-indexed*, and the *hybrid* state machine.

The *execution-indexed state machine* is the type of state machine that most people envision when they talk about a state machine, and it is the type of state machine shown in the previous examples. It has a CASE statement structure with routine for each CASE, and a state variable that controls which state is executed when the state machine is called. A good example of an execution-indexed state machine is the PB&J state machine in the previous example. The function performed by the state machine is specified by the value held in the state variable.

The other extreme is the *data-indexed state machine*. It is probably the least recognized form of a state machine, even though most designers have created several, because it doesn't use a SWITCH/CASE statement. Rather, it uses an array variable with the state variable providing the index into the array. The concept behind a data-indexed state machine is that the sequence of instructions remains constant, and the data that is acted upon is controlled by the state variable.

A *hybrid state machine* combines aspects of both the data-indexed and the execution-indexed to create a state machine with the ability to vary both its execution and the data it operates on. This hybrid approach allows the varied execution of the execution indexed with the variable data aspect of the data-indexed state machine.

We have three different formats, with different advantages and disadvantages. Execution indexed allows designers to vary the actions taken in each state, and/or respond to external sequences of events. Data indexed allows designers to vary the data acted upon in each state, but keep the execution constant. And, finally, the hybrid combines both to create a more efficient state machine that requires both the varied execution of the execution-indexed and the indexed data capability of the data-indexed state machine. Let's take a closer look at the three types and their capabilities.

1.7.1   Data-Indexed State Machines

Consider a system that uses an analog-to-digital converter, or ADC, to monitor multiple sensors. Each sensor has its own channel into the ADC, its own calibration offset/scaling factors, and its own limits. To implement these functions using a data-indexed state machine, we start by assigning a state to each input and creating an array-based storage for all the values that will be required.

Starting with the data storage, the system will need storage for the following:

1. Calibration offset and scaling values.

2. Upper and lower limit values.

3. The final, calibrated values.

Using a two-dimensional array, we can store the values in the following format. Assume that S_var is the state value associated with a specific ADC channel:

- ADC_Data[0][S_var] variable in the array holding the calibration offset values

- ADC_Data[1][S_var] variable in the array holding the calibration scaling values

- ADC_Data[2][S_var] variable in the array holding the upper limit values

- ADC_Data[3][S_var] variable in the array holding the lower limit values

- ADC_Data[4][S_var] variable in the array holding the ADC channel select command value

- ADC_Data[5][S_var] variable in the array holding the calibrated final values

The actual code to implement the state machine will look like the following:

Code Snippet 1.17

```
Void ADC(char S_var, boolean alarm)
{
   ADC_Data[4][S_var] = (ADC*ADC_Data[1][S_
     var])+ADC_Data[0][S_var];
   IF (ADC_Data[4][S_var]>ADC_Data[2][S_var]) THEN
     Alarm = true;
   IF (ADC_Data[4][S_var]<ADC_Data[3][S_var]) THEN
     Alarm = true;
   S_var++;
   IF (S_var > max_channel) then S_var = 0;
   ADC_control = ADC_Data[5][S_var];
   ADC_convert_start = true;
}
```

In the example, the first line converts the raw data value held in ADC into a calibrated value by multiplying the scaling factor and adding in the offset. The result is stored into the ADC_Data array. Lines 2 and 3 perform limit testing against the upper and lower limits store in the ADC_Data array and set the error variable if there is a problem. Next, the state variable S_var is incremented, tested against the maximum number of channels to be polled, and wrapped around if it has incremented beyond the end. Finally, the configuration data selecting the next channel is loaded into the ADC control register and the conversion is initiated—a total of seven lines of code to scan as many ADC channels as the system needs, including both individual calibration and range checking.

From the example, it seems that data-indexed state machines are fairly simple constructs, so how do they justify the lofty name of state machine? Simple, by exhibiting the ability to change its operation based on internal and external influences. Consider a variation on the previous example. If we add another variable to the data array and place the next state information into that variable, we now have a state machine that can be reprogrammed "on the fly" to change its sequence of conversions based on external input.

- ADC_Data[6] [S_var] variable in the array holding the next channel to convert

Code Snippet 1.18

```
Void ADC(char S_var, boolean alarm)
{
    ADC_Data[4][S_var] = (ADC*ADC_Data[1][S_
        var])+ADC_Data[0][S_var];
    IF (ADC_Data[4][S_var]>ADC_Data[2][S_var]) THEN
        Alarm = true;
    IF (ADC_Data[4][S_var]<ADC_Data[3][S_var]) THEN
        Alarm = true;
    S_var = ADC_Data[6][S_var];
    ADC_control = ADC_Data[5][S_var];
    ADC_convert_start = true;
}
```

Now the sequence of channels is controlled by the array ADC_Data. If the system does not require data from a specific channel, it just reprograms the array to route the state machine around the unneeded channel. The state machine could also be built with two or more next channels, with the actual next channel determined by whether a fault has occurred, or an external flag is set, or a value reported by one of the channels has been exceeded.

Don't let the simplicity of the state machine deceive you; there is power and flexibility in the data-indexed state machine. All that is required is the imagination to look beyond the simplicity and see the possibilities.

1.7.2   Execution-Indexed State Machines

Execution-indexed state machines, as described previously, are often mistakenly assumed to be little more than a CASE statement with the appropriate routines inserted for the individual states. While the CASE statement, or an equivalent machine language construct, is at the heart of an execution-based state machine, there is a lot more to their design and a lot more to their capabilities.

For instance, the capability to control its own state variable lends itself to a wide variety of capabilities that rival normal linear coding. By selectively incrementing or loading the state variable, individual states within the state machine can implement:

- Sequential execution.

- Computed GOTO instructions.

- DO/WHILE instructions.

- WHILE/DO instructions.

- FOR/NEXT instructions.

- And even GOSUB/RETURN instructions.

Let's run through some examples to demonstrate some of the capabilities of the execution-indexed state machine type.

First of all, to implement a sequence of state steps, it is simply a matter of assigning the value associated with the next state in the sequence, at the end of each state. For example:

Code Snippet 1.19

```
SWITCH(State_var)
{
    CASE 0:    State_var = 1;
               Break;
    CASE 1:    State_var = 2;
               Break;
    CASE 2:    State_var = 3;
               Break;
}
```

Somewhere in each state, the next state is loaded into the state variable. As a result, each execution of the state machine results in the execution of the current state's code block and the advancement of the state variable to the next state. If the states are defined to be sequential values, the assignment can even be replaced with a simple increment. However, there is no requirement that the states be sequential, or that the state machine must sequence down the case statement on each successive call to the state machine. It is perfectly valid to have the state machine step through the case statement in whatever pattern is convenient, particularly if the pattern of values in the state variable is convenient for some other function in the system, such as the sequence of energized windings in a brushless motor. The next state can even be defined by the values in an array, making the sequence entirely programmable.

Computed GOTO instructions are just a simple extension of the basic concept used in sequential execution. The only difference is the assignment is made from the result of a calculation. For example:

Code Snippet 1.20

```
SWITCH(State_var)
{
        CASE 0:    State_var = 10 * Var_a;
            Break;
        CASE 10:   Function_A;
            State_var = 0;
            Break;
        CASE 20:   Function_B;
            State_var = 0;
            Break;
        CASE 30:   Function_C
            State_var = 0;
            Break;
}
```

Based on the value present in Var_a, the state machine will execute one of three different states the next time it is called. This essentially implements a state machine that cannot only change its sequence of execution based on data, but can also change its execution to one of several different sequences based on data.

Another construct that can be implemented is the IF/THEN/ELSE statement. Based on the result of a comparison in one of the states, the state machine can step to one of two different states, altering its sequence. If the comparison in the conditional statement is true, then the state variable is loaded with the new state value associated with the THEN part of the IF statement and the next time the state machine is executed, it will execute the new state. If the comparison results in a false, then the state variable is loaded with a different value and the state machine executes the state associated with the ELSE portion of the IF statement. For example:

Code Snippet 1.21

```
SWITCH(State_var)
{
    CASE 0:   IF (Var_A > Var_B) THEN State_var = 1;
                               ELSE State_var = 2;
            Break;
    CASE 1:   Var_B = Var_A
            State_var = 0;
            Break;
    CASE 2:   Var_A = Var_B
            State_var = 0;
            Break;
}
```

In the example, whenever the value in Var_A is larger than the value in Var_B, the state machine advances to state 1 and the value in Var_A is copied into Var_B. The state machine then returns to state 0. If the value in Var_B is greater than or equal to Var_A, then Var_B is copied into Var_A, and the state machine returns to state 0.

Now, having seen both the GOTO and the IF/THEN/ELSE, it is a simple matter to implement all three iterative statements by simply combining the GOTO and the IF/THEN/ELSE. For example, a DO/WHILE iterative statement would be implemented as follows:

Code Snippet 1.22

```
    CASE 4:   Function;
            State_var = 5;
            Break;
    CASE 5:   IF (comparison)   THEN State_var = 4;
                                ELSE State_var = 6;
            Break;
    CASE 6:
```

In the example, state 4 holds the (DO) function within the loop, and state 5 holds the (WHILE) comparison. And, a WHILE/DO iterative statement would be implemented as follows:

Code Snippet 1.23

```
CASE 4:   IF (comparison)    THEN State_var = 5;
                             ELSE State_var = 6;
          Break;
CASE 5:   Function;
          State_var = 4;
          Break;
CASE 6:
```

In this example, state 4 holds the (WHILE) comparison, and state 5 holds the (DO) function within the loop. A FOR/NEXT iterative statement would be implemented as follows:

Code Snippet 1.24

```
CASE 3:   Counter = 6;
          State_var = 4;
          Break;
CASE 4:   IF (Counter > 0)   THEN State_var = 5;
                             ELSE State_var = 6;
          Break;
CASE 5:   Function;
          Counter = Counter - 1;
          State_var = 4;
          Break;
CASE 6:
```

In the last example, the variable (Counter) in the FOR/NEXT is assigned its value in state 3, is compared to 0 in state 4 (FOR), and is then incremented and looped back in state 5 (NEXT).

These three iterative constructs are all simple combinations of the GOTO and IF/THEN/ELSE described previously. Building them into a state machine just required breaking the various parts out into separate states, and appropriately setting the state variable.

The final construct to examine in an execution-indexed state machine is the CALL/RETURN. Now, the question arises, why do designers need a subroutine construct in state machines? What possible use is it?

Well, let's take the example of a state machine that has to generate two different delays. State machine delays are typically implemented by repeatedly calling a do-nothing state, and then returning to an active state. For example, the following is a typical state machine delay:

Code Snippet 1.25

```
CASE 3:   Counter = 6;
          State_var = 4;
          Break;
CASE 4:   IF (Counter == 0) THEN State_var = 5;
          Counter = Counter - 1;
          Break;
CASE 5:
```

This routine will wait in state 4 a total of six times before moving on to state 5. If we want to create two different delays, or use the same delay twice, we would have to create two different wait states. However, if we build the delay as a subroutine state, implementing both the CALL and RETURN, we can use the same state over and over, saving program memory. For example:

Code Snippet 1.26

```
CASE 3:   Counter = 6;
          State_var = 20;
          Back_var = 4
          Break;
     |        |
     |        |
CASE 12:  Counter = 10;
          State_var = 20;
          Back_var = 13
          Break;
     |        |
     |        |
CASE 20:  IF (Counter == 0) THEN State_var = Back_var;
          Counter = Counter - 1;
          Break;
```

In the example, states 3 and 12 are calling states and state 20 is the subroutine. Both 3 and 12 loaded the delay counter with the delays they required, loaded Back_var with the state immediately following the calling state (return address), and jumped to the delay state 20 (CALL). State 20 then delayed the appropriate number of times, and transferred the return value in Back_var into the state variable (RETURN).

By providing a return state value, and setting the counter variable before changing state, a simple yet effective subroutine system was built into a state machine. With a little work and a small array for the Back_var, the subroutine could even call other subroutines.

1.7.3   Hybrid State Machines

Hybrid state machines are a combination of both formats; they have the CASE structure of an execution-based state machine, as well as the array-based data structure of a data-indexed state machine. They are typically used in applications that require the sequential nature of an execution-based state machine, combined with the ability to handle multiple data blocks.

A good example of this hybrid requirement is a software-based serial transmit function. The function must generate a start bit, 8 data bits, a parity bit, and one or more stop bits. The start, parity, and stop bits have different functionality and implementing them within an execution-based state machine is simple and straightforward. However, the transmission of the 8 data bits does not work as well within the execution-based format. It would have to be implemented as eight nearly identical states, which would be inefficient and a waste of program memory. So, a second data-driven state machine, embedded in the first state machine, is needed to handle the 8 data bits being transmitted. The following is an example of how the hybrid format would be implemented:

Code Snippet 1.27

```
SWITCH(Ex_State_var)
{
   CASE 0:                                   // waiting for new character
                  IF (Data_avail == true) THEN Ex_State_var = 1;
                  Break;

   CASE 1:                                   // begin with a start bit
                  Output(0);
                  Ex_State_var = 2;
                  DI_State_var = 0;
                  Break;

   CASE 2:                                   // sending bits 0-7
                  If ((Tx_data & (2^DI_State_var))) == 0)
                        Then    Output(0);
                        Else    Output(1);
                  DI_State_var++;
                  If (DI_State_var == 8) Then Ex_State_var = 3;
                  Break;

   CASE 3:                                   // Output Parity bit
                  Output(Parity(Tx_data));
                  Ex_State_var = 4;
                  Break;

   CASE 4:                                   // Send Stop bit to end
                  Output(1);
                  Ex_State_var = 0
}
```

Note that the example has two state variables, Ex_State_var and DI_State_var. Ex_State_var is the state variable for the execution-indexed section of the state machine, determining which of the four cases in the SWITCH statement is executed. DI_State_var is the state variable for the data-indexed section of the state machine, determining which bit in the 8-bit data variable is transmitted on each pass through state 2. Together the two types of state machine produce a hybrid state machine that is both simple and efficient.

On a side note, it should be noted that the Ex_State_var and DI_State_var can be combined into a single data variable to conserve data memory. However, this is typically not done due to the extra overhead of separating the two values. Even if the two values are combined using a Structure declaration, the compiler will still have to include additional code to mask off the two values.

1.8 Multitasking

Multitasking is the ability to execute multiple separate tasks in a fashion that is seemingly simultaneous. Note the phrase "seemingly simultaneous." Short of a multiple processor system, there is no way to make a single processor execute multiple tasks at the same time. However, there is a way to create a system that seems to execute multiple tasks at the same time. The secret is to divide up the processor's time so it can put a segment of time on each of the tasks on a regular basis. The result is the appearance that the processor is executing multiple tasks, when in actuality the processor is just switching between the tasks too quickly to be noticed.

As an example, consider four cars driving on a freeway. Each car has a driver and a desired destination, but no engine. A repair truck arrives, but it only has one engine. For each car to move toward its destination, it must use a common engine, shared with the other cars on the freeway. (See Figure 1.1.)

Now in one scenario, the engine could be given to a single car, until it reaches its destination, and then transferred to the next car until it reaches its destination, and so on until all the cars get where they are going. While this would accomplish the desired result, it does leave the other cars sitting on the freeway until the car with the engine finishes its trip. It also means that the cars would not be able to interact with each other during their trips.

A better scenario would be to give the engine to the first car for a short period of time, then move it to the second for a short period, then the third, then the fourth, and then back to first, continuing the rotation through the cars over and over. In this scenario, all of the cars make progress toward their destinations. They won't make the same rate of progress that they would if they had exclusive use of the engine, but they all do move together. This has a couple of advantages; the cars travel at a similar rate, all of the cars complete their trip at approximately the same time, and the cars are close enough during their trip to interact with each other.

This scenario is in fact, the common method for multitasking in an operating system. A task is granted a slice of execution time, then halted, and the next task begins to execute. When its time runs out, a third task begins executing, and so on.

Figure 1.1: Automotive Multitasking

While this is an over-simplification of the process, it is the basic underlying principle of a multitasking operating system: multiple programs operating within small slices of time, with a central control that coordinates the changes. The central control manages the switching between the various tasks, handles communications between the tasks, and even determines which tasks should run next. This central control is in fact the multitasking operating system. If we plan to develop software that can multitask without an operating system, then our design must include all of the same elements of an operating system to accomplish multitasking.

1.8.1 Four Basic Requirements of Multitasking

The three basic requirements of a multitasking system are: context switching, communications, and managing priorities. To these three functions, a fourth—timing control—is required to manage multitasking in a real-time environment. Functions to handle each of these requirements must be developed within a system for that system to be able to multitask in real time successfully.

To better understand the requirements, we will start with a general description of each requirement, and then examine how the two main classes of multitasking operating systems handle the requirements. Finally, we'll look at how a stand-alone system can manage the requirements without an operating system.

1.8.1.1 Context Switching

When a processor is executing a program, several registers contain data associated with the execution. They include the working registers, the program counter, the system status register, the stack pointer, and the values on the stack. For a program to operate correctly, each of these registers must have the right data and any changes caused by the execution of the program must be accurately retained. There may also be addition data, variables used by the program, intermediate values from a complex calculation, or even hidden variables used by utilities from a higher level language used to generate the program. All of this information is considered the program, or task, context.

When multiple tasks are multitasking, it is necessary to swap in and out all of this information or context, whenever the program switches from one task to another. Without the correct context, the program that is loaded will have problems, RETURNs will not go to the right address, comparisons will give faulty results, or the microcontroller could even lose its place in the program.

To make sure the context is correct for each task, a specific function in the operating system, called the *Context Switcher*, is needed. Its function is to collect the context of the previous task and save it in a safe place. It then has to retrieve the context of the next task and restore it to the appropriate registers. In addition to the context switcher, a block of data memory sufficient to hold the context of each task must also be reserved for each task operating.

When we talk about multitasking with an operating system in the next section, one of the main differentiating points of operating systems is the event that triggers context switcher, and what effect that system has on both the context switcher and the system in general.

1.8.1.2   Communications

Another requirement of a multitasking system is the ability of the various tasks in the system to reliably communicate with one another. While this may seem to be a trivial matter, it is the very nature of multitasking that makes the communications between tasks difficult. Not only are the tasks never executing simultaneously, the receiving task may not be ready to receive when the sending task transmits. The rate at which the sending task is transmitting may be faster than the receiving task can accept the data. The receiving task may not even accept the communications. These complications, and others, result in the requirement for a *communications system* between the various tasks. Note: the generic term "intertask communications" will typically be used when describing the data passed through the communications system and the various handshaking protocols used.

1.8.1.3   Managing Priorities

The *priority manager* operates in concert with the context switcher, determining which tasks should be next in the queue to have execution time. It bases its decisions on the relative priority of the tasks and the current mode of operation for the system. It is in essence an arbitrator, balancing the needs of the various tasks based on their importance *to the system* at a given moment.

In larger operating systems, system configuration, recent operational history, and even statistical analysis of the programs can be used by the priority manager to set the system's priorities. Such a complicated system is seldom required in embedded programming, but some method for shifting emphasis from one task to another is needed for the system to adapt to the changing needs of the system.

1.8.1.4   Timing Control

The final requirement for real-time multitasking is *timing control*. It is responsible for the timing of the task's execution. Now, this may sound like just a variation on the priority manager, and the timing control does interact with the priority manager to do its job. However, while the priority manager determines which tasks are next, it is the timing control that determines the order of execution, setting when the task executes.

The distinction between the roles can be somewhat fuzzy. However, the main point to remember is that the timing control determines *when* a task is executed, and it is the priority control that determines *if* the task is executed.

Balancing the requirements of the timing control and the priority manager is seldom neither simple nor easy. After all, real-time systems often have multiple asynchronous tasks, operating at different rates, interacting with each other and the asynchronous real world. However, careful design and thorough testing can produce a system with a reasonable balance between timing and priorities.

1.8.1.5   Operating Systems

To better understand the requirements of multitasking, let's take a look at how two different types of operating systems handle multitasking. The two types of operating system are *preemptive* and *cooperative*. Both utilize a context switcher to swap one task for another; the difference is the event that triggers the context switch. A *preemptive* operating system typically uses a timer-driven interrupt, which calls the context switcher through the interrupt service routine. A *cooperative* operating system relies on subroutine calls by the task to periodically invoke the context switcher. Both systems employ the stack to capture and retrieve the return address; it is just the method that differs. However, as we will see below, this creates quite a difference in the operation of the operating systems.

Of the two systems, the more familiar is the preemptive style of operating system. This is because it uses the interrupt mechanism within the microcontroller in much the same way as an interrupt service routine does.

When the interrupt fires, the current program counter value is pushed onto the stack, along with the status and working registers. The microcontroller then calls the interrupt service routine, or ISR, which determines the cause of the interrupt, handles the event, and then clears the interrupt condition. When the ISR has completed its task, the return address, status, and register values are then retrieved and restored, and the main program continues on without any knowledge of the ISR's execution.

The difference between the operation of the ISR and a preemptive operating system is that the main program that the ISR returns to is not the same program that was running when the interrupt occurred. That's because, during the interrupt, the context switcher swaps in the context for the next task to be executed. So, basically, each task is operating within the ISR of every other task. In addition, just like the program interrupted by the ISR, each task is oblivious to the execution of all the other tasks.

The interrupt driven nature of the preemptive operating system gives rise to some advantages that are unique to the preemptive operating system:

- The slice of time that each task is allocated is strictly regulated. When the interrupt fires, the current task loses access to the microcontroller and the next task is substituted. So, no one task can monopolize the system by refusing to release the microcontroller.

- Because the transition from one task to the next is driven by hardware, it is not dependent upon the correct operation of the code within the current task. A fault condition that corrupts the program counter within one task is unlikely to corrupt another current task, provided the corrupted task does not trample on another task's variable space. The other tasks in the system should still operate, and the operating system should still swap them in and out on time. Only the corrupted task should fail. While this is not a guarantee, the interrupt nature of the preemptive system does offer some protection.

- The programming of the individual tasks can be linear, without any special formatting to accommodate multitasking. This means traditional programming practices can be used for development, reducing the amount of training required to bring onboard a new designer.

However, because the context switch is asynchronous to the task timing, meaning it can occur at any time during the task execution, complex operations within the task may be interrupted before they complete, so a preemptive operating system also suffers from some disadvantages as well:

- Multibyte updates to variables and/or peripherals may not complete before the context switch, leaving variable updates and peripheral changes incomplete. This is the reason preemptive operating systems have a communications manager to handle all communications. Its job is to only pass on updates and changes that are complete, and hold any that did not complete.

- Absolute timing of events in the task cannot rely on execution time. If a context switch occurs during a timed operation, the time between actions may include the execution time of one or more other tasks. To alleviate this problem timing functions must rely on an external hardware function that is not tied to the task's execution.

- Because the operating system does not know what context variables are in use when the context switch occurs, any and all variables used by the task, including any variables specific to the high-level language, must be saved as part of the context. This can significantly increase the storage requirements for the context switcher.

While the advantages of the preemptive operating system are attractive, the disadvantages can be a serious problem in a real-time system.

The communications problems will require a communications manager to handle multibyte variables and interfaces to peripherals. Any timed event will require a much more sophisticated timing control capable of adjusting the task's timing to accommodate specific timing delays. In addition, the storage requirements for the context switcher can require upwards of 10–30 bytes, per task—no small amount of memory space as 5–10 tasks are running at the same time. All in all, a preemptive system operates well for a PC, which has large amounts of data memory and plenty of program memory to hold special communications and timing handlers. However, in real-time microcontroller applications, the advantages are quickly outweighed by the operating system's complexity.

The second form of multitasking system is the *Cooperative* operating system. In this operating system, the event triggering the context switch is a subroutine call to the operating system by the task currently executing. Within the operating system subroutine, the current context is stored and the next is retrieved. So, when the operating system returns from the subroutine, it will be to an entirely different task, which will then run until it makes a subroutine call to the operating system. This places the responsibility for timing on the tasks themselves. They determine when they will release the microcontroller by the timing of their call to the operating system, thus the name cooperative. This solves some of the more difficult problems encountered in the preemptive operating system:

- Multibyte writes to variables and peripherals can be completed prior to releasing the microcontroller, so no special communications handler is required to oversee the communications process.

- The timed events, performed *between* calls to the operating system, can be based on execution time, eliminating the need for external hardware-based delay systems, *provided* a call to the operating system is not made between the start and end of the event.

- The context storage need only save the current address and the stack. Any variables required for statement execution, status, or even task variables do not need to be saved as all statement activity is completed before the statement making the subroutine call is executed. This means that a cooperative operating system has a significantly smaller context storage requirement than a preemptive system. This also means the context switcher does not need intimate knowledge about register usage in the high-level language to provide context storage.

However, the news is not all good; there are some drawbacks to the cooperative operating system that can be just as much a problem as the preemptive operating system:

- Because the context switch requires the task to make a call to the operating system, any corruption of the task execution, due to EMI, static, or programming errors, will cause the entire system to fail. Without the voluntary call to the operating system, a context switch cannot occur. Therefore, a cooperative operating system will typically require an external watchdog function to detect and recover from system faults.

- Because the time of the context switch is dependent on the flow of execution within the task, variations in the flow of the program can introduce variations into the system's long-term timing. Any timed events that span one or more calls to the operating system will still require an external timing function.

- Because the periodic calls to the operating system are the means of initiating a context switch, it falls to the designer to evenly space the calls throughout the programming for all tasks. It also means that if a significant change is made in a task, the placement of the calls to the operating system may need to be adjusted. This places a significant overhead on the designer to insure that the execution times allotted to each task are reasonable and approximately equal.

As with the preemptive system, the cooperative system has several advantages, and several disadvantages as well. In fact, if you examine the lists closely, you will see that the two systems have some advantages and disadvantages that are mirror images of each other. The preemptive system's context system is variable within the tasks, creating completion problems. The cooperative system gives the designer the power to determine where and when the context switch occurs, but it suffers in its handling of fault conditions. Both suffer from complexity in relation to timing issues, both require some specialized routines within the operating system to execute properly, and both require some special design work by the designer to implement and optimize.

1.8.1.6 State Machine Multitasking

So, if preemptive and cooperative systems have both good and bad points, and neither is the complete answer to writing multitasking software, is there a third alternative? The answer is yes, a compromise system designed in a cooperative style with elements of the preemptive system.

Specifically, the system uses state machines for the individual tasks with the calls to the state machine regulated by a hardware-driven timing system. Priorities are managed based on the current value in the state variables and the general state of the system. Communications are handled through a simple combination of handshaking protocols and overall system design.

The flowchart of the collective system is shown in Figure 1.2. Within a fixed infinite loop, each state machine is called based on its current priority and its timing requirements. At the end of each state, the state machine executes a return and the loop continues onto the next state machine. At the end of the loop, the system pauses, waiting for the start of the next pass, based on the time-out of a hardware timer. Communications between the tasks are handled through variables, employing various protocols to guarantee the reliable communications of data.

As with both the preemptive and cooperative systems, there are also a number of advantages to a state machine-based system:

- The entry and exit points are fixed by the design of the individual states in the state machines, so partial updates to variables or peripherals are a function of the design, not the timing of the context switch.

- A hardware timer sets the timing of each pass through the system loop. Because the timing of the loop is constant, no specific delay timing subroutines are required for the individual delays within the task. Rather, counting passes through the loop can be used to set individual task delays.

- Because the individual segments within each task are accessed via a state variable, the only context that must be saved is the state variable itself.

- Because the design leaves slack time at the end of the loop and the start of the loop is tied to an external hardware timer, reasonable changes to the execution time

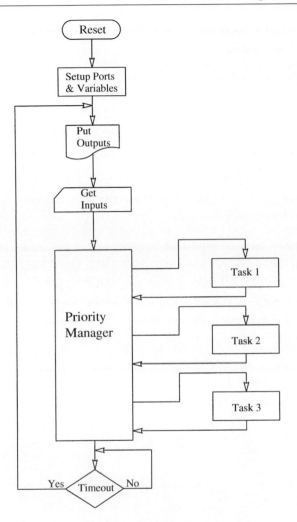

Figure 1.2: State Machine Multitasking

of individual states within the state machine do not affect the overall timing of
the system.

- The system does not require any third-party software to implement, so no license
  fees or specialized software are required to generate the system.

- Because the designer designs the entire system, it is completely scalable to whatever
  program and data memory limitation may exist. There is no minimal kernel required
  for operation.

However, just like the other operating systems, there are a few disadvantages to the state machine approach to multitasking:

- Because the system relies on the state machine returning at the end of each state, EMI, static, and programming flaws can take down all of the tasks within the system. However, because the state variable determines which state is being executed, and it is not affected by a corruption of the program counter, a watchdog timer driven reset can recover and restart uncorrupted tasks without a complete restart of the system.

- Additional design time is required to create the state machines, communications, timing, and priority control system.

The resulting state machine-based multitasking system is a collection of tasks that are already broken into function-convenient time slices, with fixed hardware-based timing and a simple priority and communication system specific to the design. Because the overall design for the system is geared specifically to the needs of the system, and not generalized for all possible designs, the operation is both simple and reliable if designed correctly.

CHAPTER 2

Device Drivers

Tammy Noergaard

2.1   In This Chapter

Defining device drivers, discussing the difference between architecture-specific and board-specific drivers, and providing several examples of different types of device drivers.

Most embedded hardware requires some type of software initialization and management. The software that directly interfaces with and controls this hardware is called a ***device driver***. All embedded systems that require software have, at the very least, device driver software in their system software layer. Device drivers are the software libraries that initialize the hardware, and manage access to the hardware by higher layers of software. Device drivers are the liaison between the hardware and the operating system, middleware, and application layers.

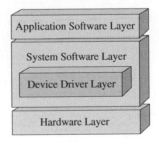 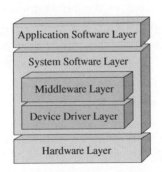 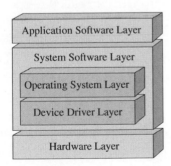

Figure 2.1: Embedded Systems Model and Device Drivers

The types of hardware components needing the support of device drivers vary from board to board, but they can be categorized according to the von Neumann model approach (see Figure 2.2). The von Neumann model can be used as a software model as well as a hardware model in determining what device drivers are required within a particular platform. Specifically, this can include drivers for the *master processor* architecture-specific functionality, *memory* and memory management drivers, *bus* initialization and transaction drivers, and I/O initialization and control drivers (such as for networking, graphics, input devices, storage devices, debugging I/O, and so on) both at the board and master CPU level.

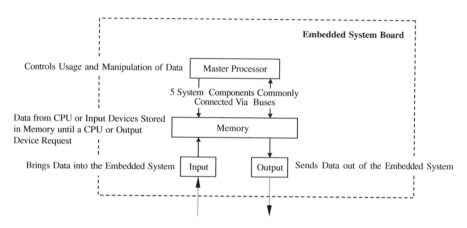

Figure 2.2: Embedded System Board Organization
Source: Based upon the von Neumann architecture model, also referred to as the Princeton architecture.

Device drivers are typically considered either **architecture-specific** or **generic**. A device driver that is architecture-specific manages the hardware that is integrated into the master processor (the architecture). Examples of architecture-specific drivers that initialize and enable components within a master processor include on-chip memory, integrated memory managers (MMUs), and floating-point hardware. A device driver that is generic manages hardware that is located on the board and not integrated onto the master processor. In a generic driver, there are typically architecture-specific portions of source code, because the master processor is the central control unit and to gain access to anything on the board usually means going through the master processor. However, the generic driver also manages board hardware that is not specific to that particular processor, which means that a generic driver can be configured to run on a variety of architectures that contain the related board hardware for which the driver is written. Generic drivers include code that initializes and manages access to the remaining major components of the board, including board buses (I2C, PCI, PCMCIA, and so on), off-chip memory (controllers, level-2+ cache, Flash, and so on), and off-chip I/O (Ethernet, RS-232, display, mouse, and so on).

Figure 2.3a shows a hardware block diagram of a MPC860-based board, and Figure 2.3b shows a systems diagram that includes examples of both MPC860 processor-specific device drivers, as well as generic device drivers.

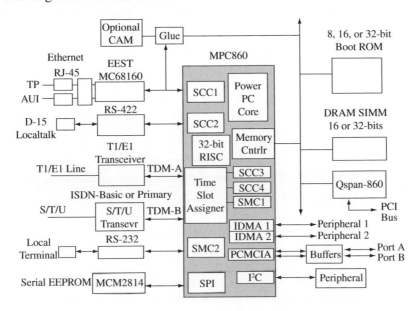

Figure 2.3a: MPC860 Hardware Block Diagram
Source: Copyright of Freescale Semiconductor, Inc. 2004. Used by permission.

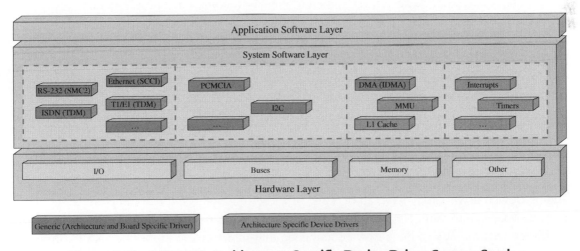

Figure 2.3b: MPC860 Architecture Specific Device Driver System Stack
Source: Copyright of Freescale Semiconductor, Inc. 2004. Used by permission.

Regardless of the type of device driver or the hardware it manages, all device drivers are generally made up of *all* or *some* combination of the following functions:

- **Hardware Startup**, initialization of the hardware upon power-on or reset.

- **Hardware Shutdown**, configuring hardware into its power-off state.

- **Hardware Disable**, allowing other software to disable hardware on-the-fly.

- **Hardware Enable**, allowing other software to enable hardware on-the-fly.

- **Hardware Acquire**, allowing other software to gain singular (locking) access to hardware.

- **Hardware Release**, allowing other software to free (unlock) hardware.

- **Hardware Read**, allowing other software to read data from hardware.

- **Hardware Write**, allowing other software to write data to hardware.

- **Hardware Install**, allowing other software to install new hardware on-the-fly.

- **Hardware Uninstall**, allowing other software to remove installed hardware on-the-fly.

Of course, device drivers may have additional functions, but some or all of the functions shown above are what device drivers inherently have in common. These functions are based upon the software's implicit perception of hardware, which is that hardware is in one of three states at any given time—*inactive, busy,* or *finished.* Hardware in the inactive state is interpreted as being either disconnected (thus the need for an install function), without power (hence the need for an initialization routine) or disabled (thus the need for an enable routine). The busy and finished states are active hardware states, as opposed to inactive; thus the need for uninstall, shutdown and/or disable functionality. Hardware that is in a busy state is actively processing some type of data and is not idle, and thus may require some type of release mechanism. Hardware that is in the finished state is in an idle state, which then allows for acquisition, read, or write requests, for example.

Again, device drivers may have all or some of these functions, and can integrate some of these functions into single larger functions. Each of these driver functions typically has code that interfaces directly to the hardware and code that interfaces to higher layers of software. In some cases, the distinction between these layers is clear, while in other drivers, the code is tightly integrated (see Figure 2.4).

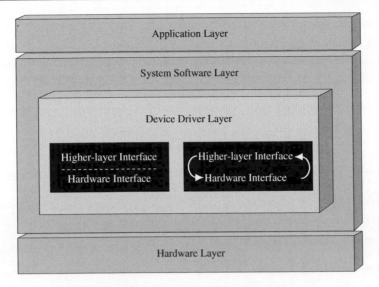

Figure 2.4: Driver Code Layers

On a final note, depending on the master processor, different types of software can execute in different modes, the most common being *supervisory* and *user* modes. These modes essentially differ in terms of what system components the software is allowed access to, with software running in supervisory mode having more access (privileges) than software running in user mode. Device driver code typically runs in supervisory mode.

The next several sections provide real-world examples of device drivers that demonstrate how device driver functions can be written and how they can work. By studying these examples, the reader should be able to look at any board and figure out relatively quickly what possible device drivers need to be included in that system, by examining the hardware and going through a checklist, using the von Neumann model as a tool for keeping track of the types of hardware that might require device drivers.

2.2   Example 1: Device Drivers for Interrupt-Handling

Interrupts are signals triggered by some event during the execution of an instruction stream by the master processor. What this means is that interrupts can be initiated asynchronously, for external hardware devices, resets, power failures, and so forth, or synchronously, for instruction-related activities such as system calls or illegal instructions. These signals cause the master processor to stop executing the current instruction stream and start the process of *handling* (processing) the interrupt.

The software that handles interrupts on the master processor and manages interrupt hardware mechanisms (i.e., the interrupt controller) consists of the *device drivers* for interrupt-handling. At least four of the ten functions from the list of device driver functionality introduced at the start of this chapter are supported by interrupt-handling device drivers, including:

- **Interrupt-Handling Startup**, initialization of the interrupt hardware (i.e., interrupt controller, activating interrupts, and so forth) on power-on or reset.

- **Interrupt-Handling Shutdown**, configuring interrupt hardware (i.e., interrupt controller, deactivating interrupts, and so forth) into its power-off state.

- **Interrupt-Handling Disable**, allowing other software to disable active interrupts on-the-fly (not allowed for nonmaskable interrupts [NMIs], which are interrupts that cannot be disabled).

- **Interrupt-Handling Enable**, allowing other software to enable inactive interrupts on-the-fly.

and one additional function unique to interrupt-handling:

- **Interrupt-Handler Servicing**, the interrupt-handling code itself, which is executed after the interruption of the main execution stream (this can range in complexity from a simple nonnested routine to nested and/or reentrant routines).

How startup, shutdown, disable, enable, and service functions are implemented in software usually depends on the following criteria:

- The types, number, and priority levels of interrupts available (determined by the interrupt hardware mechanisms on-chip and onboard).

- How interrupts are triggered.

- The interrupt policies of components within the system that trigger interrupts, and the services provided by the master CPU processing the interrupts.

The three main types of interrupts are *software, internal hardware,* and *external hardware.* Software interrupts are explicitly triggered internally by some instruction within the current instruction stream being executed by the master processor. Internal hardware interrupts, on the other hand, are initiated by an event that is a result of a problem with the current instruction stream that is being executed by the master processor because of the features (or limitations) of the hardware, such as illegal math operations (overflow, divide-by-zero), debugging (single-stepping, breakpoints), invalid instructions (opcodes), and so on.

Interrupts that are raised (requested) by some internal event to the master processor—basically, software and internal hardware interrupts—are also commonly referred to as *exceptions* or *traps*. Exceptions are internally generated hardware interrupts triggered by errors that are detected by the master processor during software execution, such as invalid data or a divide by zero. How exceptions are prioritized and processed is determined by the architecture. Traps are software interrupts specifically generated by the software, via an exception instruction. Finally, external hardware interrupts are interrupts initiated by hardware other than the master CPU—board buses and input/output (I/O) for instance.

For interrupts that are raised by external events, the master processor is either wired, through an input pin(s) called an ***IRQ*** (Interrupt Request Level) pin or port, to outside intermediary hardware (i.e., interrupt controllers), or directly to other components on the board with dedicated interrupt ports, that signal the master CPU when they want to raise the interrupt. These types of interrupts are triggered in one of two ways: ***level-triggered*** or ***edge-triggered***. A level-triggered interrupt is initiated when its interrupt request (IRQ) signal is at a certain level (i.e., HIGH or LOW—see Figure 2.5a). These interrupts are processed when the CPU finds a request for a level-triggered interrupt when sampling its IRQ line, such as at the end of processing each instruction.

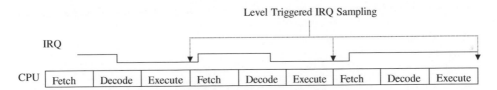

Figure 2.5a: Level-Triggered Interrupts

Edge-triggered interrupts are triggered when a change occurs on the IRQ line (from LOW to HIGH/rising edge of signal or from HIGH to LOW/falling edge of signal; see Figure 2.5b). Once triggered, these interrupts latch into the CPU until processed.

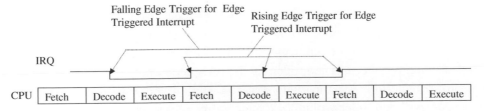

Figure 2.5b: Edge-Triggered Interrupts

Both types of interrupts have their strengths and drawbacks. With a level-triggered interrupt, as shown in the example in Figure 2.6a, if the request is being processed and has not been disabled before the next sampling period, the CPU will try to service the same interrupt again. On the flip side, if the level-triggered interrupt were triggered and then disabled before the CPU's sample period, the CPU would never note its existence and would therefore never process it. Edge-triggered interrupts could have problems if they share the same IRQ line, if they were triggered in the same manner at about the same time (say before the CPU could process the first interrupt), resulting in the CPU being able to detect only one of the interrupts (see Figure 2.6b).

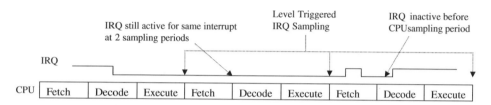

Figure 2.6a: Level-Triggered Interrupts Drawbacks

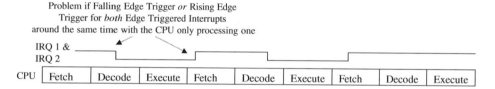

Figure 2.6b: Edge-Triggered Interrupts Drawbacks

Because of these drawbacks, level-triggered interrupts are generally recommended for interrupts that share IRQ lines, whereas edge-triggered interrupts are typically recommended for interrupt signals that are very short or very long.

At the point an IRQ of a master processor receives a signal that an interrupt has been raised, the interrupt is processed by the interrupt-handling mechanisms within the system. These mechanisms are made up of a combination of both hardware and software components. In terms of hardware, an interrupt controller can be integrated onto a board, or within a processor, to mediate interrupt transactions in conjunction with software. Architectures that include an interrupt controller within their interrupt-handling schemes include the 268/386 (x86) architectures, which use two PICs (Intel's Programmable Interrupt Controller);

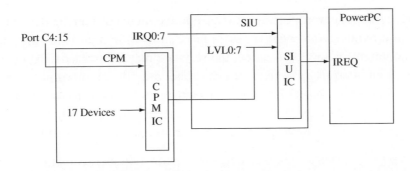

Figure 2.7a: Motorola/Freescale MPC860 Interrupt Controllers
Source: Copyright of Freescale Semiconductor, Inc. 2004. Used by permission.

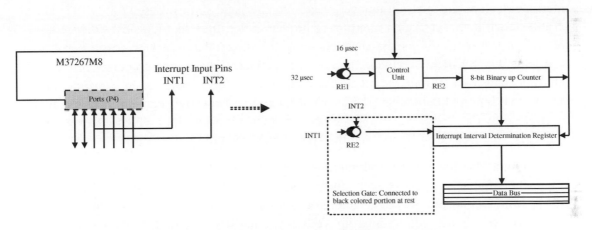

Figure 2.7b: Mitsubishi M37267M8 Circuitry

MIPS32, which relies on an external interrupt controller; and the MPC860 (shown in Figure 2.7a), which integrates two interrupt controllers, one in the CPM and one in its System Interface Unit (SIU). For systems with no interrupt controller, such as the Mitsubishi M37267M8 TV microcontroller shown in Figure 2.7b, the interrupt request lines are connected directly to the master processor, and interrupt transactions are controlled via software and some internal circuitry, such as registers and/or counters.

Interrupt acknowledgment, or *IACK*, is typically handled by the master processor when an external device triggers an interrupt. Because IACK cycles are a function of the local bus, the IACK function of the master CPU depends on interrupt policies of system buses, as well as the interrupt policies of components within the system that trigger the interrupts. With respect to an external device triggering an interrupt, the interrupt scheme depends on whether that device can provide an ***interrupt vector*** (a place in memory that holds the address of an interrupt's ISR, Interrupt Service Routine, the software that the master CPU executes after the triggering of an interrupt). For devices that cannot provide an interrupt vector, referred to as nonvectored interrupts, master processors implement an ***auto-vectored*** interrupt scheme in which one ISR is shared by the nonvectored interrupts; determining which specific interrupt to handle, interrupt acknowledgment, and so on, are all handled by the ISR software.

An *interrupt vectored* scheme is implemented to support peripherals that can provide an interrupt vector over a bus, and where acknowledgment is automatic. An IACK-related register on the master CPU informs the device requesting the interrupt to stop requesting interrupt service, and provides what the master processor needs to process the correct interrupt (such as the interrupt number, vector number, and so on). Based on the activation of an external interrupt pin, an interrupt controller's interrupt select register, a device's interrupt select register, or some combination of the above, the master processor can determine which ISR to execute. After the ISR completes, the master processor resets the interrupt status by adjusting the bits in the processor's status register or an interrupt mask in the external interrupt controller. The interrupt request and acknowledgment mechanisms are determined by the device requesting the interrupt (since it determines which interrupt service to trigger), the master processor, and the system bus protocols.

Keep in mind that this is a general introduction to interrupt-handling, covering some of the key features found in a variety of schemes. The overall interrupt-handling scheme can vary widely from architecture to architecture. For example, PowerPC architectures implement an auto-vectored scheme, with no interrupt vector base register. The 68000 architecture supports both auto-vectored and interrupt vectored schemes, whereas MIPS32 architectures have no IACK cycle, so the interrupt handler handles the triggered interrupts.

2.2.1 Interrupt Priorities

Because there are potentially multiple components on an embedded board that may need to request interrupts, the scheme that manages all of the different types of interrupts is ***priority-based***. This means that all available interrupts within a processor have an associated interrupt level, which is the priority of that interrupt within the system. Typically, interrupts starting at level "1" are the highest priority within the system, and incrementally from there (2, 3, 4, ...) the priorities of the associated interrupts decrease. Interrupts with higher levels have precedence over any instruction stream being executed by the master processor, meaning that not only do interrupts have precedence over the main program, but higher priority interrupts have priority over interrupts with lower priorities as well. When an interrupt is triggered, lower priority interrupts are typically **masked**, meaning they are not allowed to trigger when the system is handling a higher-priority interrupt. The interrupt with the highest priority is usually called a nonmaskable interrupt (NMI).

How the components are prioritized depends on the IRQ line they are connected to, in the case of external devices, or what has been assigned by the processor design. It is the master processor's internal design that determines the number of external interrupts available and the interrupt levels supported within an embedded system. In Figure 2.8a, the MPC860 CPM, SIU, and PowerPC Core all work together to implement interrupts on the MPC823 processor. The CPM allows for internal interrupts (two SCCs, two SMCs, SPI, I²C, PIP, general-purpose timers, two IDMAs, SDMA, RISC Timer) and 12 external pins of port C, and it drives the interrupt levels on the SIU. The SIU receives interrupts from 8 external pins (IRQ0-7), and 8 internal sources, for a total of 16 sources of interrupts, one of which can be the CPM, and drives the IREQ input to the Core. When the IREQ pin is asserted, external interrupt processing begins. The priority levels are shown in Figure 2.8b.

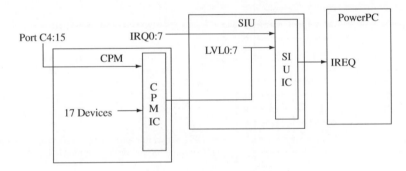

Figure 2.8a: Motorola/Freescale MPC860 Interrupt Pins and Table
Source: Copyright of Freescale Semiconductor, Inc. 2004. Used by permission.

In another processor, shown in Figures 2.9a and b, the 68000, there are eight levels of interrupts (0–7), where interrupts at level 7 have the highest priority. The 68000 interrupt table (Figure 2.9b) contains 256 32-bit vectors.

The M37267M8 architecture, shown in Figure 2.10a, allows for interrupts to be caused by 16 events (13 internal, two external, and one software) whose priorities and usages are summarized in Figure 2.10b.

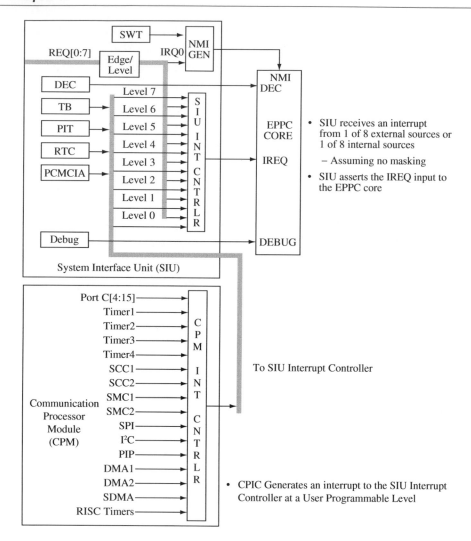

Figure 2.8b: Motorola/Freescale MPC860 Interrupt Levels
Source: Copyright of Freescale Semiconductor, Inc. 2004. Used by permission.

Several different priority schemes are implemented in the various architectures. These schemes commonly fall under one of three models: the *equal single level*, where the latest interrupt to be triggered gets the CPU; the *static multilevel*, where priorities are assigned by a priority encoder, and the interrupt with the highest priority gets the CPU; and the *dynamic multilevel*, where a priority encoder assigns priorities, and the priorities are reassigned when a new interrupt is triggered.

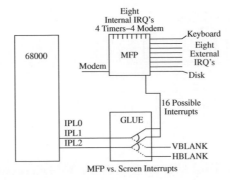

Figure 2.9a: Motorola/Freescale 68000 IRQs (There are 3 IRQ pins: IPL0, IPL1, and IPL2)

Vector Number[s]	Vector Offset (Hex)	Assignment
0	000	Reset Initial Interrupt Stack Pointer
1	004	Reset initial Program Counter
2	008	Access Fault
3	00C	Address Error
4	010	Illegal Instruction
5	014	Integer Divide by Zero
6	018	CHK, CHK2 instruction
7	01C	FTRAPcc, TRAPcc, TRAPV instructions
8	020	Privilege Violation
9	024	Trace
10	028	Line 1010 Emulator (Unimplemented A-Line Opcode)
11	02C	Line 1111 Emulator (Unimplemented F-line Opcode)
12	030	(Unassigned, Reserved)
13	034	Coprocessor Protocol Violation
14	038	Format Error
15	03C	Uninitialized Interrupt
16–23	040–050	(Unassigned, Reserved)
24	060	Spurious Interrupt
25	064	Level 1 Interrupt Autovector
26	068	Level 2 Interrupt Autovector
27	06C	Level 3 Interrupt Autovector
28	070	Level 4 Interrupt Autovector
29	074	Level 5 Interrupt Autovector
30	078	Level 6 Interrupt Autovector
31	07C	Level 7 Interrupt Autovector
32–47	080–08C	TRAP #0 D 15 Instructor Vectors
48	0C0	FP Branch or Set on Unordered Condition
49	0C4	FP Inexact Result
50	0C8	FP Divide by Zero
51	0CC	FP Underflow
52	0D0	FP Operand Error
53	0D4	FP Overflow
54	0D8	FP Signaling NAN
55	0DC	FP Unimplemented Data Type (Defined for MC68040)
56	0E0	MMU Configuration Error
57	0E4	MMU Illegal Operation Error
58	0E8	MMU Access Level Violation Error
59–63	0ECD0FC	(Unassigned, Reserved)
64-255	100D3FC	User Defined Vectors (192)

Figure 2.9b: Motorola/Freescale 68K IRQs Interrupt Table

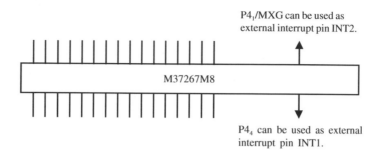

Figure 2.10a: Mitsubishi M37267M8 8-bit TV Microcontroller Interrupts

Interrupt Source	Priority	Interrupt Causes
RESET	1	(nonmaskable)
CRT	2	Occurs after character block display to CRT is completed
INT1	3	External Interrupt ** the processor detects that the level of a pin changes from 0 (LOW) to 1 (HIGH), or 1(HIGH) to 0 (LOW) and generates and interrupt request
Data Slicer	4	Interrupt occurs at end of line specified in caption position register
Serial I/O	5	Interrupt request from synchronous serial I/O function
Timer 4	6	Interrupt generated by overflow of timer 4
Xin & 4096	7	Interrupt occurs regularly with a f(Xin)/4096 period.
Vsync	8	An interrupt request synchronized with the vertical sync signal
Timer 3	9	Interrupt generated by overflow of timer 3
Timer 2	10	Interrupt generated by overflow of timer 2
Timer 1	11	Interrupt generated by overflow of timer 1
INT2	12	External Interrupt ** the processor detects that the level of a pin changes from 0 (LOW) to 1 (HIGH), or 1 (HIGH) to 0 (LOW) and generates and interrupt request
Multimaster I²C Bus interface	13	Related to I²C bus interface
Timer 5 & 6	14	Interrupt generated by overflow of timer 5 or 6
BRK instruction	15	(nonmaskable software)

Figure 2.10b: Mitsubishi M37267M8 8-bit TV Microcontroller Interrupt Table

2.2.2   Context Switching

After the hardware mechanisms have determined which interrupt to handle and have acknowledged the interrupt, the current instruction stream is halted and a context switch is performed, a process in which the master processor switches from executing the current instruction stream to another set of instructions. This alternate set of instructions being executed as the result of an interrupt is the ***interrupt service routine (ISR)*** or ***interrupt handler***. An ISR is simply a fast, short program that is executed when an interrupt is triggered. The specific ISR executed for a particular interrupt depends on whether a nonvectored or vectored scheme is in place. In the case of a nonvectored interrupt, a memory location contains the start of an ISR that the PC (program counter) or some similar mechanism branches to for all nonvectored interrupts. The ISR code then determines the source of the interrupt and provides the appropriate processing. In a vectored scheme, typically an interrupt vector table contains the address of the ISR.

The steps involved in an interrupt context switch include stopping the current program's execution of instructions, saving the context information (registers, the PC or similar mechanism that indicates where the processor should jump back to after executing the ISR) onto a stack, either dedicated or shared with other system software, and perhaps the disabling of other interrupts. After the master processor finishes executing the ISR, it context switches back to the original instruction stream that had been interrupted, using the context information as a guide.

The *interrupt services* provided by device driver code, based upon the mechanisms discussed above, include ***enabling/disabling*** interrupts through an interrupt control register on the master CPU or the disabling of the interrupt controller, ***connecting*** the ISRs to the interrupt table, providing interrupt levels and vector numbers to peripherals, providing address and control data to corresponding registers, and so on. Additional services implemented in interrupt access drivers include the ***locking/unlocking*** of interrupts, and the implementation of the actual ISRs. The pseudo code in the following example shows interrupt-handling initialization and access drivers that act as the basis of interrupt services (in the CPM and SIU) on the MPC860.

2.2.3 Interrupt Device Driver Pseudo Code Examples

The following pseudo code examples demonstrate the implementation of various interrupt-handling routines on the MPC860, specifically startup, shutdown, disable, enable, and interrupt servicing functions in reference to this architecture. These examples show how interrupt-handling can be implemented on a more complex architecture like the MPC860, and this in turn can be used as a guide to understand how to write interrupt-handling drivers on other processors that are as complex as or less complex than this one.

2.2.3.1 Interrupt-Handling Startup (Initialization) MPC860

Overview of initializing interrupts on MPC860 (in both CPM and SIU)

1. Initializing CPM Interrupts in MPC860 Example

    1.1 Setting Interrupt Priorities via CICR

    1.2 Setting individual enable bit for interrupts via CIMR

    1.3 Initializing SIU Interrupts via SIU Mask Register including setting the SIU bit associated with the level that the CPM uses to assert an interrupt.

    1.4 Set Master Enable bit for all CPM interrupts

2. Initializing SIU Interrupts on MPC860 Example

    2.1 Initializing the SIEL Register to select the edge-triggered or level-triggered interrupt-handling for external interrupts and whether processor can exit/wakeup from low power mode.

    2.2 If not done, initializing SIU Interrupts via SIU Mask Register including setting the SIU bit associated with the level that the CPM uses to assert an interrupt.

Enabling all interrupts via MPC860 "mtspr" instruction next step see Interrupt-Handling Enable.

2.2.3.2 Initializing CPM for Interrupts—4 Step Process

```
// **** Step 1 *****
// initializing the 24-bit CICR (see Figure 2.11), setting priorities
// and the interrupt levels. Interrupt Request Level, or IRL[0:2]
// allows a user to program the priority request level of the CPM
// interrupt with any number from level 0 (highest priority) through
// level 7 (lowest priority).
```

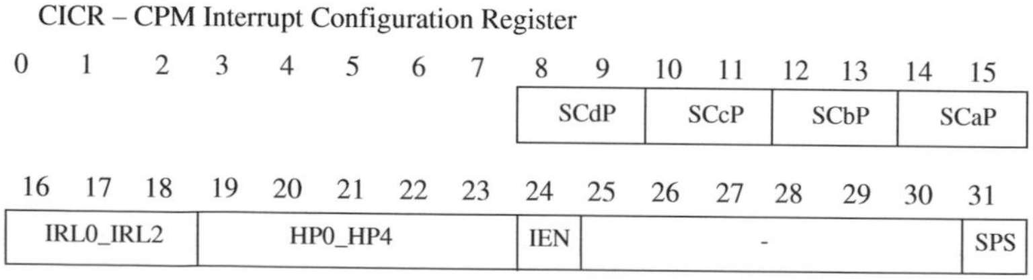

CICR – CPM Interrupt Configuration Register

0	1	2	3	4	5	6	7	8	9	10	11	12	13	14	15
								SCdP		SCcP		SCbP		SCaP	

16	17	18	19	20	21	22	23	24	25	26	27	28	29	30	31
IRL0_IRL2			HP0_HP4					IEN			-				SPS

Figure 2.11a: CICR Register

SCC	Code	Highest SCaP	SCbP	SCcP	Lowest SCdP
SCC1	00			00	
SCC2	01		01		
SCC3	10	10			
SCC4	11				11

Figure 2.11b: SCC Priorities

CIPR - CPM Interrupt Pending Register

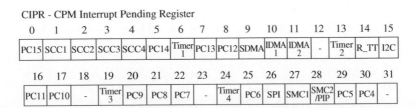

Figure 2.11c: CIPR Register

```
. . .
int RESERVED94 = 0xFF000000; // bits 0-7 reserved, all set to 1

// the PowerPC SCC's are prioritized relative to each other.
// Each SCxP field is representative of a priority for each SCC
// where SCdP is the lowest and ScaP is the highest priority.
// Each SCxP field is made up of 2-bits (0-3), one for each SCC,
// where 0d (00b) = SCC1, 1d (01b) = SCC2, 2d (10b) = SCC3, and
// 3d (11b) = SCC4. See Figure 2.11b.

int CICR.SCdP = 0x00C00000; // bits 8-9 both = 1,
                            // SCC4 = lowest priority
int CICR.SCcP = 0x00000000; // bits 10-11, both = 0, SCC1 = 2nd
                            // to lowest priority
int CICR.SCbP = 0x00040000; // bits 12-13, = 01b, SCC2 2nd
                            // highest priority
int CICR.SCaP = 0x00020000; // bits 14-15, = 10b, SCC3
                            // highest priority


// IRL0_IRL2 is a 3 bit configuration parameter called the
// Interrupt Request Level - it allows a user to program the
// priority request level of the CPM interrupt with bits 16-18
// with a value of 0 - 7 in terms of its priority mapping within
// the SIU.Iin this example, it is a priority 7 since all 3 bits
// set to 1.
```

```
int CICR.IRL0 = 0x00008000; // Interrupt request level 0
                            // (bit 16) = 1
int CICR.IRL1 = 0x00004000; // Interrupt request level 1
                            // (bit 17) = 1
int CICR.IRL2 = 0x00002000; // Interrupt request level 2
                            // (bit 18) = 1


// HP0 - HP 4 are 5 bits (19-23) used to represent one of the
// CPM Interrupt Controller interrupt sources (shown in Figure 2.8b)
// as being the highest priority source relative to their bit
// location in the CIPR register—see Figure 2.11c. In this example,
// HP0 - HP4 = 11111b (31d) so highest external priority source to
// the PowerPC core is PC15
int CICR.HP0 = 0x00001000; /* Highest priority */
int CICR.HP1 = 0x00000800; /* Highest priority */
int CICR.HP2 = 0x00000400; /* Highest priority */
int CICR.HP3 = 0x00000200; /* Highest priority */
int CICR.HP4 = 0x00000100; /* Highest priority */


// IEN bit 24 - Master enable for CPM interrupts - not enabled
// here - see step 4

int RESERVED95 = 0x0000007E; // bits 25-30 reserved, all set to 1

int CICR.SPS = 0x00000001;   // Spread priority scheme in which
                             // SCCs are spread out by priority in
                             // interrupt table, rather than grouped
                             // by priority at the top of the table



// ***** step 2 *****
// initializing the 32-bit CIMR (see Figure 2.12), CIMR bits
// correspond to CMP Interrupt Sources indicated in CIPR
// (see Figure 2.11c), by setting the bits associated with the desired
// interrupt sources in the CIMR register (each bit corresponds
// to a CPM interrupt source)
```

CIPR - CPM Interrupt Mask Register

0	1	2	3	4	5	6	7	8	9	10	11	12	13	14	15
PC15	SCC1	SCC2	SCC3	SCC4	PC14	Timer 1	PC13	PC12	SDMA	IDMA 1	IDMA 2	-	Timer 2	R_TT	I2C

16	17	18	19	20	21	22	23	24	25	26	27	28	29	30	31
PC11	PC10	-	Timer 3	PC9	PC8	PC7	-	Timer 4	PC6	SPI	SMC1	SMC2 /PIP	PC5	PC4	-

Figure 2.12: CIMR Register

```
int CIMR.PC15 = 0x80000000; // PC15 (Bit 0) set to 1, interrupt
                            // source enabled
int CIMR.SCC1 = 0x40000000; // SCC1 (Bit 1) set to 1, interrupt
                            // source enabled
int CIMR.SCC2 = 0x20000000; // SCC2 (Bit 2) set to 1, interrupt
                            // source enabled
int CIMR.SCC4 = 0x08000000; // SCC4 (Bit 4) set to 1, interrupt
                            // source enabled
int CIMR.PC14 = 0x04000000; // PC14 (Bit 5) set to 1, interrupt
                            // source enabled
int CIMR.TIMER1 = 0x02000000; // Timer1 (Bit 6) set to 1, interrupt
                              // source enabled
int CIMR.PC13 = 0x01000000; // PC13 (Bit 7) set to 1, interrupt
                            // source enabled
int CIMR.PC12 = 0x00800000; // PC12 (Bit 8) set to 1, interrupt
                            // source enabled
int CIMR.SDMA = 0x00400000; // SDMA (Bit 9) set to 1, interrupt
                            // source enabled
int CIMR.IDMA1 = 0x00200000; // IDMA1 (Bit 10) set to 1, interrupt
                             // source enabled
int CIMR.IDMA2 = 0x00100000; // IDMA2 (Bit 11) set to 1, interrupt
                             // source enabled
int RESERVED100 = 0x00080000; // Unused Bit 12
int CIMR.TIMER2 = 0x00040000; // Timer2 (Bit 13) set to 1, interrupt
                              // source enabled
int CIMR.R.TT = 0x00020000; // R-TT (Bit 14) set to 1, interrupt
                            // source enabled
int CIMR.I2C = 0x00010000; // I2C (Bit 15) set to 1, interrupt
                           // source enabled
int CIMR.PC11 = 0x00008000; // PC11 (Bit 16) set to 1, interrupt
                            // source enabled
int CIMR.PC10 = 0x00004000; // PC11 (Bit 17) set to 1, interrupt
                            // source enabled
```

```
int RESERVED101 = 0x00002000; // Unused bit 18
int CIMR.TIMER3 = 0x00001000; // Timer3 (Bit 19) set to 1, interrupt
                             // source enabled
int CIMR.PC9 = 0x00000800; // PC9 (Bit 20) set to 1, interrupt
                             // source enabled
int CIMR.PC8 = 0x00000400; // PC8 (Bit 21) set to 1, interrupt
                             // source enabled
int CIMR.PC7 = 0x00000200; // PC7 (Bit 22) set to 1, interrupt
                             // source enabled
int RESERVED102 = 0x00000100; // unused bit 23
int CIMR.TIMER4 = 0x00000080; // Timer4 (Bit 24) set to 1, interrupt
                             // source enabled
int CIMR.PC6 = 0x00000040; // PC6 (Bit 25) set to 1, interrupt
                             // source enabled
int CIMR.SPI = 0x00000020; // SPI (Bit 26) set to 1, interrupt
                             // source enabled
int CIMR.SMC1 = 0x00000010; // SMC1 (Bit 27) set to 1, interrupt
                             // source enabled
int CIMR.SMC2-PIP = 0x00000008; // SMC2/PIP (Bit 28) set to 1,
                             // interrupt source enabled
int CIMR.PC5 = 0x00000004; // PC5 (Bit 29) set to 1, interrupt
                             // source enabled
int CIMR.PC4 = 0x00000002; // PC4 (Bit 30) set to 1, interrupt
                             // source enabled
int RESERVED103 = 0x00000001; // unused bit 31
```

```
// ***** step 3 *****
// Initializing the SIU Interrupt Mask Register (see Figure 2.13)
// including setting the SIU bit associated with the level that
// the CPM uses to assert an interrupt.
```

SIMASK - SIU Mask Register

0	1	2	3	4	5	6	7	8	9	10	11	12	13	14	15
IRM0	LVM0	IRM1	LVM1	IRM2	LVM2	IRM3	LVM3	IRM4	LVM4	IRM5	LVM5	IRM6	LVM6	IRM7	LVM7

16	17	18	19	20	21	22	23	24	25	26	27	28	29	30	31
							Reserved								

Figure 2.13: SIMASK Register

```
int SIMASK.IRM0 = 0x80000000; // enable external interrupt
                              // input level 0
int SIMASK.LVM0 = 0x40000000; // enable internal interrupt
                              // input level 0
int SIMASK.IRM1 = 0x20000000; // enable external interrupt
                              // input level 1
int SIMASK.LVM1 = 0x10000000; // enable internal interrupt
                              // input level 1
int SIMASK.IRM2 = 0x08000000; // enable external interrupt
                              // input level 2
int SIMASK.LVM2 = 0x04000000; // enable internal interrupt
                              // input level 2
int SIMASK.IRM3 = 0x02000000; // enable external interrupt
                              // input level 3
int SIMASK.LVM3 = 0x01000000; // enable internal interrupt
                              // input level 3
int SIMASK.IRM4 = 0x00800000; // enable external interrupt
                              // input level 4
int SIMASK.LVM4 = 0x00400000; // enable internal interrupt
                              // input level 4
int SIMASK.IRM5 = 0x00200000; // enable external interrupt
                              // input level 5
int SIMASK.LVM5 = 0x00100000; // enable internal interrupt
                              // input level 5
int SIMASK.IRM6 = 0x00080000; // enable external interrupt
                              // input level 6
int SIMASK.LVM6 = 0x00040000; // enable internal interrupt
                              // input level 6
int SIMASK.IRM7 = 0x00020000; // enable external interrupt
                              // input level 7
int SIMASK.LVM7 = 0x00010000; // enable internal interrupt
                              // input level 7
int RESERVED6 = 0x0000FFFF; // unused bits 16-31
```

```
// ***** step 4 *****
// IEN bit 24 of CICR register- Master enable for CPM interrupts
// int CICR.IEN = 0x00000080;
// interrupts enabled IEN = 1
// Initializing SIU for interrupts - 2 step process
```

SIEL - SIU Interrupt Edge Level Mask Register

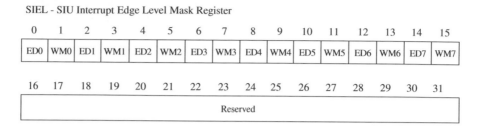

Figure 2.14: SIEL Register

```
// ***** step 1 *****
// Initializing the SIEL Register (see Figure 2.14) to select the
// edge-triggered (set to 1 for falling edge indicating interrupt
// request) or level-triggered (set to 0 for a 0 logic level
// indicating interrupt request) interrupt-handling for external
// interrupts (bits 0,2,4,6,8,10,12,14) and whether processor can
// exit/wakeup from low power mode (bits 1,3,5,7,9,11,13,15). Set
// to 0 is NO, set to 1 is Yes
```

```
int SIEL.ED0 = 0x80000000; // interrupt level 0 (falling)
                           // edge-triggered
int SIEL.WM0 = 0x40000000; // IRQ at interrupt level 0 allows CPU to
                           // exit from low power mode
int SIEL.ED1 = 0x20000000; // interrupt level 1 (falling)
                           // edge-triggered
int SIEL.WM1 = 0x10000000; // IRQ at interrupt level 1 allows CPU to
                           // exit from low power mode
int SIEL.ED2 = 0x08000000; // interrupt level 2 (falling)
                           // edge-triggered
int SIEL.WM2 = 0x04000000; // IRQ at interrupt level 2 allows CPU to
                           // exit from low power mode
int SIEL.ED3 = 0x02000000; // interrupt level 3 (falling)
                           // edge-triggered
int SIEL.WM3 = 0x01000000; // IRQ at interrupt level 3 allows CPU to
                           // exit from low power mode
int SIEL.ED4 = 0x00800000; // interrupt level 4 (falling)
                           // edge-triggered
int SIEL.WM4 = 0x00400000; // IRQ at interrupt level 4 allows CPU to
                           // exit from low power mode
int SIEL.ED5 = 0x00200000; // interrupt level 5 (falling)
                           // edge-triggered
```

```
int SIEL.WM5 = 0x00100000; // IRQ at interrupt level 5 allows CPU to
                           // exit from low power mode
int SIEL.ED6 = 0x00080000; // interrupt level 6 (falling)
                           // edge-triggered
int SIEL.WM6 = 0x00040000; // IRQ at interrupt level 6 allows CPU to
                           // exit from low power mode
int SIEL.ED7 = 0x00020000; // interrupt level 7 (falling)
                           // edge-triggered
int SIEL.WM7 = 0x00010000; // IRQ at interrupt level 7 allows CPU to
                           // exit from low power mode
int RESERVED7 = 0x0000FFFF; // bits 16-31 unused
```

```
// ***** step 2 *****
// Initializing SIMASK register - done in step intializing CPM
```

2.2.3.3 Interrupt-Handling Shutdown on MPC860

There essentially is no shutdown process for interrupt-handling on the MPC860, other than perhaps disabling interrupts during the process.

```
// Essentially disabling all interrupts via IEN bit 24 of
// CICR - Master disable for CPM interrupts
CICR.IEN = "CICR.IEN" AND "0"; // interrupts disabled IEN = 0
```

2.2.3.4 Interrupt-Handling Disable on MPC860

```
// To disable specific interrupt means modifying the SIMASK,
// so disabling the external
// interrupt at level 7 (IRQ7) for example is done by
// clearing bit 14
SIMASK.IRM7 = "SIMASK.IRM7" AND "0"; // disable external interrupt
                                     // input level 7

// disabling of all interrupts takes effect with the mtspr
// instruction.
// mtspr 82,0; // disable interrupts via mtspr (move
//                to special purpose register) instruction
```

2.2.3.5 Interrupt-Handling Enable on MPC860

```
// specific enabling of particular interrupts done in initialization
// section of this example - so the interrupt enable of all interrupts
// takes effect with the mtspr instruction.

mtspr 80,0;     // enable interrupts via mtspr (move to special purpose
                // register) instruction

// in review, to enable specific interrupt means modifying the SIMASK,
// so enabling the external interrupt at level 7 (IRQ7) for example
// is done my setting bit 14

SIMASK.IRM7 = "SIMASK.IRM7" OR "1"; // enable external interrupt
                                    // input level 7
```

2.2.3.6 Interrupt-Handling Servicing on MPC860

In general, this ISR (and most ISRs) essentially disables interrupts first, saves the context information, processes the interrupt, restores the context information, and then enables interrupts.

```
InterruptServiceRoutineExample ()
{
  . . .
  // disable interrupts
  disableInterrupts(); // mtspr 82,0;
  // save registers

   saveState();
  // read which interrupt from SI Vector Register (SIVEC)
  interruptCode = SIVEC.IC;

  // if IRQ 7 then execute
  if (interruptCode = IRQ7) {
  . . .

  // If an IRQx is edge-triggered, then clear the service bit in
  // the SI Pending Register by putting a "1".
  SIPEND.IRQ7 = SIPEND.IRQ7 OR "1";
```

```
// main process
. . .
} // endif IRQ7

// restore registers
restoreState();
// re-enable interrupts
enableInterrupts(); // mtspr 80,0;
}
```

2.2.4   *Interrupt-Handling and Performance*

The performance of an embedded design is affected by the ***latencies*** (delays) involved with the interrupt-handling scheme. The interrupt *latency* is essentially the time from when an interrupt is triggered until its ISR starts executing. The master CPU, under normal circumstances, accounts for a lot of overhead for the time it takes to process the interrupt request and acknowledge the interrupt, obtaining an interrupt vector (in a vectored scheme), and context switching to the ISR. In the case when a lower-priority interrupt is triggered during the processing of a higher priority interrupt, or a higher priority interrupt is triggered during the processing of a lower priority interrupt, the interrupt latency for the original lower priority interrupt increases to include the time in which the higher priority interrupt is handled (essentially how long the lower priority interrupt is disabled). Figure 2-15 summarizes the variables that impacts interrupt latency.

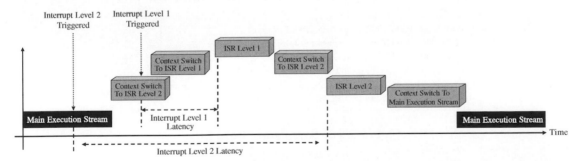

Figure 2.15: Interrupt Latency

Within the ISR itself, additional overhead is caused by the context information being stored at the start of the ISR and retrieved at the end of the ISR. The time to context switch back to the original instruction stream that the CPU was executing before the interrupt was triggered also adds to the overall interrupt execution time. While the hardware aspects of interrupt-handling—the context switching, processing interrupt requests, and so on—are beyond the software's control, the overhead related to when the context information is saved, as well as how the ISR is written both in terms of the programming language used and the size, are under the software's control. Smaller ISRs, or ISRs written in a lower-level language like assembly, as opposed to larger ISRs or ISRs written in higher-level languages like Java, or saving/retrieving less context information at the start and end of an ISR, can all decrease the interrupt-handling execution time and increase performance.

2.3   Example 2: Memory Device Drivers

While in reality all types of physical memory are two-dimensional arrays (matrices) made up of cells addressed by a unique row and column, the master processor and programmers view memory as a large one-dimensional array, commonly referred to as the ***Memory Map*** (see Figure 2.16). In the memory map, each cell of the array is a row of bytes (8 bits) and the number of bytes per row depends on the width of the data bus (8-bit, 16-bit, 32-bit, 64-bit, and so on). This, in turn, depends on the width of the registers of the master architecture. When physical memory is referenced from the software's point-of-view, it is commonly referred to as logical memory, and its most basic unit is the byte. Logical memory is made up of all the physical memory (registers, ROM, and RAM) in the entire embedded system.

Address Range	Accessed Device	Port Width
0x00000000 - 0x003FFFFF	Flash PROM Bank 1	32
0x00400000 - 0x007FFFFF	Flash PROM Bank 2	32
0x04000000 - 0x043FFFFF	DRAM 4 Mbyte (1Meg x 32-bit)it)	32
0x09000000 - 0x09003FFF	MPC Internal Memory Map	32
0x09100000 - 0x09100003	BCSR - Board Control & Status Register	32
0x10000000 - 0x17FFFFFF	PCMCIA Channel	16

Figure 2.16: Sample Memory Map

The software must provide the processors in the system with the ability to access various portions of the memory map. The software involved in managing the memory on the master processor and on the board, as well as managing memory hardware mechanisms, consists of the device drivers for the management of the overall memory subsystem. The memory subsystem includes all types of memory management components, such as memory controllers and MMU, as well as the types of memory in the memory map, such as registers, cache, ROM, DRAM, and so on. All or some combination of six of the ten device driver functions from the list of device driver functionality introduced at the start of this chapter are commonly implemented, including:

- **Memory Subsystem Startup**, initialization of the hardware on power-on or reset (initialize TLBs for MMU, initialize/configure MMU).

- **Memory Subsystem Shutdown**, configuring hardware into its power-off state. Note that under the MPC860 there is no necessary shutdown sequence for the memory subsystem, so pseudo code examples are not shown.

- **Memory Subsystem Disable**, allowing other software to disable hardware on-the-fly (disabling cache).

- **Memory Subsystem Enable**, allowing other software to enable hardware on-the-fly (enable cache).

- **Memory Subsystem Write**, storing in memory a byte or set of bytes (i.e., in cache, ROM, and main memory).

- **Memory Subsystem Read**, retrieving from memory a "copy" of the data in the form of a byte or set of bytes (i.e., in cache, ROM, and main memory).

Regardless of what type of data is being read or written, all data within memory is managed as a sequence of bytes. While one memory access is limited to the size of the data bus, certain architectures manage access to larger ***blocks*** (a contiguous set of bytes) of data, called ***segments***, and thus implement a more complex address translation scheme in which the logical address provided via software is made up of a *segment number* (address of start of segment) and *offset* (within a segment) which is used to determine the physical address of the memory location.

The order in which bytes are retrieved or stored in memory depends on the ***byte ordering*** scheme of an architecture. The two possible byte ordering schemes are ***little endian*** and ***big endian***. In little endian mode, bytes (or "bits" with 1 byte [8-bit] schemes) are retrieved and stored in the order of the lowest byte first, meaning the lowest byte is furthest to the left. In big endian mode bytes are accessed in the order of the highest byte first, meaning that the lowest byte is furthest to the right (see Figure 2.17).

Odd Bank		Even Bank	
F	90	87	E
D	E9	11	C
8	F1	24	A
9	01	46	8
7	76	DE	6
5	14	33	4
3	55	12	2
1	AB	FF	0

Data Bus (15:8)	Data Bus (7:0)

In <u>little-endian</u> mode if a byte is read from address "0", an "FF" is returned, if 2 bytes are read from address 0, then (reading from the lowest byte which is furthest to the LEFT in little-endian mode) an "ABFF" is returned. If 4 bytes (32-bits) are read from address 0, then a "5512ABFF" is returned.

In <u>big-endian</u> mode if a byte is read from address "0", an "FF" is returned, if 2 bytes are read from address 0, then (reading from the lowest byte which is furthest to the RIGHT in big-endian mode) an "FFAB" is returned. If 4 bytes (32-bits) are read from address 0, then a "1255FFAB" is returned.

Figure 2.17: Endianess

What is important regarding memory and byte ordering is that performance can be greatly impacted if data requested is not aligned in memory according to the byte ordering scheme defined by the architecture. As shown in Figure 2.17, memory is either soldered into or plugged into an area on the embedded board, called memory ***banks***. While the configuration and number of banks can vary from platform to platform, memory addresses are aligned in an odd or even bank format. If data is aligned in little endian mode, data taken from address "0" in an even bank is "ABFF," and as such is an aligned memory access. So, given a 16-bit data bus, only one memory access is needed. But if data were to be taken from address "1" (an odd bank) in a memory aligned as shown in Figure 2.17, the little endian ordering scheme should retrieve "12AB" data. This would require two memory accesses, one to read the AB, the odd byte, and one to read "12," the even byte, as well as some mechanism within the processor or in driver code to perform additional work to align them as "12AB." Accessing data in memory that is aligned according to the byte ordering scheme can result in access times at least twice as fast.

Finally, how memory is actually accessed by the software will, in the end, depend on the programming language used to write the software. For example, assembly language has various architecture-specific addressing modes that are unique to an architecture, and Java allows modifications of memory through objects.

2.3.1   *Memory Management Device Driver Pseudo Code Examples*

The following pseudo code demonstrates implementation of various memory management routines on the MPC860, specifically startup, disable, enable, and writing/erasing functions in reference to the architecture. These examples demonstrate how memory management can be implemented on a more complex architecture, and this in turn can serve as a guide to understanding how to write memory management drivers on other processors that are as complex as or less complex than the MPC860 architecture.

2.3.1.1 Memory Subsystem Startup (Initialization) on MPC860

In the sample memory map in Figure 2.18, the first two banks are 8 MB of Flash, then 4 MB of DRAM, followed by 1 MB for the internal memory map and control/status registers. The remainder of the map represents 4 MB of an additional PCMCIA card. The main memory subsystem components that are initialized in this example are the physical memory chips themselves (i.e., Flash, DRAM), which in the case of the MPC860 is initialized via a memory controller, configuring the internal memory map (registers and dual-port RAM), as well as configuring the MMU.

Address Range	Accessed Device	Port Width
0x00000000 - 0x003FFFFF	Flash PROM Bank 1	32
0x00400000 - 0x007FFFFF	Flash PROM Bank 2	32
0x04000000 - 0x043FFFFF	DRAM 4 Mbyte (1Meg x 32-bit)it)	32
0x09000000 - 0x09003FFF	MPC Internal Memory Map	32
0x09100000 - 0x09100003	BCSR - Board Control & Status Register	32
0x10000000 - 0x17FFFFFF	PCMCIA Channel	16

Figure 2.18: Sample Memory Map

1. Initializing the Memory Controller and Connected ROM/RAM

The MPC860 memory controller (shown in Figure 2.19) is responsible for the control of up to eight memory banks, interfacing to SRAM, EPROM, flash EPROM, various DRAM devices, and other peripherals (i.e., PCMCIA). Thus, in this example of the MPC860, onboard memory (Flash, SRAM, DRAM, and so on) is initialized by initializing the memory controller.

The memory controller has two different types of subunits, the general-purpose chip-select machine (GPCM) and the user-programmable machines (UPMs), which exist to connect to certain types of memory. The GPCM is designed to interface to SRAM, EPROM, Flash EPROM, and other peripherals (such as PCMCIA), whereas the UPMs are designed to interface to a wide variety of memory, including DRAMs. The pinouts of the MPC860's memory controller reflect the different signals that connect these subunits to the various types of memory (see Figures 2.20a, b, and c). For every chip select (CS), there is an associated memory bank.

With every new access request to external memory, the memory controller determines whether the associated address falls into one of the eight address ranges (one for each bank)

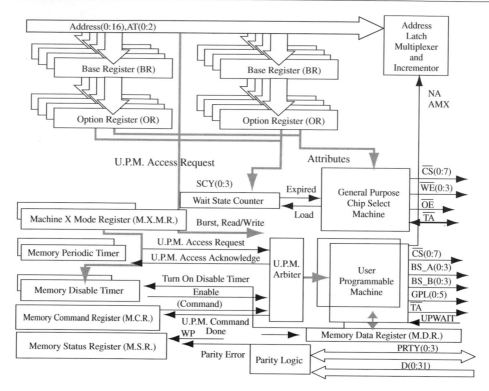

Figure 2.19: MPC860 Integrated Memory Controller
Source: Copyright of Freescale Semiconductor, Inc. 2004. Used by permission.

defined by the eight base registers (which specify the start address of each bank) and option registers (which specify the bank length) pairs (see Figure 2.21). If it does, the memory access is processed by either the GPCM or one of the UPMs, depending on the type of memory located in the memory bank that contains the desired address.

Because each memory bank has a pair of base and option registers (BR0/OR0–BR7/OR7), they need to be configured in the memory controller initialization drivers. The base register (BR) fields are made up of a 16-bit start address BA (bits 0-16); AT (bits 17-19) specifies the address type (allows sections of memory space to be limited to only one particular type of data), a port size (8, 16, 32-bit); a parity checking bit; a bit to write protect the bank (allowing for read-only or read/write access to data); a memory controller machine selection set of bits (for GPCM or one of the UPMs); and a bit indicating if the bank is valid. The option register (OR) fields are made up of bits of control information for configuring the GPCM and UPMs accessing and addressing scheme (i.e., burst accesses, masking, multiplexing, and so on).

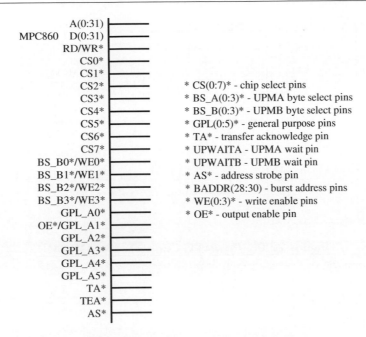

Figure 2.20a: Memory Controller Pins

Source: Copyright of Freescale Semiconductor, Inc. 2004. Used by permission.

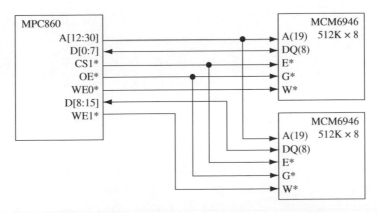

Figure 2.20b: PowerPC Connected to SRAM

Source: Copyright of Freescale Semiconductor, Inc. 2004. Used by permission.

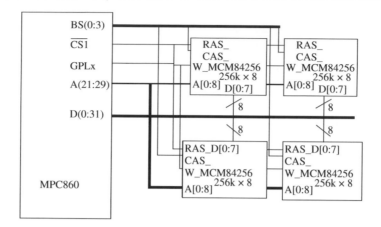

Figure 2.20c: PowerPC Connected to DRAM
Source: Copyright of Freescale Semiconductor, Inc. 2004. Used by permission.

BRx - Base Register

0	1	2	3	4	5	6	7	8	9	10	11	12	13	14	15
BA0 – BA15															

16	17	18	19	20	21	22	23	24	25	26	27	28	29	30	31
BA16	AT0_AT2			PS0_PS1		PARE		WP		MS0_MS1			Reserved		V

ORx - Option Register

0	1	2	3	4	5	6	7	8	9	10	11	12	13	14	15
AM0 – AM15															

16	17	18	19	20	21	22	23	24	25	26	27	28	29	30	31
AM16	ATM0_ATM2			CSNT/SAM	ACS0_ACS1		BI	SCY0_SCY3				SETA	TRLX	EHTR	Res

Figure 2.21: Base and Option Registers

The type of memory located in the various banks, and connected to the appropriate CS, can then be initialized for access via these registers. So, given the memory map example in Figure 2.18, the pseudo code for configuring the first two banks (of 4 MB of Flash each), and the third bank (4 MB of DRAM) would be as follows:

Note: Length initialized by looking up the length in the table below, and entering 1's from bit 0 to bit position indicating that length, and entering 0's into the remaining bits.

0	1	2	3	4	5	6	7	8	9	10	11	12	13	14	15	16
2 G	1 G	512 M	256 M	128 M	64 M	32 M	16 M	8 M	4 M	2 M	1 M	512 K	256 K	128 K	64 K	32 K

```
// OR for Bank 0 - 4 MB of flash , 0x1FF8 for bits AM (bits 0-16)
// OR0 = 0x1FF80954;
// Bank 0 - Flash starting at address 0x00000000 for bis BA
// (bits 0-16), configured for GPCM, 32-bit
BR0 = 0x00000001;

// OR for Bank 1 - 4 MB of flash , 0x1FF8 for bits AM (bits 0-16)
// OR1 = 0x1FF80954; Bank 1 - 4 MB of Flash on CS1 starting at
// address 0x00400000, configured for GPCM, 32-bit
BR1 = 0x00400001;

// OR for Bank 2 - 4 MB of DRAM , 0x1FF8 for bits AM (bits 0-16)
// OR2 = 0x1FF80800; Bank 2 - 4 MB of DRAM on CS2 starting at
// address 0x04000000, configured for UPMA, 32-bit
BR2 = 0x04000081;

// OR for Bank 3 for BCSR OR3 = 0xFFFF8110; Bank 3 - Board Control
// and Status Registers from address 0x09100000
BR3 = 0x09100001;
. . .
```

So, to initialize the memory controller, the base and option registers are initialized to reflect the types of memory in its banks. While no additional GPCM registers need initialization, for memory managed by the UPMA or UPMB, at the very least, the memory periodic timer prescaler register (MPTPR) is initialized for the required refresh time-out (i.e., for DRAM), and the related memory mode register (MAMR or MBMR) for configuring the UPMs needs initialization. The core of every UPM is a (64 × 32 bit) RAM array that specifies the specific type of accesses (logical values) to be transmitted to the UPM managed memory chips for a given clock cycle. The RAM array is initialized via the memory command register (MCR), which is specifically used during initialization to read from and write to the RAM array, and the memory data register (MDR), which stores the data the MCR uses to write to or read from the RAM array (see sample pseudo code below).

```
. . .
// set periodic timer prescaler to divide by 8
MPTPR = 0x0800; // 16 bit register

// periodic timer prescaler value for DRAM refresh period
// (see the PowerPC manual for  calculation), timer enable,...
MAMR = 0xC0A21114;

// 64-Word UPM RAM Array content example --the values in this
// table were generated using the UPM860 software available on
// the Motorola/Freescale Netcomm Web site.

UpmRamARRY:
// 6 WORDS - DRAM 70ns - single read. (offset 0 in upm RAM)
.long 0x0fffcc24, 0x0fffcc04, 0x0cffcc04, 0x00ffcc04,
.long 0x00ffcc00, 0x37ffcc47

// 2 WORDs - offsets 6-7 not used
.long 0xffffffff, 0xffffffff
// 14 WORDs - DRAM 70ns - burst read. (offset 8 in upm RAM)
.long 0x0fffcc24, 0x0fffcc04, 0x08ffcc04, 0x00ffcc04,0x00ffcc08,
.long 0x0cffcc44, 0x00ffec0c, 0x03ffec00, 0x00ffec44, 0x00ffcc08,
.long 0x0cffcc44, 0x00ffec04, 0x00ffec00, 0x3fffec47
// 2 WORDs - offsets 16-17 not used
.long 0xffffffff, 0xffffffff
// 5 WORDs - DRAM 70ns - single write. (offset 18 in upm RAM)
.long 0x0fafcc24, 0x0fafcc04, 0x08afcc04, 0x00afcc00,0x37ffcc47
// 3 WORDs - offsets 1d-1f not used
.long 0xffffffff, 0xffffffff, 0xffffffff
// 10 WORDs - DRAM 70ns - burst write. (offset 20 in upm RAM)
.long 0x0fafcc24, 0x0fafcc04, 0x08afcc00, 0x07afcc4c, 0x08afcc00
.long, 0x07afcc4c, 0x08afcc00, 0x07afcc4c, 0x08afcc00, 0x37afcc47
// 6 WORDs - offsets 2a-2f not used
.long 0xffffffff, 0xffffffff, 0xffffffff, 0xffffffff
.long 0xffffffff, 0xffffffff
// 7 WORDs - refresh 70ns. (offset 30 in upm RAM)
.long 0xe0ffcc84, 0x00ffcc04, 0x00ffcc04, 0x0fffcc04, 0x7fffcc04,
.long 0xffffcc86, 0xffffcc05
// 5 WORDs - offsets 37-3b not used
.long 0xffffffff, 0xffffffff, 0xffffffff, 0xffffffff,0xffffffff
// 1 WORD - exception. (offset 3c in upm RAM)
.long 0x33ffcc07
```

```
// 3 WORDs - offset 3d-3f not used
.long 0xffffffff, 0xffffffff, 0x40004650
UpmRAMArrayEnd:

// Write To UPM Ram Array
Index = 0
Loop While Index < 64
{
MDR = UPMRamArray[Index]; // Store data to MDR
MCR = 0x0000; // Issue "Write" command to MCR register to
// store what is in MDR in RAM Array
Index = Index + 1;
} // end loop
. . .
```

2. Initializing the Internal Memory Map on the MPC860

The MPC860's internal memory map contains the architecture's special purpose registers (SPRs), as well as dual-port RAM, also referred to as parameter RAM, that contain the buffers of the various integrated components, such as Ethernet or I2C, for example. On the MPC860, it is simply a matter of configuring one of these SPRs, the Internal Memory Map Register (IMMR) shown in Figure 2.22, to contain the base address of the internal memory map, as well as some factory-related information on the specific MPC860 processor (part number and mask number).

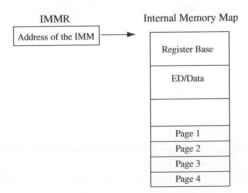

Figure 2.22: IMMR

In the case of the sample memory map used in this section, the internal memory map starts at 0x09000000, so in pseudo code form, the IMMR would be set to this value via the "mfspr" or "mtspr" commands:

```
mtspr 0x090000FF // the top 16 bits are the address,
                 // bits 16-23 are the part number
// (0x00 in this example), and bits 24-31 is the mask number
// (0xFF in this example).
```

3. Initializing the MMU on the MPC860

The MPC860 uses the MMUs to manage the board's virtual memory management scheme, providing logical/effective to physical/real address translations, cache control (instruction MMU and instruction cache, data MMU and data cache), and memory access protections. The MPC860 MMU (shown in Figure 2.23a) allows support for a 4 GB uniform (user) address space that can be divided into pages of a variety of sizes, specifically 4 kB, 16 kB, 512 kB, or 8 MB, that can be individually protected and mapped to physical memory.

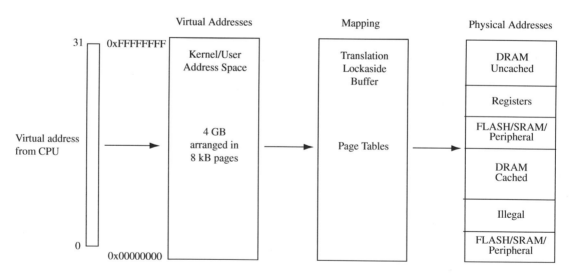

Figure 2.23a: TLB within VM Scheme

Using the smallest page size a virtual address space can be divided into on the MPC860 (4 kB), a translation table—also commonly referred to as the *memory map* or *page table*—would contain a million address translation entries, one for each 4 kB page in the

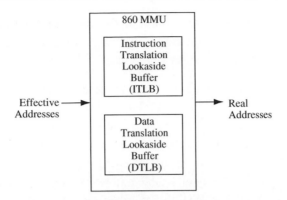

Figure 2.23b: TLB

4 GB address space. The MPC860 MMU does not manage the entire translation table at one time (in fact, most MMUs do not). This is because embedded boards do not typically have 4 GB of physical memory that needs to be managed at one time. It would be very time consuming for an MMU to update a million entries with every update to virtual memory by the software, and an MMU would need to use a lot of faster (and more expensive) on-chip memory in order to store a memory map of such a size. So, as a result, the MPC860 MMU contains small caches within it to store a subset of this memory map. These caches are referred to as translation look aside buffers (TLBs, shown in Figure 2.23b—one instruction and one data), and are part of the MMU's initialization sequence. In the case of the MPC860, the TLBs are 32-entry and fully associative caches. The entire memory map is stored in cheaper off-chip main memory as a two-level tree of data structures that define the physical memory layout of the board and their corresponding effective memory address.

The TLB is how the MMU translates (maps) logical/virtual addresses to physical addresses. When the software attempts to access a part of the memory map not within the TLB, a *TLB miss* occurs, which is essentially a trap requiring the system software (through an exception handler) to load the required translation entry into the TLB. The system software that loads the new entry into the TLB does so through a process called a ***tablewalk***. This is basically the process of traversing the MPC860's two-level memory map tree in main memory to locate the desired entry to be loaded in the TLB. The first level of the PowerPC's multilevel translation table scheme (its translation table structure uses one level-1 table and one or more level-2 tables) refers to a page table entry in the page table of the second level. There are 1024 entries, where each entry is 4 bytes (24 bits), and represents a segment of virtual memory that is 4 MB in size. The format of an entry in the level 1 table is made up of a valid bit field (indicating that the 4 MB respective segment is valid), a level-2 base address field (if valid bit is set, pointer to base address of the level-2 table that represents the associated 4 MB segment of virtual memory), and several attribute fields describing the various attributes of the associated memory segment.

Within each level-2 table, every entry represents the pages of the respective virtual memory segment. The number of entries of a level-2 table depends on the defined virtual memory page size (4 kB, 16 kB, 512 kB, or 8 MB) see Table 2.1. The larger the virtual memory page size, the less memory used for level-2 translation tables, since there are fewer entries in the translation tables—for example, a 16 MB physical memory space can be mapped using 2×8 MB pages (2048 bytes in the level 1 table and a 2×4 in the level-2 table for a total of 2056 bytes), or 4096×4 kB pages (2048 bytes in the level-1 table and a 4×4096 in the level-2 table for a total of 18,432 bytes).

Table 2.1 Level 1 and 2 Entries

Page Size	No. of Pages per Segment	Number of Entries in L2T	L2T Size (Bytes)
8 MB	.5	1	4
512 kB	8	8	32
16 kB	256	1024*	4096
4 kB	1024	1024	4096

In the MPC860's TLB scheme, the desired entry location is derived from the incoming effective memory address. The location of the entry within the TLB sets is specifically determined by the index field(s) derived from the incoming logical memory address. The format of the 32-bit logical (effective) address generated by the PowerPC Core differs depending on the page size. For a 4 kB page, the effective address is made up of a 10-bit level-1 index, a 10-bit level-2 index, and a 12-bit page offset (see Figure 2.24a). For a 16 kB page, the page offset becomes 14 bits, and the level-2 index is 8-bits (see Figure 2.24b). For a 512 kB page, the page offset is 19 bits, and the level-2 index is then 3 bits long (Figure 2.24c)—and for an 8 MB page, the page offset is 23 bits long, there is no level-2 index, and the level-1 index is 9-bits long (Figure 2.24d).

Figure 2.24a: 4 kB Effective Address Format

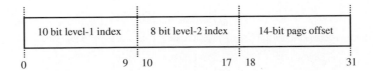

Figure 2.24b: 16 kB Effective Address Format

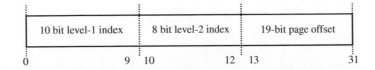

Figure 2.24c: 512 kB Effective Address Format

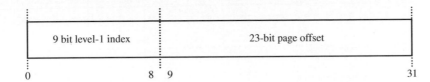

Figure 2.24d: 8 MB Effective Address Format

The page offset of the 4 kB effective address format is 12 bits wide to accommodate the offset within the 4 kB (0x0000 to 0x0FFF) pages. The page offset of the 16 kB effective address format is 14 bits wide to accommodate the offset within the 16 kB (0x0000 to 0x3FFF) pages. The page offset of the 512 kB effective address format is 19 bits wide to accommodate the offset within the 512 kB (0x0000 to 0x7FFFF) pages, and the page offset of the 8 MB effective address format is 23 bits wide to accommodate the offset within the 8 MB (0x0000 to 0x7FFFF8) pages.

In short, the MMU uses these effective address fields (level-1 index, level-2 index, and offset) in conjunction with other registers, TLB, translation tables, and the tablewalk process to determine the associated physical address (see Figure 2.25).

The MMU initialization sequence involves initializing the MMU registers and translation table entries. The initial steps include initializing the MMU Instruction Control Register (MI_CTR) and the Data Control Registers (MD_CTR) shown in Figures 2.26a and b. The fields in both registers are generally the same, most of which are related to memory protection.

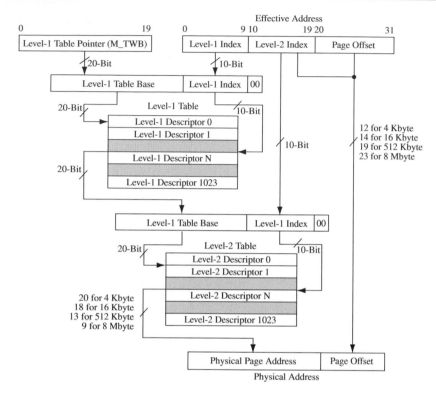

Figure 2.25: 2-Level Translation Table for 4 kB Page Scheme

MI_CTR - MMU Instruction Control Register

0	1	2	3	4	5	6	7	8	9	10	11	12	13	14	15
GPM	PPM	CI DEF	Res	RS V4I	Res	PPCS					Reserved				

16	17	18	19	20	21	22	23	24	25	26	27	28	29	30	31
Res			ITLB_INDX					Reserved							

Figure 2.26a: MI_CTR

MD_CTR - MMU Data Control Register

0	1	2	3	4	5	6	7	8	9	10	11	12	13	14	15
GPM	PPM	CI DEF	WT DEF	RS V4D	TW AM	PPCS					Reserved				

16	17	18	19	20	21	22	23	24	25	26	27	28	29	30	31
Res			DTLB_INDX					Reserved							

Figure 2.26b: MD_CR

Initializing translation table entries is a matter of configuring two memory locations (level 1 and level 2 descriptors), and three register pairs, one for data and one for instructions, in each pair, for a total of six registers. This equals one each of an Effective Page Number (EPN) register, Tablewalk Control (TWC) register, and Real Page Number (RPN) register.

The level 1 descriptor (see Figure 2.27a) defines the fields of the level-1 translation table entries, such as the Level-2 Base Address (L2BA), the access protection group, page size, and so on. The level-2 page descriptor (see Figure 2.27b) defines the fields of the level-2 translation table entries, such as: the physical page number, page valid bit, page protection, and so on. The registers shown in Figures 2.27c–e are essentially TLB source registers used to load entries into the TLBs. The Effective Page Number (EPN) registers contain the effective address to be loaded into a TLB entry. The Tablewalk Control (TWC) registers contain the attributes of the effective address entry to be loaded into the TLB (i.e., page size, access protection, and so on), and the Real Page Number (RPN) registers contain the physical address and attributes of the page to be loaded into the TLB.

Level 1 Descriptor Format

0	1	2	3	4	5	6	7	8	9	10	11	12	13	14	15
L2BA															

16	17	18	19	20	21	22	23	24	25	26	27	28	29	30	31
L2BA				Reserved			Access Prot Group				G	PS		WT	V

Figure 2.27a: L1 Descriptor

Level 2 Descriptor Format

0	1	2	3	4	5	6	7	8	9	10	11	12	13	14	15
RPN															

16	17	18	19	20	21	22	23	24	25	26	27	28	29	30	31
RPN				PP		E	C	TLBH				SPS	SH	CI	V

Figure 2.27b: L2 Descriptor

Mx_EPN - Effective Page Number Register   X = 1, P. 11–15; x = D

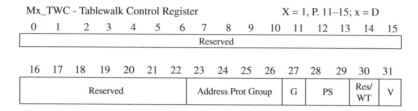

Figure 2.27c: Mx-EPN

Mx_TWC - Tablewalk Control Register   X = 1, P. 11–15; x = D

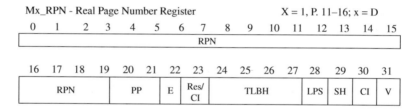

Figure 2.27d: Mx-TWC

Mx_RPN - Real Page Number Register   X = 1, P. 11–16; x = D

0	1	2	3	4	5	6	7	8	9	10	11	12	13	14	15
RPN

16	17	18	19	20	21	22	23	24	25	26	27	28	29	30	31
RPN				PP		E	Res/CI	TLBH				LPS	SH	CI	V

Figure 2.27e: Mx-RPN

An example of a MMU initialization sequence on the MPC860 is pseudo coded below.

```
// Invalidating TLB entries
tlbia ;   // the MPC860's instruction to invalidate entries
          // within the TLBs, also the "tlbie" can be used

// Initializing the MMU Instruction Control Register
. . .
MI_CTR.fld.all = 0; // clear all fields of register so group
                    // protection mode = PowerPC mode, page
                    // protection mode is page resolution, etc.
MI_CTR.fld.CIDEF = 1; // instruction cache inhibit default when
                      // MMU disabled
. . .
```

```
// Initializing the MMU Data Control Register
. . .

MD_CTR.fld.all = 0; // clear all fields of register so
                    // group protection mode = PowerPC mode,
                    // page protection mode is page resolution, etc.

MD_CTR.fl d.TWAM = 1; // tablewalk assist mode = 4kbyte page
                      // hardware assist
MD_CTR.fld.CIDEF = 1; // data cache inhibit default when MMU disabled
. . .
```

Move to Exception Vector Table the Data and Instruction TLB Miss and Error ISRs (MMU interrupt vector locations shown in table below).

Offset (hex)	Interrupt Type
01100	Implementation Dependent Instruction TLB Miss
01200	Implementation Dependent Data TLB Miss
01300	Implementation Dependent Instruction TLB Error
01400	Implementation Dependent Data TLB Error

With a TLB miss, an ISR loads the descriptors into the MMU. Data TLB Reload ISR example:

```
. . .
// put next code into address, incrementing vector by 4 after
// each line i.e., "mtspr M_TW,r0" = "07CH, 011H, 013H, 0A6H",
// so put integer 0x7C1113A6H at vector 0x1200 and increment
// vector by 4;
install start of ISR at vector address offset = 0x1200;

// save general purpose register into MMU tablewalk special register
mtspr M_TW, GPR;

mfspr GPR, M_TWB; // load GPR with address of level one descriptor
lwz GPR, (GPR);   // Load level one page entry

// save level two base pointer and level one # attributes into DMMU
// tablewalk control register
mtspr MD_TWC,GPR;
```

```
// load GPR with level two pointer while taking into account
// the page size mfspr GPR, MD_TWC;

lwz GPR, (GPR); // Load level two page entry
mtspr MD_RPN, GPR; // Write TLB entry into real page number register

// restore GPR from tablewalk special register return to main
// execution stream;
mfspr GPR, M_TW;
...
```

Instruction TLB Reload ISR Example:

```
// put next code into address, incrementing vector by 4 after each
// line i.e., "mtspr M_TW,r0" = "07CH, 011H, 013H, 0A6H", so put
// integer 0x7C1113A6H at vector 0x1100 and increment vector by 4;

install start of ISR at vector address offset = 0x1100;

...

// save general purpose register into MMU tablewalk special
// register
mtspr M_TW, GPR;

mfspr GPR, SRR0     // load GPR with instruction miss effective
                    // address
mtspr MD_EPN, GPR   // save instruction miss effective address
                    // in MD_EPN
mfspr GPR, M_TWO    // load GPR with address of level one
                    // descriptor
lwz GPR, (GPR)      // Load level one page entry
mtspr MI_TWC,GPR    // save level one attributes
mtspr MD_TWC,GPR    // save level two base pointer

// load R1 with level two pointer while taking into account the
// page size
mfspr GPR, MD_TWC

lwz GPR, (GPR)      // Load level two page entry
mtspr MI_RPN, GPR   // Write TLB entry
mfspr GPR, M_TW     // restore R1
```

```
return to main execution stream;

//Initialize L1 table pointer and clear L1 Table i.e., MMU
//tables/TLBs 043F0000 - 043FFFFF
Level1_Table_Base_Pointer = 0x043F0000;

index:= 0;
WHILE ((index MOD 1024) is NOT = 0) DO
Level1 Table Entry at Level1_Table_Base_Pointer + index = 0;
index = index + 1;

end WHILE;
...
```

Initialize translation table entries and map in desired segments in level1 table and pages in level2 tables. For example, given the physical memory map below, the l1 and l2 descriptors would need to be configured for Flash, DRAM, and so on.

Address Range	Accessed Device	Port Width
0x00000000 - 0x003FFFFF	Flash PROM Bank 1	32
0x00400000 - 0x007FFFFF	Flash PROM Bank 2	32
0x04000000 - 0x043FFFFF	DRAM 4 Mbyte (1Meg x 32-bit)it)	32
0x09000000 - 0x09003FFF	MPC Internal Memory Map	32
0x09100000 - 0x09100003	SCSR - Board Control & Status Register	32
0x10000000 - 0x17FFFFFF	PCMCIA Channel	16

Figure 2.28a: Physical Memory Map

PS	#	Used for...	Address Range	CI	WT	S/U	R/W	SH
8M	1	Monitor & trans. tbls	0x0 - 0x7FFFFF	N	Y	S	R/O	Y
512K	2	Stack & scratchpad	0x40000000 - 0x40FFFFF	N	N	S	R/W	Y
512K	1	CPM data buffers	0x4100000 - 0x417FFFF	Y	-	S	R/W	Y
512K	5	Prob. prog. & data	0x4180000 - 0x43FFFFF	N	N	S/U	R/W	Y
16K	1	MPC int mem. map	0x9000000 -	Y	-	S	R/W	Y
16K	1	Board config. regs	0x9100000 - 0x9103FFF	Y	-	S	R/W	Y
8M	16	PCMCIA	0x10000000 - 0x17FFFFFF	Y	-	S	R/W	Y

Figure 2.28b: L1/L2 Configuration

```
// i.e., Initialize entry for and Map in 8 MB of Flash at 0x00000000,
// adding entry into L1 table, and adding a level2 table for every
// L1 segment—as shown in Figure 2.28b, page size is 8 MB,
// cache is not inhibited, marked as write-through, used in
// supervisor mode, read only, and shared

// 8 MB Flash               .

. . .

Level2_Table_Base_Pointer = Level1_Table_Base_Pointer +
size of L1 Table (i.e.,1024);
L1desc(Level1_Table_Base_Pointer + L1Index).fld.BA = Level2_Table_
Base_Pointer;
L1desc(Level1_Table_Base_Pointer + L1Index).fld.PS = 11b;
// page size = 8MB


// Writethrough attribute = 1 writethrough cache policy region
L1desc.fld(Level1_Table_Base_Pointer + L1Index).WT = 1 ;

L1desc(Level1_Table_Base_Pointer + L1Index).fld.PS = 1;
// page size = 512K


// level-one segment valid bit = 1 segment valid
L1desc(Level1_Table_Base_Pointer + L1Index).fld.V = 1;


// for every segment in L1 table, there is an entire level2 table
L2index:=0;
WHILE (L2index < # Pages in L1Table Segment) DO
L2desc[Level2_Table_Base_Pointer + L2index * 4].fld.RPN = physical
page number;
L2desc[Level2_Table_Base_Pointer + L2index * 4].fld.CI = 0;
// Cache Inhibit Bit = 0
. . .

L2index = L2index + 1 ;
end WHILE;


// i.e., Map in 4 MB of DRAM at 0x04000000, as shown in Figure 2.29b,
// divided into eight 512 Kb pages. Cache is enabled, and is in
// copy-back mode, supervisormode, supports reading and writing,
// and it is shared
```

```
. . .

Level2_Table_Base_Pointer = Level2_Table_Base_Pointer +
Size of L2Table for 8MB Flash;
L1desc(Level1_Table_Base_Pointer + L1Index).fld.BA =
Level2_Table_Base_Pointer;
L1desc(Level1_Table_Base_Pointer + L1Index).fld.PS = 01b;
// page size = 512KB

// Writethrough Attribute = 0 copyback cache policy region
L1desc.fld(Level1_Table_Base_Pointer + L1Index).WT = 0;
L1desc(Level1_Table_Base_Pointer + L1Index).fld.PS = 1;
// page size = 512K

// level-one segment valid bit = 1 segment valid
L1desc(Level1_Table_Base_Pointer + L1Index).fld.V = 1;
. . .

// Initializing Effective Page Number Register
loadMx_EPN(mx_epn.all);

// Initializing the Tablewalk Control Register Descriptor
load Mx_TWC(L1desc.all);

// Initializing the Mx_RPN Descriptor
load Mx_RPN (L2desc.all);
. . .
```

... At this point the MMU and caches can be enabled (see memory subsystem enable section).

2.3.1.2 Memory Subsystem Disable on MPC860

```
// Disable MMU -- The MPC860 powers up with the MMUs in disabled
// mode, but to disable translation IR and DR bits need to be
// cleared.
. . .
rms msr ir 0; rms msr dr 0; // disable translation
. . .
```

```
// Disable Caches
. . .

// disable caches (0100b in bits 4-7, IC_CST[CMD] and DC_CST[CMD]
// registers)
addis r31,r0,0x0400
mtspr DC_CST,r31
mtspr IC_CST,r31
. . .
```

2.3.1.3   Memory Subsystem Enable on MPC860

```
// Enable MMU via setting IR and DR bits and "mtmsr" command
// on MPC860

. . .
ori r3,r3,0x0030;       // set the IR and DR bits
mtmsr r3;               // enable translation
isync;
. . .
// enable caches
. . .
addis r31,r0,0x0a00     // unlock all in both caches
mtspr DC_CST,r31
mtspr IC_CST,r31
addis r31,r0,0x0c00     // invalidate all in both caches
mtspr DC_CST,r31
mtspr IC_CST,r31

// enable caches (0010b in bits 4-7,IC_CST[CMD] and DC_CST[CMD]
// registers)
addis r31,r0,0x0200
mtspr DC_CST,r31
mtspr IC_CST,r31
. . .
```

2.3.1.4   Memory Subsystem Writing/Erasing Flash

While reading from Flash is the same as reading from RAM, accessing Flash for writing or erasing is typically much more complicated. Flash memory is divided into blocks, called sectors, where each sector is the smallest unit that can be erased. While flash chips differ in the process required to perform a write or erase, the general handshaking is similar to the pseudo code examples below for the Am29F160D Flash chip. The Flash erase function notifies the Flash chip of the impending operation, sends the command to erase the sector, and then loops, polling the Flash chip to determine when it completes. At the end of the erase function, the Flash is then set to standard read mode. The write routine is similar to that of the erase function, except the command is transmitted to perform a write to a sector, rather than an erase.

```
. . .
// The address at which the Flash devices are mapped
int FlashStartAddress = 0x00000000;

int FlashSize = 0x00800000; // The size of the flash devices in
bytes – i.e., 8MB.

// flash memory block offset table from the flash base of the various
sectors, as well as, the corresponding sizes.
BlockOffsetTable = {{ 0x00000000, 0x00008000 },{ 0x00008000,
                      0x00004000 },
  { 0x0000C000, 0x00004000 }, { 0x00010000, 0x00010000 },
  { 0x00020000, 0x00020000 }, { 0x00040000, 0x00020000 },
  { 0x00060000, 0x00020000 }, { 0x00080000, 0x00020000 }, . . . .};

// Flash write pseudo code example
FlashErase (int startAddress, int offset) {
. . . .
// Erase sector commands
Flash [startAddress + (0x0555 << 2)] = 0x00AA00AA;
// unlock 1 flash command
Flash [startAddress + (0x02AA << 2)] = 0x00550055;
// unlock 2 flash command
Flash [startAddress + (0x0555 << 2)] = 0x00800080);
// erase setup flash command
Flash [startAddress + (0x0555 << 2)] = 0x00AA00AA;
// unlock 1 flash command
Flash [startAddress + (0x02AA << 2)] = 0x00550055;
// unlock 2 flash command
Flash [startAddress + offset] = 0x00300030;
// set flash sector erase command
```

```
// Poll for completion: avg. block erase time is 700msec,
// worst-case block erase time is 15sec
int poll;
int loopIndex = 0;
while (loopIndex < 500) {
for (int i = 0; i < 500 * 3000; i++);
poll = Flash(startAddr + offset);
if ((poll AND 0x00800080) = 0x00800080 OR
(poll AND 0x00200020) = 0x00200020) {
exit loop;
}
loopIndex++;
}

// exit
Flash (startAddr) = 0x00F000F0; // read reset command
Flash(startAddr + offset) == 0xFFFFFFFF;
}
```

2.4   Example 3: Onboard Bus Device Drivers

Associated with every bus is (1) some type of protocol that defines how devices gain access to the bus (arbitration), (2) the rules attached devices must follow to communicate over the bus (handshaking), and (3) the signals associated with the various bus lines. Bus protocol is supported by the bus device drivers, which commonly include all or some combination of all of the 10 functions from the list of device driver functionality introduced at the start of this chapter, including:

- **Bus Startup**, initialization of the bus upon power-on or reset.

- **Bus Shutdown**, configuring bus into its power-off state.

- **Bus Disable**, allowing other software to disable bus on-the-fly.

- **Bus Enable**, allowing other software to enable bus on-the-fly.

- **Bus Acquire**, allowing other software to gain singular (locking) access to bus.

- **Bus Release**, allowing other software to free (unlock) bus.

- **Bus Read**, allowing other software to read data from bus.

- **Bus Write**, allowing other software to write data to bus.

- **Bus Install**, allowing other software to install new bus device on-the-fly for expandable buses.

- **Bus Uninstall**, allowing other software to remove installed bus device on-the-fly for expandable buses.

Which of the routines are implemented and how they are implemented depends on the actual bus. The pseudo code below is an example of an I2C bus initialization routine provided as an example of a bus startup (initialization) device driver on the MPC860.

2.4.1   *Onboard Bus Device Driver Pseudo Code Examples*

The following pseudo code gives an example of implementing a bus initialization routine on the MPC860, specifically the startup function in reference to the architecture. These examples demonstrate how bus management can be implemented on a more complex architecture, and this can be used as a guide to understand how to write bus management drivers on other processors of equal or lesser complexity than the MPC860 architecture. Other driver routines have not been pseudo coded, because the same concepts apply here as in Sections 2.2 and 2.3— essentially, looking in the architecture and bus documentation for the mechanisms that enable a bus, disable a bus, acquire a bus, and so on.

2.4.1.1   *I^2C Bus Startup (Initialization) on the MPC860*

The I^2C (inter-integrated circuit) protocol is a serial bus with one serial data line (SDA) and one serial clock line (SCL). With the I^2C protocol, all devices attached to the bus have a unique address (identifier), and this identifier is part of the data stream transmitted over the SDL line.

The components on the master processor that support the I^2C protocol are what need initialization. In the case of the MPC860, there is an integrated I^2C controller on the master processor (see Figure 2.29). The I^2C controller is made up of transmitter registers, receiver registers, a baud rate generator, and a control unit. The baud rate generator generates the clock signals when the I^2C controller acts as the I^2C bus master—if in slave mode, the controller uses the clock signal received from the master. In reception mode, data is transmitted from the SDA line into the control unit, through the shift register, which in turn transmits the data to the receive data register. The data that will be transmitted over the I^2C bus from the PPC is initially stored in the transmit data register and transferred out through the shift register to the control unit and over the SDA line. Initializing the I^{2}C bus on the MPC860 means initializing the I^{2}C SDA and SCL pins, many of the I^{2}C registers, some of the parameter RAM, and the associated buffer descriptors.

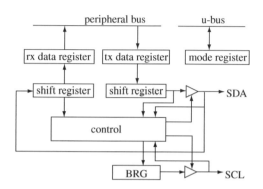

Figure 2.29: I^{2}C Controller on MPC860
Source: Copyright of Freescale Semiconductor, Inc. 2004. Used by permission.

The MPC860 I^{2}C SDA and SCL pins are configured via the Port B general purpose I/O port (see Figures 2.30a and b). Because the I/O pins can support multiple functions, the specific function a pin will support needs to be configured via port B's registers (shown in Figure 2.30c). Port B has four read/write (16-bit) control registers: the Port B Data Register (PBDAT), the Port B Open Drain Register (PBODR), the Port B Direction Register (PBDIR), and the Port B Pin Assignment Register (PBPAR). In general, the PBDAT register contains the data on the pin, the PBODR configures the pin for open drain or active output, the PBDIR configures the pin as either an input or output pin, and the PBPAR assigns the pin its function (I^{2}C, general purpose I/O, and so on).

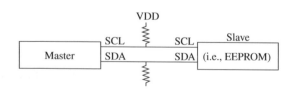

Figure 2.30a: SDA and SCL Pins on MPC860
Source: Copyright of Freescale Semiconductor, Inc. 2004. Used by permission.

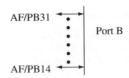

Figure 2.30b: MPC860 Port B Pins
Source: Copyright of Freescale Semiconductor, Inc. 2004. Used by permission.

	14	15	16	17	18	19	20	21	22	23	24	25	26	27	28	29	30	31
PBDAT	D	D	D	D	D	D	D	D	D	D	D	D	D	D	D	D	D	D
PBODR	OD	OD	OD	OD	OD	OD	OD	OD	OD	OD	OD	OD	OD	OD	OD	OD	OD	OD
PBDIR	DR	DR	DR	DR	DR	DR	DR	DR	DR	DR	DR	DR	DR	DR	DR	DR	DR	DR
PBPAR	DD	DD	DD	DD	DD	DD	DD	DD	DD	DD	DD	DD	DD	DD	DD	DD	DD	DD

Figure 2.30c: MPC860 Port B Register
Source: Copyright of Freescale Semiconductor, Inc. 2004. Used by permission.

An example of initializing the SDA and SCL pins on the MPC860 is given in the pseudo code below.

```
. . .
immr = immr & 0xFFFF0000; // MPC8xx internal register map
// Configure Port B pins to enable SDA and SCL
immr->pbpar = (pbpar) OR (0x00000030); // set to dedicated I2C
immr->pbdir = (pbdir) OR (0x00000030); // Enable I2CSDA and I2CSCL
as outputs
. . . .
```

The I²C registers that need initialization include the I²C Mode Register (I2MOD), I²C Address Register (I2ADD), the Baud Rate Generator Register (I2BRG), the I²C Event Register (I2CER), and the I²C Mask Register (I2CMR) (shown in Figures 2.31a–e).

I²C Mode Register (I2MOD)

0	1	2	3	4	5	6	7
—	—	REVD	GCD	FLT	PDIV		EN

Figure 2.31a: I2MOD

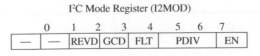

I²C Address Register (I2ADD)

0	1	2	3	4	5	6	7
			SAD[0:6]				—

SAD0 – SAD6 = Slave Address

Figure 2.31b: I2ADD

I²C BRG Register (I2BRG)

0	1	2	3	4	5	6	7
			DIV[0:7]				

Figure 2.31c: I2BRG

I²C Event Register (I2CER)

0	1	2	3	4	5	6	7
—	—	—	TXE	—	BSY	TXB	RXB

Figure 2.31d: I2CER

I²C Mask Register (I2CMR)

0	1	2	3	4	5	6	7
—	—	—	TXE	—	BSY	TXB	RXB

Figure 2.31e: I2CMR

An example of I²C register initialization pseudo code is as follows:

```
/* I2C Registers Initialization Sequence */
. . .
// Disable I2C before initializing it, LSB character order for
// transmission and reception, I2C clock not filtered, clock
// division factor of 32, etc.
immr->i2mod = 0x00;

immr->i2add = 0x80;   // I2C MPC860 address = 0x80
immr->i2brg = 0x20;   // divide ratio of BRG divider
immr->i2cer = 0x17;   // Clear out I2C events by setting relevant
                      // bits to "1"
immr->i2cmr = 0x17;   // Enable interrupts from I2C in
                      // corresponding I2CER
immr->i2mod = 0x01;   // Enable I2C bus
. . .
```

Five of the 15 field I^2C parameter RAM need to be configured in the initialization of I^2C on the MPC860. They include the receive function code register (RFCR), the transmit function code register (TFCR), and the maximum receive buffer length register (MRBLR), the base value of the receive buffer descriptor array (Rbase), and the base value of the transmit buffer descriptor array (Tbase) shown in Figure 2.32.

Offset[1]	Name	Width	Description
0x00	RBASE	Hword	Rx/TxBD table base address. Indicate where the BD tables begin in the dual-port RAM. Setting Rx/TxBD[W] in the last BD in each BD table determines how many BDs are allocated for the Tx and Rx sections of the I^2C. Initialize RBASE/TBASE before enabling the I^2C. Furthermore, do not configure BD tables of the I^2C to overlap any other active controller's parameter RAM. RBASE and TBASE should be divisible by eight.
0x02	TBASE	Hword	
0x04	RFCR	Byte	Rx/Tx function code. Contains the value to appear on AT[1–3] when the associated SDMA channel accesses memory. Also controls the byte-ordering convention for transfers.
0x05	TFCR	Byte	
0x06	MRBLR	Hword	Maximum receive buffer length. Defines the maximum number of bytes the I^2C receiver writes to a receive buffer before moving to the next buffer. The receiver writes fewer bytes to the buffer than the MRBLR value if an error or end-of-frame occurs. Receive buffers should not be smaller than MRBLR. Transmit buffers are unaffected by MRBLR and can vary in length; the number of bytes to be sent is specified in TxBD[Data Length]. MRBLR is not intended to be changed while the I^2C is operating. However it can be changed in a single bus cycle with one 16-bit move (not two 8-bit bus cycles back-to-back). The change takes effect when the CP moves control to the next RxBD. To guarantee the exact RxBD on which the change occurs, change MRBLR only while the I^2C receiver is disabled. MRBLR should be greater than zero.
0x08	RSTATE	Word	Rx internal state. Reserved for CPM use.
0x0C	RPTR	Word	Rx internal data pointer[2] is updated by the SDMA channels to show the next address in the buffer to be accessed.
0x10	RBPTR	Hword	RxBD pointer. Points to the next descriptor the receiver transfers data to when it is in an idle state or to the current descriptor during frame processing for each I^2C channel. After a reset or when the end of the descriptor table is reached, the CP initializes RBPTR to the value in RBASE. Most applications should not write RBPTR, but it can be modified when the receiver is disabled or when no receive buffer is used.
0x12	RCOUNT	Hword	Rx internal byte count[2] is a down-count value that is initialized with the MRBLR value and decremented with every byte the SDMA channels write.
0x14	RTEMP	Word	Rx temp. Reserved for CPM use.
0x18	TSTATE	Word	Tx internal state. Reserved for CPM use.
0x1C	TPTR	Word	Tx internal data pointer[2] is updated by the SDMA channels to show the next address in the buffer to be accessed.

(Continued)

Figure 2.32: I^2C Parameter RAM

Offset[1]	Name	Width	Description
0x20	TBPTR	Hword	TxBD pointer. Points to the next descriptor that the transmitter transfers data from when it is in an idle state or to the current descriptor during frame transmission. After a reset or when the end of the descriptor table is reached, the CPM initialized TBPTR to the value in TBASE. Most applications should not write TBPTR, but it can be modified when the transmitter is disabled or when no transmit buffer is used.
0x22	TCOUNT	Hword	Tx internal byte count[2] is a down-count value initialized with TxBD[Data Length] and decremented with every byte read by the SDMA channels.
0x24	TTEMP	Word	Tx temp. Reserved for CP use.
0x28-0x 2F	–	–	Used for I²C/SPI relocation.

[1] As programmed in I²C_BASE, the default value is IMMR + 0x3C80.
[2] Normally, these parameters need not be accessed.

Figure 2.32: Continued

See the following pseudo code for an example of I²C parameter RAM initialization:

```
// I2C Parameter RAM Initialization
. . .

// specifies for reception big endian or true little endian byte
// ordering and channel # 0

immr->I2Cpram.rfcr = 0x10;

// specifies for reception big endian or true little endian byte
// ordering and channel # 0
immr->I2Cpram.tfcr = 0x10;
immr->I2Cpram.mrblr  = 0x0100; // the maximum length of
                               // I2C receive buffer
immr->I2Cpram.rbase = 0x0400; // point RBASE to first RX BD
immr->I2Cpram.tbase = 0x04F8; // point TBASE to TX BD

. . .
```

Data to be transmitted or received via the I²C controller (within the CPM of the PowerPC) is input into buffers that the transmit and receive buffer descriptors refer to. The first half-word (16 bits) of the transmit and receive buffer contain status and control bits (as shown in Figures 2.33a and b). The next 16 bits contain the length of the buffer.

In both buffers the Wrap (W) bit indicates whether this buffer descriptor is the final descriptor in the buffer descriptor table (when set to 1, the I²C controller returns to the first buffer in the buffer descriptor ring). The Interrupt (I) bit indicates whether the I²C controller issues an interrupt when this buffer is closed. The last bit (L) indicates whether this buffer contains the last character of the message. The CM bit indicates whether the I²C controller clears the empty (E) bit of the reception buffer or ready (R) bit of the transmission buffer when it is finished with this buffer. The continuous mode (CM) bit refers to continuous mode in which, if a single buffer descriptor is used, continuous reception from a slave I²C device is allowed.

In the case of the transmission buffer, the ready bit indicates whether the buffer associated with this descriptor is ready for transmission. The transmit start condition (S) bit indicates whether a start condition is transmitted before transmitting the first byte of this buffer. The NAK bit indicates that the I²C aborted the transmission because the last transmitted byte did not receive an acknowledgment. The underrun condition (UN) bit indicates that the controller encountered an underrun condition while transmitting the associated data buffer. The collision (CL) bit indicates that the I²C controller aborted transmission because the transmitter lost while arbitrating for the bus. In the case of the reception buffer, the empty blanks are reserved blanks are reserved bit indicates if the data buffer associated with this buffer descriptor is empty and the overrun (OV) bit indicates whether an overrun occurred during data reception.

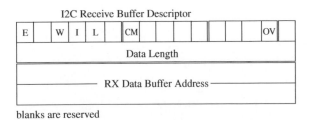

Figure 2.33a: Receive Buffer Descriptor

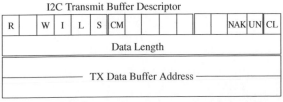

Figure 2.33b: Transmit Buffer Descriptor

An example of I²C buffer descriptor initialization pseudo code would look as follows:

```
// I2C Buffer Descriptor Initialization
. . .
// 10 reception buffers initialized
index = 0;
While (index<9) do
{
// E = 1, W = 0, I = 1, L = 0, OV = 0
  immr->udata_bd ->rxbd[index].cstatus = 0x9000;
  immr->bd ->rxbd[index].length = 0; // buffer empty
  immr->bd ->rxbd[index].addr = . . . index = index + 1;
}
// last receive buffer initialized
immr->bd->rxbd[9].cstatus = 0xb000; // E = 1, W = 1, I = 1,
                                     // L = 0, OV =0
  immr->bd ->rxbd[9].length = 0; // buffer empty
immr->udata_bd ->rxbd[9].addr = . . .;

// transmission buffer
immr->bd ->txbd.length = 0x0010;    // transmission buffer 2 bytes
                                    // long

// R = 1, W = 1, I = 0, L = 1, S = 1, NAK = 0, UN = 0, CL = 0
immr->bd->txbd.cstatus = 0xAC00;

immr->udata_bd ->txbd.bd_addr = . . .;

/* Put address and message in TX buffer */
. . .

// Issue Init RX & TX Parameters Command for I2C via CPM command
// register CPCR.
while(immr->cpcr & (0x0001)); // Loop until ready to issue command
immr->cpcr = (0x0011);        // Issue Command
while(immr->cpcr & (0x0001)); // Loop until command proecessed
. . .
```

2.5   Board I/O Driver Examples

The board I/O subsystem components that require some form of software management include the components integrated on the master processor, as well as an I/O slave controller, if one exists. The I/O controllers have a set of status and control registers used to control the processor and check on its status. Depending on the I/O subsystem, commonly all or some combination of all of the 10 functions from the list of device driver functionality introduced at the start of this chapter are typically implemented in I/O drivers, including:

- **I/O Startup**, initialization of the I/O on power-on or reset.

- **I/O Shutdown**, configuring I/O into its power-off state.

- **I/O Disable**, allowing other software to disable I/O on-the-fly.

- **I/O Enable**, allowing other software to enable I/O on-the-fly.

- **I/O Acquire**, allowing other software gain singular (locking) access to I/O.

- **I/O Release**, allowing other software to free (unlock) I/O.

- **I/O Read**, allowing other software to read data from I/O.

- **I/O Write**, allowing other software to write data to I/O.

- **I/O Install**, allowing other software to install new I/O on-the-fly.

- **I/O Uninstall**, allowing other software to remove installed I/O on-the-fly.

The Ethernet and RS232 I/O initialization routines for the PowerPC and ARM architectures are provided as examples of I/O startup (initialization) device drivers. These examples are to demonstrate how I/O can be implemented on more complex architectures, such as PowerPC and ARM, and this in turn can be used as a guide to understand how to write I/O drivers on other processors that are as complex as or less complex than the PowerPC and ARM architectures. Other I/O driver routines were not pseudo coded in this chapter, because the same concepts apply here as in Sections 2.2 and 2.3 In short, it is up to the responsible developer to study the architecture and I/O device documentation for the mechanisms used to read from an I/O device, write to an I/O device, enable an I/O device, and so on.

2.5.1   Example 4: Initializing an Ethernet Driver

The example used here will be the widely implemented LAN protocol Ethernet, which is primarily based upon the IEEE 802.3 family of standards.

As shown in Figure 2.34, the software required to enable Ethernet functionality maps to the lower section of Ethernet the OSI data-link layer. The hardware components can all be mapped to the physical layer of the OSI model, but will not be discussed in this section.

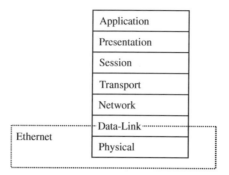

Figure 2.34: OSI Model

The Ethernet component that can be integrated onto the master processor is called the ***Ethernet Interface***. The only firmware (software) that is implemented is in the Ethernet interface. The software is dependent on how the hardware supports two main components of the IEEE802.3 Ethernet protocol: the ***media access management*** and ***data encapsulation***.

2.5.1.1   Data Encapsulation [Ethernet Frame]

In an Ethernet LAN, all devices connected via Ethernet cables can be set up as a bus or star topology (see Figure 2.35).

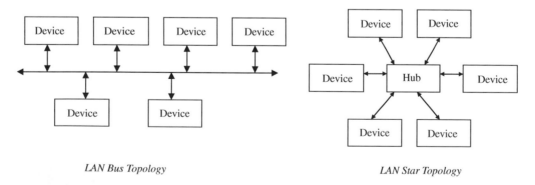

Figure 2.35: Ethernet Topologies

In these topologies, all devices share the same signaling system. After a device checks for LAN activity and determines after a certain period there is none, the device then transmits its Ethernet signals serially. The signals are then received by all other devices attached to the LAN—thus the need for an "Ethernet frame," which contains the data as well as the information needed to communicate to each device which device the data is actually intended for.

Ethernet devices encapsulate data they want to transmit or receive into what are called "Ethernet frames." The Ethernet frame (as defined by IEEE 802.3) is made of up a series of bits, each grouped into fields. Multiple Ethernet frame formats are available, depending on the features of the LAN. Two such frames (see the IEEE 802.3 specification for a description of all defined frames) are shown in Figure 2.36.

Basic Ethernet Frame

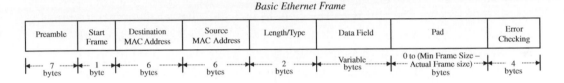

Basic Ethernet Frame with VLAN Tagging

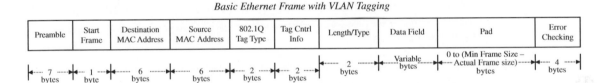

Figure 2.36: Ethernet Frames

The ***preamble*** bytes tell devices on the LAN that a signal is being sent. They are followed by "10101011" to indicate the ***start*** of a ***frame***. The media access control (***MAC***) ***addresses*** in the Ethernet frame are physical addresses unique to each Ethernet interface in a device, so every device has one. When the frame is received by a device, its data-link layer looks at the destination address of the frame. If the address does not match its own MAC address, the device disregards the rest of the frame.

The ***data*** field can vary in size. If the data field is less than or equal to 1500 then the ***Length/Type*** field indicates the number of bytes in the data field. If the data field is greater than 1500, then the type of MAC protocol used in the device that sent the frame is defined in ***Length/ Type***. While the data field size can vary, the MAC Addresses, the Length/Type, the Data, Pad, and Error checking fields must add up to be at least 64 bytes long. If not, the ***pad*** field is used to bring up the frame to its minimum required length.

The *error checking* field is created using the MAC Addresses, Length/Type, Data Field, and Pad fields. A 4-byte *CRC* (cyclical redundancy check) value is calculated from these fields and stored at the end of the frame before transmission. At the receiving device, the value is recalculated, and if it does not match the frame is discarded.

Finally, remaining frame formats in the Ethernet specification are extensions of the basic frame. The VLAN (virtual local-area network) tagging frame shown above is an example of one of these extended frames, and contains two additional fields: *802.1Q tag type* and *Tag Control Information*. The *802.1Q tag type* is always set to 0x8100 and serves as an indicator that there is a VLAN tag following this field, and not the Length/Type field that in this format is shifted 4 bytes over within the frame. The *Tag Control Information* is actually made up of three fields: the *user priority field* (UPF), the *canonical format indicator* (CFI), and the *VLAN identifier* (VID). The UPF is a 3-bit field that assigns a priority level to the frame. The CFI is a 1-bit field to indicate whether there is a Routing Information Field (RIF) in the frame, while the remaining 12 bits is the VID, which identifies which VLAN this frame belongs to. Note that while the VLAN protocol is actually defined in the IEEE 802.1Q specification, it is the IEEE 802.3ac specification that defines the Ethernet-specific implementation details of the VLAN protocol.

2.5.1.2 Media Access Management

Every device on the LAN has an equal right to transmit signals over the medium, so there have to be rules that ensure every device gets a fair chance to transmit data. Should more than one device transmit data at the same time, these rules must also allow the device a way to recover from the data colliding. This is where the two MAC protocols come in: the IEEE 802.3 *Half-Duplex* Carrier Sense Multiple Access/Collision Detect (CDMA/CD) and the IEEE 802. 3x *Full-Duplex Ethernet* protocols. These protocols, implemented in the Ethernet interface, dictate how these devices behave when sharing a common transmission medium.

Half-Duplex CDMA/CD capability in an Ethernet device means that a device can either receive or transmit signals over the same communication line, but not do both (transmit and receive) at the same time. Basically, a Half-Duplex CDMA/CD (also, known as the MAC sublayer) in the device can both transmit and receive data, from a higher layer or from the physical layer in the device. In other words, the MAC sublayer functions in two modes: transmission (data received from higher layer, processed, then passed to physical layer) or reception (data received from physical layer, processed, then passed to higher layer). The transmit data encapsulation (TDE) component and the transmit media access management (TMAM) components provide the transmission mode functionality, while the receive media access management (RMAM) and the receive data decapsulation (RDD) components provide the reception mode functionality.

2.5.1.3 CDMA/CD (MAC Sublayer) Transmission Mode

When the MAC sublayer receives data from a higher layer to transmit to the physical layer, the TDE component first creates the Ethernet frame, which is then passed to the TMAM component. Then, the TMAM component waits for a certain period of time to ensure the transmission line is quiet, and that no other devices are currently transmitting. When the TMAM component has determined that the transmission line is quiet, it then transmits (via the physical layer) the data frame over the transmission medium, in the form of bits, one bit at a time (serially). If the TMAM component of this device learns that its data has collided with other data on the transmission line, it transmits a series of bits for a predefined period to let all devices on the system know that a collision has occurred. The TMAM component then stops all transmission for another period of time, before attempting to retransmit the frame again.

Figure 2.37 is a high-level flowchart of the MAC layer processing a MAC client's (an upper layer) request to transmit a frame.

2.5.1.4 CDMA/CD (MAC Sublayer) Reception Mode

When the MAC sublayer receives the stream of bits from the physical layer, to be later transmitted to a MAC client, the MAC sublayer RMAM component receives these bits from the physical layer as a "frame." Note that, as the bits are being received by the RMAM component, the first two fields (preamble and start frame delimiter) are disregarded. When the physical layer ceases transmission, the frame is then passed to the RDD component for processing. It is this component that compares the MAC Destination Address field in this frame to the MAC Address of the device. The RDD component also checks to ensure the fields of the frame are properly aligned, and executes the CRC Error Checking to ensure the frame was not damaged in route to the device (the Error Checking field is stripped from the frame). If everything checks out, the RDD component then transmits the remainder of the frame, with an additional status field appended, to the MAC Client.

Figure 2.38 is a high-level flowchart of the MAC layer processing incoming bits from the physical layer.

It is not uncommon to find that half-duplex capable devices are also full-duplex capable. This is because only a subset of the MAC sublayer protocols implemented in half-duplex are

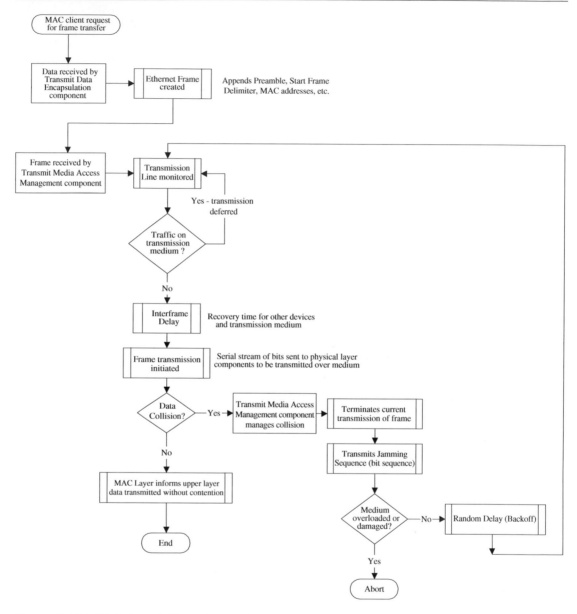

Figure 2.37: High-Level Flowchart of MAC Layer Processing a MAC Client's Request to Transmit a Frame

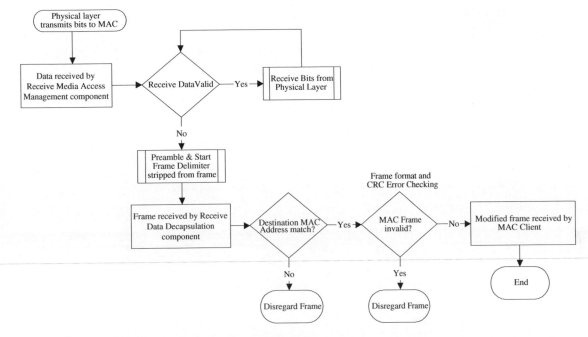

Figure 2.38: High-Level Flowchart of MAC Layer Processing Incoming Bits from the Physical Layer

needed for full-duplex operation. Basically, a full-duplex capable device can receive and transmit signals over the same communication media line at the same time. Thus, the throughput in a full-duplex LAN is double that of a half-duplex system.

The transmission medium in a full-duplex system must also be capable of supporting simultaneous reception and transmission without interference. For example: 10Base-5, 10Base-2, and 10Base-FX are cables that *do not* support full-duplex, while 10/100/1000Base-T and 100Base-FX meet full-duplex media specification requirements.

Full-duplex operation in a LAN is restricted to connecting only two devices, and both devices must be capable and configured for full duplex operation. While it is restricting to only allow point to point links, the efficiency of the link in a full-duplex system is actually improved. Having only two devices eliminates the potential for collisions, and eliminates any need for the CDMA/CD algorithms implemented in a half-duplex capable device. Thus, while the reception algorithm is the same for both full and half duplex, Figure 2.39 flowcharts the high-level functions of full-duplex in transmission mode.

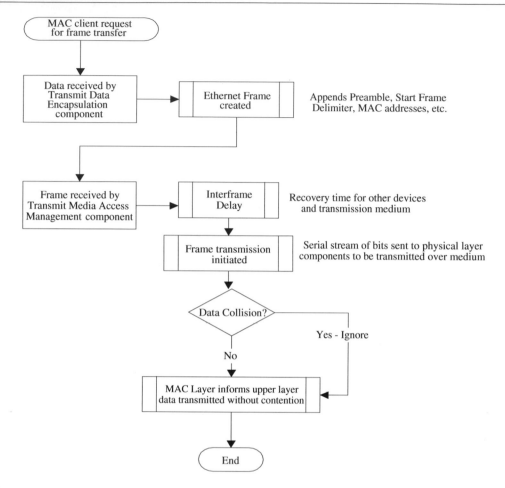

Figure 2.39: Flowchart of High-Level Functions of Full-Duplex in Transmission Mode

Now that you have a definition of all components (hardware and software) that make up an Ethernet system, let's take a look at how architecture-specific Ethernet components are implemented via software on various reference platforms.

2.5.1.5   Motorola/Freescale MPC823 Ethernet Example

Figure 2.40 is a diagram of a MPC823 connected to Ethernet hardware components on the board.

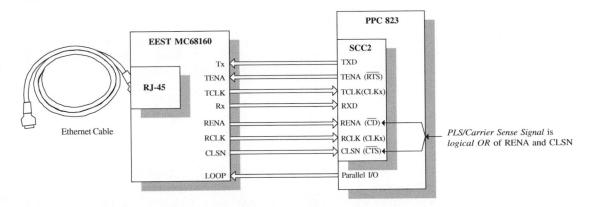

Figure 2.40: MPC823 Ethernet Block Diagram
Source: Copyright of Freescale Semiconductor, Inc. 2004. Used by permission.

A good starting point for understanding how Ethernet runs on a MPC823 is Section 16 in the 2000 *MPC823 User's Manual* on the MPC823 component that handles networking and communications, called the CPM (Communication Processor Module). It is here that we learn that configuring the MPC823 to implement Ethernet is done through serial communication controllers (SCCs).

From the 2000 *MPC823 User's Manual*

16.9 THE SERIAL COMMUNICATION CONTROLLERS

The MPC823 has two serial communication controllers (SCC2 and SCC3) that can be configured independently to implement different protocols. They can be used to implement bridging functions, routers, gateways, and interface with a wide variety of standard WANs, LANs, and proprietary networks. . . .

The serial communication controllers do not include the physical interface, but it is the logic that formats and manipulates the data obtained from the physical interface. Many functions of the serial communication controllers are common to (among other protocols) the Ethernet controller. The serial communication controller's main features include support for full 10Mbps Ethernet/IEEE 802.3.

Section 16.9.22 in the *MPC823 User's Manual* discusses in detail the features of the Serial Communication Controller in Ethernet mode, including full-duplex operation support. In fact, what actually needs to be implemented in software to initialize and configure Ethernet on the PPC823 can be based on the Ethernet programming example in Section 16.9.23.7.

From the 2000 *MPC823 User's Manual*

16.9.23.7 SCC2 ETHERNET PROGRAMMING EXAMPLE

The following is an example initialization sequence for the SCC2 in Ethernet mode. The CLK1 pin is used for the Ethernet receiver and the CLK2 pin is used for the transmitter.

1. Configure the port A pins to enable the TXD1 and RXD1 pins. Write PAPAR bits 12 and 13 with ones, PADIR bits 12 and 13 with zeros, and PAODR bit 13 with zero.

2. Configure the Port C pins to enable CTS2(CLSN) and CD2 (RENA). Write PCPAR and PCDIR bits 9 and 8 with zeros and PCSO bits 9 and 8 with ones.

3. Do not enable the RTS2(TENA) pin yet because the pin is still functioning as RTS and transmission on the LAN could accidentally begin.

4. Configure port A to enable the CLK1 and CLK2 pins. Write PAPAR bits 7 and 6 with ones and PADIR bits 7 and 6 with zeros.

5. Connect the CLK1 and CLK2 pins to SCC2 using the serial interface. Write the R2CS field in the SICR to 101 and the T2CS field to 100.

6. Connect SCC2 to the NMSI and clear the SC2 bit in the SICR.

7. Initialize the SDMA configuration register (SDCR) to 0x0001.

8. Write RBASE and TBASE in the SCC2 parameter RAM to point to the RX buffer descriptor and TX buffer descriptor in the dual-port RAM. Assuming one RX buffer descriptor at the beginning of dual-port RAM and one TX buffer descriptor following that RX buffer descriptor, write RBASE with 0x2000 and TBASE with 0x2008.

9. Program the CPCR to execute the INIT RX BD PARAMETER command for this channel.

10. Write RFCR and TFCR with 0x18 for normal operation.

11. Write MRBLR with the maximum number of bytes per receive buffer. For this case assume 1520 bytes, so MRBLR = 0x05F0. In this example, the user wants to receive an entire frame into one buffer, so the MRBLR value is chosen to be the first value larger than 1518 that is evenly divisible by four.

12. Write C_PRES with 0xFFFFFFFF to comply with 32-bit CCITT-CRC.

13. Write C_MASK with 0xDEBB20E3 to comply with 32-bit CDITT-CRC.

14. Clear CRCEC, ALEC, and DISFC for clarity.

15. Write PAD with 0x8888 for the pad value.

16. Write RET_LIM with 0x000F.

17. Write MFLR with 0x05EE to make the maximum frame size 1518 bytes.

18. Write MINFLR with 0x0040 to make the minimum frame size 64 bytes.

19. Write MAXD1 and MAXD2 with 0x005EE to make the maximum DMA count 1518 bytes.

20. Clear GADDR1-GADDR4. The group hash table is not used.

21. Write PADDR1_H with 0x0380, PADDR1_M with 0x12E0, and PADDR1_L with 0x5634 to configure the physical address 8003E0123456.

22. Write P_Per with 0x000. It is not used.

23. Clear IADDR1-IADDR4. The individual hash table is not used.

24. Clear TADDR_H, TADDR_M, and TADDR_L for clarity.

25. Initialize the RX buffer descriptor and assume the RX data buffer is at 0x00001000 main memory. Write 0xB000 to Rx_BD_Status, 0x0000 to Rx_BD_Length (optional), and 0x00001000 to Rx_BD_Pointer.

26. Initialize the TX buffer descriptor and assume the TX data frame is at 0x00002000 main memory and contains fourteen 8-bit characters (destination and source addresses plus the type field. Write 0xFC00 to Tx_BD_Status, add PAD to the frame and generate a CRC. Then write 0x000D to Tx_BD_Length and 0x00002000 to Tx_BD_Pointer.

27. Write 0xFFFF to the SCCE-Ethernet to clear any previous events.

28. Write 0x001A to the SCCM-Ethernet to enable the TXE, RXF, and TXB interrupts.

29. Write 0x20000000 to the CIMR so that SCC2 can generate a system interrupt. The CICR must also be initialized.

30. Write 0x00000000 to the GSMR_H to enable normal operation of all modes.

31. Write 0x1088000C to the GSMR_L to configure the CTS2 (CLSN) and CD2 (RENA) pins to automatically control transmission and reception (DIAG field) and the Ethernet mode. TCI is set to allow more setup time for the EEST to receive the MPC82 transmit data. TPL and TPP are set for Ethernet requirements. The DPLL is not used with Ethernet. Notice that the transmitter (ENT) and receiver (ENR) have not been enabled yet.

32. Write 0xD555 to the DSR.

33. Set the PSMR-SCC Ethernet to 0x0A0A to configure 32-bit CRC, promiscuous mode and begin searching for the start frame delimiter 22 bits after RENA.

34. Enable the TENA pin (RTS2). Since the MODE field of the GMSR_L is written to Ethernet, the TENA signal is low. Write PCPAR bit 14 with a one and PCDIR bit 14 with a zero.

35. Write 0x1088003C to the GSMR_L register to enable the SCC2 transmitter and receiver. This additional write ensures that the ENT and ENR bits are enabled last.

NOTE: After 14 bytes and the 46 bytes of automatic pad (plus the 4 bytes of CRC) are transmitted, the TX buffer descriptor is closed. Additionally, the receive buffer is closed after a frame is received. Any data received after 1520 bytes or a single frame causes a busy (out-of-buffer) condition since only one RX buffer descriptor is prepared.

It is from Section 16.9.23.7 that the Ethernet initialization device driver source code can be written. It is also from this section that it can be determined how Ethernet on the MPC823 is configured to be ***interrupt driven***. The actual initialization sequence can be divided into seven major functions: disabling SCC2, configuring ports for Ethernet transmission and reception, initializing buffers, initializing parameter RAM, initializing interrupts, initializing registers, and starting Ethernet (see pseudo code below).

2.5.1.6 MPC823 Ethernet Driver Pseudo Code

```
// disabling SCC2
   // Clear GSMR_L[ENR] to disable the receiver
   GSMR_L = GSMR_L & 0x00000020
   // Issue Init Stop TX Command for the SCC
   Execute Command (GRACEFUL_STOP_TX)
   // clear GSLM_L[ENT] to indicate that transmission has stopped
   GSMR_L = GSMR_L & 0x00000010

-=-=-=-=

// Configure port A to enable TXD1 and RXD1 - step 1 from user's
// manual
PADIR = PADIR & 0xFFF3   // Set PAPAR[12,13]
PAPAR = PAPAR | 0x000C   // clear PADIR[12,13]
PAODR = PAODR & 0xFFF7   // clear PAODR[12]

// Configure port C to enable CLSN and RENA - step 2 from
// user's manual
PCDIR = PCDIR & 0xFF3F   // clear PCDIR[8,9]
PCPAR = PCPAR & 0xFF3F   // Clear PCPAR[8,9]
PCSO = PCSO |0x00C0      // set PCSO[8,9]
```

```
// step 3 - do nothing now
// configure port A to enable the CLK2 and CLK4 pins.- step 4 from
// user's manual
PAPAR = PAPAR | 0x0A00   // set PAPAR[6] (CLK2) and PAPAR[4] (CLK4).
PADIR = PADIR & 0xF5FF   // Clear PADIR[4] and PADIR[6]. (All 16-bit)


// Initializing the SI Clock Route Register (SICR) for SCC2.
// Set SICR[R2CS] to 111 and Set SICR[T2CS] to 101, Connect SCC2 to
// NMSI and Clear
SICR[SC2] - steps 5 & 6 from user's manual
SICR = SICR & 0xFFFFBFFF
SICR = SICR | 0x00003800
SICR = (SICR & 0xFFFFF8FF) | 0x00000500


// Initializing the SDMA configuration register - step 7
SDCR = 0x01 // Set SDCR to 0x1 (SDCR is 32-bit) - step 7 from
            // user's manual


// Write RBASE in the SCC1 parameter RAM to point to the RxBD table
// and the TxBD table in the dual-port RAM and specify the size of
// the respective buffer descriptor pools.  - step 8 user's manual
RBase = 0x00 (for example)
RxSize = 1500 bytes (for example)
TBase = 0x02 (for example)
TxSize = 1500 bytes (for example)
Index = 0
While (index < RxSize) do
{
// Set up one receive buffer descriptor that tells the communication
// processor that the next packet is ready to be received - similar
// to step 25
// Set up one transmit buffer descriptor that tells the communication
// processor that the next packet is ready to be transmitted -
// similar step 26
index = index+1}


// Program the CPCR to execute the INIT_RX_AND_TX_PARAMS - deviation
// from step 9 in user's guide
execute Command(INIT_RX_AND_TX_PARAMS)


 // write RFCR and TFCR with 0x10 for normal operation (All 8-bits)
 // or 0x18 for normal operation and Motorola/Freescale byte
 // ordering - step 10 from user's manual
```

```
 RFCR = 0x10
 TFCR = 0x10

 // Write MRBLR with the maximum number of bytes per receive buffer
 // and assume 16 bytes – step 11 user's manual
 MRBLR = 1520

 // write C_PRES with 0xFFFFFFFF to comply with the 32 bit CRC-CCITT
 // – step 12 user's manual
 C_PRES = 0xFFFFFFFF

 // write C_MASK with 0xDEBB20E3 to comply with the 16 bit CRC-CCITT
 // – step 13 user's manual
 C_MASK = 0xDEBB20E3

 // Clear CRCEC, ALEC, and DISFC for clarity – step 14 user's manual
 CRCEC = 0x0
 ALEC = 0x0
 DISFC = 0x0

 // Write PAD with 0x8888 for the PAD value – step 15 user's manual
 PAD = 0x8888

 // Write RET_LIM to specify how many retries (with 0x000F for
 // example)-step 16
 RET_LIM = 0x000F

 // Write MFLR with 0x05EE to make the maximum frame size 1518 bytes
 // – step 17
 MFLR = 0x05EE

 // Write MINFLR with 0x0040 to make the minimum frame size 64 bytes
 // – step 18
 MINFLR = 0x0040

 // Write MAXD1 and MAXD2 with 0x05F0 to make the maximum DMA count
 // 1520 bytes – step 19
 MAXD1 = 0x05F0
 MAXD2 = 0x05F0

 // Clear GADDR1-GADDR4. The group hash table is not used – step 20
 GADDR1 = 0x0
 GADDR2 = 0x0
 GADDR3 = 0x0
 GADDR4 = 0x0
```

```
// Write PADDR1_H, PADDR1_M and PADDR1_L with the 48-bit station
address - step 21
stationAddr = "embedded device's Ethernet address" =
(for example) 8003E0123456
PADDR1_H = 0x0380 ["80 03" of the station address]
PADDR1_M = 0x12E0 ["E0 12" of the station address]
PADDR1_L = 0x5634 ["34 56" of the station address]

// Clear P_PER. It is not used - step 22
P_PER = 0x0

// Clear IADDR1-IADDR4. The individual hash table is not used -
// step 23
IADDR1 = 0x0
IADDR2 = 0x0
IADDR3 = 0x0
IADDR4 = 0x0

// Clear TADDR_H, TADDR_M and TADDR_L for clarity - step 24
groupAddress = "embedded device's group address" = no group address
for example
TADDR_H = 0 [similar as step 21 high byte reversed]
TADDR_M = 0 [middle byte reversed]
TADDR_L = 0 [low byte reversed]

// Initialize the RxBD and assume that Rx data buffer is at
// 0x00001000. Write 0xB000 to RxBD[Status and Control]
// Write 0x0000 to RxBD[Data Length] Write 0x00001000 to
// RxDB[BufferPointer] - step 25
RxBD[Status and Control] is the status of the buffer = 0xB000
Rx data buffer is the byte array the communication processor can
use to store the incoming packet in. = 0x00001000
Save Buffer and Buffer Length in Memory, Then Save Status

// Initialize the TxBD and assume that Tx data buffer is at
// 0x00002000 Write 0xFC00 to TxBD[Status and Control]
// Write 0x0000 to TxBD[Data Length]

// Write 0x00002000 to TxDB[BufferPointer] - step 26


TxBD[Status and Control] is the status of the buffer = 0xFC00 Tx data
buffer is the byte array the communication processor can use to store
```

```
the outgoing packet in. = 0x00002000
Save Buffer and Buffer Length in Memory, Then Save Status


// Write 0xFFFF to the SCCE-Transparent to clear any previous events
// - step 27 user's manual SCCE = 0xFFFF


// Initialize the SCCM-Transparent (SCC mask register) depending
// on the interrupts required of the SCCE[TXB, TXE, RXB, RXF]
// interrupts possible. - step 28 user's manual
 // Write 0x001B to the SCCM for generating TXB, TXE, RXB, RXF
 // interrupts (all events).
Write 0x0018 to the SCCM for generating TXE and RXF
// Interrupts (errors). Write 0x0000 to the SCCM in order to mask all
// interrupts.
SCCM = 0x0000


// Initialize CICR, and Write to the CIMR so that SCC2 can generate
// a system interrupt.- step 29
CIMR = 0x200000000
CICR = 0x001B9F80


// write 0x00000000 to the GSMR_H to enable normal operation of
// all modes - step 30 user's manual
GSMR_H = 0x0


// GSMR_L: 0x1088000C: TCI = 1, TPL = 0b100, TPP = 0b01, MODE = 1100
// to configure the CTS2 and CD2 pins to automatically control
// transmission and reception (DIAG field). Normal operation of the
// transmit clock is used. Notice that the transmitter (ENT) and
// receiver (ENR) are not enabled yet. - step 31 user's manual
GSMR_L = 0x1088000C


// Write 0xD555 to the
DSR - step 32 DSR = 0xD555


// Set PSMR-SCC Ethernet to configure 32-bit CRC - step 33
   // 0x080A: IAM = 0, CRC = 10 (32-bit), PRO = 0, NIB = 101
   // 0x0A0A: IAM = 0, CRC = 10 (32-bit), PRO = 1, NIB = 101
   // 0x088A: IAM = 0, CRC = 10 (32-bit), PRO = 0, SBT = 1, NIB = 101
   // 0x180A: HBC = 1, IAM = 0, CRC = 10 (32-bit), PRO = 0, NIB = 101
PSMR = 0x080A
```

```
// Enable the TENA pin (RTS2) since the MODE field of the GSMR_L is
// written to Ethernet, the TENA signal is low. Write PCPAR bit 14
// with a one and PCDIR bit 14 with a zero - step 34
PCPAR = PCPAR | 0x0001
PCDIR = PCDIR & 0xFFFE
// Write 0x1088003C to the GSMR_L register to enable the SCC2
// transmitter and receiver. - step 35
GSMR_L = 0x1088003C

-=-=-=-
// start the transmitter and the receiver
// After initializing the buffer descriptors, program the
// CPCR to execute an INIT RX AND TX PARAMS command for this channel.
        Execute Command(Cp.INIT_RX_AND_TX_PARAMS)

// Set GSMR_L[ENR] and GSMR_L[ENT] to enable the receiver and the
// Transmitter
        GSMR_L = GSMR_L | 0x00000020 | 0x00000010

// END OF MPC823 ETHERNET INITIALIZATION SEQUENCE - now when
// appropriate interrupt triggered, data is moved to or from
// transmit/receive buffers
```

2.5.1.7   NetSilicon NET+ARM40 Ethernet Example

Figure 2.41 is a diagram of a NET+ARM connected to Ethernet hardware components on the board.

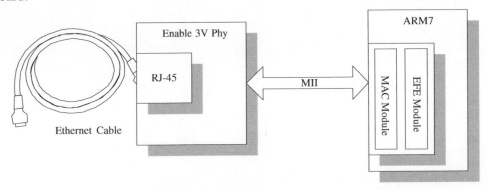

Figure 2.41: NET+ARM Ethernet Block Diagram

Like the MPC823, the NET+ARM40 Ethernet protocol is configured to have full-duplex support, as well as be *interrupt driven*. However, unlike the MPC823, the NET+ARM's initialization sequence is simpler and can be divided into three major functions: performing reset of Ethernet processor, initializing buffers, and enabling DMA channels (see *NET+ARM Hardware User's Guide for NET+ARM 15/40* and pseudo code below).

NET+ARM40 Pseudo Code

```
. . .
// Perform a low level reset of the NCC ethernet chip
// determine MII type
MIIAR = MIIAR & 0xFFFF0000 | 0x0402
MIICR = MIICR | 0x1
// wait until current PHY operation completes

if using MII
{
// set PCSCR according to poll count - 0x00000007 (>= 6),
// 0x00000003 (< 6) enable autonegotiation
}
else { //ENDEC MODE
EGCR = 0x0000C004
// set PCSCR according to poll count - 0x00000207 (>= 6),
// 0x00000203 (< 6) set EGCR to correct mode if automan jumper
// removed from board
}

// clear transfer and receive registers by reading values
get LCC
get EDC
get MCC
get SHRTFC
get LNGFC
get AEC
get CRCEC
get CEC

// Inter-packet Gap Delay = 0.96usec for MII and 9.6usec for
10BaseT
if using MII then {
B2BIPGGTR = 0x15
NB2BIPGGTR = 0x0C12
} else {
B2BIPGGIR = 0x5D
NB2BIPGGIR = 0x365a);
}

MACCR = 0x0000000D
```

```
// Perform a low level reset of the NCC ethernet chip continued

// Set SAFR = 3: PRO Enable Promiscuous Mode(receive ALL packets),
// 2: PRM Accept ALL multicast packets,  1: PRA Accept multicast
// packets using Hash Table, 0: BROAD Accept ALL broadcast packets
SAFR = 0x00000001

// load Ethernet address into addresses 0xFF8005C0 - 0xFF8005C8
// load MCA hash table into addresses 0xFF8005D0 - 0xFF8005DC

STLCR = 0x00000006

If using MII {
  // Set EGCR according to what rev - 0xC0F10000 (rev < 4),
  0xC0F10000 (PNA support disabled)

else {
  // ENDEC mode
   EGCR = 0xC0C08014}

 // Initialize buffer descriptors
    // setup Rx and Tx buffer descriptors
    DMABDP1A = "receive buffer descriptors"
        DMABDP2 = "transmit buffer descriptors"

 // enable Ethernet DMA channels
  // setup the interrupts for receive channels
  DMASR1A = DMASR1A & 0xFF0FFFFF | (NCIE | ECIE | NRIE | CAIE)

  // setup the interrupts for transmit channels
  DMASR2 = DMASR2 & 0xFF0FFFFF | (ECIE | CAIE)

    // Turn each channel on

    If MII is 100Mbps then {
            DMACR1A = DMACR1A & 0xFCFFFFFF | 0x02000000
            }
 DMACR1A = DMACR1A & 0xC3FFFFFF | 0x80000000

 If MII is 100Mbps then {
      DMACR2 = DMACR2 & 0xFCFFFFFF | 0x02000000
      }
```

```
   else if MII is 10Mbps{
        DMACR2 = DMACR2 & 0xFCFFFFFF
        }
  DMACR2 = DMACR2 & 0xC3FFFFFF | 0x84000000

  // Enable the interrupts for each channel
  DMASR1A = DMASR1A | NCIP | ECIP | NRIP | CAIP
  DMASR2 = DMASR2 | NCIP | ECIP | NRIP | CAIP


  // END OF NET+ARM ETHERNET INITIALIZATION SEQUENCE - now
  // when appropriate interrupt triggered, data is moved to
  // or from transmit/receive buffers
```

2.5.2   Example 5: Initializing an RS-232 Driver

One of the most widely implemented asynchronous serial I/O protocols is the **RS-232** or EIA-232 (Electronic Industries Association-232), which is primarily based upon the Electronic Industries Association family of standards. These standards define the major components of any RS-232 based system, which is implemented almost entirely in hardware.

The firmware (software) required to enable RS-232 functionality maps to the lower section of the OSI data-link layer. The hardware components can all be mapped to the physical layer of the OSI model, but will not be discussed in this section.

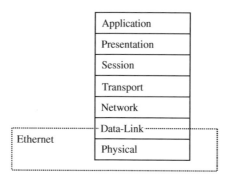

Figure 2.42: OSI Model

The RS-232 component that can be integrated on the master processor is called the *RS-232 Interface*, which can be configured for synchronous or asynchronous transmission. For example, in the case of asynchronous transmission, the only firmware (software) that is

RS-232 System Model

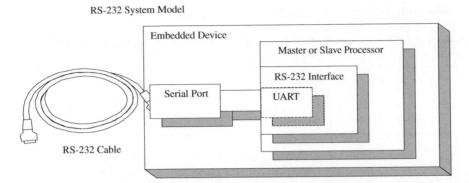

Figure 2.43: RS-232 Hardware Diagram

implemented for RS-232 is in a component called the UART (universal asynchronous transmitter receiver), which implements the serial data transmission.

Data is transmitted asynchronously over RS-232 in a stream of bits that are traveling at a constant rate. The frame processed by the UART is in the format shown in Figure 2.44.

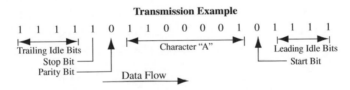

Figure 2.44: RS-232 Frame Diagram

The RS232 protocol defines frames as having: 1 start bit, 7-8 data its, 1 parity bit, and 1-2 stop bits.

2.5.2.1   Motorola/Freescale MPC823 RS-232 Example

Figure 2.45 is a MPC823 connected to RS-232 hardware components on the board.

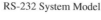
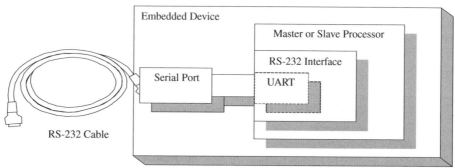

Figure 2.45: MPC823 RS-232 Block Diagram
Source: Copyright of Freescale Semiconductor, Inc. 2004. Used by permission.

There are different integrated components on a MPC823 that can be configured into UART mode, such as SCC2 and the SMCs (serial management controllers). SCC2 was discussed in the previous section as being enabled for Ethernet, so this example will look at configuring an SMC for the serial port. Enabling RS-232 on a MPC823 through the serial management controllers (SMCs) is discussed in Section 16.11, The Serial Management Controllers in the 2000 *MPC823 User's Manual.*

From the 2000 *MPC823 User's Manual*

16.11 THE SERIAL MANAGEMENT CONTROLLERS

The serial management controllers (SMCs) consist of two full-duplex ports that can be independently configured to support any one of three protocols—UART, Transparent, or general-circuit interface (GCI). Simple UART operation is used to provide a debug/monitor port in an application, which allows a serial communication controller (SCCx) to be free for other purposes. The serial management controller clock can be derived from one of four internal baud rate generators or from a 16x external clock pin . . .

The software for configuring and initializing RS-232 on the MPC823 can be based upon the SMC1 UART controller programming example in Section 16.11.6.15.

From the 2000 *MPC823 User's Manual*

16.11.6.15 SMC1 UART CONTROLLER PROGRAMMING EXAMPLE.

The following is an initialization sequence for 9600 baud, 8 data bits, no parity, and 1 stop bit operation of an SMC1 UART controller assuming a 25MHz system frequency. BRG1 and SMC1 are used.

1. Configure the port B pins to enable SMTXD1 and SMRXD1. Write PBPAR bits 25 and 24 with ones and then PBDIR and PBODR bits 25 and 24 with zeros.

2. Configure the BRG1. Write 0x010144 to BRGC1. The DIV16 bit is not used and divider is 162 (decimal). The resulting BRG1 clock is 16x the preferred bit rate of SMC1 UART controller.

3. Connect the BRG1 clock to SMC1 using the serial interface. Write the SMC1 bit SIMODE with a D and the SMC1CS field in SIMODE register with 0x000.

4. Write RBASE and TBASE in the SMC1 parameter RAM to point to the RX buffer descriptor and TX buffer descriptor in the dual-port RAM. Assuming one RX buffer descriptor at the beginning of dual-port RAM and one TX buffer descriptor following that RX buffer descriptor, write RBASE with 0x2000 and TBASE with 0x2008.

5. Program the CPCR to execute the INIT RX AND TX PARAMS command. Write 0x0091 to the CPCR.

6. Write 0x0001 to the SDCR to initialize the SDMA configuration register.

7. Write 0x18 to the RFCR and TFCR for normal operation.

8. Write MRBLR with the maximum number of bytes per receive buffer. Assume 16 bytes, so MRBLR = 0x0010.

9. Write MAX_IDL with 0x0000 in the SMC1 UART parameter RAM for the clarity.

10. Clear BRKLN and BRKEC in the SMC1 UART parameter RAM for the clarity.

11. Set BRKCR to 0x0001, so that if a STOP TRANSMIT command is issued, one bit character is sent.

12. Initialize the RX buffer descriptor. Assume the RX data buffer is at 0x00001000 in main memory. Write 0xB000 to RX_BD_Status. 0x0000 to RX_BD_Length (not required), and 0x00001000 to RX_BD_Pointer.

13. Initialize the TX buffer descriptor. Assume the TX data buffer is at 0x00002000 in main memory and contains five 8-bit characters. Then write 0xB000 to TX_BD_Status, 0x0005 to TX_BD_Length, and 0x00002000 to TX_BD_Pointer.

14. Write 0xFF to the SMCE-UART register to clear any previous events.

15. Write 0x17 to the SMCM-UART register to enable all possible serial management controller interrupts.

16. Write 0x00000010 to the CIMR to SMC1 can generate a system interrupt. The CICR must also be initialized.

17. Write 0x4820 to SMCMR to configure normal operation (not loopback), 8-bit characters, no parity, 1 stop bit. Notice that the transmitter and receiver are not enabled yet.

18. Write 0x4823 to SMCMR to enable the SMC1 transmitter and receiver. This additional write ensures that the TEN and REN bits are enabled last.

NOTE: After 5 bytes are transmitted, the TX buffer descriptor is closed. The receive buffer is closed after 16 bytes are received. Any data received after 16 bytes causes a busy (out-of-buffers) condition since only one RX buffer descriptor is prepared.

Similar to the Ethernet implementation, MPC823 serial driver is configured to be *interrupt driven*, and its initialization sequence can also be divided into seven major functions: disabling SMC1, setting up ports and the baud rate generator, initializing buffers, setting up parameter RAM, initializing interrupts, setting registers, and enabling SMC1 to transmit/receive (see the following pseudo code).

2.5.2.2 MPC823 Serial Driver Pseudo Code

```
. . .

// disabling SMC1

// Clear SMCMR[REN] to disable the receiver
     SMCMR = SMCMR & 0x0002

   // Issue Init Stop TX Command for the SCC
               execute command(STOP_TX)

   // clear SMCMR[TEN] to indicate that transmission has stopped
            SMCMR = SMCMR & 0x0002

-=-=-

// Configure port B pins to enable SMTXD1 and SMRXD1. Write PBPAR
// bits 25 and 24 with ones and then PBDIR bits 25 and 24 with
// zeros - step 1 user's manual
```

```
PBPAR = PBPAR | 0x000000C0
PBDIR = PBDIR & 0xFFFFFF3F
PBODR = PBODR & 0xFFFFFF3F

// Configure BRG1 - BRGC: 0x10000 - EN = 1 -  25 MHZ: BRGC:
// 0x010144 - EN = 1, CD = 162 (b10100010), DIV16 = 0 (9600)
// BRGC: 0x010288 - EN = 1, CD = 324 (b101000100), DIV16 = 0
// (4800) 40 Mhz:  BRGC: 0x010207 - EN = 1, CD =
// 259 (b1 0000 0011), DIV16 = 0
(9600) - step 2 user's manual

BRGC= BRGC | 0x010000

// Connect the BRG1 (Baud rate generator) to the SMC. Set the
// SIMODE[SMCx] and the SIMODE[SMC1CS] depending on baud rate
// generator where SIMODE[SMC1] = SIMODE[16], and SIMODE[SMC1CS] =
// SIMODE[17-19] - step 3
// user's manual

SIMODE = SIMODE & 0xFFFF0FFF | 0x1000

// Write RBASE and TBASE in the SCM parameter RAM to point to
// the RxBD table and the TxBD table in the dual-port RAM - step 4

RBase = 0x00 (for example)
RxSize = 128 bytes (for example)
TBase = 0x02 (for example)
TxSize = 128 bytes (for example)
Index = 0
While (index < RxSize) do
{

// Set up one receive buffer descriptor that tells the
// communication processor that the next packet is ready to be
// received - similar to  step 12 Set up one transmit buffer
// descriptor that tells the communication processor that the
// next packet is
// ready to be transmitted - similar step 13
index = index+1}
// Program the CPCR to execute the INIT RX AND TX PARAMS
// command. - step 5
execute Command(INIT_RX_AND_TX_PARAMS)
```

```
// Initialize the SDMA configuration register,  Set SDCR to 0x1
// (SDCR is 32-bit) - step 6 user's manual
SDCR =0x01

// Set RFCR,TFCR -- Rx,Tx Function Code, Initialize to 0x10 for
// normal operation (All 8-bits), Initialize to 0x18 for normal
// operation and Motorola/Freescale byte ordering - step 7
RFCR = 0x10
TFCR = 0x10

// Set MRBLR -- Max. Receive Buffer Length, assuming 16 bytes
// (multiple of 4) - step 8
MRBLR = 0x0010

// Write MAX_IDL (Maximum idle character) with 0x0000 in the SMC1
// UART parameter RAM to disable the MAX_IDL functionality - step 9
MAX_IDL = 0

// Clear BRKLN and BRKEC in the SMC1 UART parameter RAM for
// clarity - step 10
BRKLN = 0
BRKEC = 0

// Set BRKCR to 0x01 - so that if a STOP TRANSMIT command is issued,
// one break character is
// sent - step 11 BRKCR = 0x01
```

2.6   Summary

This chapter discussed device drivers, the type of software needed to manage the hardware in an embedded system. This chapter also introduced a general set of device driver routines, which make up most device drivers. Interrupt handling (on the PowerPC platform), memory management (on the PowerPC platform), I^2C bus (on a PowerPC-based platform), and I/O (Ethernet and RS-232 on PowerPC and ARM-based platforms) were real-world examples provided, along with pseudo code to demonstrate how device driver functionality can be implemented.

CHAPTER 3

Embedded Operating Systems

Tammy Noergaard
Jean Labrosse

3.1   In This Chapter

Define operating system; discuss process management, scheduling, and intertask communication; introduce memory management at the OS level; and discuss I/O management in operating systems.

An operating system (OS) is an optional part of an embedded device's system software stack, meaning that not all embedded systems have one. OSes can be used on any processor (ISA) to which the OS has been ported. As shown in Figure 3.1, an OS either sits over the hardware, over the device driver layer or over a BSP (Board Support Package, which will be discussed in Section 3.5 of this chapter).

The OS is a set of software libraries that serves two main purposes in an embedded system: providing an abstraction layer for software on top of the OS to be less dependent on

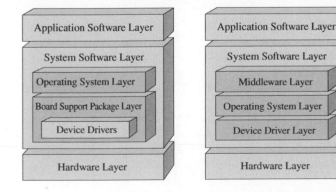
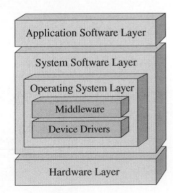

Figure 3.1: OSes and the Embedded Systems Model

hardware, making the development of middleware and applications that sit on top of the OS easier, and managing the various system hardware and software resources to ensure the entire system operates efficiently and reliably. While embedded OSes vary in what components they possess, all OSes have a ***kernel*** at the very least. The kernel is a component that contains the main functionality of the OS, specifically all or some combination of features and their interdependencies, shown in Figures 3.2a–e, including:

- **Process Management**. How the OS manages and views other software in the embedded system (via *processes*—more in Section 3.3, Multitasking and Process Management). A subfunction typically found within process management is *interrupt and error detection management*. The multiple interrupts and/or traps generated by the various processes need to be managed efficiently so that they are handled correctly and the processes that triggered them are properly tracked.

- **Memory Management**. The embedded system's memory space is shared by all the different processes, so access and allocation of portions of the memory space need to be managed (more in Section 3.4, Memory Management). Within memory management, other subfunctions such as *security system management* allow for portions of the embedded system sensitive to disruptions that can result in the disabling of the system, to remain secure from unfriendly, or badly written, higher-layer software.

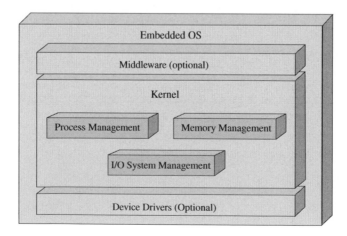

Figure 3.2a: General OS Model

Figure 3.2b: Kernel Subsystem Dependencies

- **I/O System Management**. I/O devices also need to be shared among the various processes and so, just as with memory, access and allocation of an I/O device need to be managed (more in Section 3.5, I/O and File System Management). Through I/O system management, *file system management* can also be provided as a method of storing and managing data in the forms of files.

Because of the way in which an operating system manages the software in a system, using processes, the process management component is the most central subsystem in an OS. All other OS subsystems depend on the process management unit.

Figure 3.2c: Kernel Subsystem Dependencies

Since all code must be loaded into main memory (RAM or cache) for the master CPU to execute, with boot code and data located in nonvolatile memory (ROM, Flash, and so on), the process management subsystem is equally dependent on the memory management subsystem.

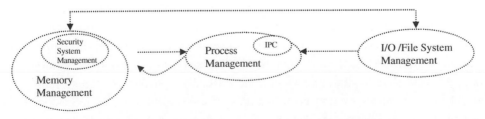

Figure 3.2d: Kernel Subsystem Dependencies

I/O management, for example, could include networking I/O to interface with the memory manager in the case of a network file system (NFS).

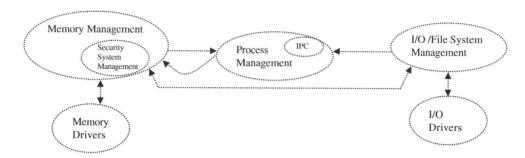

Figure 3.2e: Kernel Subsystem Dependencies

Outside the kernel, the Memory Management and I/O Management subsystems then rely on the device drivers, and vice versa, to access the hardware.

Whether inside or outside an OS kernel, OSes also vary in what other system software components, such as device drivers and middleware, they incorporate (if any). In fact, most embedded OSes are typically based on one of three models, the ***monolithic, layered,*** or ***microkernel*** (client-server) design. In general, these models differ according to the internal design of the OS's kernel, as well as what other system software has been incorporated into the OS. In a monolithic OS, middleware and device driver functionality is typically integrated into the OS along with the kernel. This type of OS is a single executable file containing all of these components (see Figure 3.3).

Monolithic OSes are usually more difficult to scale down, modify, or debug than their other OS architecture counterparts, because of their inherently large, integrated, cross-dependent nature. Thus, a more popular algorithm, based on the monolithic design, called the ***monolithic-modularized*** algorithm, has been implemented in OSes to allow for easier debugging, scalability, and better performance over the standard monolithic approach. In a monolithic-modularized OS, the functionality is integrated into a single executable file that is made up of ***modules***, separate pieces of code reflecting various OS functionality. The embedded Linux operating system is an example of a monolithic-based OS, whose main modules are shown in Figure 3.4. The Jbed RTOS, MicroC/OS-II, and PDOS are all examples of embedded monolithic OSes.

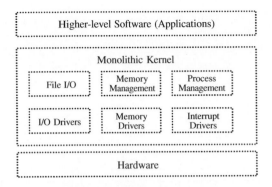

Figure 3.3: Monolithic OS

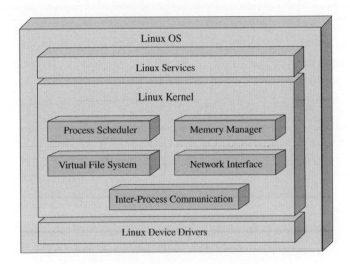

Figure 3.4: Linux OS Block Diagram

In the *layered* design, the OS is divided into hierarchical layers (0 . . . N), where upper layers are dependent on the functionality provided by the lower layers. Like the monolithic design, layered OSes are a single large file that includes device drivers and middleware (see Figure 3.5). While the layered OS can be simpler to develop and maintain than a monolithic design, the APIs provided at each layer create additional overhead that can impact size and performance. DOS-C (FreeDOS), DOS/eRTOS, and VRTX are all examples of a layered OS.

An OS that is stripped down to minimal functionality, commonly only process and memory management subunits as shown in Figure 3.6, is called a ***client-server*** OS, or a **microkernel**.

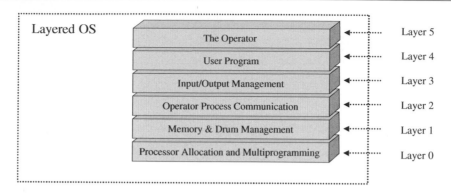

Figure 3.5: Layered OS Block Diagram

(Note: a subclass of micro-Management Management kernels are stripped down even further to only process management functionality, and Device Drivers are commonly referred to as *nanokernels*.) The remaining functionality typical of other I/O Memory Interrupt kernel algorithms is abstracted out of the kernel, while device drivers, for instance, are usually abstracted out of a microkernel Hardware entirely, as shown in Figure 3.6. A microkernel also typically differs in its process management implementation over other types of OSes. This is discussed in more detail in Section 3.3, Multitasking and Process Management: Intertask Communication and Synchronization.

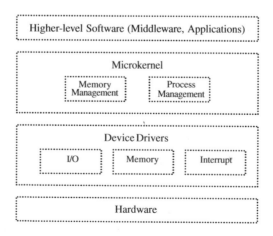

Figure 3.6: Microkernel-based OS Block Diagram

The microkernel OS is typically a more scalable (modular) and debuggable design, since additional components can be dynamically added in. It is also more secure since much of the functionality is now independent of the OS, and there is a separate memory space for client and server functionality. It is also easier to port to new architectures. However, this model may be slower than other OS architectures, such as the monolithic, because of the communication paradigm between the microkernel components and other "kernel-like" components. Overhead is also added when switching between the kernel and the other OS components and non-OS components (relative to layered and monolithic OS designs). Most of the off-the-shelf embedded OSes—and there are at least a hundred of them—have kernels that fall under the microkernel category, including OS-9, C Executive, vxWorks, CMX-RTX, Nucleus Plus, and QNX.

3.2   What Is a Process?

To understand how OSes manage an embedded device's hardware and software resources, the reader must first understand how an OS views the system. An OS differentiates between a program and the executing of a program. A program is simply a passive, static sequence of instructions that could represent a system's hardware and software resources. The actual execution of a program is an active, dynamic event in which various properties change relative to time and the instruction being executed. A ***process*** (commonly referred to as a ***task*** in many embedded OSes) is created by an OS to encapsulate all the information that is involved in the executing of a program (i.e., stack, PC, the source code and data, and so on). This means that a program is only part of a task, as shown in Figure 3.7.

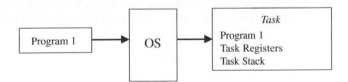

Figure 3.7: OS Task

Embedded OSes manage all embedded software using tasks, and can either be ***unitasking*** or ***multitasking***. In unitasking OS environments, only one task can exist at any given time, whereas in a multitasking OS, multiple tasks are allowed to exist simultaneously. Unitasking OSes typically do not require as complex a task management facility as a multitasking OS. In a multitasking environment, the added complexity of allowing multiple existing tasks requires that each process remain independent of the others and not affect any other without the specific programming to do so. This multitasking model provides each process with more security, which is not needed in a unitasking environment. Multitasking can actually provide a more organized way for a complex embedded system to function. In a multitasking environment, system activities are divided up into simpler, separate components, or the same activities can be running in multiple processes simultaneously, as shown in Figure 3.8.

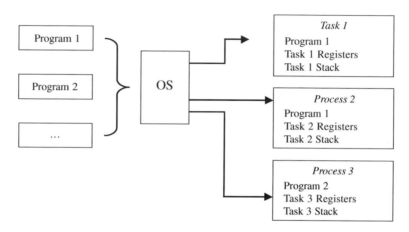

Figure 3.8: Multitasking OS

Some multitasking OSes also provide ***threads*** (*lightweight processes*) as an additional, alternative means for encapsulating an instance of a program. Threads are created within the context of a task (meaning a thread is bound to a task), and, depending on the OS, the task can own one or more threads. A thread is a sequential execution stream within its task. Unlike tasks, which have their own independent memory spaces that are inaccessible to other tasks, threads of a task share the same resources (working directories, files, I/O devices, global data, address space, program code, and so on), but have their own PCs, stack, and scheduling information (PC, SP, stack, registers, and so on) to allow for the instructions they are executing to be scheduled independently. Since threads are created within the context of the same task and can share the same memory space, they can allow for simpler communication and coordination relative to tasks. This is because a task can contain at least one thread executing one program in one address space, or can contain many threads executing different portions of one program in one address space (see Figure 3.9), needing no intertask communication mechanisms. Also, in the case of shared resources, multiple threads are typically less expensive than creating multiple tasks to do the same work.

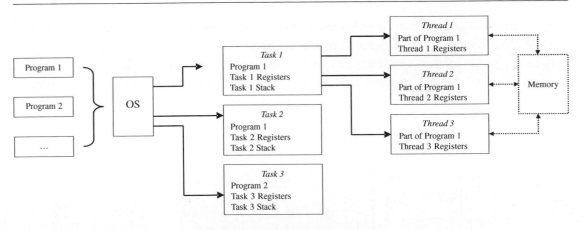

Figure 3.9: Tasks and Threads

Usually, programmers define a separate task (or thread) for each of the system's distinct activities to simplify all the actions of that activity into a single stream of events, rather than a complex set of overlapping events. However, it is generally left up to the programmer as to how many tasks are used to represent a system's activity and, if threads are available, if and how they are used within the context of tasks.

DOS-C is an example of a unitasking embedded OS, whereas vxWorks (Wind River), embedded Linux (Timesys), and Jbed (Esmertec) are examples of multitasking OSes. Even within multitasking OSes, the designs can vary widely. vxWorks has one type of task, each of which implements one "thread of execution." Timesys Linux has two types of tasks, the Linux fork and the periodic task, whereas Jbed provides six different types of tasks that run alongside threads: *OneshotTimer Task* (which is a task that is run only once), *PeriodicTimer Task* (a task that is run after a particular set time interval), *HarmonicEvent Task* (a task that runs alongside a periodic timer task), *JoinEvent Task* (a task that is set to run when an associated task completes), *InterruptEvent Task* (a task that is run when a hardware interrupt occurs), and the *UserEvent Task* (a task that is explicitly triggered by another task). More details on the different types of tasks are given in the next section.

3.3   Multitasking and Process Management

Multitasking OSes require an additional mechanism over unitasking OSes to manage and synchronize tasks that can exist simultaneously. This is because, even when an OS allows multiple tasks to coexist, one master processor on an embedded board can only execute one task or thread at any given time. As a result, multitasking embedded OSes must find some way of allocating each task a certain amount of time to use the master CPU, and switching the master processor between the various tasks. It is by accomplishing this through task *implementation, scheduling, synchronization,* and *intertask communication* mechanisms that an OS successfully gives the illusion of a single processor simultaneously running multiple tasks (see Figure 3.10).

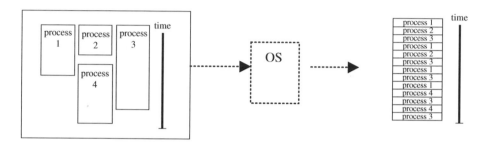

Figure 3.10: Interleaving Tasks

3.3.1   Process Implementation

In multitasking embedded OSes, tasks are structured as a hierarchy of parent and child tasks, and when an embedded kernel starts up only one task exists (as shown in Figure 3.11). It is from this first task that all others are created (note: the first task is also created by the programmer in the system's initialization code).

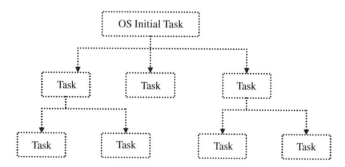

Figure 3.11: Task Hierarchy

Task creation in embedded OSes is primarily based on two models, *fork/exec* (which is derived from the IEEE /ISO POSIX 1003.1 standard) and *spawn* (which is derived from fork/exec). Since the spawn model is based on the fork/exec model, the methods of creating tasks under both models are similar. All tasks create their child tasks through fork/exec or spawn system calls. After the system call, the OS gains control and creates the **Task Control Block** (TCB), also referred to as a *Process Control Block* (PCB) in some OSes, that contains OS control information, such as task ID, task state, task priority, error status, and CPU context information, such as registers, for that particular task. At this point, memory is allocated for the new child task, including for its TCB, any parameters passed with the system call, and the code to be executed by the child task. After the task is set up to run, the system call returns and the OS releases control back to the main program.

The main difference between the fork/exec and spawn models is how memory is allocated for the new child task. Under the fork/exec model, as shown in Figure 3.12, the "fork" call

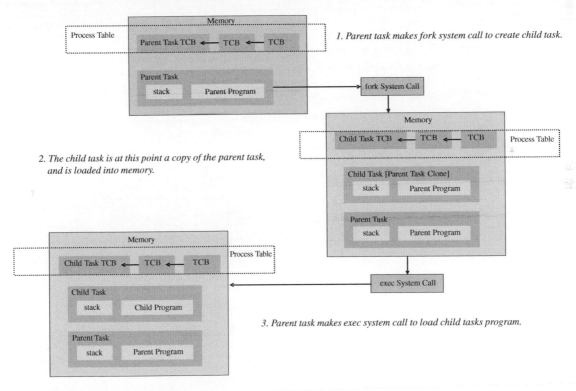

< < Task creation based upon fork/exec involve 4 major steps > >

Figure 3.12: FORK/EXEC Process Creation

creates a copy of the parent task's memory space in what is allocated for the child task, thus allowing the child task to **inherit** various properties, such as program code and variables, from the parent task. Because the parent task's entire memory space is duplicated for the child task, two copies of the parent task's program code are in memory, one for the parent, and one belonging to the child. The "exec" call is used to explicitly remove from the child task's memory space any references to the parent's program and sets the new program code belonging to the child task to run.

The spawn model, on the other hand, creates an entirely new address space for the child task. The spawn system call allows for the new program and arguments to be defined for the child task. This allows the child task's program to be loaded and executed immediately at the time of its creation.

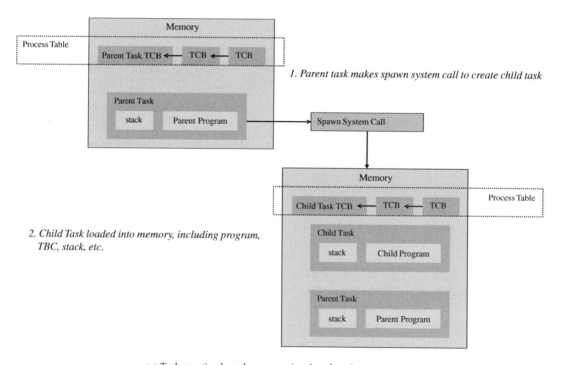

<< Task creation based on spawn involves 2 major steps >>

Figure 3.13: Spawn Process Creation

Both process creation models have their strengths and drawbacks. Under the spawn approach, there are no duplicate memory spaces to be created and destroyed, and then new space allocated, as is the case with the fork/exec model. The advantages of the fork/exec model, however, include the efficiency gained by the child task inheriting properties from the parent task, and then having the flexibility to change the child task's environment afterwards. In Figures 3.1–3.3, real-world embedded OSes are shown along with their process creation techniques.

Example 3.1: Creating a Task in vxWorks

The two major steps of spawn task creation form the basis of creating tasks in vxWorks. The vxWorks system call "taskSpawn" is based on the POSIX spawn model, and it is what creates, initializes, and activates a new (child) task.

```
int taskSpawn(
   {Task Name},
   {Task Priority 0-255, related to scheduling and will be
        discussed in the next section},
   {Task Options - VX_FP_TASK, execute with floating point
        coprocessor
        VX_PRIVATE_ENV, execute task with private environment
        VX_UNBREAKABLE, disable breakpoints for task
        VX_NO_STACK_FILL, do not fill task stack with 0xEE}
   {Stack Size}
   {Task address of entry point of program in memory-
        initial PC value}
   {Up to 10 arguments for task program entry routine})
```

After the spawn system call, an image of the child task (including TCB, stack, and program) is allocated into memory. Below is a pseudo code example of task creation in the vxWorks RTOS where a parent task "spawns" a child task software timer.

Task Creation vxWorks Pseudo Code

```
// parent task that enables software timer
void parentTask(void)
{
. . .

if sampleSoftware Clock NOT running {
        /"newSWClkId" is a unique integer value assigned by
        kernel when task is created newSWClkId = taskSpawn
        ("sampleSoftwareClock", 255, VX_NO_STACK_FILL, 3000,
        (FUNCPTR) minuteClock, 0, 0, 0, 0, 0, 0, 0, 0, 0, 0);
        . . .
}


//child task program Software Clock
void minuteClock (void) {
        integer seconds;
        while (softwareClock is RUNNING) {
            seconds = 0;
            while (seconds < 60) {
                seconds = seconds+1;
            }
. . .
}
```

Example 3.2: Jbed RTOS and Task Creation

In Jbed, there is more than one way to create a task, because in Java there is more than one way to create a Java thread—and in Jbed, tasks are extensions of Java threads. One of the most common methods of creating a task in Jbed is through the "task" routines, one of which is: public Task (long duration, long allowance, long deadline, RealtimeEvent event).

Task creation in Jbed is based on a variation of the spawn model, called *spawn threading*. Spawn threading is spawning, but typically with less overhead and with tasks sharing the same memory space. Below is a pseudo code example of task creation of an OneShot task, one of Jbed's six different types of tasks, in the Jbed RTOS where a parent task "spawns" a child task software timer that runs only one time.

Task Creation Jbed Pseudo Code

```
// Define a class that implements the Runnable interface for
   the software clock
public class ChildTask implements Runnable{

      //child task program Software Clock
      public void run () {
          integer seconds;

          while (softwareClock is RUNNING) {
              seconds = 0;
              while (seconds < 60) {
                  seconds = seconds+1;
              }
              . . .
          }
 }

// parent task that enables software timer
void parentTask(void)
{
. . .
if sampleSoftware Clock NOT running {

      try{
            DURATION,
            ALLOWANCE,
            DEADLINE,
            OneshotTimer );
      }catch( AdmissionFailure error ){
            Print Error Message ( "Task creation failed" );
      }
}
. . .
  }
```

The creation and initialization of the Task object is the Jbed (Java) equivalent of a TCB. The task object, along with all objects in Jbed, are located in Jbed's heap (in Java, there is only one heap for all objects). Each task in Jbed is also allocated its own stack to store primitive data types and object references.

Example 3.3: Embedded Linux and fork/exec

In embedded Linux, all process creation is based on the fork/exec model:

```
int fork (void) void exec (...)
```

In Linux, a new "child" process can be created with the fork system call (shown above), which creates an almost identical copy of the parent process. What differentiates the parent task from the child is the process ID—the process ID of the child process is returned to the parent, whereas a value of "0" is what the child process believes its process ID to be.

```
#include <sys/types.h>
#include <unistd.h>

void program(void)
{

    processId child_processId;

      /* create a duplicate: child process */
      child_processId = fork();

      if (child_processId == -1) {
              ERROR;
      }
      else if (child_processId == 0) {
            run_childProcess();
      }
       else {
       run_parentParent();
       }
```

The exec function call can then be used to switch to the child's program code.

```
int program (char* program, char** arg_list)
{
  processed child_processId;

/* Duplicate this process */
child_processId = fork ();

if (child_pId != 0)
  /* This is the parent process */
  return child_processId;
  else
  {
  /* Execute PROGRAM, searching for it in the path */
  execvp (program, arg_list);

  /* execvp returns only if an error occurs */
  fprintf (stderr, "Error in execvp\n");
  abort ();
  }
}
```

Tasks can terminate for a number of different reasons, such as normal completion, hardware problems such as lack of memory, and software problems such as invalid instructions. After a task has been terminated, it must be removed from the system so that it does not waste resources, or even keep the system in limbo. In *deleting* tasks, an OS de-allocates any memory allocated for the task (TCBs, variables, executed code, and so on). In the case of a parent task being deleted, all related child tasks are also deleted or moved under another parent, and any shared system resources are released.

When a task is deleted in vxWorks, other tasks are not notified, and any resources, such as memory allocated to the task are not freed—it is the responsibility of the programmer to manage the deletion of tasks using the subroutines below.

In Linux, processes are deleted with the ***void exit (int status)*** system call, which deletes the process and removes any kernel references to process (updates flags, removes processes from queues, releases data structures, updates parent-child relationships, and so on). Under Linux, child processes of a deleted process become children of the main init parent process.

Call	Description
exit()	Terminates the calling task and frees memory (task stacks and task control blocks only).
taskDelete()	Terminates a specified task and frees memory (task stacks and task control blocks only).*
taskSafe()	Protects the calling task from deletion.
taskUnsafe()	Undoes a taskSafe() (makes the calling task available for deletion).

* Memory that is allocated by the task during its execution is *not* freed when the task is terminated.

```
void vxWorksTaskDelete (int taskId)
{
    int localTaskId = taskIdFigure (taskId);

    /* no such task ID */
    if (localTaskId == ERROR)
        printf ("Error: ask not found.\n");
    else if (localTaskId == 0)
        printf ("Error: The shell can't delete itself.\n");
    else if (taskDelete (localTaskId) != OK)
        printf ("Error");
}
```

Figure 3.14a: vxWorks and Spawn Task Deleted

```
#include <stdio.h>
#include <stdlib.h>

main ()
{...
if (fork == 0)
  exit (10);
...
}
```

Figure 3.14b: Embedded Linux and fork/exec Task Deleted

Because Jbed is based on the Java model, a garbage collector is responsible for deleting a task and removing any unused code from memory once the task has stopped running. Jbed uses a nonblocking mark-and-sweep garbage collection algorithm that marks all objects still being used by the system and deletes (sweeps) all unmarked objects in memory.

In addition to creating and deleting tasks, an OS typically provides the ability to **suspend** a task (meaning temporarily blocking a task from executing) and **resume** a task (meaning any blocking of the task's ability to execute is removed). These two additional functions are provided by the OS to support task **states**. A task's state is the activity (if any) that is going on with that task once it has been created, but has not been deleted. OSes usually define a task as being in one of three states:

- **Ready:** The process is ready to be executed at any time, but is waiting for permission to use the CPU.

- **Running:** The process has been given permission to use the CPU, and can execute.

- **Blocked** or **Waiting:** The process is waiting for some external event to occur before it can be "ready" to "run."

OSes usually implement separate READY and BLOCK/WAITING "queues" containing tasks (their TCBs) that are in the relative state (see Figure 3.15). Only one task at any one time can be in the RUNNING state, so no queue is needed for tasks in the RUNNING state.

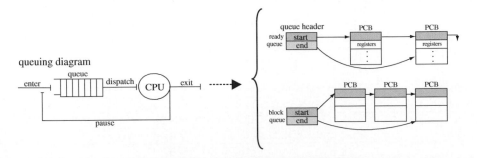

Figure 3.15: Task States and Queues

Based on these three states (Ready, Blocked, and Running), most OSes have some process state transition model similar to the state diagram in Figure 3.16. In this diagram, the "New" state indicates a task that has been created, the "Exit" state is a task that has terminated (suspended or stopped running). The other three states are defined above (Ready, Running, and Blocked). The state transitions (according to Figure 3.16) are New → Ready (where a task has entered the ready queue and can be scheduled for running), Ready → Running (based on the kernel's scheduling algorithm, the task has been selected to run), Running → Ready (the task has finished its turn with the CPU, and is returned to the ready queue for the next time around), Running → Blocked (some event has occurred to move the task into the blocked queue, not to run until the event has occurred or been resolved), and Blocked → Ready (whatever blocked task was waiting for has occurred, and task is moved back to ready queue).

Figure 3.16: Task State Diagram

When a task is moved from one of the queues (READY or BLOCKED/WAITING) into the RUNNING state, it is called a ***context switch***. Examples 3.4–3.6 give real-world examples of OSes and their state management schemes.

Example 3.4: vxWorks Wind Kernel and States

Other than the RUNNING state, VxWorks implements nine variations of the READY and BLOCKED/WAITING states, as shown in the following table and state diagram.

State	Description
STATE + 1	The state of the task with an inherited priority
READY	Task in READY state
DELAY	Task in BLOCKED state for a specific time period
SUSPEND	Task is BLOCKED usually used for debugging
DELAY + S	Task is in 2 states: DELAY & SUSPEND
PEND	Task in BLOCKED state due to a busy resource
PEND + S	Task is in 2 states: PEND & SUSPEND
PEND + T	Task is in PEND state with a time-out value
PEND + S + T	Task is in 2 states: PEND state with a time-out value and SUSPEND

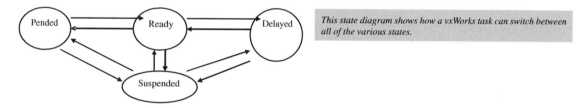

This state diagram shows how a vxWorks task can switch between all of the various states.

Figure 3.17a1: State Diagram for vxWorks Tasks

Under vxWorks, separate ready, pending, and delay state queues exist to store the TCB information of a task that is within that respective state (see Figure 3.17a2).

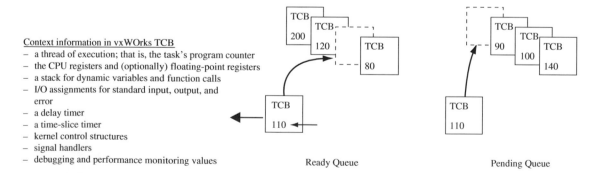

Context information in vxWOrks TCB
– a thread of execution; that is, the task's program counter
– the CPU registers and (optionally) floating-point registers
– a stack for dynamic variables and function calls
– I/O assignments for standard input, output, and error
– a delay timer
– a time-slice timer
– kernel control structures
– signal handlers
– debugging and performance monitoring values

Ready Queue

Pending Queue

Figure 3.17a2: vxWorks Tasks and Queues

A task's TCB is modified and is moved from queue to queue when a context switch occurs. When the Wind kernel context switches between two tasks, the information of the task currently running is saved in its TCB, while the TCB information of the new task to be executed is loaded for the CPU to begin executing. The Wind kernel contains two types of context switches: synchronous, which occurs when the running task blocks itself (through pending, delaying, or suspending), and asynchronous, which occurs when the running task is blocked due to an external interrupt.

Example 3.5: Jbed Kernel and States

In Jbed, some states of tasks are related to the type of task, as shown in the table and state diagrams below. Jbed also uses separate queues to hold the task objects that are in the various states.

State	Description
RUNNING	For all types of tasks, task is currently executing
READY	For all types of tasks, task in READY state
STOP	In Oneshot Tasks, task has completed execution
AWAIT TIME	For all types of tasks, task in BLOCKED state for a specific time period
AWAIT EVENT	In Interrupt and Joined tasks, BLOCKED while waiting for some event to occur

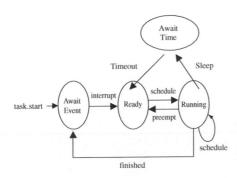

This state diagram shows some possible states for Interrupt tasks. Basically, an interrupt task is in an Await Event state until a hardware interrupt occurs—at which point the Jbed scheduler moves an Interrupt task into the Ready state to await its turn to run. At any time, the Joined Task can enter a timed waiting period.

Figure 3.17b1: State Diagram for Jbed Interrupt Tasks

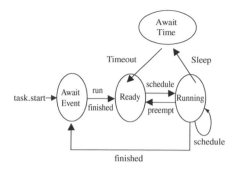

This state diagram shows some possible states for Joined tasks. Like the Interrupt task, the Joined task is in an Await Event state until an associated task has finished running—at which point the Jbed scheduler moves a Joined task into the Ready state to await its turn to run. At any time, the Joined Task can enter a timed waiting period.

Figure 3.17b2: State Diagram for Jbed Joined Tasks

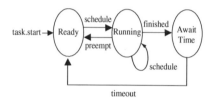

This state diagram shows some possible states for Periodic tasks. A Periodic task runs continuously at certain intervals and gets moved into the Await Time state after every run to await that interval before being put into the ready state.

Figure 3.17b3: State Diagram for Periodic Tasks

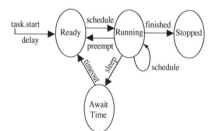

This state diagram shows some possible states for Oneshot tasks. A Oneshot task can either run once and then end (stop), or be blocked for a period of time before actually running.

Figure 3.17b4: State Diagram for Oneshot Tasks

Example 3.6: Embedded Linux and States

In Linux, RUNNING combines the traditional READY and RUNNING states, while there are three variations of the BLOCKED state.

State	Description
RUNNING	Task is either in the RUNNING or READY state
WAITING	Task in BLOCKED state waiting for a specific resource or event
STOPPED	Task is BLOCKED, usually used for debugging
ZOMBIE	Task is BLOCKED and no longer needed

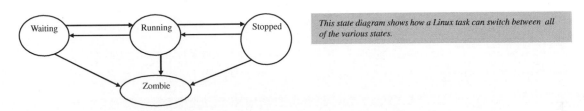

This state diagram shows how a Linux task can switch between all of the various states.

Figure 3.17c1: State Diagram for Linux Tasks

Under Linux, a process's context information is saved in a PCB called the task_struct shown in Figure 3.17c2 below. Shown boldface in the figure is an entry in the task_struct containing a Linux process's state. In Linux there are separate queues that contain the task_struct (PCB) information for the process with that respective state.

3.3.2 Process Scheduling

In a multitasking system, a mechanism within an OS, called a *scheduler* (shown in Figure 3.18), is responsible for determining the order and the duration of tasks to run on the CPU. The scheduler selects which tasks will be in what states (ready, running, or blocked), as well as loading and saving the TCB information for each task. On some OSes the same scheduler allocates the CPU to a process that is loaded into memory and ready to run, while in other OSes a *dispatcher* (a separate scheduler) is responsible for the actual allocation of the CPU to the process.

There are many scheduling algorithms implemented in embedded OSes, and every design has its strengths and trade-offs. The key factors that impact the effectiveness and performance of a scheduling algorithm include its *response time* (time for scheduler to make the context switch to a ready task and includes waiting time of task in ready queue), *turnaround time* (the time it takes for a process to complete running), *overhead* (the time and data needed to determine which tasks will run next), and *fairness* (what are the determining factors as to which processes get to run). A scheduler needs to balance utilizing

```
struct task_struct
{
                      ...
// -1 unrunnable, 0 runnable, >0 stopped
volatile long      state;
// number of clock ticks left to run in this scheduling slice, decremented by a timer.
   long           counter;
   // the process' static priority, only changed through well-known system calls like nice, POSIX.1b
   // sched_setparam, or 4.4BSD/SVR4 setpriority.
   long           priority;
   unsigned       long signal;
   // bitmap of masked signals
   unsigned       long blocked;
// per process flags, defined below
   unsigned       long flags;
   int errno;
   // hardware debugging registers
   long           debugreg[8];
   struct exec_domain *exec_domain;
   struct linux_binfmt *binfmt;
   struct task_struct *next_task, *prev_task;
   struct task_struct *next_run, *prev_run;
   unsigned long      saved_kernel_stack;
   unsigned long      kernel_stack_page;
   int            exit_code, exit_signal;
    unsigned long      personality;
   int            dumpable:1;
   int            did_exec:1;
   int            pid;
   int            pgrp;
   int            tty_old_pgrp;
   int            session;
// boolean value for session group leader
   int            leader;
   int            groups[NGROUPS];
   // pointers to (original) parent process, youngest child, younger sibling, older sibling, respectively.
   // (p->father can be replaced with p->p_pptr->pid)
   struct task_struct *p_opptr, *p_pptr, *p_cptr,
                *p_ysptr, *p_osptr;
   struct wait_queue  *wait_chldexit;
   unsigned short     uid,euid,suid,fsuid;
   unsigned short     gid,egid,sgid,fsgid;
   unsigned long      timeout;
// the scheduling policy, specifies which scheduling class the task belongs to, such as : SCHED_OTHER
//(traditional UNIX process), SCHED_FIFO(POSIX.1b FIFO realtime process - A FIFO realtime process
//will run until either a) it blocks on I/O, b) it explicitly yields the CPU or c) it is preempted by another
//realtime process with a higher p->rt_priority  value.) and SCHED_RR(POSIX round-robin realtime
//process – SCHED_RRis the same as SCHED_FIFQ except that when its timeslice expires it goes back
//to the end of the run queue).
   unsigned long      policy;

//realtime priority
   unsigned long      rt_priority;
   unsigned long      it_real_value, it_prof_value, it_virt_value;
   unsigned long      it_real_incr, it_prof_incr, it_virt_incr;
   struct timer_list  real_timer;
   long           utime, stime, cutime, cstime, start_time;
   // mm fault and swap info: this can arguably be seen as either
   // mm-specific or thread-specific */
   unsigned long      min_flt, maj_flt, nswap, cmin_flt, cmaj_flt,
   // cnswap;
   int swappable:1;
   unsigned long      swap_address;
   // old value of maj_flt
   unsigned long      old_maj_flt;
   // page fault count of the last time
   unsigned long      dec_flt;
   // number of pages to swap on next pass
   unsigned long      swap_cnt;
//limits
   struct rlimit      rlim[RLIM_NLIMITS];
   unsigned short     used_math;
   char           comm[16];
// file system info
   int            link_count;
// NULL if no tty
   struct tty_struct  *tty;
// ipc stuff
   struct sem_undo    *semundo;
   struct sem_queue   *semsleeping;
// ldt for this task - used by Wine.  If NULL, default_ldt is used
   struct desc_struct *ldt;
// tss for this task
   struct thread_struct tss;
// filesystem information
   struct fs_struct   *fs;
// open file information
   struct files_struct *files;
// memory management info
   struct mm_struct   *mm;
// signal handlers
   struct signal_struct *sig;
#ifdef __SMP__
   int            processor;
   int            last_processor;
   int            lock_depth;    /* Lock depth.
                        We can context switch in and out
                        of holding a syscall kernel lock . . . */
#endif
}
```

Figure 3.17c2: Task Structure

the system's resources—keeping the CPU, I/O, as busy as possible—with task ***throughput***, processing as many tasks as possible in a given amount of time. Especially in the case of fairness, the scheduler has to ensure that task ***starvation***, where a task never gets to run, does not occur when trying to achieve a maximum task throughput.

In the embedded OS market, scheduling algorithms implemented in embedded OSes typically fall under two approaches: ***nonpreemptive*** and ***preemptive*** scheduling. Under nonpreemptive scheduling, tasks are given control of the master CPU until they have finished execution, regardless of the length of time or the importance of the other tasks that are waiting. Scheduling algorithms based on the nonpreemptive approach include:

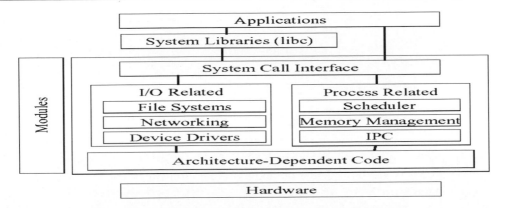

Figure 3.18: OS Block Diagram and the Scheduler

- **First-Come-First-Serve (FCFS)/Run-to-Completion,** where tasks in the READY queue are executed in the order they entered the queue, and where these tasks are run until completion when they are READY to be run (see Figure 3.19). Here, nonpreemptive means there is no BLOCKED queue in an FCFS scheduling design.

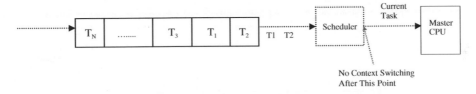

Figure 3.19: First-Come-First-Serve Scheduling

The response time of a FCFS algorithm is typically slower than other algorithms (i.e., especially if longer processes are in front of the queue requiring that other processes wait their turn), which then becomes a fairness issue since short processes at the end of the queue get penalized for the longer ones in front. With this design, however, starvation is not possible.

- **Shortest Process Next (SPN)/Run-To-Completion,** where tasks in the READY queue are executed in the order in which the tasks with the shortest execution time are executed first (see Figure 3.20).

The SPN algorithm has faster response times for shorter processes. However, then the longer processes are penalized by having to wait until all the shorter processes in the queue have run. In this scenario, starvation can occur to longer processes if the

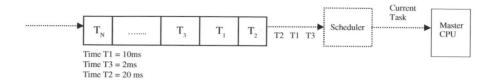

Time T1 = 10ms
Time T3 = 2ms
Time T2 = 20 ms

Figure 3.20: Shortest Process Next Scheduling

ready queue is continually filled with shorter processes. The overhead is higher than that of FCFS, since the calculation and storing of run times for the processes in the ready queue must occur.

- **Cooperative**, where the tasks themselves run until they tell the OS when they can be context switched (for example, for I/O). This algorithm can be implemented with the FCFS or SPN algorithms, rather than the run-to-completion scenario, but starvation could still occur with SPN if shorter processes were designed not to "cooperate," for example.

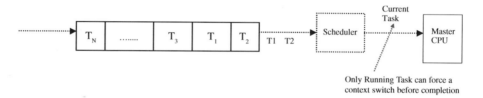

Only Running Task can force a
context switch before completion

Figure 3.21: Cooperative Scheduling

Nonpreemptive algorithms can be riskier to support since an assumption must be made that no one task will execute in an infinite loop, shutting out all other tasks from the master CPU. However, OSes that support nonpreemptive algorithms do not force a context-switch before a task is ready, and the overhead of saving and restoration of accurate task information when switching between tasks that have not finished execution is only an issue if the nonpreemptive scheduler implements a cooperative scheduling mechanism. In *preemptive scheduling,* on the other hand, the OS forces a context-switch on a task, whether or not a running task has completed executing or is cooperating with the context switch. Common scheduling algorithms based upon the preemptive approach include:

- **Round Robin/FIFO (First In, First Out) Scheduling**

    The Round Robin/FIFO algorithm implements a FIFO queue that stores ready processes (processes ready to be executed). Processes are added to the queue at the end of the queue, and are retrieved to be run from the start of the queue. In the FIFO system, all processes are treated equally regardless of their workload or interactivity. This is mainly due to the possibility of a single process maintaining control of the processor, never blocking to allow other processes to execute.

Under round-robin scheduling, each process in the FIFO queue is allocated an equal time slice (the duration each process has to run), where an interrupt is generated at the end of each of these intervals to start the preemption process. (Note: scheduling algorithms that allocate time slices, are also referred to as time-sharing systems.) The scheduler then takes turns rotating among the processes in the FIFO queue and executing the processes consecutively, starting at the beginning of the queue. New processes are added to the end of the FIFO queue, and if a process that is currently running is not finished executing by the end of its allocated time slice, it is preempted and returned to the back of the queue to complete executing the next time its turn comes around. If a process finishes running before the end of its allocated time slice, the process voluntarily releases the processor, and the scheduler then assigns the next process of the FIFO queue to the processor (see Figure 3.22).

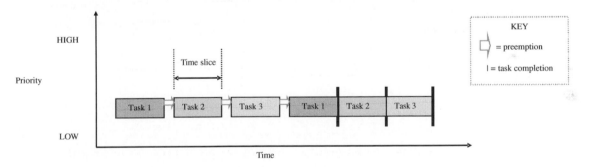

Figure 3.22: Round-Robin/FIFO Scheduling

While Round Robin/FIFO scheduling ensures the equal treatment of processes, drawbacks surface when various processes have heavier workloads and are constantly preempted, thus creating more context switching overhead. Another issue occurs when processes in the queue are interacting with other processes (such as when waiting for the completion of another process for data), and are continuously preempted from completing any work until the other process of the queue has finished its run. The throughput depends on the time slice. If the time slice is too small, then there are many context switches, while too large a time slice is not much different from a nonpreemptive approach, like FCFS. Starvation is not possible with the round-robin implementation.

- **Priority (Preemptive) Scheduling**

    The priority preemptive scheduling algorithm differentiates between processes based on their relative importance to each other and the system. Every process is assigned a priority, which acts as an indicator of orders of precedence within the system. The processes with the highest priority always preempt lower priority processes when they want to run, meaning a running task can be forced to block by the scheduler if a higher priority task becomes ready to run. Figure 3.23 shows three tasks (1, 2, 3—where task 1 is the lowest priority task and task 3 is the highest), and task 3 preempts task 2, and task 2 preempts task 1.

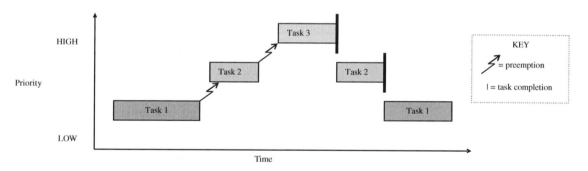

Figure 3.23: Preemptive Priority Scheduling

While this scheduling method resolves some of the problems associated with round-robin/FIFO scheduling in dealing with processes that interact or have varying workloads, new problems can arise in priority scheduling including:

— *Process starvation*, where a continuous stream of high priority processes keep lower priority processes from ever running. Typically resolved by aging lower priority processes (as these processes spend more time on queue, increase their priority levels).

— *Priority inversion*, where higher priority processes may be blocked waiting for lower priority processes to execute, and processes with priorities in between have a higher priority in running, thus both the lower priority as well as higher priority processes do not run (see Figure 3.24).

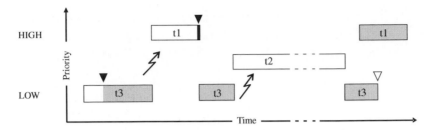

Figure 3.24: Priority Inversion

— How to *determine the priorities* of various processes. Typically, the more important the task, the higher the priority it should be assigned. For tasks that are equally important, one technique that can be used to assign task priorities is the **Rate Monotonic Scheduling** (RMS) scheme, in which tasks are assigned a priority based on how often they execute within the system. The premise behind this model is that, given a preemptive scheduler and a set of tasks that are completely independent (no shared data or resources) and are run periodically (meaning run at regular time intervals), the more often a task is executed within this set, the higher its priority should be. The RMS Theorem says that if the above assumptions are met for a scheduler and a set of "n" tasks, all timing deadlines will be met if the inequality $\sum E_i/T_i <= n(2^{1/n} - 1)$ is verified, where

> i = periodic task
> n = number of periodic tasks
> T_i = the execution period of task i
> E_i = the worst-case execution time of task i
> E_i/T_i = the fraction of CPU time required to execute task i

So, given two tasks that have been prioritized according to their periods, where the shortest period task has been assigned the highest priority, the "$n(2^{1/n} - 1)$" portion of the inequality would equal approximately .828, meaning the CPU utilization of these tasks should not exceed about 82.8% in order to meet all hard deadlines. For 100 tasks that have been prioritized according to their periods, where the shorter period tasks have been assigned the higher priorities, CPU utilization of these tasks should not exceed approximately 69.6% (100 * (21/100 – 1)) in order to meet all deadlines.

Real-World Advice

To Benefit Most from a Fixed-Priority Preemptive OS

Algorithms for assigning priorities to OS tasks are typically classified as fixed-priority where tasks are assigned priorities at design time and do not change through the life cycle of the task, dynamic-priority where priorities are assigned to tasks at run-time, or some combination of both algorithms. Many commercial OSes typically support only the fixed-priority algorithms, since it is the least complex scheme to implement. The key to utilizing the fixed-priority scheme is:

- *to assign the priorities of tasks according to their periods, so that the shorter the periods, the higher the priorities.*

- *to assign priorities using a fixed-priority algorithm (like the Rate Monotonic Algorithm, the basis of RMS) to assign fixed priorities to tasks, and as a tool to quickly to determine if a set of tasks is schedulable.*

- *to understand that in the case when the inequality of a fixed-priority algorithm, like RMS, is not met, an analysis of the specific task set is required. RMS is a tool that allows for assuming that deadlines would be met in most cases if the total CPU utilization is below the limit ("most" cases meaning there are tasks that are not schedulable via any fixed-priority scheme). It is possible for a set of tasks to still be schedulable in spite of having a total CPU utilization above the limit given by the inequality. Thus, an analysis of each task's period and execution time needs to be done in order to determine if the set can meet required deadlines.*

- *to realize that a major constraint of fixed-priority scheduling is that it is not always possible to completely utilize the master CPU 100%. If the goal is 100% utilization of the CPU when using fixed priorities, then tasks should be assigned harmonic periods, meaning a task's period should be an exact multiple of all other tasks with shorter periods.*

Based on the article "Introduction to Rate Monotonic Scheduling" by Michael Barr, *Embedded Systems Programming*, February 2002.

- **EDF (Earliest Deadline First)/Clock Driven Scheduling**

  As shown in Figure 3.25, the EDF/Clock Driven algorithm schedules priorities to processes according to three parameters: ***frequency*** (number of times process is run), ***deadline*** (when processes execution needs to be completed), and ***duration*** (time it takes to execute the process). While the EDF algorithm allows for timing constraints to be verified and enforced (basically guaranteed deadlines for all tasks), the difficulty is defining an exact duration for various processes. Usually, an average estimate is the best that can be done for each process.

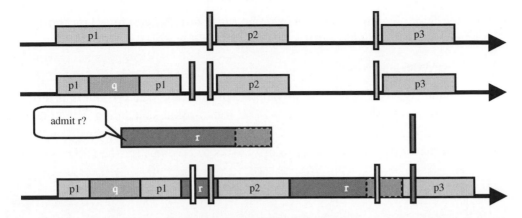

Figure 3.25: EDF Scheduling

3.3.2.1   Preemptive Scheduling and the Real-Time Operating System (RTOS)

One of the biggest differentiators between the scheduling algorithms implemented within embedded operating systems is whether the algorithm guarantees its tasks will meet execution time deadlines. If tasks always meet their deadlines (as shown in the first two graphs in Figure 3.26), and related execution times are predictable (deterministic), the OS is referred to as a **Real-Time Operating System** (RTOS).

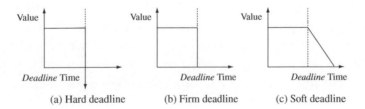

Figure 3.26: OSes and Deadlines

Preemptive scheduling must be one of the algorithms implemented within RTOS schedulers, since tasks with real-time requirements have to be allowed to preempt other tasks. RTOS schedulers also make use of their own array of *timers*, ultimately based on the system clock, to manage and meet their hard deadlines.

Whether an RTOS or a nonreal-time OS in terms of scheduling, all will vary in their implemented scheduling schemes. For example, vxWorks (Wind River) is a priority-based and round-robin scheme, Jbed (Esmertec) is an EDF scheme, and Linux (Timesys) is a priority-based scheme. Examples 3.7–3.9 examine further the scheduling algorithms incorporated into these embedded off-the-shelf operating systems.

Example 3.7: vxWorks Scheduling

The Wind scheduler is based upon both preemptive priority and round-robin real-time scheduling algorithms. As shown in Figure 3.27a1, round-robin scheduling can be teamed with preemptive priority scheduling to allow for tasks of the *same priority* to share the master processor, as well as allow higher priority tasks to preempt for the CPU.

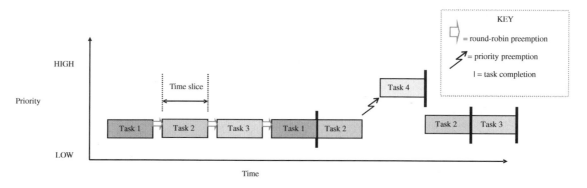

Figure 3.27a1: Preemptive Priority Scheduling Augmented with Round-Robin Scheduling

Without round-robin scheduling, tasks of equal priority in vxWorks would never preempt each other, which can be a problem if a programmer designs one of these tasks to run in an infinite loop. However, the preemptive priority scheduling allows vxWorks its real-time capabilities, since tasks can be programmed never to miss a deadline by giving them the higher priorities to preempt all other tasks. Tasks are assigned priorities via the "taskSpawn" command at the time of task creation:

```
int taskSpawn(
{Task Name},
{Task Priority 0-255, related to scheduling and will be discussed
   in the next section},
{Task Options – VX_FP_TASK, execute with floating point coprocessor
     VX_PRIVATE_ENV, execute task with private environment
     VX_UNBREAKABLE,   disable breakpoints for task
     VX_NO_STACK_FILL, do not fill task stack with 0xEE}
{Task address of entry point of program in memory– initial PC value}
{Up to 10 arguments for task program entry routine})
```

Example 3.8: Jbed and EDF Scheduling

Under the Jbed RTOS, all six types of tasks have the three variables "duration," "allowance," and "deadline" when the task is created for the EDF scheduler to schedule all tasks, as shown in the method (java subroutine) calls below.

```
public Task(
            long duration,
            long allowance,
            long deadline,
            RealtimeEvent event)
     Throws AdmissionFailure
```

```
Public Task (java.lang.String name,
            long duration,
            long allowance,
            long deadline,
            RealtimeEvent event)
        Throws AdmissionFailure
```

```
Public Task (java.lang.Runnable target,
            java.lang.String name,
            long duration,
            long allowance,
            long deadline,
            RealtimeEvent event)
        Throws AdmissionFailure
```

Example 3.9: TimeSys Embedded Linux Priority-based Scheduling

As shown in Figure 3.27b1, the embedded Linux kernel has a scheduler that is made up of four modules:

- *System call interface module*, which acts as the interface between user processes and any functionality explicitly exported by the kernel.

- *Scheduling policy module*, which determines which processes have access to the CPU.

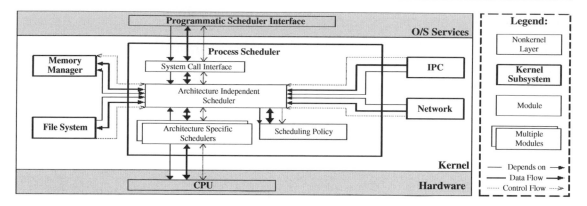

Figure 3.27b1: Embedded Linux Block Diagram

- *Architecture specific scheduler module*, which is an abstraction layer that interfaces with the hardware (i.e., communicating with CPU and the memory manager to suspend or resume processes).

- *Architecture independent scheduler module*, which is an abstraction layer that interfaces between the scheduling policy module and the architecture specific module.

The scheduling policy module implements a "priority-based" scheduling algorithm. While most Linux kernels and their derivatives (2.2/2.4) are nonpre-emptable, have no rescheduling, and are not real-time, Timesys' Linux scheduler is priority-based, but has been modified to allow for real-time capabilities. Timesys has modified the traditional Linux's standard software timers, which are too coarsely grained to be suitable for use in most real-time applications because they rely on the kernel's jiffy timer, and implements high-resolution clocks and timers based on a hardware timer. The scheduler maintains a table listing all of the tasks within the entire system and any state information associated with the tasks. Under Linux, the total number of tasks allowed is only limited to the size of physical memory available. A dynamically allocated linked list of a task structure, whose fields that are relevant to scheduling are highlighted in Figure 3.27b2, represents all tasks in this table.

```
struct task_struct
{                        . . .
                   // -1 unrunnable, 0 runnable, >0 stopped
  volatile long     state;
```

// number of clock ticks left to run in this scheduling slice, decremented
by a timer.
```
  long               counter;
```

// the process' static priority, only changed through well-known system
calls like nice, POSIX.1b
// sched_setparam, or 4.4BSD/SVR4 setpriority.
```
  long               priority;

  unsigned           long signal;
```

// bitmap of masked signals
```
  unsigned           long blocked;
```

// per process flags, defined below
```
  unsigned           long flags;
  int errno;
```

// hardware debugging registers
```
  long               debugreg[8];
  struct exec_domain *exec_domain;
  struct linux_binfmt *binfmt;
  struct task_struct *next_task, *prev_task;
  struct task_struct *next_run, *prev_run;
  unsigned long      saved_kernel_stack;
  unsigned long      kernel_stack_page;
  int                exit_code, exit_signal;
  unsigned long      personality;
  int                dumpable:1;
  int                did_exec:1;
  int                pid;
  int                pgrp;
  int                tty_old_pgrp;
  int                session;
// Boolean value for session group leader
  int                leader;
  int                groups[NGROUPS];
```

// pointers to (original) parent process, youngest child, younger sibling,
// older sibling, respectively. (p->father can be replaced with p->p_pptr->pid)
```
  struct task_struct *p_opptr, *p_pptr, *p_cptr,
                     *p_ysptr, *p_osptr;
  struct wait_queue  *wait_chldexit;
  unsigned short     uid,euid,suid,fsuid;
  unsigned short     gid,egid,sgid,fsgid;
  unsigned long      timeout;
```

// the scheduling policy, specifies which scheduling class the task belongs to,
// such as : SCHED_OTHER (traditional UNIX process), SCHED_FIFO
// (POSIX.1b FIFO realtime process - A FIFO realtime process will
//run until either a) it blocks on I/O, b) it explicitly yields the CPU or c) it is
// preempted by another realtime process with a higher p->rt_priority value.)
// and SCHED_RR (POSIX round-robin realtime process –
//SCHED_RR is the same as SCHED_FIFO, except that when its timeslice
// expires it goes back to the end of the run queue).
```
  unsigned long      policy;
```

//realtime priority
```
  unsigned long      rt_priority;
```

```
  unsigned long      it_real_value, it_prof_value, it_virt_value;
  unsigned long      it_real_incr, it_prof_incr, it_virt_incr;
  struct timer_list  real_timer;
  long               utime, stime, cutime, cstime, start_time;
```

// mm fault and swap info: this can arguably be seen as either mm-
specific or thread-specific */
```
  unsigned long      min_flt, maj_flt, nswap, cmin_flt, cmaj_flt,
cnswap;
  int swappable:1;
  unsigned long      swap_address;
```

// old value of maj_flt
```
  unsigned long      old_maj_flt;
```

// page fault count of the last time
```
  unsigned long      dec_flt;
```

// number of pages to swap on next pass
```
  unsigned long      swap_cnt;
```

//limits
```
  struct rlimit      rlim[RLIM_NLIMITS];
  unsigned short     used_math;
  char               comm[16];
```

// file system info
```
  int                link_count;
```

// NULL if no tty
```
  struct tty_struct  *tty;
```

// ipc stuff
```
  struct sem_undo    *semundo;
  struct sem_queue   *semsleeping;
```

// ldt for this task - used by Wine. If NULL, default_ldt is used
```
  struct desc_struct *ldt;
```

// tss for this task
```
  struct thread_struct tss;
```

// filesystem information
```
  struct fs_struct   *fs;
```

// open file information
```
  struct files_struct *files;
```

// memory management info
```
  struct mm_struct   *mm;
```

// signal handlers
```
  struct signal_struct *sig;
#ifdef __SMP__
  int                processor;
  int                last_processor;
  int                lock_depth;    /* Lock depth.
                                    We can context switch in and out
                                    of holding a syscall kernel lock . . . */
#endif
                     . . .
}
```

Figure 3.27b2: Task Structure

After a process has been created in Linux, through the fork or fork/exec commands, for instance, its priority is set via the setpriority command.

```
int setpriority(int which, int who, int prio);
   which = PRIO_PROCESS, PRIO_PGRP, or PRIO_USER
   who  = interpreted relative to which
   prio = priority value in the range -20 to 20
```

3.3.3 *Intertask Communication and Synchronization*

Different tasks in an embedded system typically must share the same hardware and software resources, or may rely on each other in order to function correctly. For these reasons, embedded OSes provide different mechanisms that allow for tasks in a multitasking system to intercommunicate and synchronize their behavior so as to coordinate their functions, avoid problems, and allow tasks to run simultaneously in harmony.

Embedded OSes with multiple intercommunicating processes commonly implement interprocess communication (IPC) and synchronization algorithms based on one or some combination of ***memory sharing, message passing,*** and ***signaling*** mechanisms.

With the ***shared data*** model shown in Figure 3.28, processes communicate via access to shared areas of memory in which variables modified by one process are accessible to all processes.

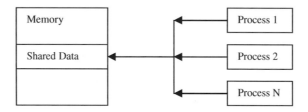

Figure 3.28: Memory Sharing

While accessing shared data as a means to communicate is a simple approach, the major issue of ***race conditions*** can arise. A race condition occurs when a process that is accessing shared variables is preempted before completing a modification access, thus affecting the integrity of shared variables. To counter this issue, portions of processes that access shared data, called *critical sections*, can be earmarked for *mutual exclusion* (or *Mutex* for short). Mutex mechanisms allow shared memory to be locked up by the process accessing it, giving that process exclusive access to shared data. Various mutual exclusion mechanisms can be implemented not only for coordinating access to shared memory, but for coordinating access to other shared system resources as well. Mutual exclusion techniques for synchronizing tasks that wish to concurrently access shared data can include:

- *Processor assisted locks* for tasks accessing shared data that are scheduled such that no other tasks can preempt them; the only other mechanisms that could force a context switch are interrupts. Disabling interrupts while executing code in the critical section would avoid a race condition scenario if the interrupt handlers access the same data. Figure 3.29 demonstrates this processor-assisted lock of disabling interrupts as implemented in vxWorks.

VxWorks provides an interrupt locking and unlocking function for users to implement in tasks.

Another possible processor-assisted lock is the "test-and-set-instruction" mechanism (also referred to as the condition variable scheme). Under this mechanism, the setting and testing of a register flag (condition) is an atomic function, a process that cannot be interrupted, and this flag is tested by any process that wants to access a critical section.

In short, both the interrupt disabling and the condition variable type of locking schemes guarantee a process exclusive access to memory, where nothing can preempt the access to shared data and the system cannot respond to any other event for the duration of the access.

```
FuncA ()
    {
        int lock = intLock ();
        .
        . critical region that cannot be interrupted
        .
        intUnlock (lock);
    }
```

Figure 3.29: vxWorks Processor Assisted Locks

- *Semaphores*, which can be used to lock access to shared memory (mutual exclusion), and also can be used to coordinate running processes with outside events (synchronization). The semaphore functions are *atomic* functions, and are usually invoked through system calls by the process. Example 3.10 demonstrates semaphores provided by vxWorks.

Example 3.10: vxWorks Semaphores

VxWorks defines three types of semaphores:

1. *Binary* semaphores are binary (0 or 1) flags that can be set to be available or unavailable. Only the associated resource is affected by the mutual exclusion when a binary semaphore is used as a mutual exclusion mechanism (whereas processor assisted locks, for instance, can affect other unrelated resources within the system). A binary semaphore is initially set = 1 (full) to show the resource is available. Tasks check the binary semaphore of a resource when wanting access, and if available, then take the associated semaphore when accessing a resource (setting the binary semaphore = 0), and then give it back when finishing with a resource (setting the binary semaphore = 1).

   When a binary semaphore is used for task synchronization, it is initially set equal to 0 (empty), because it acts as an event other tasks are waiting for. Other tasks that need to run in a particular sequence then wait (block) for the binary semaphore to be equal to 1 (until the event occurs) to take the semaphore from the original task and set it back to 0. The vxWorks pseudo code example below demonstrates how binary semaphores can be used in vxWorks for task synchronization.

```
#include "vxWorks.h"
#include "semLib.h"
#include "arch/arch/ivarch.h"  /* replace arch with /*
* architecture type */

SEM_ID syncSem; /* ID of sync semaphore */

init (int someIntNum)
{
 /* connect interrupt service routine */
 intConnect (INUM_TO_IVEC (someIntNum), eventInterruptSvcRout, 0);

 /* create semaphore */
 syncSem = semBCreate (SEM_Q_FIFO, SEM_EMPTY);

 /* spawn task used for synchronization. */
 taskSpawn ("sample", 100, 0, 20000, task1, 0,0,0,0,0,0,0,0,0,0);
}

task1 (void)
{
 ...
 semTake (syncSem, WAIT_FOREVER); /* wait for event to occur */
 printf ("task 1 got the semaphore\n");
 ... /* process event */
}

eventInterruptSvcRout (void)
{
 ...
 semGive (syncSem); /* let task 1 process event */
 ...
}
```

2. *Mutual Exclusion* semaphores are binary semaphores that can only be used for mutual exclusion issues that can arise within the vxWorks scheduling model, such as: priority inversion, deletion safety (insuring that tasks that are accessing a critical section and blocking other tasks aren't unexpectedly deleted), and recursive access to resources. Below is a pseudo code example of a mutual exclusion semaphore used recursively by a task's subroutines.

```
/* Function A requires access to a resource which it acquires
* by taking
* mySem;
* Function A may also need to call function B, which also
* requires mySem:
*/
/* includes */
#include "vxWorks.h"
#include "semLib.h"
SEM_ID mySem;

/* Create a mutual-exclusion semaphore. */
init ()
{
 mySem = semMCreate (SEM_Q_PRIORITY);
}

funcA ()
{
 semTake (mySem, WAIT_FOREVER);
 printf ("funcA: Got mutual-exclusion semaphore\n");
 ...
 funcB ();
 ...
 semGive (mySem);
 printf ("funcA: Released mutual-exclusion semaphore\n");
}
funcB ()
{
 semTake (mySem, WAIT_FOREVER);
 printf ("funcB: Got mutual-exclusion semaphore\n");
 ...
 semGive (mySem);
 printf ("funcB: Releases mutual-exclusion semaphore\n");
}
```

3. *Counting* semaphores are positive integer counters with two related functions: incrementing and decrementing. Counting semaphores are typically used to manage multiple copies of resources. Tasks that need access to resources decrement the value of the semaphore, when tasks relinquish a resource, the value of the semaphore is

incremented. When the semaphore reaches a value of "0," any task waiting for the related access is blocked until another task gives back the semaphore.

```
/* includes */
#include "vxWorks.h"
#include "semLib.h"
SEM_ID mySem;

/* Create a counting semaphore. */
init ()
{
 mySem = semCCreate (SEM_Q_FIFO,0);
}
. . .
```

On a final note, with mutual exclusion algorithms, only one process can have access to shared memory at any one time, basically having a lock on the memory accesses. If more than one process blocks waiting for their turn to access shared memory, and relying on data from each other, a ***deadlock*** can occur (such as priority inversion in priority based scheduling). Thus, embedded OSes have to be able to provide deadlock-avoidance mechanisms as well as deadlock-recovery mechanisms. As shown in the examples above, in vxWorks, semaphores are used to avoid and prevent deadlocks.

Intertask communication via ***message passing*** is an algorithm in which messages (made up of data bits) are sent via message queues between processes. The OS defines the protocols for process addressing and authentication to ensure that messages are delivered to processes reliably, as well as the number of messages that can go into a queue and the message sizes. As shown in Figure 3.30, under this scheme, OS tasks send messages to a message queue, or receive messages from a queue to communicate.

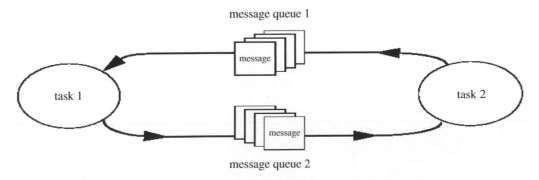

Figure 3.30: Message Queues

Microkernel-based OSes typically use the message passing scheme as their main synchronization mechanism. Example 3.11 demonstrates message passing in more detail, as implemented in vxWorks.

Example 3.11: Message Passing in vxWorks

VxWorks allows for intertask communication via message passing queues to store data transmitted between different tasks or an ISR. VxWorks provides the programmer four system calls to allow for the development of this scheme:

Call	Description
msgQCreate( )	Allocates and initializes a message queue.
msgQDelete( )	Terminates and frees a message queue.
msgQSend( )	Sends a message to a message queue.
msgQReceive( )	Receives a message from a message queue.

These routines can then be used in an embedded application, as shown in the source code example below, to allow for tasks to intercommunicate:

```
/* In this example, task t1 creates the message queue and
* sends a message
* to task t2. Task t2 receives the message from the queue
* and simply displays the message. */

/* includes */
#include "vxWorks.h"
#include "msgQLib.h"

/* defines */
#define MAX_MSGS (10)
#define MAX_MSG_LEN (100)
MSG_Q_ID myMsgQId;

task2 (void)
{
 char msgBuf[MAX_MSG_LEN];
 /* get message from queue; if necessary wait until msg is
 available */
 if (msgQReceive(myMsgQId, msgBuf, MAX_MSG_LEN, WAIT_FOREVER)
== ERROR)
 return (ERROR);
 /* display message */
 printf ("Message from task 1:\n%s\n", msgBuf);
}
```

```
#define MESSAGE "Greetings from Task 1"
task1 (void)
{
 /* create message queue */
 if ((myMsgQId = msgQCreate (MAX_MSGS, MAX_MSG_LEN,
 * MSG_Q_PRIORITY)) ==
NULL)
  return (ERROR);
  /* send a normal priority message, blocking if queue is full */
  if (msgQSend (myMsgQId, MESSAGE, sizeof (MESSAGE),
WAIT_FOREVER,MSG_PRI_NORMAL) ==
ERROR)
 return (ERROR);
 }
```

3.3.3.1 Signals and Interrupt Handling (Management) at the Kernel Level

Signals are indicators to a task that an asynchronous event has been generated by some external event (other processes, hardware on the board, timers, and so on) or some internal event (problems with the instructions being executed among others). When a task receives a signal, it suspends executing the current instruction stream and context switches to a signal handler (another set of instructions). The signal handler is typically executed within the task's context (stack) and runs in the place of the signaled task when it is the signaled task's turn to be scheduled to execute.

BSD 4.3	POSIX 1003.1
sigmask( )	sigemptyset( ), sigfillset( ), sigaddset( ), sigdelset( ), sigismember( )
sigblock( )	sigprocmask( )
sigsetmask( )	sigprocmask( )
pause( )	sigsuspend( )
sigvec( )	sigaction( )
(none)	sigpending( )
signal( )	signal( )
kill( )	kill( )

Figure 3.31: vxWorks Signaling Mechanism (The wind kernel supports two types of signal interface: UNIX BSD-style and POSIX-compatible signals.)

Signals are typically used for interrupt handling in an OS, because of their asynchronous nature. When a signal is raised, a resource's availability is unpredictable. However, signals can be used for general intertask communication, but are implemented so that the possibility of a signal handler blocking or a deadlock occurring is avoided. The other intertask communication mechanisms (shared memory, message queues, and so on), along with signals, can be used for ISR-to-Task level communication, as well.

When signals are used as the OS abstraction for interrupts and the signal handling routine becomes analogous to an ISR, the OS manages the interrupt table, which contains the interrupt and information about its corresponding ISR, as well as provides a system call (subroutine) with parameters that that can be used by the programmer. At the same time, the OS protects the integrity of the interrupt table and ISRs, because this code is executed in kernel/supervisor mode. The general process that occurs when a process receives a signal generated by an interrupt and an interrupt handler is called OS interrupt subroutine and is shown in Figure 3.32.

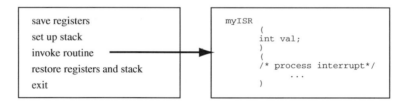

Figure 3.32: OS Interrupt Subroutine

The architecture determines the interrupt model of an embedded system (that is, the number of interrupts and interrupt types). The interrupt device drivers initialize and provide access to interrupts for higher layer of software. The OS then provides the signal inter-process communication mechanism to allow for its processes to work with interrupts, as well as can provide various interrupt subroutines that abstracts out the device driver.

While all OSes have some sort of interrupt scheme, this will vary depending on the architecture they are running on, since architectures differ in their own interrupt schemes. Other variables include *interrupt latency/response*, the time between the actual initiation of an interrupt and the execution of the ISR code, and *interrupt recovery*, the time it takes to switch back to the interrupted task. Example 9-12 shows an interrupt scheme of a real-world embedded RTOS.

Example 3.12: Interrupt Handling in vxWorks

Except for architectures that do not allow for a separate interrupt stack (and thus the stack of the interrupted task is used), ISRs use the same interrupt stack, which is initialized and configured at system start-up, outside the context of the interrupting task. Table 3.1

summarizes the interrupt routines provided in vxWorks, along with a pseudo code example of using one of these routines.

Table 3.1: Interrupt Routines in vxWorks

Call	Description
intConnect( )	Connects a C routine to an interrupt vector.
intContext( )	Returns TRUE if called from interrupt level.
intCount( )	Gets the current interrupt nesting depth.
intLevelSet( )	Sets the processor interrupt mask level.
intLock( )	Disables interrupts.
intUnlock( )	Re-enables interrupts.
intVecBaseSet( )	Sets the vector base address.
intVecBaseGet( )	Gets the vector base address.
intVecSet( )	Sets an exception vector.
intVecGet( )	Gets an exception vector.

```
/* This routine intializes the
 * serial driver, sets up interrupt
 * vectors, and performs hardware
 * intialization of the serial ports.
 */
voidInitSerialPort(void)
{
initSerialPort();
(void) intConnect(INUM_TO_IVEC(INT_
NUM_SCC), serialInt, 0);
...
}
```

3.4   Memory Management

As mentioned earlier in this chapter, a kernel manages program code within an embedded system via tasks. The kernel must also have some system of loading and executing tasks within the system, since the CPU only executes task code that is in cache or RAM. With multiple tasks sharing the same memory space, an OS needs a security system mechanism to protect task code from other independent tasks. Also, since an OS must reside in the same memory space as the tasks it is managing, the protection mechanism needs to include managing its own code in memory and protecting it from the task code it is managing. It is these functions, and more, that are the responsibility of the memory management components of an OS. In general, a kernel's memory management responsibilities include:

- Managing the mapping between logical (physical) memory and task memory references.

- Determining which processes to load into the available memory space.

- Allocating and de-allocating of memory for processes that make up the system.

- Supporting memory allocation and de-allocation of code requests (within a process) such as the C language "alloc" and "dealloc" functions, or specific buffer allocation and de-allocation routines.

- Tracking the memory usage of system components.

- Ensuring cache coherency (for systems with cache).

- Ensuring process memory protection.

Physical memory is composed of two-dimensional arrays made up of cells addressed by a unique row and column, in which each cell can store 1 bit. Again, the OS treats memory as one large one-dimensional array, called a ***memory map***. Either a hardware component integrated in the master CPU or on the board does the conversion between logical and physical addresses (such as an MMU), or it must be handled via the OS.

How OSes manage the logical memory space differs from OS to OS, but kernels generally run kernel code in a separate memory space from processes running higher level code (i.e., middleware and application layer code). Each of these memory spaces (kernel containing kernel code and user containing the higher-level processes) are managed differently. In fact, most OS processes typically run in one of two modes: ***kernel mode*** and ***user mode***, depending on the routines being executed. Kernel routines run in kernel mode (also referred to as supervisor mode), in a different memory space and level than higher layers of software such as middleware or applications. Typically, these higher layers of software run in user mode, and can only access anything running in kernel mode via ***system calls***, the higher-level interfaces to the kernel's subroutines. The kernel manages memory for both itself and user processes.

3.4.1   User Memory Space

Because multiple processes are sharing the same physical memory when being loaded into RAM for processing, there also must be some protection mechanism so processes cannot inadvertently affect each other when being swapped in and out of a single physical memory space. These issues are typically resolved by the operating system through memory "swapping," where partitions of memory are swapped in and out of memory at run-time. The most common partitions of memory used in swapping are ***segments*** (fragmentation of processes from within) and ***pages*** (fragmentation of logical memory as a whole). Segmentation and paging not only simplify the swapping—memory allocation and de-allocation—of tasks in memory, but allow for code reuse and memory protection, as well as providing the foundation for virtual memory. Virtual memory is a mechanism managed by the OS to allow a device's limited memory space to be shared by multiple competing "user" tasks, in essence enlarging the device's actual physical memory space into a larger "virtual" memory space.

3.4.1.1 *Segmentation*

As mentioned in an earlier section of this chapter, a process encapsulates all the information that is involved in executing a program, including source code, stack, data, and so on. All of the different types of information within a process are divided into "logical" memory units of variable sizes, called *segments*. A segment is a set of logical addresses containing the same type of information. Segment addresses are logical addresses that start at 0, and are made up of a *segment number*, which indicates the base address of the segment, and a *segment offset*, which defines the actual physical memory address. Segments are independently protected, meaning they have assigned accessibility characteristics, such as shared (where other processes can access that segment), Read-Only, or Read/Write.

Most OSes typically allow processes to have all or some combination of five types of information within segments: text (or code) segment, data segment, bss (block started by symbol) segment, stack segment, and the heap segment. A *text* segment is a memory space containing the source code. A *data* segment is a memory space containing the source code's initialized variables (data). A *bss* segment is a statically allocated memory space containing the source code's uninitialized variable (data). The data, text, and bss segments are all fixed in size at compile time, and are as such *static* segments; it is these three segments that typically are part of the *executable file*. Executable files can differ in what segments they are composed of, but in general they contain a header, and different sections that represent the types of segments, including name, permissions, and so on, where a segment can be made up of one or more sections. The OS creates a task's image by *memory mapping* the contents of the executable file, meaning loading and interpreting the segments (sections) reflected in the executable into memory. There are several executable file formats supported by embedded OSes, the most common including:

- *ELF* (Executable and Linking Format): UNIX-based, includes some combination of an ELF header, the program header table, the section header table, the ELF sections, and the ELF segments. Linux (Timesys) and vxWorks (WRS) are examples of OSes that support ELF. (See Figure 3.33.)

- *Class* (Java Byte Code): A class file describes one java class in detail in the form of a stream of 8-bit bytes (hence the name "byte code"). Instead of segments, elements of the class file are called items. The Java class file format contains the class description, as well as how that class is connected to other classes. The main

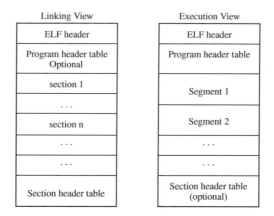

Figure 3.33: ELF Executable File Format

components of a class file are a symbol table (with constants), declaration of fields, method implementations (code) and symbolic references (where other classes references are located). The Jbed RTOS is an example that supports the Java Byte Code format. (See Figure 3.34.)

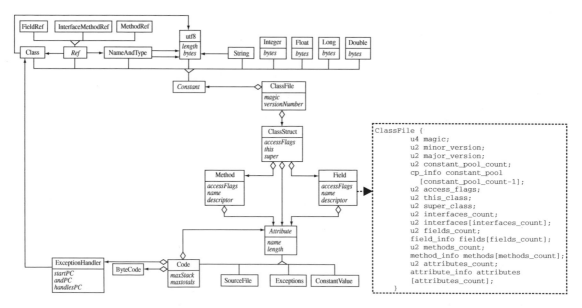

Figure 3.34: Class Executable File Format

- *COFF* (Common Object File Format); A class file format which (among other things) defines an image file that contains file headers that include a file signature, COFF Header, an Optional Header, and also object files that contain only the COFF Header. Figure 3.35 shows an example of the information stored in a COFF header. WinCE[MS] is an example of an embedded OS that supports the COFF executable file format.

Offset	Size	Field	Description
0	2	Machine	Number identifying type of target machine.
2	2	Number of Sections	Number of sections; indicates size of the Section Table, which immediately follows the headers.
4	4	Time/Date Stamp	Time and date the file was created.
8	4	Pointer to Symbol	Offset, within the COFF file, of the symbol table.
12	4	Number of Symbols	Number of entries in the symbol table. This data can be used in locating the string table, which immediately follows the symbol table.
16	2	Optional Header	Size of the optional header, which is Size included for executable files but not object files. An object file should have a value of 0 here.
18	2	Characteristics	Flags indicating attributes of the file.

Figure 3.35: Class Executable File Format

The *stack* and *heap* segments, on the other hand, are not fixed at compile time, and can change in size at runtime and so are dynamic allocation components. A *stack* segment is a section of memory that is structured as a LIFO (last in, first out) queue, where data is "Pushed" onto the stack, or "Popped" off of the stack (push and pop are the only two operations associated with a stack). Stacks are typically used as a simple and efficient method within a program for allocating and freeing memory for data that is predictable (i.e., local variables, parameter passing, and so on). In a stack, all used and freed memory space is located consecutively within the memory space. However, since "push" and "pop" are the only two operations associated with a stack, a stack can be limited in its uses.

A *heap* segment is a section of memory that can be allocated in blocks at runtime, and is typically set up as a free linked-list of memory fragments. It is here that a kernel's memory management facilities for allocating memory come into play to support the "malloc" C function (for example) or OS-specific buffer allocation functions. Typical memory allocation schemes include:

- *FF* (first fit) algorithm, where the list is scanned from the beginning for the first "hole" that is large enough.

- *NF* (next fit), where the list is scanned from where the last search ended for the next "hole" that is large enough.

- *BF* (best fit), where the entire list is searched for the hole that best fits the new data.

- *WF* (worst fit), which is placing data in the largest available "hole."

- *QF* (quick fit), where a list is kept of memory sizes and allocation is done from this information.

- *The buddy system*, where blocks are allocated in sizes of powers of 2. When a block is de-allocated, it is then merged with contiguous blocks.

The method by which memory that is no longer needed within a heap is freed depends on the OS. Some OSes provide a garbage collector that automatically reclaims unused memory (garbage collection algorithms include generational, copying, and mark and sweep; see Figures 3.36a–c). Other OSes require that the programmer explicitly free memory through a system call (i.e., in support of the "free" C function). With the latter technique, the programmer has to be aware of the potential problem of memory leaks, where memory is lost because it has been allocated but is no longer in use and has been forgotten, which is less likely to happen with a garbage collector.

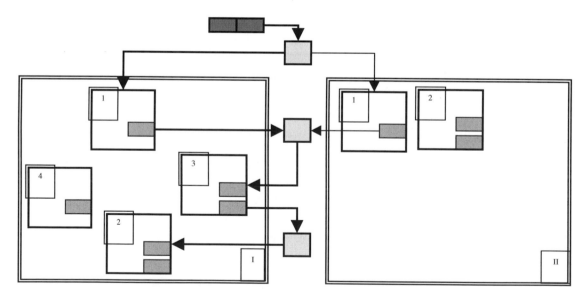

Figure 3.36a: Copying Garbage Collector Diagram

Another problem occurs when allocated and freed memory cause memory fragmentation, where available memory in the heap is spread out in a number of holes, making it more difficult to allocate memory of the required size. In this case, a memory compaction algorithm must be implemented if the allocation/de-allocation algorithms causes a lot of fragmentation. This problem can be demonstrated by examining garbage collection algorithms.

The copying garbage collection algorithm works by copying referenced objects to a different part of memory, and then freeing up the original memory space. This algorithm uses a larger memory area to work, and usually cannot be interrupted during the copy (it blocks the systems). However, it does ensure that what memory is used, is used efficiently by compacting objects in the new memory space.

The mark and sweep garbage collection algorithm works by "marking" all objects that are used, and then "sweeping" (de-allocating) objects that are unmarked. This algorithm is usually nonblocking, so the system can interrupt the garbage collector to execute other functions when necessary. However, it doesn't compact memory the way a Copying garbage collector would, leading to memory fragmentation with small, unusable holes possibly existing where de-allocated objects used to exist. With a mark and sweep garbage collector, an additional memory compacting algorithm could be implemented making it a mark (sweep) and compact algorithm.

Mark and Sweep

With Compact (for Mark and Compact)

Figure 3.36b: Mark and Sweep and Mark and Compact Garbage Collector Diagram

Finally, the generational garbage collection algorithm separates objects into groups, called *generations*, according to when they were allocated in memory. This algorithm assumes that most objects that are allocated are short-lived, thus copying or compacting the remaining objects with longer lifetimes is a waste of time. So, it is objects in the younger generation group that are cleaned up more frequently than objects in the older generation groups. Objects can also be moved from a younger generation to an older generation group. Each generational garbage collector also may employ different algorithms to de-allocate objects within each generational group, such as the copying algorithm or mark and sweep algorithms described above. Compaction algorithms would be needed in both generations to avoid fragmentation problems.

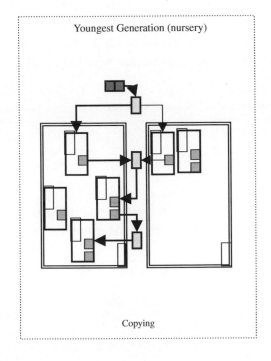

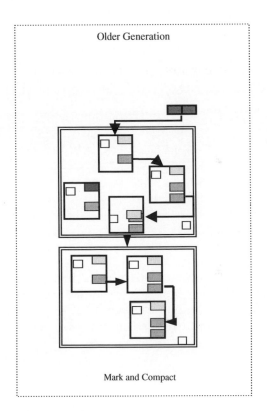

Figure 3.36c: Generational Garbage Collector Diagram

Finally, heaps are typically used by a program when allocation and deletion of variables are unpredictable (linked lists, complex structures, and so on). However, heaps aren't as simple or as efficient as stacks. As mentioned, how memory in a heap is allocated and de-allocated is typically affected by the programming language the OS is based on, such as a C-based OS using "malloc" to allocate memory in a heap and "free" to de-allocate memory or a Java-based OS having a garbage collector. Pseudo code examples 3.13–3.15 demonstrate how heap space can be allocated and de-allocated under various embedded OSes.

Example 3.13: vxWorks Memory Management and Segmentation

VxWorks tasks are made up of text, data, and bss static segments, as well as each task having its own stack.

The vxWorks system called "taskSpawn" is based on the POSIX spawn model, and is what creates, initializes, and activates a new (child) task. After the spawn system call, an image of the child task (including TCB, stack, and program) is allocated into memory. In the pseudo code below, the code itself is the text segment, data segments are any initialized variables, and the bss segments are the uninitialized variables (for example, seconds). In the taskSpawn system call, the task stack size is 3000 bytes, and is not filled with 0xEE because of the VX_NO_ STACK_FILL parameter in the system call.

Task Creation vxWorks Pseudo Code

```
// parent task that enables software timer
void parentTask(void)
{
. . .
if sampleSoftware Clock NOT running {

    /"newSWClkId" is a unique integer value assigned by kernel
    when task is created newSWClkId = taskSpawn
    ("sampleSoftwareClock", 255,  VX_NO_STACK_FILL, 3000,
    (FUNCPTR) minuteClock, 0, 0, 0, 0, 0, 0, 0, 0, 0, 0);
    . . .

}


//child task program Software Clock
void minuteClock (void) {
        integer seconds;

        while (softwareClock is RUNNING) {
            seconds = 0;
            while (seconds < 60) {
                    seconds = seconds+1;
        }
    . . .
}
```

Heap space for vxWorks tasks is allocated by using the C-language malloc/new system calls to dynamically allocate memory. There is no garbage collector in vxWorks, so the programmer must de-allocate memory manually via the free() system call.

```
/* The following code is an example of a driver that performs
 * address translations. It attempts to allocate a cache-safe buffer,
 * fill it, and then write it out to the device. It uses
 * CACHE_DMA_FLUSH to make sure the data is current. The driver then
 * reads in new data and uses CACHE_DMA_INVALIDATE to guarantee
   cache coherency. */
#include "vxWorks.h"
#include "cacheLib.h"
#include "myExample.h"
STATUS myDmaExample (void)
{
void * pMyBuf;
void * pPhysAddr;
/* allocate cache safe buffers if possible */
if ((pMyBuf = cacheDmaMalloc (MY_BUF_SIZE)) == NULL)
return (ERROR);
... fill buffer with useful information ...
/* flush cache entry before data is written to device */
CACHE_DMA_FLUSH (pMyBuf, MY_BUF_SIZE);
/* convert virtual address to physical */
pPhysAddr = CACHE_DMA_VIRT_TO_PHYS (pMyBuf);
/* program device to read data from RAM */
myBufToDev (pPhysAddr);
... wait for DMA to complete ...
... ready to read new data ...
/* program device to write data to RAM */
myDevToBuf (pPhysAddr);
... wait for transfer to complete ...
/* convert physical to virtual address */
pMyBuf = CACHE_DMA_PHYS_TO_VIRT (pPhysAddr);
/* invalidate buffer */
CACHE_DMA_INVALIDATE (pMyBuf, MY_BUF_SIZE);
... use data ...
/* when done free memory */
if (cacheDmaFree (pMyBuf) == ERROR)
return (ERROR);
return (OK);
}
```

Example 3.14: Jbed Memory Management and Segmentation

In Java, memory is allocated in the Java heap via the "new" keyword (unlike the "malloc" in C, for example). However, there are a set of interfaces defined in some Java standards, called JNI or Java Native Interface, which allows for C and/or assembly code to be integrated within Java code, so in essence, the "malloc" is available if JNI is supported. For memory de-allocation, as specified by the Java standard, is done via a garbage collector.

Jbed is a Java-based OS, and as such supports "new" for heap allocation.

```
public void CreateOneshotTask(){
  // Task execution time values
  final long DURATION = 100L; // run method takes < 100us
  final long ALLOWANCE = 0L; // no DurationOverflow handling
  final long DEADLINE = 1000L;// complete within 1000us
  Runnable target; // Task's executable code
  OneshotTimer taskType;
  Task task;

  // Create a Runnable object
  target = new MyTask();         ◄------------------------┐
                                                          ┊
  // Create oneshot tasktype with no delay                ┊
  taskType = new OneshotTimer( 0L );     ┊------- Memory
                                         ┊         allocation in Java
  // Create the task                     ┊
  try {                                  ┊
  task = new Task( target,    ◄----------------------------┘
  DURATION, ALLOWANCE, DEADLINE,
  taskType );
  }catch( AdmissionFailure e ){
  System.out.println( "Task creation failed" );
  return;
  }
```

Memory de-allocation is handled automatically in the heap via a Jbed garbage collector based on the mark and sweep algorithm (which is nonblocking and is what allows Jbed

to be an RTOS). The GC can be run as a reoccurring task, or can be run by calling a
"runGarbageCollector" method.

Example 3.15: Linux Memory Management and Segmentation

Linux processes are made up of text, data, and bss static segments, as well as each process
has its own stack (which is created with the fork system call). Heap space for Linux tasks
are allocated via the C-language malloc/new system calls to dynamically allocate memory.
There is no garbage collector in Linux, so the programmer must de-allocate memory
manually via the free() system call.

```c
void *mem_allocator (void *arg)
{
   int i;
   int thread_id = *(int *)arg;
   int start = POOL_SIZE * thread_id;
   int end = POOL_SIZE * (thread_id + 1);

   if(verbose_flag) {
     printf("Releaser %i works on memory pool %i to %i\n",
        thread_id, start, end);
     printf("Releaser %i started ...\n", thread_id);
   }

   while(!done_flag) {
       /* find first NULL slot */
       for (i = start; i < end; ++i){
         if (NULL == mem_pool[i]) {
               mem_pool[i] = malloc(1024);
               if (debug_flag)
                 printf("Allocate %i: slot %i\n",
                     thread_id, i);
               break;
         }
       }
   }
   pthread_exit(0);
}
```

```
void *mem_releaser(void *arg)
{
   int i;
   int loops = 0;
   int check_interval = 100;
   int thread_id = *(int *)arg;
   int start = POOL_SIZE * thread_id;
   int end = POOL_SIZE * (thread_id + 1);

   if(verbose_flag) {
     printf("Allocator %i works on memory pool %i to %i\n",
         thread_id, start, end);
     printf("Allocator %i started ...\n", thread_id);
}

    while(!done_flag) {

        /* find non-NULL slot */
        for (i = start; i < end; ++i) {
          if (NULL != mem_pool[i]) {
                  void *ptr = mem_pool[i];
                  mem_pool[i] = NULL;
                  free(ptr);
                  ++counters[thread_id];
                  if (debug_flag)
                    printf("Releaser %i: slot %i\n",
                         thread_id, i);
                  break;
          }
        }
        ++loops;
        if((0 == loops % check_interval)&&
            (elapsed_time(&begin)>run_time)){
                  done_flag = 1;
                  break;
        }
   }
   pthread_exit(0);
}
```

3.4.1.2   *Paging and Virtual Memory*

Either with or without segmentation, some OSes divide logical memory into some number of fixed-size partitions, called ***blocks, frames, pages*** or ***some combination of a few or all of these***. For example, with OSes that divide memory into frames, the logical address is comprised of a frame number and offset. The user memory space can then, also, be divided into pages, where page sizes are typically equal to *frame* sizes.

When a process is loaded in its entirety into memory (in the form of pages), its pages may not be located within a contiguous set of frames. Every process has an associated process table that tracks its pages, and each page's corresponding frames in memory. The logical address spaces generated are unique for each process, even though multiple processes share the same physical memory space. Logical address spaces are typically made up of a page-frame number, which indicates the start of that page, and an offset of an actual memory location within that page. In essence, the logical address is the sum of the page number and the offset.

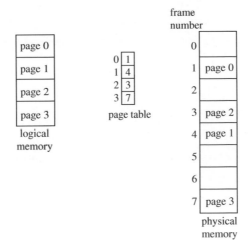

Figure 3.37: Paging

An OS may start by *prepaging*, or loading the pages needed to get started, and then implementing the scheme of ***demand paging*** where processes have no pages in memory, and pages are only loaded into RAM when a ***page fault*** (an error occurring when attempting to access a page not in RAM) occurs. When a page fault occurs, the OS takes over and loads the needed page into memory, updates page tables, and then the instruction that triggered the page fault in the first place is re-executed. This scheme is based on Knuth's Locality of Reference theory, which estimates that 90% of a system's time is spent on processing just 10% of code.

Dividing up logical memory into pages aids the OS in more easily managing tasks being relocated in and out physical of various types of memory in the memory hierarchy, memory a process called *swapping*. Common page selection and replacement schemes to determine which pages are swapped include:

- *Optimal*, using future reference time, swapping out pages that won't be used in the near future.

- *Least Recently Used* (LRU), which swaps out pages that have been used the least recently.

- *FIFO* (First-In-First-Out), which as its name implies, swaps out the pages that are the oldest (regardless of how often it is accessed) in the system. While a simpler algorithm then LRU, FIFO is much less efficient.

- *Not Recently Used* (NRU), swaps out pages that were not used within a certain time period.

- *Second Chance*, FIFO scheme with a reference bit, if "0" will be swapped out (a reference bit is set to "1" when access occurs, and reset to "0" after the check).

- *Clock Paging*, pages replaced according to clock (how long they have been in memory), in clock order, if they haven't been accessed (a reference bit is set to "1" when access occurs, and reset to "0" after the check).

While every OS has its own swap algorithm, all are trying to reduce the possibility of *thrashing*, a situation in which a system's resources are drained by the OS constantly swapping in and out data from memory. To avoid thrashing, a kernel may implement a *working set* model, which keeps a fixed number of pages of a process in memory at all times. Which pages (and the number of pages) that comprise this working set depends on the OS, but typically it is the pages accessed most recently. A kernel that wants to prepage a process also needs to have a working set defined for that process before the process's pages are swapped into memory.

3.4.1.3 Virtual Memory

Virtual memory is typically implemented via demand segmentation (fragmentation of processes from within, as discussed in a previous section) and/or demand paging (fragmentation of logical user memory as a whole) memory fragmentation techniques. When virtual memory is implemented via these "demand" techniques, it means that only the pages and/or segments that are currently in use are loaded into RAM.

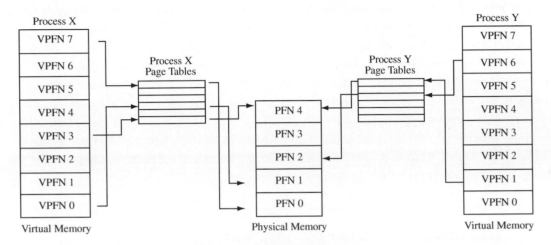

Figure 3.38: Virtual Memory

As shown in Figure 3.38, in a virtual memory system, the OS generates *virtual* addresses based on the logical addresses, and maintains *tables* for the sets of logical addresses into virtual addresses conversions (on some processors table entries are cached into TLBs). The OS (along with the hardware) then can end up managing more than one different address space for each process (the physical, logical, and virtual). In short, the software being managed by the OS views memory as one continuous memory space, whereas the kernel actually manages memory as several fragmented pieces that can be segmented and paged, segmented and unpaged, unsegmented and paged, or unsegmented and unpaged.

3.4.2 Kernel Memory Space

The kernel's memory space is the portion of memory in which the kernel code is located, some of which is accessed via system calls by higher-level software processes, and is where the CPU executes this code from. Code located in the kernel memory space includes required IPC mechanisms, such as those for message passing queues. Another example is when tasks are creating some type of fork/exec or spawn system calls. After the task creation system call, the OS gains control and creates the *Task Control Block* (TCB), also referred to as a *Process Control Block* (PCB) in some OSes, within the kernel's memory space that contains

OS control information and CPU context information for that particular task. Ultimately, what is managed in the kernel memory space, as opposed to in the user space, is determined by the hardware, as well as the actual algorithms implemented within the OS kernel.

As previously mentioned, software running in user mode can only access anything running in kernel mode via *system calls*. System calls are the higher-level (user mode) interfaces to the kernel's subroutines (running in kernel mode). Parameters associated with system calls that need to be passed between the OS and the system callee running in user mode are then passed via registers, a stack, or in the main memory heap. The types of system calls typically fall under the types of functions being supported by the OS, so they include file systems management (i.e., opening/modifying files), process management (i.e., starting/ stopping processes), I/O communications, and so on. In short, where an OS running in kernel mode views what is running in user mode as processes, software running in user mode views and defines an OS by its system calls.

3.5   I/O and File System Management

Some embedded OSes provide memory management support for a temporary or permanent file system storage scheme on various memory devices, such as Flash, RAM, or hard disk. File systems are essentially a collection of files along with their management protocols (see Table 3.2). File system algorithms are middleware and/or application software that is *mounted* (installed) at some mount point (location) in the storage device.

In relation to file systems, a kernel typically provides file system management mechanisms for, at the very least:

- *Mapping* files onto secondary storage, Flash, or RAM (for instance).

- Supporting the primitives for manipulating files and directories.

    — *File Definitions and Attributes*: Naming Protocol, Types (i.e., executable, object, source, multimedia, and so on), Sizes, Access Protection (Read, Write, Execute, Append, Delete, and so on), Ownership, and so on.

    — *File Operations*: Create, Delete, Read, Write, Open, Close, and so on.

    — *File Access Methods*: Sequential, Direct, and so on.

    — Directory Access, Creation and Deletion.

Table 3.2: Middleware File System Standards

File System	Summary
FAT32 (File Allocation Table)	Where memory is divided into the smallest unit possible (called sectors). A group of sectors is called a cluster. An OS assigns a unique number to each cluster, and tracks which files use which clusters. FAT32 supports 32-bit addressing of clusters, as well as smaller cluster sizes than that of the FAT predecessors (FAT, FAT16, and so on).
NFS (Network File System)	Based on RPC (Remote Procedure Call) and XDR (Extended Data Representation), NFS was developed to allow external devices to mount a partition on a system as if it were in local memory. This allows for fast, seamless sharing of files across a network.
FFS (Flash File System)	Designed for Flash memory.
DosFS	Designed for real-time use of block devices (disks) and compatible with the MS-DOS file system.
RawFS	Provides a simple *raw file system* that essentially treats an entire disk as a single large file.
TapeFS	Designed for tape devices that do not use a standard file or directory structure on tape. Essentially treats the tape volume as a raw device in which the entire volume is a large file.
CdromFS	Allows applications to read data from CD-ROMs formatted according to the ISO 9660 standard file system.

OSes vary in terms of the primitives used for manipulating files (i.e., naming, data structures, file types, attributes, operations, and so on), what memory devices files can be mapped to, and what file systems are supported. Most OSes use their standard I/O interface between the file system and the memory device drivers. This allows for one or more file systems to operate in conjunction with the operating system.

I/O Management in embedded OSes provides an additional abstraction layer (to higher level software) away from the system's hardware and device drivers. An OS provides a uniform interface for I/O devices that perform a wide variety of functions via the available kernel system calls, providing protection to I/O devices since user processes can only access I/O via these system calls, and managing a fair and efficient I/O sharing scheme among the multiple processes. An OS also needs to manage synchronous and asynchronous communication coming from I/O to its processes, in essence be event-driven by responding to requests from both sides (the higher level processes and low-level hardware), and manage the data transfers. In order to accomplish these goals, an OS's I/O management scheme is typically made up of a generic device-driver interface both to user processes and device drivers, as well as some type of buffer-caching mechanism.

Device driver code controls a board's I/O hardware. In order to manage I/O, an OS may require all device driver code to contain a specific set of functions, such as startup, shutdown, enable, disable, and so on. A kernel then manages I/O devices, and in some OSes file systems as well, as "black boxes" that are accessed by some set of generic APIs by higher-layer processes. OSes can vary widely in terms of what types of I/O APIs they provide to upper layers. For example, under Jbed, or any Java-based scheme, all resources (including I/O) are viewed and structured as objects. VxWorks, on the other hand, provides a communications mechanism, called *pipes*, for use with the vxWorks I/O subsystem. Under vxWorks, pipes are virtual I/O devices that include underlying message queue associated with that pipe. Via the pipe, I/O access is handled as either a stream of bytes (*block* access) or one byte at any given time (*character* access).

In some cases, I/O hardware may require the existence of OS buffers to manage data transmissions. Buffers can be necessary for I/O device management for a number of reasons. Mainly they are needed for the OS to be able to capture data transmitted via block access. The OS stores within buffers the stream of bytes being transmitted to and from an I/O device independent of whether one of its processes has initiated communication to the device. When performance is an issue, buffers are commonly stored in cache (when available), rather than in slower main memory.

3.6   OS Standards Example: POSIX (Portable Operating System Interface)

Standards may greatly impact the design of a system component—and operating systems are no different. One of the key standards implemented in off-the-shelf embedded OSes today is portable operating system interface (POSIX). POSIX is based upon the IEEE (*1003.1-2001*) and The Open Group (*The Open Group Base Specifications Issue 6*) set of standards that define a standard operating system interface and environment. POSIX provides OS-related standard APIs and definitions for process management, memory management, and I/O management functionality (see Table 3.3).

Table 3.3: POSIX Functionality

OS Subsystem	Function	Definition
Process Management	Threads	Functionality to support multiple flows of control within a process. These flows of control are called threads and they share their address space and most of the resources and attributes defined in the operating system for the owner process. The specific functional areas included in threads support are: • Thread management: the creation, control, and termination of multiple flows of control that share a common address space. • Synchronization primitives optimized for tightly coupled operation of multiple control flows in a common, shared address space.
	Semaphores	A minimum synchronization primitive to serve as a basis for more complex synchronization mechanisms to be defined by the application program.
	Priority scheduling	A performance and determinism improvement facility to allow applications to determine the order in which threads that are ready to run are granted access to processor resources.
	Real-time signal extension	A determinism improvement facility to enable asynchronous signal notifications to an application to be queued without impacting compatibility with the existing signal functions.
	Timers	A mechanism that can notify a thread when the time as measured by a particular clock has reached or passed a specified value, or when a specified amount of time has passed.
	IPC	A functionality enhancement to add a high-performance, deterministic interprocess communication facility for local communication.
Memory Management	Process memory locking	A performance improvement facility to bind application programs into the high-performance random access memory of a computer system. This avoids potential latencies introduced by the operating system in storing parts of a program that were not recently referenced on secondary memory devices.
	Memory mapped files	A facility to allow applications to access files as part of the address space.
	Shared memory objects	An object that represents memory that can be mapped concurrently into the address space of more than one process.

Table 3.3: POSIX Functionality (continued)

OS Subsystem	Function	Definition
I/O Management	Synchronionized I/O	A determinism and robustness improvement mechanism to enhance the data input and output mechanisms, so that an application can ensure that the data being manipulated is physically present on secondary mass storage devices.
	Asynchronous I/O	A functionality enhancement to allow an application process to queue data input and output commands with asynchronous notification of completion.
. . .	. . .	. . .

How POSIX is translated into software is shown in Examples 3.16 and 3.17, examples in Linux and vxWorks of POSIX threads being created (note the identical interface to the POSIX thread create subroutine).

Example 3.16: Linux POSIX Example

Creating a Linux POSIX thread:

```
if(pthread_create(&threadId, NULL, DEC threadwork, NULL)) {
printf("error");
   . . .
   }
```

Here, threadId is a parameter for receiving the thread ID. The second argument is a thread attribute argument that supports a number of scheduling options (in this case NULL indicates the default settings will be used). The third argument is the subroutine to be executed upon creation of the thread. The fourth argument is a pointer passed to the subroutine (i.e., pointing to memory reserved for the thread, anything required by the newly created thread to do its work, and so on).

Example 3.17: vxWorks POSIX Example

Creating a POSIX thread in vxWorks:

```
...
pthread_t tid;
int ret;

/* create the pthread with NULL attributes to designate
* default values */
ret = pthread_create(&threadId, NULL, entryFunction, entryArg);

...
```

Here, threadId is a parameter for receiving the thread ID. The second argument is a thread attribute argument that supports a number of scheduling options (in this case NULL indicates the default settings will be used). The third argument is the subroutine to be executed upon creation of the thread. The fourth argument is a pointer passed to the subroutine (i.e., pointing to memory reserved for the thread, anything required by the newly created thread to do its work, and so on).

Essentially, the POSIX APIs allow for software that is written on one POSIX-compliant OS to be easily ported to another POSIX OS, since by definition the APIs for the various OS system calls must be identical and POSIX compliant. It is up to the individual OS vendors to determine how the internals of these functions are actually performed. This means that, given two different POSIX compliant OSes, both probably employ very different internal code for the same routines.

3.7   OS Performance Guidelines

The two subsystems of an OS that typically impact OS performance the most, and differentiate the performance of one OS from another, are the memory management scheme (specifically the process swapping model implemented) and the scheduler. The performance of one virtual memory-swapping algorithm over another can be compared by the number of page faults they produce, given the same set of memory references—that is, the same number of page frames assigned per process for the exact same process on both OSes. One algorithm can be further tested for performance by providing it with a variety of different memory references and noting the number of page faults for various number of page frames per process configurations.

While the goal of a scheduling algorithm is to select processes to execute in a scheme that maximizes overall performance, the challenge OS schedulers face is that there are a number of performance indicators. Furthermore, algorithms can have opposite effects on an indicator, even given the exact same processes. The main performance indicators for scheduling algorithms include:

- *Throughput*, which is the number of processes being executed by the CPU at any given time. At the OS scheduling level, an algorithm that allows for a significant number of larger processes to be executed before smaller processes runs the risk of having a lower throughput. In a SPN (shortest process next) scheme, the throughput may even vary on the same system depending on the size of processes being executed at the moment.

- *Execution time*, the average time it takes for a running process to execute (from start to finish). Here, the size of the process affects this indicator. However, at the scheduling level, an algorithm that allows for a process to be continually preempted allows for significantly longer execution times. In this case, given the same process, a comparison of a nonpre-emptable versus pre-emptable scheduler could result in two very different execution times.

- *Wait time*, the total amount of time a process must wait to run. Again this depends on whether the scheduling algorithm allows for larger processes to be executed before slower processes. Given a significant number of larger processes executed (for whatever reason), any subsequent processes would have higher wait times. This indicator is also dependent on what criteria determines which process is selected to run in the first place—a process in one scheme may have a lower or higher wait time than if it is placed in a different scheduling scheme.

On a final note, while scheduling and memory management are the leading components impacting performance, to get a more accurate analysis of OS performance one must measure the impact of both types of algorithms in an OS, as well as factor in an OS's *response time* (essentially the time from when a user process makes the system call to when the OS starts processing the request). While no one factor alone determines how well an OS performs, OS performance in general can be *implicitly* estimated by how hardware resources in the system (the CPU, memory, and I/O devices) are utilized for the variety of processes. Given the right processes, the more time a resource spends executing code as opposed to sitting idle *can be* indicative of a more efficient OS.

3.8   OSes and Board Support Packages (BSPs)

The *board support package* (BSP) is an optional component provided by the OS provider, the main purpose of which is simply to provide an abstraction layer between the operating system and generic device drivers.

A BSP allows for an OS to be more easily ported to a new hardware environment, because it acts as an integration point in the system of hardware dependent and hardware independent source code. A BSP provides subroutines to upper layers of software that can customize the hardware, and provide flexibility at compile time. Because these routines point to separately compiled device driver code from the rest of the system application software, BSPs provide run-time portability of generic device driver code. As shown in Figure 3.39, a BSP provides architecture-specific device driver configuration management, and an API for the OS (or higher layers of software) to access generic device drivers. A BSP is also responsible for managing the initialization of the device driver (hardware) and OS in the system.

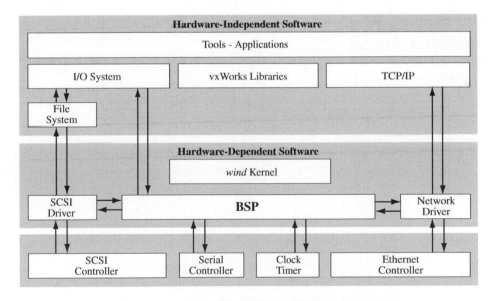

Figure 3.39: BSP within Embedded Systems Model

The device configuration management portion of a BSP involves architecture-specific device driver features, such as constraints of a processor's available addressing modes, endianess, and interrupts (connecting ISRs to interrupt vector table, disabling/enabling, control registers) and so on, and is designed to provide the most flexibility in porting generic device drivers to a new architecture-based board, with its differing endianess, interrupt scheme, and other architecture-specific features.

3.8.1   *Advantages and Disadvantages of Real-Time Kernels*

A real-time kernel, also called a *Real-Time Operating System* (RTOS), allows real-time applications to be designed and expanded easily; functions can be added without requiring major changes to the software. In fact, if you add low priority tasks to your system, the responsiveness of your system to high priority tasks is almost not affected! The use of an RTOS simplifies the design process by splitting the application code into separate tasks. With a preemptive RTOS, all time-critical events are handled as quickly and as efficiently as possible. An RTOS allows you to make better use of your resources by providing you with valuable services, such as semaphores, mailboxes, queues, time delays, and time-outs.

You should consider using a real-time kernel if your application can afford the extra requirements: extra cost of the kernel, more ROM/RAM, and 2–4% additional CPU overhead.

The one factor I haven't mentioned so far is the cost associated with the use of a real-time kernel. In some applications, cost is everything and would preclude you from even considering an RTOS.

Currently about 150+ RTOS vendors exist. Products are available for 8-, 16-, 32-, and even 64-bit microprocessors. Some of these packages are complete operating systems and include not only the real-time kernel but also an input/output manager, windowing systems (display), a file system, networking, language interface libraries, debuggers, and cross-platform compilers. The development cost to use an RTOS varies from $70 (U.S. Dollars) to well over $30,000. The RTOS vendor might also require *royalties* on a per-target-system basis. Royalties are like buying a chip from the RTOS vendor that you include with each unit sold. The RTOS vendors call this *silicon software*. The royalty fee varies between $5 to more than $500 per unit. μC/OS-II is not free software and needs to be licensed for commercial use. Like any other software package these days, you also need to consider the maintenance cost, which can set you back another 15% of the development cost of the RTOS per year!

3.9   Summary

This chapter introduced the different types of embedded OSes, as well as the major components that make up most embedded OSes. This included discussions of process management, memory management, and I/O system management. This chapter also discussed the POSIX standard and its impact on the embedded OS market in terms of what function requirements are specified. The impact of OSes on system performance was discussed, as well as OSes that supply a board-independent software abstraction layer, called a board support package (BSP).

CHAPTER 4

Networking

Kamal Hyder
Bob Perrin

Just as silicon has advanced, so have software development techniques. The old days of writing code on punch cards, toggling in binary bootstrap loaders or keying in hexadecimal opcodes are long gone. The tried, true, and tiresome technique of "burn and learn" is still with us, but in a greatly reduced capacity. Most applications are developed using assemblers, compilers, linkers, loaders, simulators, emulators, EPROM programmers, and debuggers.

Selecting software development tools suited to a particular project is important and complex. Bad tool choices can greatly extend development times. Tools can cost thousands of dollars per developer, but the payoff can be justifiable because of increased productivity. On the other hand, initial tool choice can adversely affect the product's maintainability years down the road.

For example, deciding to use JAVA to develop code for a PIC® microcontroller in a coffeemaker is a poor choice. While there are tools available to do this, and programmers willing to do this, code maintenance is likely to be an issue. Once the JAVA-wizard programmer moves to developing code for web sites, it may be difficult to find another JAVA-enabled programmer willing to sustain embedded code for a coffeemaker. Equally silly would be to use an assembler to write a full-up GUI (graphical user interface)-based MMI.

A quick trip to the Embedded Systems Conference will reveal a wide array of development tools. Many of these are ill suited for embedded development, if not for reasons of scale or cost, then for reasons of code maintainability or tool stability.

The two time-tested industry-approved solutions for embedded development are assembly and C. Forth, BASIC, JAVA, PLM, Pascal, UML, XML and a plethora of other obscure languages have been used to produce functioning systems. However, for low-level fast code, such as Interrupt Service Routines (ISRs), assembly is the only real option. For high-level coding, C is the best choice due to the availability of software engineers that know the language and the wide variety of available libraries.

Selecting a tool vendor is almost as important as selecting a language. Selecting a tool vendor without a proven track record is a risk. If the tool proves problematic, good tech-support will be required.

Public domain tools have uncertain histories and no guarantee of support. The idea behind open source tools is that if support is needed, the user can tweak the tool's code-base to force the tool to behave as desired. For some engineers, this is a fine state of affairs. On the other hand, many embedded software engineers may not know, or even desire to know, how to tweak, for example, a backend code generator on a compiler.

Rabbit Semiconductor and Z-World offer a unique solution to the tool dilemma facing embedded systems designers. Rabbit Semiconductor designs ICs and core modules. Z-World designs board-level and packaged controllers based on Rabbit chips. Both companies share the development and maintenance of Dynamic C™.

Dynamic C offers the developer an integrated development environment (IDE) where C and assembly can be written and blended. Once an application is coded, Dynamic C will download the executable image to the target system over a serial cable. Debugging tools such as single stepping, break points, and watch-windows are provided within the IDE, without the need for an expensive In-Circuit Emulator (ICE).

Between Z-World and Rabbit Semiconductor, all four classes of controllers are available as well as a complete set of highly integrated development tools. Libraries support a file system, Compact Flash interfaces, TCP/IP, IrDA, SDLC/HDLC, SPI, I2C, AES, FFTs, and the uCOS/II RTOS.

One of the most attractive features of Dynamic C is that the TCP/IP stack is royalty free. This is unusual in the embedded industry, where companies are charging thousands of dollars for TCP/IP support. If TCP/IP is required for an application, the absence of royalties makes Dynamic C a very attractive tool.

For these reasons, we have chosen the Rabbit core module and Dynamic C for our networking development example.

Before considering embedded networks, we will start this chapter with a brief description of the RCM3200 Rabbit core and then get into the Rabbit development environment. We will cover development and debugging aspects of Dynamic C and we will highlight some of the differences between Dynamic C and ANSI C. Then we will move on to our networking examples, which are based on the Rabbit core and make use of the Dynamic C development system.

4.1   Introduction to the RCM3200 Rabbit Core

A processor does not mean a lot by itself. The designer has to select the right support components, such as memory, external peripherals, interface components, and so on. The designer has to interface these components to the CPU, and design the timing and the glue logic to make them all work together. There are design risks involved in undertaking such a task, not to mention the time in designing, prototyping, and testing such a system.

Using a core module solves most of these issues. Buying a low-cost module that integrates all these peripherals means someone has already taken the design through the prototyping, debugging, and assembly phases. In addition, core manufacturers generally take EMI issues into account. This allows the embedded system builder to focus on interface issues and application code.

There are several advantages to using cores. The greatest advantage is reduced time-to-market. Instead of putting together the fundamental building blocks such as CPU, RAM, and ROM, the designer can quickly start coding and focus instead on the application they are trying to develop.

To illustrate how to use a core module, we will set up an RCM3200 core module and step through the code development process.

The RCM3200 core offers the following features:

- The Rabbit 3000 CPU running at 44.2 MHz

- 512 K of Flash memory for code

- 512 K of fast SRAM for program execution

- 256 K of battery backed SRAM for data storage

- Built in real-time clock

- 10/100Base-T Ethernet

- Six serial ports

- 52 bits of digital I/O

- Operation from 3.15 V to 3.45 V

During development, cores mount on prototyping boards supplied by Rabbit Semiconductor. An RCM3200 prototyping board contains connectors for power and I/O, level shifters for serial I/O, a reset switch, and a pair of switches and LEDs connected to I/O pins. A useful feature of the prototyping board is the prototyping area that has both through-holes and SMT pads. This is where designers can populate their own devices and interface them with the core.

The Rabbit Semiconductor prototyping boards are designed to allow a system developer to build preliminary designs and write code on the prototyping board. This allows initial system development to occur even if the application's target hardware is not available.

Once final hardware is complete, the core module can be moved from the prototyping board to the target hardware and the system software can then be finalized and tested.

4.2   *Introduction to the Dynamic C Development Environment*

The Dynamic C development system includes an editor, compiler, downloader, and in-circuit debugger. The development tools allow users to write and compile their code on a Windows platform, and download the executable code to the core. Dynamic C is a powerful platform for development and debugging.

4.2.1   *Development*

- Dynamic C includes an integrated development environment (IDE). Users do not need to buy or use separate editors, compilers, assemblers or linkers.

- Dynamic C has an extensive library of drivers. For most applications, designers do not need to write low-level peripheral interface code. They simply need to make the right API calls. Designers can focus on developing the higher-level application rather than spend their time writing low-level drivers.

- Dynamic C uses a serial port to download code into the target core. There is no need to use an expensive CPU or ROM emulator. Users of most cores load and run code from flash.

- Dynamic C is not ANSI C. We will highlight some of the differences as we move along.

4.2.2   Debugging

Dynamic C has a host of debugging features. In a traditional development environment, a CPU emulator performs these functions. However, Dynamic C performs these functions, saving the developer hundreds or thousands of dollars in emulator costs. Dynamic C's debugging features include:

- Breakpoints—Set breakpoints that can stop program flow where required, so that the programmer can examine and change the state of variables and registers or figure out how the program got to a certain part of the code

- Single stepping—Step into or over functions at a source or machine code level. Single stepping will let the programmer examine program flow, or values of CPU registers, program variables, or memory locations.

- Code disassembly—The disassembly window displays addresses, opcodes, mnemonics, and machine cycle times. This can help the programmer examine how C code got converted into assembly language, as well as calculate how many machine cycles it may take to execute a section of code.

- Switch between debugging at machine code level and source code level by simply opening or closing the disassembly window.

- Watch expressions—This window displays values of selected variables or even complex expressions, including function calls. The programmer can therefore examine or evaluate values of selected variables during program execution. Watch expressions can be updated with or without stopping program execution and can be used to trigger the operation of hardware devices in the target. Use the mouse to "hover over" a variable name to examine its value.

- Register window—All processor registers and flags are displayed. The contents of registers may be modified as needed.

- Stack window—Shows the contents of the top of the stack.

- Hex memory dump—Displays the contents of memory at any address.

- STDIO window—**printf** outputs to this window, and keyboard input on the host PC can be detected for debugging purposes.

4.3   Brief Introduction to Dynamic C Libraries

Dynamic C provides extensive libraries of drivers. Low-level drivers have already been written and provided for common devices. For instance, Dynamic C drivers for I2C, SPI, various LCD displays, keypads, file systems on flash memory devices, and even GPS interfaces are already provided. A complete TCP stack is also included for cores that support networking.

There are some differences between Dynamic C and ANSI C. This will be especially important to programmers porting code to a Rabbit environment. As we cover various aspects of code development, we will highlight differences between Dynamic C and ANSI C.

Source code for Dynamic C libraries is supplied with the Dynamic C distribution. Although the Dynamic C library files end with a ".LIB" extension, these are actually source files that can be opened with a text editor.

For example, let us examine the LCD library. If Dynamic C is installed into its default directories, we find an LCD library file at DynamicC\Lib\Displays\LCD122KEY7.LIB: The library file defines various variables and functions. Because it is an LCD library, we find functions that initialize a display and allow the programmer to write to an LCD.

Looking at the function descriptions, the programmer can quickly understand how Rabbit's engineers implemented each function. The embedded systems designer can tailor the library functions to suit particular applications and save them in separate libraries.

Quick Summary

- Dynamic C is not ANSI C

- Dynamic C library files end with a ".LIB" extension, and are source files that can be opened with a text editor

---

4.4 Memory Spaces in Dynamic C

Here we will see how Dynamic C manipulates the MMU to provide an optimal memory usage for the application.

The Rabbit has an external 8-bit data bus. This allows the processor to interface to inexpensive 8-bit memory devices. The trade-off with a small data bus is the multiple bus accesses required to read large amounts of data. To minimize the time required to fetch operands containing addresses while still providing a useful amount of address space, the Rabbit uses a 16-bit address for all instruction operands.

A 16-bit address requires two read cycles over the data bus to acquire an address as an operand. This implies an address space limited to 216 (65,536) bytes. A 16-bit address space, while usable, is somewhat limiting.

To achieve a usable memory space larger than 216 bytes the Rabbit's designers gave the microprocessor a memory management unit (MMU). This device maps a 16-bit logical address to a 20-bit physical address.

The Rabbit designers could have simply made the Rabbit's instructions accept 20-bit address operands. This would require 3 bytes to contain the operands and would therefore require three fetches over the 8-bit data bus to pull in the complete 20-bit address. This is a 50% penalty over the 2 fetches required to gather a 16-bit address.

Many programs fit quite handily in a 16-bit address space. The performance penalty incurred by making all the instructions operate on a 20-bit address is not desirable. The MMU offers a compromise between a large address space and an efficient bus utilization. Good speed and code density are achieved by minimizing the instruction length. The MMU makes available a large address space to applications requiring more than a 16-bit address space.

The *Rabbit 3000™ Designer's Handbook* covers the MMU in exacting detail. However, most engineers using the Rabbit only need understand the rudimentary details of how Dynamic C uses the feature-rich Rabbit MMU.

4.4.1 Rabbit's Memory Segments

The Rabbit 3000's MMU maps four segments from the 16-bit logical address space into the 20-bit physical address space accessible on the chip's address pins. These segments are shown in Figure 4.1.

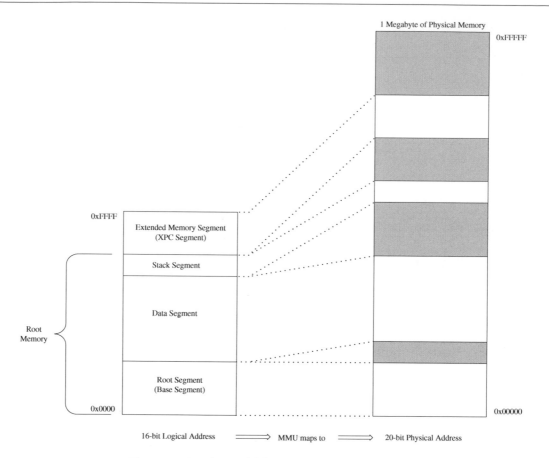

Figure 4.1: The Rabbit 3000 MMU Segments

Dynamic C uses the available segments differently depending on whether separate instruction and data space is enabled. First, we will consider the case without separate I & D space enabled.

4.4.2   Dynamic C's Memory Usage without Separate I & D Space

Dynamic C's support of separate I & D space allows much better memory utilization than the older model without separate I & D space. This section is included for the benefit of engineers who may have to maintain code written for the older memory model. New applications should be developed using separate I & D space. The newer memory model almost doubles the amount of root memory available to an application.

Dynamic C uses each of the four segments for specific purposes. The Root Segment and Data Segment hold the most frequently accessed program code and data. The Stack Segment is where the system stack resides. The Extended Memory Segment is used to access code or data that is placed outside of the memory mapped into the lower three segments.

A bit of Rabbit terminology worth remembering is the term ***root memory***. Root memory contains the memory pointed to by the Root segment, the Data Segment, and the Stack Segment (per *Rabbit 3000 Microprocessor Designer's Handbook*). This can be seen in Figure 4.1.

Another bit of nomenclature to keep in mind is the word ***segment***. When we use the word segment we are referring to the logical address space that the MMU maps to physical memory. This is a function of the Rabbit 3000 chip. Of course, Dynamic C sets up the MMU registers, but a segment is a slice of logical address space and correspondingly a reference to the physical memory mapped.

Segments can be remapped during runtime. The XPC segment gets remapped frequently to access extended memory, but most applications do not remap the other segments while running.

The semantics may seem a little picky, but this attention to detail will help to enforce the logical abstractions between Dynamic C's usage of the Rabbit's hardware resources and the resources themselves.

An example is the phrase ***Stack Segment*** and the word ***stack***. The Stack Segment is just a mapping of a slice of physical memory into logical address space. There is no intrinsic hardware requirement that the system stack be located in this segment. The Stack Segment was so named because Dynamic C happens to use this third MMU segment to hold the system stack. The Stack Segment is a piece of memory mapped by the MMU's third segment. The stack is a data structure that could be placed in any segment.

The Root Segment is sometimes referred to as the Base Segment. The Root Segment maps to BIOS code, application code, and Dynamic C constants. In most designs the Root Segment is mapped to flash memory. The BIOS is placed at address 0x00000 and grows upward. The application code is placed above the BIOS and grows to the top of the segment. Constants are intermixed with the application code.

Dynamic C refers to executable code placed in the Root Segment as ***Root Code***. The Dynamic C constants are called ***Root Constants*** and are also stored in the Root Segment.

The Data Segment is used by Dynamic C primarily to hold C variables. The Rabbit 3000 microprocessor can actually execute code from any segment; however, Dynamic C uses the

Data Segment primarily for data. Application data placed in the Data Segment is called *Root Data*.

Some versions of Dynamic C do squeeze a few extra goodies into the Data Segment that one might not normally associate with being program data. These items are nonetheless critically important to the proper functioning of an embedded system. A quick glance at Figure 4.2 will reveal that at the top 1024 bytes of the data segment are allocated to hold watch-code for debugging and interrupt vectors. Future versions of Dynamic C may use more or less space and may place different items in this space.

Dynamic C begins placing C variables (Root Data) just below the watch-code and grows them downward toward the Root Segment. All static variables, even those local to functions placed in the extended memory, are located in Data Segment. This is important to keep in mind as the Data Segment can fill up quickly.

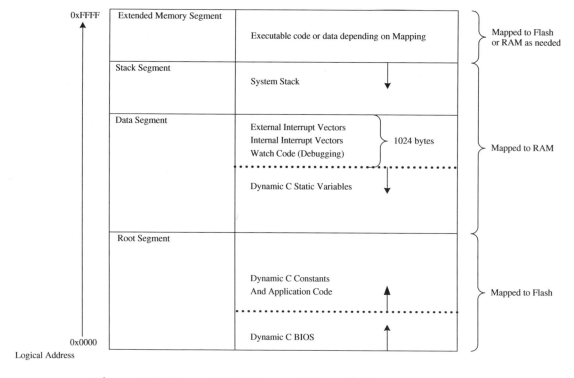

Figure 4.2: Dynamic C's Usage of the Rabbit 3000 Segments

Dynamic C's default settings allocate approximately 28 K bytes for the Data Segment and 24 K bytes for the Root Segment spaces. The macro DATAORG, found in `RabbitBios.c`, can be modified, in steps of 0x1000, to change the boundary between these two spaces. Each increase of 0x1000 will gain 0x1000 bytes for code with an attendant loss of 0x1000 for data. Each incremental decrease of 0x1000 will have the opposite effect.

The Stack Segment, as the name implies, holds the system stack. The stack is used by Dynamic C to keep track of return addresses as well as to pass some variables between functions. Variables of type *auto* also reside on the stack. The system stack starts at the top of the stack segment and grows downward.

The XPC Segment, sometimes called the Extended Memory Segment, allows access to code and data that is stored in the physical memory devices outside of the areas pointed to by the three segments in **Root Memory**. Root Memory is comprised of the Root Segment, the Data Segment, and the Stack Segment.

The system's **extended memory** is all of the memory not mapped into the Root Memory as shown in Figure 4.1. Extended Memory includes not only the physical memory mapped into the XPC segment, but all the other physical memory shown in Figure 4.1 in gray.

When we refer to extended memory, we are not referring just to memory mapped into the XPC Segment. The XPC segment is the tool (MMU segment) that Dynamic C uses to access all of the system's extended memory. We will use XMEM interchangeably with extended memory to mean all physical memory not mapped into Root Memory.

Generally, functions can be placed in XMEM or in root code space interchangeably. The only reason a function must be placed in root memory is if the function is an interrupt service routine (ISR) or if the function modifies the MMU mapping of the XPC register.

If an application grows large, moving functions to XMEM is a good choice for increasing the available root code space. Rabbit Semiconductor has an excellent technical note TN219, "Root Memory Usage Reduction Tips." For engineers with large applications, this technical note is a must read.

An easy method to gain more space for Root Code is simply to enable separate I & D space, but for when that is not an option, moving function code to XMEM is the best alternative.

4.4.3 Placing Functions in XMEM

Assembly or C functions may be placed in root memory or extended memory. Access to variables in C statements is not affected by the placement of the function, since all variables are in the Data Segment of root memory. Dynamic C will automatically place C functions in extended memory as root memory fills.

Functions placed in extended memory will incur a slight 12 machine cycle execution penalty on call and return. This is because the assembly instructions LCALL and LRET take longer to execute than the assembly instructions CALL and RET. If execution speed is important, consider leaving frequently called functions in the root segment.

Short, frequently used functions may be declared with the root keyword to force Dynamic C to load them in Root Memory. Functions that have embedded assembly that modifies the MMU's special function register called XPC must also be located in Root Memory. It is always a good idea to use the root keyword to explicitly tell Dynamic C to locate functions in root memory if the functions must be placed in root memory.

Interrupt service routines (ISRs) must always be located in root memory.

Dynamic C provides the keyword xmem to force a function into extended memory. If the application program is structured such that it really matters where functions are located, the keywords root and xmem should be used to tell the compiler explicitly where to locate the functions. If Dynamic C is left to its own devices, there is no guarantee that different versions of Dynamic C will locate functions in the same memory segments. This can sometimes be an issue for code maintenance.

For example, say an application is released with one version of Dynamic C, and a year later the application must be modified. If the xmem and root keywords are contained in the application code, it does not matter what version of Dynamic C the second engineer uses to modify the application. The compiler will place the functions in the intended memory—XMEM or Root Memory.

4.4.4 Separate Instruction and Data Memory

The Rabbit 3000 microprocessor supports a separate memory space for instructions and data. By enabling separate I & D spaces, Dynamic C is essentially given double the amount of root memory for both code and data. This is a powerful feature, and one that separates the Rabbit 3000 processors and Dynamic C from many other processor/tool combinations on the market.

The application developer has control over whether Dynamic C uses separate instruction and data space (I & D space). In the Dynamic C integrated development environment (IDE) the engineer need only navigate the OPTIONS ⇨ PROJECT OPTIONS ⇨ COMPILER menu and use the check box labeled "enable separate instruction and data spaces."

When Separate I & D space is enabled, some of the terms Z-World uses to describe MMU segments and their contents are slightly altered from the older memory model without separate I & D spaces. Likewise, some of the macro definitions in RabbitBios.c have altered meanings.

For example, the DATAORG macro in the older memory model tells the compiler how much memory to allocate to the Data Segment (used for Root Data) and the Root Segment (used for Root Code) and Root Constants. In a separate I & D space model, the DATAORG macro has no effect on the amount of memory allocated to code (instructions), but instead, tells the compiler how to split the data space between Root Data and Root Constants. With separate I & D space enabled, each increase of 0x1000 will decrease Root Data and increase Root Constant spaces by 0x1000 each.

The reason for the difference in function is an artifact of how Dynamic C uses the segments and how the MMU maps memory when separate I & D space is enabled. For most software engineers, it is enough to know that enabling separate I & D space will usually map 44 K of SRAM and flash for use as Root Data and Root Constants and 52 K of flash for use as Root Code.

The more inquisitive developer may wish to delve deeper into the memory mapping scheme. To accommodate this, we will briefly cover how separate I & D space works, but the nitty-gritty details are to be found on the accompanying CD.

When separate I & D space is enabled, the lower two MMU segments are mapped to different address spaces in the physical memory depending on whether the fetch is for an instruction or data. Dynamic C treats the lower MMU two segments (the Root Segment and the Data Segment) as one combined larger segment for Root Code during instruction fetches. During data fetches, Dynamic C uses the lowest MMU segment (the Root Segment) to access Root Constants. During data fetches the second MMU segment (the Data Segment) is used to access Root Data.

When separate I & D space is enabled, the lower two MMU segments are both mapped to flash for instruction fetches, while for data fetches the lower MMU segment is mapped to flash (to store Root Constants) and the second MMU segment is mapped to SRAM (to store Root Data).

This is an area where it is easy to become lost or misled by nomenclature. When separate I & D space is enabled, the terms Root Code and Root Data mean more or less the same thing to the compiler in that *code* and *data* are being manipulated. However, the underlying segment mapping is very different than when separate I & D space is not enabled.

When separate I & D space is not enabled, the Root Code is only to be found in the physical memory mapped into the lowest MMU segment (the Root Segment).

When separate I & D space is enabled, the Root Code is found in both the lower MMU segments (named *Root Segment* and *Data Segment*). Dynamic C knows that the separate I & D feature on the Rabbit 3000 allows both of the lower MMU segments to map to alternate places in physical memory depending on the type of CPU fetch. Dynamic C sets up the lower MMU segments so that they BOTH map to flash when an instruction is being fetched. Therefore Root Code can be stored in physical memory such that Dynamic C can use the two lower MMU segments to access Root Code.

This may seem contrary to the segment name of the second MMU segment, the Data Segment. The reader must bear in mind that the MMU segments were named based on the older memory model without separate I & D space. In that model, the CPU segment names were descriptive of how Dynamic C used the MMU segments. When the Rabbit 3000 came out and included the option for separate I & D space, the MMU segments were still given their legacy names. When separate I & D space was enabled, Dynamic C used the MMU segments differently, but the segment names on the microprocessor remained the same.

This brings us to how Dynamic C uses the lower two MMU segments when separate I & D space is enabled and a data fetch (or write) occurs. We are already familiar with the idea of Root Data, and this is mapped into physical memory (SRAM) through the second MMU segment—the Data Segment.

Constants are another type of data with which Dynamic C must contend. In the older memory model without separate I & D space enabled, constants (Root Constants) were intermixed with the code and accessed by Dynamic C through the lowest MMU segment (the Root Segment). In the new memory model with separate I & D space enabled, Dynamic C still uses the lower MMU segment (the root segment) to access Root Constants. However, with separate I & D space enabled, when data accesses occur, the lowest MMU segment (root segment) is mapped to a space where code is not stored. This means there is more space to store Root Constants as they are not sharing memory with Root Code.

Root Constants must be stored in flash. This implies that the lowest MMU segment is mapped into physical flash memory for both instruction and data accesses. Root Code resides in flash, as do Root Constants.

Given this overview, we can consider the effect of DATAORG again. DATAORG is used to specify the size of the first two MMU segments. Since Dynamic C maps the first two MMU segments to Root Code for instruction accesses, and treats the first two MMU segments as one big logical address space for Root Code, changing DATAORG has no effect on the space available for Root Code.

Now consider the case when separate I & D space is enabled and data is being accessed. The lowest MMU segment (the Root Segment) is mapped into flash and is used to access Root Constants. The second MMU segment (the Data Segment) is mapped into SRAM and is used to access Root Data.

Changing DATAORG can increase or decrease the size of the first two segments. For data accesses, this means the size of flash mapped to the MMU's first segment is either made larger or smaller while the second segment is oppositely affected. This means there will be more or less flash memory mapped (through the first MMU segment) for Dynamic C to use for Root Constants with a corresponding decrease or increase in SRAM mapped (through the second MMU segment) for Dynamic C to use as Root Data.

When separate I & D spaces are enabled, the stack segment and extended memory segment are unaffected. This means that the same system stack is mapped regardless of whether instructions or data are being fetched. Likewise, extended memory can still be mapped anywhere in physical memory to accommodate storing/retrieving either executable code or application data.

For most engineers it is enough just to know that using separate I & D space gives the developer the most Root Memory for the application. In the rare circumstance in which the memory model needs to be tweaked, the DATAORG macro is easily used to adjust the ratio of Root Data to Root Constant space available. For the truly hardcore, the Rabbit documentation has all the details.

4.4.5   Putting It All Together

We have spent a considerable amount of time going over segments.

Quick Summary

- Logical addresses are 16-bits

- Physical addresses exist outside the CPU in a 20-bit space

- The MMU maps logical addresses to physical addresses through segments

- Depending on application requirements such as speed and space, it may be important to control where code and data are placed. Dynamic C's defaults can be overridden, allowing the programmer to decide where to place these code elements in memory

---

4.5   How Code Is Compiled and Run

Let's look at the traditional build process and contrast it with how Dynamic C builds code:

4.5.1   How Code Is Built in Traditional Development Environments

- The programmer edits the code in an editor, often part of the IDE; the editor saves the source file in a text format.

- The programmer compiles the code, from within the IDE, from command line parameters, or by using a make utility. The programmer can either do a *Compile All*, which will compile all modules; or the make utility or IDE can only compile the modules that were changed since the last time the code was built. The compiler generates object code and a list file that shows how each line of C code got compiled into one or more lines of assembly code. Unless specified, each object module has relative memory references and is relocatable within the memory space, meaning it can reside anywhere in memory. Similarly, each assembly module gets assembled and generates its own relocatable object module and list file.

- If there are no compilation or assembly errors, the linker executes next, putting the various object modules together into a single binary file. The linker converts relative addresses into absolute addresses, and creates a single binary file of the entire program. Almost all linkers nowadays also have a built-in locator that locates code into specific memory locations. The linker generates a map file that shows a number of useful things, including where each object module resides in memory, how much space does the whole program take, and so on. If library modules are utilized, the linker simply links in precompiled object code from the libraries.

- The programmer can download the binary file into the target system using a monitor utility, a bootstrap loader, using an EPROM emulator, or by simply burning the image into an EPROM and plugging in the device into the prototyping board. If a CPU emulator is being used, the programmer can simply download the code into the emulator.

Figure 4.3 illustrates how code is built on most development environments.

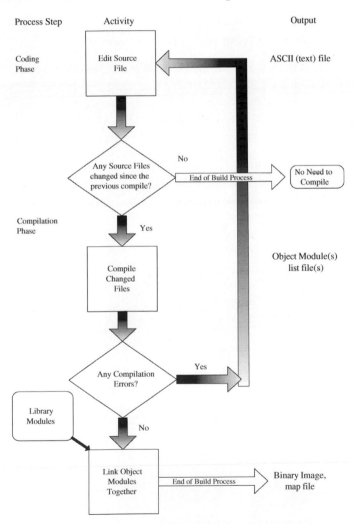

Figure 4.3: The Traditional Build Process

4.5.2   *How Code Is Built with Dynamic C*

• The programmer edits the code in the Dynamic C IDE, and saves the source file in a text format.

• The Dynamic C IDE compiles the code. If needed, the programmer can compile from command line parameters. Unlike most other development environments, Dynamic C prefers to compile every source file and every library file for each build. There is an option that allows the user to define precompiled functions.

• There is no separate linker. Each build results in a single binary file (with the ".BIN" extension) and a map file (with the ".MAP" extension).

• The Dynamic C IDE downloads the executable binary file into the target system using the programming cable.

Figure 4.4 illustrates how code is built and run with Dynamic C:

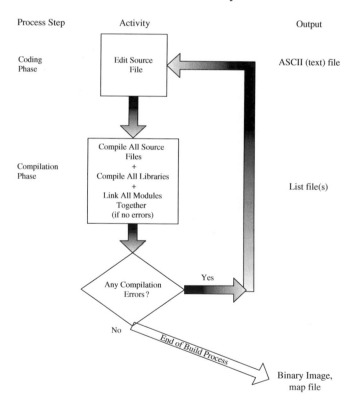

Figure 4.4: How Dynamic C Builds Code

Quick Summary

- Dynamic C builds code differently from the traditional edit/compile/link/download cycle
- Each time code is built, Dynamic C always compiles each library file and each source file
- Each time code is run, Dynamic C does a complete build
- Within the Dynamic C IDE, executable images can be downloaded to a target system through a simple programming cable

---

4.6   Setting Up a PC as an RCM3200 Development System

Before we start using Dynamic C to write code, we need to set up an RCM3200 core module and prototyping board. This simple process only takes a few minutes.

Setting up an RCM3200 development system requires fulfilling the following steps:

1.   Using the CD-ROM found in the development kit, install Dynamic C on your system.

2.   Choose a COM (serial) port on your PC to connect to the RCM3200 .

3.   Attach the RCM3200 to the prototyping board.

4.   Connect the serial programming cable between the PC and the core module.

5.   Provide power to the prototyping board.

Now that the hardware is setup, we need to configure Dynamic C. Some Rabbit core modules are able to run c ⇨ Project Options ⇨ Compiler menu. The RCM 3200 will run programs from fast SRAM instead of flash.

For our simple examples, it really doesn't matter whether we configure Dynamic C to generate code that will run from fast SRAM or from flash. However, for the sake of consistency, we always configure Dynamic C to enable code to be run from fast SRAM for the examples in this text that use the RCM3200.

4.7   Time to Start Writing Code!

Now that the RCM3200 system is ready for software development, it is time to roll up the sleeves and start writing code. The first program is very simple. The intent of this exercise is to make sure the computer (the host PC) is able to talk to the RCM3200. Once we are able to successfully compile and run a program, we will explore some of Dynamic C's debugging features, as well as some differences between Dynamic C and ANSI C.

4.7.1 Project: Everyone's First Rabbit Program

It has been customary for computer programmers to start familiarizing themselves with a new language or a development environment by writing a program that simply prints a string ("Hello World") on the screen. We do just that—here's the program listing:

```
main()
{
        printf ("Hello World"); // output a string
}
```

Program 4.1: helloWorld.c

Here's how to compile and run the Rabbit program:

1. Launch Dynamic C through the Windows Start Menu—or the Dynamic C Desktop Icon.

2. Click "File" and "Open" to load the source file "HELLOWORLD.C." This program is found on the CD-ROM accompanying this book.

3. Press the **F9** function key to run the code.

After compiling the code, the IDE loads it into the Rabbit Core, opens a serial window on the screen, titled the "STDIO window," and runs the program. The text "Hello World" appears in the STDIO window. When the program terminates, the IDE shows a dialog box that reads "Program Terminated. Exit Code 0."

If this doesn't work, the following troubleshooting tips maybe helpful:

- The target should be ready, indicated by the message "BIOS successfully compiled . . . " If this message did not appear or a communication error occurred, recompile the BIOS by typing <Ctrl+Y> or select **Reset Target/Compile BIOS** from the **Compile** menu.

- If the message "No Rabbit Processor Detected" appears, verify the target system has power and the programming cable is connected between the PC and the target.

- The programming cable must be connected to the controller. The colored wire on the programming cable is closest to pin 1 on the programming header on the controller. Make sure you use the connector labeled as PROG and not the connector labeled DIAG. The other end of the programming cable must be connected to the PC serial port. The COM port specified in the Dynamic C **Options** menu must be the same as the one to which the programming cable is connected.

- To verify the correct serial port is connected to the target, select **Compile**, then **Compile BIOS**, or press **<Ctrl+Y>**. If the "BIOS successfully compiled ..." message does not display, try a different serial port using the Dynamic C **Options** menu. Don't change anything in this menu except the COM number. The baud rate should be 115,200 bps and the stop bits should be 1.

A Useful Dynamic C Shortcut

"**F9**" causes Dynamic C to do the following:

- Compiles the project source code

- Assuming that there were no compilation errors,

  — Loads the code into flash on the target board

  — Begins execution of the application code on the target system

---

Although the program terminates, the IDE is still controlling the target. In this mode, called debug or run mode, the IDE will not let the programmer edit the code. For the IDE to release the target and allow editing, we need to close the debug session by clicking on "Edit" and "Edit Mode." Alternatively, pressing **F4** will enter Edit Mode.

Auxiliary Files Created by Dynamic C during Compilation

helloWorld.BDL is the binary download image of the program

helloWorld.BRK stores breakpoint information. It can be opened with a text editor to see the number of breakpoints and the position of each breakpoint in the source file

helloWorld.HDL is a simple Intel format Hex download file of the program image

helloWorld.MAP shows what the labels (variables and code references) resolve to. In addition, it shows the length and origin of each module and which memory space (Root Code, Root Data, or XMEM Code) in which the module resides

helloWorld.ROM is a program image in a proprietary format

4.7.2 Dynamic C's Debugging Features

Dynamic C offers powerful debugging features. This innovation eliminates the need for an expensive hardware emulator. This section covers the basics of using Dynamic C's debugging features.

Program 4.2 (watchDemo.C on the enclosed CD-ROM) is the simple program that will be used to illustrate Dynamic C's debugging features.

```
void delay ()
{
  int j;
  for (j=0; j<20000; j++);            // create a delay
}
main() {
    int count;                        // define variable
    count = 0;                        // initialize counter
    while (1)                         // start an endless loop
    {
        count++;                      // increment counter
        delay();                      // wait a bit
        printf("count = %d\n", count);// print something useful
    } // end of endless loop
} // end of program
```

Program 4.2: watchDemo.c

The code will print increasing values for count in the STDIO window.

4.7.3 Dynamic C Help

Dynamic C makes obtaining help for its library functions simple. Placing the cursor over a function name in the source file and pressing **<Ctrl+H>** will bring up a documentation box for that function.

4.7.4 Single Stepping

To step through a program, we need to compile and download the program to the RCM3200 without running the program. This can be accomplished by any of the following methods:

- pressing <F5>

- selecting Compile from the menu bar

- selecting the lightning bolt icon on the Tool Bar

The IDE highlights (in green) the first character of the program's first executable statement. This green highlighting is always used to indicate the current execution point.

The **F7** and **F8** keys will single step the statements—each time one of these keys is pressed, one program statement will be executed. The difference is that if the statement is a call to another function, pressing **F8** will completely execute the called function, while pressing **F7** will execute the called function one statement at a time.

To summarize, **F8** allows the user to "step over" functions, while **F7** allows the user to "step into" functions.

Pressing **F7** to single step watchDemo.c will execute the following sequence of statements,

```
count = 0;      The first statement of the program
while           The "while" statement
(1)             The statement that must be evaluated for the "while" branch
count++;        The first statement in the "while" body
for (j=0; j<20000; j++);  The first statement in delay();
```

The "for" loop in delay() has a conditional statement (j<20000), the loop control variable adjustment statement (j++), and a null statement (;) in the loop body. The programmer would have to press **F7** another 60,000 times to complete all of the statements in the delay() function.

Using step-over, **F8**, Dynamic C would execute the following sequence of statements:

```
count = 0;    The first statement of the program
while         The "while" statement
 (1)          The statement that must be evaluated for the "while" branch
count++;      The first statement in the "while" body
delay();      The delay() function (all 20,000 loops)
printf("count = %d\n", count); The function that prints to STDOUT
 (1)          The statement that must be evaluated for the "while" branch
count++;      The first statement in the "while" body
delay();      The delay() function (all 20,000 loops)
printf("count = %d\n", count); The function that prints to STDOUT
```

and so on.

F8 is useful for stepping over functions that are believed to be good, while allowing the programmer to carefully examine the execution path of suspect code.

4.7.5 Adding a Breakpoint

Breakpoints are useful for controlling program execution during debugging. Placing the cursor over the statement where the breakpoint is desired and pressing **F2** will assign a breakpoint to the statement. **F2** can be used to remove existing breakpoints. Placing the cursor over an existing breakpoint and pressing **F2** will remove the breakpoint. Dynamic C indicates that a breakpoint exists by changing the background color of the first character in the statement to red. Alternatively, breakpoints may be placed or removed using the **Toggle Breakpoint** option from the **Run** menu.

Breakpoints are used to halt program execution so the programmer can inspect the state of the target system's hardware, CPU's registers, and program variables. When the program halts due to a breakpoint, step-into and step-over can be used to observe the execution path of suspect code.

Once breakpoints are set, the program is run (using **F9**). The program will advance until it hits a breakpoint. While the program is paused, the IDE allows breakpoints to be added or removed from the target.

Breakpoints may also be set in a program running at full speed. This will cause the program to break if the execution thread hits a breakpoint.

Another technique for using breakpoints will allow software developers to determine if a particular segment of code is being executed. A breakpoint can be placed in the segment of interest and program execution started. If the program never halts, then the segment of interest was not executed. This is a useful technique for determining if code branches occur as expected.

"Normal" breakpoints allow interrupts to continue being serviced even when the breakpoint has been reached. "Hard" breakpoints can be used to disable all execution while the breakpoint is being serviced. This type can be especially useful when debugging ISRs.

4.7.6   Watch Expressions

Once a program is halted, examining the contents of variables is simple. The easiest way to examine a variable is to hover the mouse over the variable of interest. Dynamic C will pop up a small box showing the value. An expression may be evaluated in a similar manner by highlighting the expression and then hovering over it.

Dynamic C also provides a tool called a watch window. The programmer can view expressions added to the watch window.

For example, adding the integer "count" from watchDemo.c to a watch window allows the developer to observe how "count" changes as the code is single stepped.

Expressions in watch-windows are updated whenever code execution is halted. Breakpoints halt code execution, as do the step-into and step-over tools.

Expressions can be updated while code is running at full speed by pressing **<Ctrl+U>**. The Dynamic C help menu for "Watch Windows" explains that runWatch() should be periodically executed in the program to support **<Ctrl+U>** requests for updating the IDE Watch Window.

Expressions can be added to a watch-window by selecting **Add/Del Watch Expression** from the **Inspect** menu, or by using the **<Ctrl+W>** shortcut.

You can use the Watch Window to change the values of variables as well as execute functions.

Summary of Some Dynamic C Shortcut Keys:

Edit, Compile, and Run
F4: Enter Edit Mode
F5: Compile code and load executable on target, but don't begin execution
F9: Compile and load executable on target if not already done and then begin execution

Debugging Functions
Ctrl+W: Add or Delete a Watch Expression
Ctrl+U: Update Watch Expression Window
F2: Set or Remove a Breakpoint
Ctrl+A: Clear All Breakpoints
F7: Single Step walking into functions (step-into)
F8: Single Step walking over functions (step-over)

Fast Help
Ctrl+H: Provide help on a function
Alt+F1: Instruction Set Reference

4.7.7 Dynamic C Is Not ANSI C

The American National Standards Institute (ANSI) maintains a standard for the C programming language. Code that conforms to the ANSI standard is theoretically more easily ported between operating systems.

Embedded systems have unique performance demands. Different companies have adopted different approaches to meeting the unique challenges presented by embedded systems.

Z-World has enjoyed nearly two decades of success with the approach of carefully adapting the C programming language to suit the demands of embedded controllers. Language extensions and creative interpretation of standard C syntax make Dynamic C a desirable environment for embedded systems development.

As universities teach ANSI C with a bent toward desktop application development, it is worth pointing out some of the differences between Dynamic C and ANSI C. This will help newcomers to Dynamic C avoid common pitfalls while being able to take full advantage of the enhancements Dynamic C offers.

Let's examine a pitfall that newcomers often fall into—declarations with initializations.

Most programmers will look at the code presented in Program 4.3A and expect "count" to be an integer that is initialized to be 0. Dynamic C takes a slightly different view of the code.

```
void main( void ) {
    int count=0;
    printf("count = %d\n", count);
}
```

Program 4.3A: oops.c

Dynamic C assumes a variable that is initialized when declared is a constant. This is a common enough situation that Dynamic C will generate a warning when the above code is compiled. The warning states, "line 2 : WARNING oops.c : Initialized variables are placed in flash as constants. Use keyword 'const' to remove this warning."

Changing the declaration to, `const int count=0;` causes the compiler to generate the same executable code, but without the warning.

If the desired result is to declare an integer named count and initialize it to zero then the code shown in Program 4.3B should be used.

```
void main( void ) {
    int count;
    count=0;
    printf("count = %d\n", count);
}
```

Program 4.3B: better.c

Initializing global variables may at first glance appear impossible. Dynamic C offers a ready solution with the compiler directive #GLOBAL_INIT.

The Dynamic C documentation explains, "`#GLOBAL_INIT` sections are blocks of code that are run once before `main()` is called."

The code shown in Program 4.3C shows how a global variable can be initialized.

```
int GlobalVarCount;
void main( void ) {
#GLOBAL_INIT
{
 GlobalVarCount = 1;
}
      printf("GlobalVarCount = %d\n", GlobalVarCount);
}
```

Program 4.3C: GlobalVarInit.c

The #GLOBAL_INIT directive can be used to initialize any static variables. If a function declares a static variable and needs that variable initialized only once, then #GLOBAL_INIT can be used to accomplish this. Program 4.3D shows how to do it. The output generated is,

> LocalStaticVar = 1
> LocalStaticVar = 2
> LocalStaticVar = 3

The static integer LocalStaticVar is initialized to 1 before main() is executed.

```
void xyzzy (void) {
static int LocalStaticVar;
#GLOBAL_INIT
{
 LocalStaticVar = 1;
}
printf("LocalStaticVar = %d\n", LocalStaticVar++);
}
void main( void ) {
xyzzy();
xyzzy();
xyzzy();
}
```

Program 4.3D: LocalVarInitializedOnce.c

Program 4.3E shows how "not to" initialize a static variable. The static integer LocalStaticVar is assigned the value of 1 every time xyzzy() is called. This is generally not a desired behavior for static variables, which are intended to retain their value between function calls. The output generated is,

    LocalStaticVar = 1
    LocalStaticVar = 1
    LocalStaticVar = 1

```
void xyzzy (void)
{
static int LocalStaticVar;
LocalStaticVar = 1
printf("LocalStaticVar = %d\n", LocalStaticVar++);
}
void main( void )
{
xyzzy();
xyzzy();
xyzzy();
}
```

Program 4.3E: LocalVarAlwaysInitialized.c

As useful as the compiler directive #GLOBAL_INIT is, it can become a source of confusion.

The key point to remember when using #GLOBAL_INIT is that the order of execution of #GLOBAL_INIT sections is not guaranteed!

All #GLOBAL_INIT code sections are chained together and executed before main() is executed.

Global variables can be modified in multiple #GLOBAL_INIT code segments. If this is done, the compiler will not generate any warnings or errors. If the coder is careless, a global initialization may be overwritten by a subsequent #GLOBAL_INIT code segment. Since the order of execution of #GLOBAL_INIT sections is not guaranteed, the global variable is not guaranteed to have been initialized by the intended #GLOBAL_INIT code segment. To further complicate matters, a source file may have the order of execution of #GLOBAL_INIT sections altered by different versions of the compiler.

#GLOBAL_INIT is a useful and reliable compiler directive. Like any tool, the powerful #GLOBAL_INIT directive must be used within the compiler's constraints. Do not initialize global variables in multiple #GLOBAL_INIT sections. Realize that the order of execution of #GLOBAL_INIT sections is not guaranteed. Know that different versions of Dynamic C are free to reorder the execution of #GLOBAL_INIT sections. Code accordingly.

> ### Summary: Other Differences between Dynamic C and ANSI C
>
> Variables initialized upon declaration are constants to Dynamic C and placed in flash memory.
>
> GLOBAL_INIT is a convenient way to initialize variables in functions.

4.7.8   Dynamic C Memory Spaces

Section 4.4 discussed where Dynamic C places variables. We will now reexamine the placement of code and data and how we can force Dynamic C to put code and data where we want.

We compiled Program 4.4 (memory1.c) using Dynamic C version 7.33, and then examined the associated map file (memory1.map). Here are the pertinent excerpts from the source code and the map file:

```
int my_function(int data)
{
    static int var_func_static;
    int var_func;

    var_func_static = 3;
    var_func = var_func_static*data;

    printf ("%d multiplied by %d is %d\n",data,var_func_static,
 var_func);

    return var_func;
}

void main()
{
    static int var_static1;
    int var_not_static1;
    static const char my_string[]="I like what I have seen
 so far!\n";

    var_static1 = 0xA;
    var_not_static1 = 0x5;

    var_not_static1 = my_function (var_static1);

    printf ("%s",my_string);
}
```

Program 4.4: memory1.c

The top section of the map file shows origin and sizes of various segments:

```
// Segment        Origin          Size
Root Code         00:0000         0055d5
Root Data         00:bfff         000899
Xmem Code         ff:e200         001716
```

Excerpts of the map file show us where in memory we will find my_string, my_function(), and main():

```
// Global/static data symbol mapping and source reference.
//   Addr       Size  Symbol
     b857          2  my_function:var_func_static
     b855          2  main:var_static1
  10:022c         33  main:my_string
// Parameter and local auto symbol mapping and source reference.
// Offset Rel. to        Size     Symbol
        4       SP          2     my_function:data
        0       SP          2     my_function:var_func_not_static
        0       SP          2     main:var_not_static1
// Function mapping and source reference.
//   Addr     Size        Function
     1c26       63        my_function
     1c65       58        main
```

Looking at the addresses above and comparing them to the global static data symbol addresses, we can see that the static variables got placed in the Root Code space, while the string got placed in XMEM.

We can see that Dynamic C lumped together the static variables from main() and my_function with the string constant, and kept the nonstatic variables in the stack. Notice that the stack has reserved two bytes for my_function:data; this is how the lone integer parameter gets passed from main() to my_function().

Also notice that the function and main got placed in Root Code.

Now that we are beginning to get comfortable with where Dynamic C places code and data by default, let's play with it a little—let's try to save root space and move as much as we can to XMEM. We think the program may take just a little longer to execute since Dynamic C and the MMU will have to convert all physical memory accesses to the internal logical representation, but we will save on the precious root space. Changing the above program to work differently, we get memory2.c in Program 4.5:

```
xmem int my_function(int data)
{
    static int var_func_static;
    int var_func_not_static;

    var_func_static = 3;
    var_func_not_static = var_func_static*data;

    printf ("%d multiplied by %d is %d\n",data,var_func_static,
var_func_not_static);

    return var_func_not_static;

}

xmem void main()
{
    static int var_static1;
    int var_not_static1;
    static const char my_string[]="I like what I have seen
so far!\n";

    var_static1 = 0xA;
    var_not_static1 = 0x5;

    var_not_static1 = my_function (var_static1);

    printf ("%s",my_string);
}
```

Program 4.5: memory2.c

We can expect to see some differences in the map file; we should find the code for main() and my_function() in xmem space. Let's look at the map file and find out if that is the case:

```
// Segment    Origin        Size
Root Code     00:0000       005561
Root Data     00:bfff       00089f
Xmem          ff:e200       001792


// Function mapping and source reference.
//   Addr      Size          Function
     e420       64           my_function
     e460       60           main

// Global/static data symbol mapping and source reference.
//   Addr      Size          Symbol
     b857        2           my_function:var_func_static
     b855        2           my_function:var_func_not_static
     b853        2           main:var_static1
     b851        2           main:var_not_static1
 10:022c        33  main:my_string


// Parameter and local auto symbol mapping and source reference.
// Offset Rel. to      Size    Symbol
         3     SP        2      my_function:data
```

This time we can see that the function and main got placed in XMEM space. The variables, except for the one used for parameter passing between main and the function, got placed in Root Code.

The keywords xmem and root allow the engineer to force the compiler to locate functions in specific areas of memory. The map file can be used to verify that Dynamic C did what the engineer intended.

Dynamic C versions 8.01 and higher are quite smart about how they locate functions. Most software engineers need not worry about manually locating functions. This is especially true when separate I & D space is enabled, as that gives plenty of root space for both code and data for most applications. However, in the cases when engineers want to tweak the compiler's choices, xmem and root give the engineer full control.

Now we are ready to work with networks.

4.8   Embedded Networks

Networks are ubiquitous, and now exist in places where they did not exist five years ago. Broadband home networks and public WiFi networks are being deployed globally at a great pace. The Internet is the most identifiable form for networking for the layperson—we can now find Internet access in large and small offices, homes, hotel rooms, restaurants, airports, coffee shops, cruise ships, and commercial airplanes. We take for granted more and more services that use networking to improve our daily lives. Credit card transactions, email, online banking, e-Commerce, online delivery tracking, and online movie rentals are just a few examples of commonplace services that did not exist a decade ago.

Networked embedded devices are finding uses in diverse areas from building access controls to smart homes to wireless cameras and entertainment appliances. A growing number of industrial devices, as well as consumer and enterprise-grade devices now use embedded web servers for configuration, management, monitoring, and diagnostics. Industry analysts are predicting the use of embedded devices in the near future that converge media, entertainment, and productivity. Networking is one of the key enablers to that vision.

In this chapter, we will look at networking from the perspective of Rabbit core modules. We will discuss common networking protocols at a high level, examining not how they work or how they are implemented, but how they can be used on Rabbit core modules.

Networking is a very broad field, including local area networks (LANs), wide area networks (WANs), metropolitan area networks (MANs), and wireless technologies. Each of these areas has its own protocols and interfaces. One can get into a lot of detail with networking protocols—DHCP and TCP, for example, have been described in thicker books than this one. Consider the subject of socket programming that we have covered in just one section here—entire books have been written on this subject and the reader is advised to look for more detailed coverage elsewhere. Dynamic C libraries support just a core set of protocols from these technologies, and these are enough for most embedded applications. The goal of this chapter is not to educate the user on networking protocols, but to examine how a networked application can be built using Dynamic C's networking features. Refer to the enclosed CD-ROM for some excellent technical papers on Dynamic C's TCP/IP implementation.

Networking is not limited to Ethernet. RS-485 is a physical interface widely used for building networks. With RS-485, programmers often have to write their own protocol. Although Dynamic C libraries provide strong support for RS-485 and other physical network interfaces, this chapter will focus only on Ethernet-based connectivity.

We will first examine a number of networking protocols that are supported by Dynamic C, and then build some projects that use some of these protocols. We will also build some applications with C++ and Java that will help us control some of these projects. Later in this chapter we will bring it all together—hardware characterization and interfacing, user interface design, and embedded web server programming.

4.9 Dynamic C Support for Networking Protocols

In this section, we will briefly describe some of the networking protocols supported by Dynamic C. The authors assume that readers are familiar with the seven-layer OSI model.

From the programmer's perspective, there isn't much for Dynamic C to do at the presentation and session layers; most of the action happens at the transport and application layers. Embedded applications are most likely to use the layers in the application layer (FTP, HTTP, and so on) or TCP and UDP directly in the transport layer. Most deployed networking uses 4 layers of the OSI; protocols operating at these layers are the ones most likely to be used in applications. Figure 4.5 shows the four layers most relevant to embedded developers.

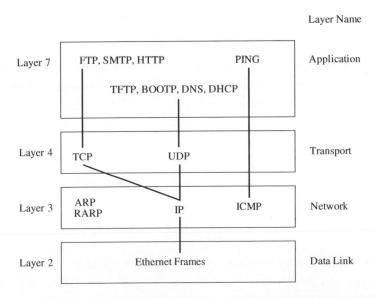

Figure 4.5: The Four-Layer Networking Model and Related Protocols

4.9.1 Common Networking Protocols

Dynamic C provides support for the following protocols:

- IP: The **Internet Protocol** is where the magic starts. The Data Link layer deals with switching Ethernet frames, based on MAC addresses, while the *Network* layer uses IP addresses to describe sources and destinations on a network.

- ARP: The **Address Resolution Protocol** allows a device to discover a MAC address, given an IP address. This forms a bridge between the TCP/IP protocol suite and any link level address, of which Ethernet is an example.

- RARP: The **Reverse ARP** does the opposite of ARP—it provides us with the IP address that is associated with the given MAC address.

- ICMP: The **Internet Control Message Protocol** implements various messages not encapsulated by the other TCP/IP protocols. In particular, the well-known "ping" command uses ICMP messages to determine if a network device is reachable.

- TCP and UDP are two major transport protocols. TCP is connection oriented, while UDP is connectionless. TCP provides reliable data transfer, while UDP provides best-effort service. These will be described here in some detail, and we will cover some more detail in Section 4.15.

- TCP: The **Transmission Control Protocol** is the building block for a host of networking services. The main purpose of TCP is to provide reliable connection-oriented data transfer, and it uses various methods for flow control and error detection to accomplish its mission. Routing and congestion cause the timing of packet arrivals to be nondeterministic, which does not guarantee that packets will arrive in the same sequence in which they were transmitted. TCP uses a sequencing mechanism to line up packets for upper layers in the same order the packets were sent.

As shown in Figure 4.5, the following applications use TCP as the underlying transport:

- FTP: The **File Transfer Protocol** allows us to do just that—transfer files over a network. Internet users often use FTP as a mechanism to download files from a remote host.

- SMTP: The **Simple Mail Transfer Protocol** is used to send and receive email. A number of popular email clients use SMTP for the underlying mail transport.

- HTTP: The **Hypertext Transfer Protocol** is commonly used with browsers. HTTP defines how web pages are formatted and transmitted, and certain commands that the browser must respond to. The actual formatting of the web pages is defined by HTML (**Hypertext Markup Language**).

- UDP: Unlike TCP, the **User Datagram Protocol** does not guarantee data reliability. In fact, there is no guarantee whether a packet sent via UDP will get to its destination (that is why the UDP transport is often called a "best effort" datagram service). Moreover, UDP does not reassemble packets to get them lined up in the same order they were delivered. UDP's connectionless nature results in simplicity of implementation code[1] and lower housekeeping overhead. Unlike a TCP connection, which must be synchronized and maintained through the network, UDP requires neither initialization handshake between the source and the destination, nor the networking resources that are tied up in maintaining a reliable connection.

As shown in Figure 4.5, the following applications use UDP as the underlying transport:

- TFTP: The **Trivial File Transfer Protocol** is a simpler version of FTP, and uses UDP to transfer files; it does not use TCP's reliable delivery mechanisms.

- DNS: The **Domain Name System** is a mapping scheme between domain names and their associated IP addresses. For example, every time a browser tries to access http://www.google.com/, the domain name server translates the "google" domain name into its associated IP address: 216.239.39.99, and the browser accesses the IP address without going through the DNS translation. If the user types "216.239.39.99" into a browser window, the browser will access the "google" web server without going through a domain name server.

- DHCP: The **Dynamic Host Configuration Protocol** allows dynamic assignment of IP addresses to devices. In the embedded systems context, this means that an embedded system can boot without an IP address, and can negotiate with a DHCP server to receive an IP address. Dynamic assignment of IP addresses is common,

---

[1] In the protocol layer, not necessarily in the user's application.

since it eases the burden on network administrators to statically assign and manage IP addresses. It is common to have DHCP servers built into routers for home networking.

TCP and UDP ensure integrity of the payload with checksums (up to a certain extent, since the checksum mechanism is not perfect).

TCP is a point-to-point connection protocol, whereas UDP is connectionless and therefore allows for other possibilities, such as broadcast messages.

Some additional applications are of interest to us—we can use the following utilities to debug networked applications, and some others are listed in Section 4.20.

- Telnet: this utility uses TCP to perform remote logins. It is commonly used to log into networking devices such as routers and switches. For security reasons, some network administrators block telnet access to their devices. Moreover, telnet is not secure because it sends unencrypted data across networks.

- Ping: As Figure 4.5 shows, Ping uses ICMP messages to determine whether a networked device is reachable via its IP address.

4.9.2   *Optional Modules Available for Dynamic C*

In addition to providing support for the networking protocols listed in Section 4.9.1, Dynamic C supports the following protocols, provided as add-on modules:

- PPP: The **Point-to-Point Protocol** allows a device to perform TCP/IP networking over a serial connection. Most dialup Internet connections use PPP.

- AES: The **Advanced Encryption Standard** is meant as a replacement for the aging DES (**Data Encryption Standard**). While the DES provided a key size of 56 bits, AES supports 128, 192, and 256 bit keys. Although triple DES can be used for added security, it is not as efficient as AES. Dynamic C supports 128, 192, and 256 bit AES encryption and decryption through the `aes_crypt.lib` library. AES is not itself a protocol but is used by other protocols.

- SSL/HTTPS: The **Secure Socket Layer/Secure HTTP** module allows users to have an encrypted web server (HTTPS). This is useful for creating a secure interface to a networked embedded device, and should always be used when the web server is accessible from the Internet, especially when the embedded systems control physical, potentially dangerous, devices.

- SNMP: The **Simple Network Management Protocol** uses a messaging mechanism to manage networking devices. A request/response mechanism is typically deployed for device management and maintenance of status data. Dynamic C supports SNMP through the snmp.lib library.

4.10   Typical Network Setup

Before looking at the Rabbit core module's network connectivity, we will present a "big picture" view of where the core module will exist in a networked development environment. We will highlight both a corporate network and a home network, since a software developer may work in either or both of these environments.

4.10.1   Typical Corporate Network

A corporate network generally has a lot of redundant devices to offer a high degree of availability. The network is used for internal operations, as well as for customer-facing activities such as the corporate web site, e-Commerce, and remote connectivity with partners, customers, and employees. Having various network elements in active and standby mode provides for quick failover and recovery. The firewalls secure the internal corporate network against unprivileged access, and there is a "demilitarized zone," the DMZ, that exists outside the corporate firewalls.

The web servers can be on either side of the firewall. If they are on the internal network behind the firewall, port 80 is opened for them to allow web requests to go through the firewall.

These details vary, depending on the company's security policies and infrastructure.

Access switches inside this simple corporate set up connect users in the corporate intranet. In addition, application servers in the secure intranet run corporate applications for email, databases, inventory, management, and so on. The corporate intranet, shown in Figure 4.6, is partitioned into various Virtual Local Area Networks (VLANS) that separate functional access. For example, corporate users and administrators use separate VLANs, while engineers use one called a "Test VLAN" that provides access to the Internet but not to

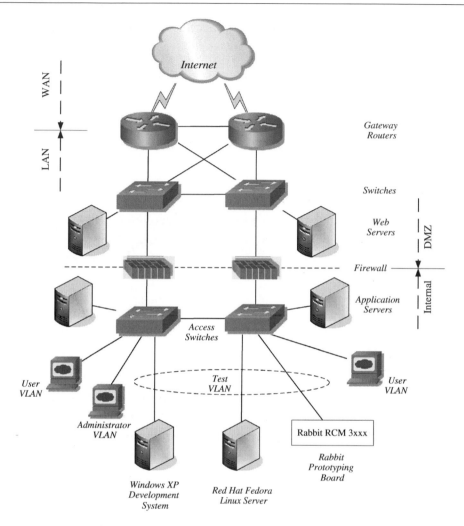

Figure 4.6: Rabbit Core Module in a Corporate Network

corporate applications. Various other networking elements, for example, that implement storage, content caching, and intrusion detection, are not shown here.

The overall network can use a mix of fiber optic, Gigabit Ethernet, 10/100 Mbit Ethernet, serial, and frame relay technologies in different parts of the network.

During development phase, a Rabbit core module will likely connect to an internal "Test VLAN" that development engineers will use to test networked applications. Dynamic C will run on a Windows workstation in the same VLAN, and the development engineer may use a Linux machine to test connectivity with the Rabbit core module. The engineer will connect these devices to an access switch that provides 10/100 Mbps connectivity to the engineer's office or lab bench.

It is not necessary to have the development system on the same network as the core module. Since the Rabbit core module is programmed via a serial link, as long as the programmer does not need to test the device over the network from the development machine, the core module can work on a separate network. The core module can be tested via a test system on the test network.

4.10.2   Typical Home Network

Conceptually, the home network consists of a router that connects the home LAN to an external WAN. A cable modem serves as the link between the cable-based broadband service and the router. The Internet service provider (ISP) dynamically assigns an IP address to the router's WAN connection. The router usually implements NAT (Network Address Translation) to allow multiple computers to connect to the Internet with only one address provided by the ISP. The router also provides DHCP to assign IP addresses to devices on the home network. A private addressing scheme can be used on the LAN that is separate from that on the WAN. The router connects to a switch that provides layer 2 switching to all the home devices.

Figure 4.7 breaks out the functional pieces of a multifunction home networking device. The router, firewall, Ethernet switch, and even the wireless access point, can be in a single box. Almost everything in the diagram is connected with 10/100 Mbps Ethernet. Unlike the corporate environment, there is no redundancy, and little or no management capability in the networking devices.

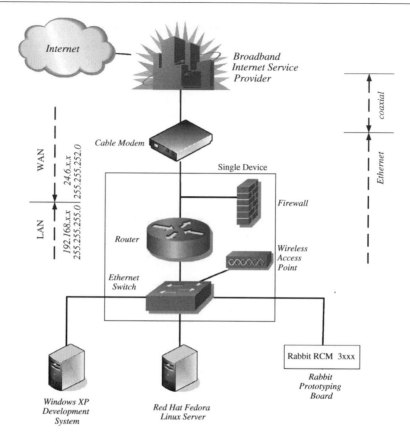

Figure 4.7: Networked Environment for the Home

4.11   *Setting Up a Core Module's Network Configuration*

Before a Rabbit core module can be used for networking, several decisions have to be made, that include where the core module will fit in the overall network, how it will be addressed, whether the networking configuration will be hard coded, or whether it may change at runtime. While this section will not cover all these issues in detail, it will introduce readers to various aspects of networking that need to be considered when bringing up a core module.

4.11.1   *Setting Up the IP Address*

As is the case with any network device, the Rabbit core module needs to have an IP address that is consistent with the overall network addressing scheme where the module will be used. If this is done incorrectly, various network elements will not recognize the core module and will not respond to queries from the device.

Two methods are commonly used to set up an IP address in networking devices:

- *Static* addressing uses hard-coded or manually-configured IP addresses. These addresses do not change unless manually reconfigured. A module that uses static addressing will work on the specific network segment for which it is configured, but will need reconfiguration to work in other networking segments.

- *Dynamic* addressing uses an external entity, a DHCP server that uses a request/response mechanism to assign IP addresses dynamically. How this works most often is each time a network device power cycles, it requests an IP address from the DHCP server.[2] After both devices negotiate and agree upon an IP address, the DHCP server leases that IP address and makes an entry in the DHCP *Client Table*. Since the addressing is dynamic, the device can renew its IP address at any time by requesting the DHCP server for a renewal. Dynamic addressing makes it easy to configure network elements and create a "plug and play" environment where devices can be plugged into different network segments without the need for manual reconfiguration. While static addressing is useful for certain applications, such as router ports or web servers, dynamic addressing may be convenient for instances when the device is acting as a client.[3]

In previous versions of Dynamic C, the programmer often had to use static IP addressing. With version 7.2, Rabbit Semiconductor improved the DHCP implementation. In previous versions, DHCP used to be blocking (that is, when a DHCP negotiation was happening, no other code on the device could be run), but now it isn't. Whether a programmer uses static or dynamic addressing (or which ones will be supported) is a design-time decision. There are good arguments for either, depending on the purpose of the device. For example, if there are multiple embedded devices running the same firmware and static addressing, they would boot up with the same IP address and cause conflict on the network. In such a scheme, DHCP would be preferable.

---

[2] Not only is the address dynamic, but its duration is time limited (and configurable and negotiable).
[3] When initially "bringing up" target hardware, use of a static address can make life easier, since the IP address is one thing that isn't changing.

Two steps are required to set up an addressing scheme with Dynamic C:

1. The programmer should create a custom configuration library, say, `custom_config.lib`, based on the `\lib\tcpip\tcp_config.lib` library. The programmer should make all necessary changes to the custom configuration library. For this chapter, we will create a library called `custom_config.lib` for custom network configurations.

2. The `TCPCONFIG` macro describes various parameters of the physical interface that will be used in the user program. The `TCPCONFIG` macro works as follows:

    *   If the macro has a value of 0, the network configuration is done in the main program, not in any libraries.

    *   If the macro has a value less than 100, the network configuration is done by `tcp_config.lib`.

    *   If the macro has a value at or higher than 100, the network configuration is done in the custom library, such as `custom_config.lib`.

When using the `TCPCONFIG` macro in a custom library, the programmer must copy the appropriate library code from `tcp_config.lib` to custom library. For example, when using static IP configuration with Ethernet, the programmer will define `TCPCONFIG` to be 100, and copy the appropriate code from `tcp_config.lib` to `custom_config.lib`, so that the static IP address will be defined in `custom_config.lib`. The programmer does not have to replicate all of `tcp_config.lib` into `custom_config.lib`, but to simply use enough of it to do the custom network configuration.

The top of `custom_config.lib` must contain the following definition:

```
#ifndef CUSTOM_CONFIG_H
#define CUSTOM_CONFIG_H
```

The programmer should set up the `TCPCONFIG` macro according to Table 4.1. The top of `tcp_config.lib` will describe the steps for creating custom networking configurations.

Instead of creating a separate library for custom configuration, the other option for the programmer is to modify `tcp_config.lib` directly (to set the IP address, gateway, network mask, and DNS server), and to set the `TCPCONFIG` macro accordingly.

In addition, we can use the `ifconfig()` function to make changes at run-time.

Table 4.1 Table Used to Set the TCPCONFIG Macro*

TCPCONFIG	Ethernet	PPP	DHCP	Runtime	Comments
0					Do not do any configuration in the library; this will be done in the main program
1	Yes	No	No	No	
2	No	Yes	No	No	
3	Yes	No	Yes	No	
4	Yes	Yes	No	No	
5	Yes	No	Yes	No	Like #3, but no optional flags
6	Yes	No	No	Yes	
7	Yes	No	Yes	No	DHCP, with static IP fallback

*This table is subject to change, since Rabbit Semiconductor is expected to continue development on the libraries.

4.11.2   Link Layer Selection

We also need to tell the core module which physical interface we are using. Three interfaces are supported:

- Ethernet: This is the main focus of this chapter, and we will build all our code to support this protocol. The macro USE_ETHERNET should be defined to support this protocol, and the macro USING_ETHERNET can be queried to find out whether the core module is using Ethernet.

- Point-to-Point Protocol (PPP): This protocol is commonly used over serial ports; it uses encapsulation and transports other protocols over a point-to-point link. For example, a Rabbit core module's serial ports can be enabled to work with PPP. The macro USE_PPPSERIAL should be defined to support this protocol and the macro USING_PPPSERIAL can be queried to find out whether the core module is using PPP.

- PPP Over Ethernet (PPPOE): With this protocol, an Ethernet frame transports the PPP frame. The macro USE_PPPOE should be defined to support this protocol and the macro USING_PPPOE can be queried to find out whether the core module is using PPPOE.

We will not cover PPP and PPPOE; readers can find more information about these in the Dynamic C TCP/IP manual.

The link layer selection is part of the setup in `tcp_config.lib`. This should not be done in the application; some of the predefined configurations in `tcp_config.lib` should instead be used.

4.11.3   TCP/IP Definitions at Compile Time

Programmers should use the `#use dcrtcp.lib` directive to choose the networking library.

It is critical to call the `sock_init()` function before proceeding with networking or calling any functions that relate to networking. Moreover, the code should check to insure that the call to `sock_init()` has been successful; a return value of 0 indicates that the call to `sock_ init()` was successful.

The network interface takes some time to come up after a call to `sock_init()`. This is especially the case when dynamic addressing is being used, because negotiation with a DHCP server can take time.

A short piece of code will tell us if `sock_init()` has been successful, the interface has come up, and we are ready to proceed:

```
    sock_init();
 while (ifpending(IF_DEFAULT) == IF_COMING_UP) tcp_tick(NULL);
```

Once we are able to proceed, we need to make sure that the function `tcp_tick()` gets called periodically. This can be done within the "big loop" of the program or within a separate costate.

In certain cases, the following definitions will be useful. These will help us allocate enough TCP and UDP socket buffers, respectively:

```
#define                                  1
MAX_TCP_SOCKET_BUFFERS
#define                                  1
MAX_UDP_SOCKET_BUFFERS
#define TCP_BUF_SIZE                     2048
#define UDP_BUF_SIZE                     2048
```

These are described in more detail in the *Rabbit TCP/IP User Manual*. These macros can be modified from the defaults to fit the resource profile of the user application. For example, we might want to increase TCP_BUF_SIZE to increase performance, but at the cost of more memory usage. Note that any definitions of these macros must come before the "#use dcrtcp.lib" line.

4.11.4   TCP/IP Definitions at Runtime

The ifconfig() function is used to make changes at run-time. This function is similar to the TCPCONFIG macro, except that it sets network parameters at runtime. In addition, the programmer can use ifconfig() to retrieve runtime information. The function allows us to set an arbitrary number of parameters in one call.

A number of other functions are available to look at run-time information:

- ifup(): attempts to activate the specified interface.

- ifdown(): attempts to deactivate the specified interface.

- ifstatus(): returns the status of the specified interface, whether it is up or down.

- ifpending(): returns indication of whether the specified interface is up, down, pending up or pending down. This reveals more information than ifstatus(), which only indicates the current state (up or down).

- is_valid_iface(): returns a Boolean indicator of whether the given interface number is valid for the configuration.

These functions are described in detail in the *Rabbit TCP/IP User Manual*, provided on the CD-ROM.

4.11.5   Debugging Macros for Networking

Dynamic C provides a numbers of useful macros for debugging networked applications:

- #define DCRTCP_DEBUG: turns on network-related debugging

- #define DCRTCP_VERBOSE: turns on all verbosity

- #define DCRTCP_STATS: turns on statistics counters

The above macros enable debugging and verbosity for functions related to ARP, IP, UDP, TCP, BOOTP, ICMP, DNS, SNMP, PPP, IGMP, and so on.

When the VERBOSE macros are defined, setting the variable debug_on to a number 0 through 6 will enable various levels of TCP-related messages. The higher debug_on is set, the more messages.

4.12   Project 1: Bringing Up a Rabbit Core Module for Networking

Here we take our first baby steps. Before learning to do something more exciting over a network, we must first insure that the core module comes up and is accessible via the network. In the following examples, we will bring up the RCM3400 prototyping board. We will verify network connectivity by using the ping command, to make sure we can reach the core module on our network segment.

4.12.1   Configuration for Static Addressing

Program 4.6 brings up the prototyping board with a static IP address. In order to do this, we needed to take the following steps:

Table 4.1 tells us that the TCPCONFIG macro needs to be set to a "1" for static addressing. If we use the default values in TCP_CONFIG.LIB, the board will come up with a private Class A address of "10.10.6.100." Assuming that we are going to make the device work in a Class C private address space, as shown in Figure 4.6, we will define a static IP address of "192.168.1.50." Therefore, we need to define a custom library and set up the TCPCONFIG macro accordingly.

We will create a custom configuration library, CUSTOM_CONFIG.LIB, and will store it in the same folder as TCP_CONFIG.LIB. We will copy the following definitions from TCP_CONFIG.LIB to CUSTOM_CONFIG.LIB and will modify them to suit the addressing in our environment:

```
#define _PRIMARY_STATIC_IP              "192.168.1.50"
#define _PRIMARY_NETMASK                "255.255.255.0"

#ifndef MY_NAMESERVER
#define MY_NAMESERVER                   "192.168.1.1"
#endif

#ifndef MY_GATEWAY
#define MY_GATEWAY                      "192.168.1.1"
#endif
```

The above network addresses will need to be modified if we use the board in other network segments.

Next, since we are defining our own configuration, we will define a custom value for the TCPCONFIG macro. Since values above 100 are read from CUSTOM_CONFIG.LIB instead of TCP_CONFIG.LIB, we will use a value of 100; this is consistent with static IP configuration from Table 4.1. Except for the line checking for the value of TCPCONFIG, everything else is just copied from TCP_CONFIG.LIB:

```
#if TCPCONFIG == 100
        #define USE_ETHERNET                 1
        #define IFCONFIG_ETH0 \
                    IFS_IPADDR,aton(_PRIMARY_STATIC_IP), \
                    IFS_NETMASK,aton(_PRIMARY_NETMASK), \
                    IFS_UP
#endif
```

Finally, we need to make sure that the master library file MYLIBS.DIR has an entry for CUSTOM_CONFIG.LIB. Thus, the top two lines of MYLIBS.DIR contain the two libraries we have defined so far in the book:

```
CUSTOMLIBS\MYLIB.LIB
LIB\TCPIP\CUSTOM_CONFIG.LIB
```

Each time Dynamic C gets reinstalled, the LIB.DIR file gets rewritten. Therefore, the file needs to be modified each time a Dynamic C upgrade is performed.

Program 4.6 includes the code needed to bring up the RCM3400 prototyping board with static IP addressing. Once the relevant initialization code has been run, the board displays its IP address in the stdio window.

```
// basicStatic.c

#define PORTA_AUX_IO

#define TCPCONFIG 100

#memmap xmem
#use "dcrtcp.lib"

/********************************/
void main()
{
char buffer[100];

    // debug_on = 5;

    brdInit();

    printf("\nWaiting to bring up TCP with static
addressing...\n");

    sock_init();

    // wait until the interface comes up
    while (ifpending(IF_DEFAULT) == IF_COMING_UP) tcp_tick(NULL);

    /* Print who we are ... */
    printf("My IP address is %s\n\n", inet_ntoa(buffer,
gethostid()) );

    while (1)
    {
            tcp_tick(NULL);
    } // while

} // main
```

Program 4.6: Configuration for Static Addressing

To state the obvious, we need to make sure that the static IP address is not already in use in the network segment. Two devices on the same network segment, using the same IP address, can cause all kinds of conflicts. Moreover, this can look like the beginning of a network attack to managed switches and intrusion detection systems in a corporate environment and these switch ports may get shut down.

To verify connectivity, we should ping the core module from a workstation to make sure we can reach that IP address on the network. Figure 4.8 shows the output of the `ping` utility:

```
C:\>ping 192.168.1.50

Pinging 192.168.1.50 with 32 bytes of data:

Reply from 192.168.1.50: bytes=32 time=1ms TTL=64
Reply from 192.168.1.50: bytes=32 time<1ms TTL=64
Reply from 192.168.1.50: bytes=32 time<1ms TTL=64
Reply from 192.168.1.50: bytes=32 time<1ms TTL=64

Ping statistics for 192.168.1.50:
    Packets: Sent = 4, Received = 4, Lost = 0 (0% loss),
Approximate round trip times in milli-seconds:
    Minimum = 0ms, Maximum = 1ms, Average = 0ms
```

Figure 4.8: Verifying Connectivity to the Core Module

4.12.2   *Configuration for Dynamic Addressing*

In order to support dynamic address allocation through DHCP, we need to change just one macro definition in Program 4.6. The code fragment shown in Program 4.7 does just that:

```
// basicDHCP.c
#define TCPCONFIG 5
```

Program 4.7: Configuration for Dynamic Addressing

We do not need to define anything in CUSTOM_CONFIG.LIB, since we are using a predefined configuration that will be read from TCP_CONFIG.LIB.

Once we compile and run the program, and after the relevant initialization code has been run, the core module will display its IP address in the stdio window. At this point, we can ping the core module to make sure we can reach that IP address on the network.

What happens if there is no DHCP server present? The programmer can set up hard-coded (or configured) "fallback" IP addresses to use in case the Rabbit core module is unable to dynamically receive an IP address. The core module tries to contact the DHCP server several times over a period of about twelve seconds (this can be configured). If there is no response, then it falls back to using a fixed IP address and network mask.

The fallback address should be specified using:

```
ifconfig(IF_DEFAULT, IFS_DHCP_FB_IPADDR, <ipaddr>, IFS_END);
```

See the function description for `ifconfig()` for details.

There is also an `IFS_DHCP_FALLBACK` that tells DHCP whether to allow any fallback, plus `IFG_DHCP_FELLBACK` to test whether the stack is currently using a fallback.

If there is no fallback address, the network port will not be usable, since no host is allowed to have a zero IP address.

4.12.3 A Special Case for Dynamic Addressing

A group of embedded controllers, working together in an environment, can often have the same firmware running in them. This brings us to a special case with dynamic addressing: what happens if all these devices power up and look for a DHCP server, and a DHCP server is not found? Are these devices going to fall back on a default IP address? This will not work, because if they are running the same firmware, they may all default to the same IP address, which will cause conflicts in the network.

When designing for such an environment, the systems designer must consider cases where networked devices have to "look within" for determining an IP address, instead of relying on an external DHCP server. In fact, the Internet Engineering Task Force (IETF)[4] has devoted a working group to this area, called "Zero Configuration Networking."[5]   Among other things that involve small network connectivity, the working group looks at address resolution without DHCP.

---

[4] Look at http://www.ietf.org.
[5] Look at http://www.zeroconf.org/.

4.13 The Client Server Paradigm

Before we explore network programming in greater depth, it is important to understand the client/server paradigm. This is a common approach to network programming, including the Berkeley Socket API, which we will examine in the next section.

The word "server" may make us think of rack-mounted enterprise-grade machines with multiple CPUs, terabytes of storage and redundant power supplies, running complex applications. A server is completely different in the network paradigm, and, for the most part, is similar to a client. From a networking perspective, the main differences between a client and a server are:

- A client initiates the connection. The client requests services when needed, and the server responds. For the most part, the server is listening for incoming requests and only then does it take action.

- A client generally handles one communication at a time while a server can be communicating with multiple clients. A client may have multiple connections with the same or different servers as part of the communication.

Depending on the connection protocol and the programming interface used, the server and the client need to do certain things in sequence in order to establish communication. For example, we will later examine which calls a TCP client has to make and in what order so that it can establish a connection with a server.

How a client and server communicate depends entirely on the application, and both parties must follow a given set of rules for things to work.[6] For example, a browser on a personal computer knows what to do once a user has logged into a brokerage account, and the server on the other side knows that it will now be accepting encrypted communication that it has to act upon. From an embedded systems perspective, both parties have to use the same protocols and have to know what connections to talk to. Section 4.14 describes a well-known interface that helps us keep the communications in order.

---

[6] The set of rules is called a protocol.

4.13.1 *What to Be Careful about in Client Server Programming*

There are certain special cases that we have to be careful about in client server programming:

- Byte Order: although two systems may send or receive the same data, one architecture may send out the most significant byte first while the other may start conversation with the least significant byte. We need to understand how the client and server communicate so that the data bytes do not get swapped unexpectedly. "Big Endian," which specifies "the most significant byte first," is the general rule for network communication.

Ideally, we need to be independent from the platform byte order. This helps make the code portable to platforms with different byte orders, and the code does not depend on the other side to have a specific byte order.

It is often enough to use the standard Berkeley macros htonl, htons, ntohl, and ntohs to convert values between host and network byte order:

- Higher Level Protocol Definition: we should be clear about various aspects of the communication, such as who starts communicating, what data needs to be exchanged and the format of the data, how the request and response are structured, what are the time-outs, and so on. It is critical to avoid conditions where both the client and the server are indefinitely waiting for a response from the other party.

- Disconnection: both parties should agree upon termination of the communication. In certain cases, it may be enough to assume that communication may be terminated at any time, while in other cases, a given sequence of bytes sent from party to another may indicate that the party wishes to disconnect. Either way, the system should be designed accordingly so that it can recover from expected and unexpected disconnection.

4.14 *The Berkeley Sockets Interface*

The sockets interface is yet another subject that several books have been written about, and an exhaustive discussion of Berkeley Sockets is outside the scope of this book. We are presenting the sockets interface here only to introduce the reader to the code on the PC side. Note that the Dynamic C TCP/IP API (Application Programming Interface) is not the Berkeley sockets API, although there are some similarities.

The Berkeley Sockets Interface was developed at UC Berkeley in the early 1980s for BSD (Berkeley Software Distribution) UNIX. It is commonly referred to as the Sockets API or just Sockets.

The sockets API provides an abstraction from the underlying network stack, where the programmer interfaces with the sockets API, without thinking about how the network stack is implemented on a given hardware platform or operating system. The sockets API also helps with platform independence, because it helps make the application code portable between languages and operating systems. Almost all of the popular languages of today have adopted the Berkeley sockets API; the programs presented in this chapter use this API for Java, C#, and C++.

A fundamental concept in the Sockets API is that the destination of a message is a port in the destination machine; the port being a virtual connection spot to plug into or to send a message to. There are "well known port numbers" used by applications in the network stack, and unused port numbers can be dynamically assigned by the application. Therefore, a socket is a combination of an IP address and a port number. Port numbers range from 0 through 65,535, and most operating systems reserve port numbers below 1024. We have to make sure that our applications use port numbers that are allowed by the operating systems we work with. In Section 4.18.1, we will illustrate ports needing to be explicitly opened for I/O since they are blocked by an operating system's built-in firewall.

While the sockets concept is generally applied to two separate computers on a network, sockets can also be used to communicate between processes running on the same computer. For example, there exists an internal loopback address (127.0.0.1) on most systems that can be used as a software interface.

Another advantage of abstracting the network layer is that the client and server applications do not have to be written in the same language or development platform. For example, we will use the same Rabbit program to work with servers on both the Windows and Linux platforms in various languages. Using an agreed-upon interface and networking protocols makes platform and language abstraction possible.

The Berkeley socket API uses two key types of sockets: Datagram and Stream:[7]

---

[7] There is a third type, called raw, used for custom protocol development. It bypasses the TCP and UDP layer and works with raw IP packets.

DGRAM (datagram)[8]   sockets are used for UDP communication. The following functions work with DGRAM sockets:

* `sendto()`: sends data.

* `recvfrom()`: receives data. This can happen only after a `bind()` operation has been performed.

* `bind()`: attaches the socket to a local port address. This needs to be done before data can be sent or received.

STREAM sockets are used for TCP communications. Since TCP is a connection-oriented protocol, a connection needs to be established before a STREAM socket can send or receive data. There are two ways to create a TCP connection:

* An *active* socket can be created by using the `connect()` function. This requires us to specify the TCP/IP port address of the other party and, once connected (i.e., after the TCP three-way handshake), we can carry out two-way communication. A client usually connects to a server through an active connection.

* A *passive* socket can be created by using the `bind()` and `listen()` functions. Servers often start out with a passive socket, where they wait and listen for incoming connections. Once that has taken place, making a call to the `accept()` function creates an active socket. Multiple active sockets can be created from a single passive socket, since TCP allows multiple connections on a single port, as long as the ip_address: port_number combination is unique. For example, each of the following is a separate connection:

  10.0.0.1:5555 — 10.0.0.2:80
  10.0.0.1:5556 — 10.0.0.2:80
  10.0.0.3:5555 — 10.0.0.2:80

Thinking one level higher, a server creates a socket and "names" it, so that it is identifiable and clients can find it, and then the server waits, listening for service requests. Clients create a socket, find a server socket by its name or IP address and port, and then try to connect to the server socket. Once the basic conversation has taken place, both parties can continue with two-way communication.

---

[8] A datagram is to UDP what a packet is to TCP.

4.14.1 Making a Socket Connection

We will examine the socket interface both from the server side and the client side. In either case, before a client or server can talk to anyone, it needs to create a socket. For our purposes the most important part of this step is specifying the type of socket, whether TCP or UDP. Applications need to make the socket() call first to create the socket.

If the server is creating a socket, it must then use the bind() function to attach the socket to an IP address and port number. If using TCP, the server can then get into a listen state and listen for incoming connections. Otherwise, if using UDP, the server can block until it receives a UDP datagram.

If using TCP, the client tries to connect to a server using `connect()` and can then `read()` and `write()` data to that socket. The client does not need to call `bind()`; it gets an arbitrary local port number, which for TCP clients is usually just fine. In case of UDP, the client can simply use `sendto()` and `recvfrom()` to talk to a UDP server.

These operations are summarized in Figures 4.9a and b.

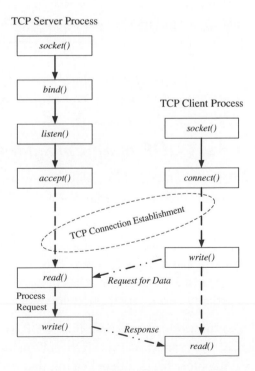

Figure 4.9a: TCP socket Operation

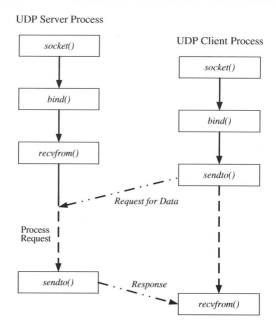

Figure 4.9b: UDP Socket Operation

We will keep these socket operations in mind in developing code on the Windows and Linux platforms. We will use the appropriate calls in C++ and Java to use the sockets interface.

4.15   Using TCP versus UDP in an Embedded Application

UDP is geared towards fast delivery while TCP is focused on guaranteed delivery, albeit at the cost of additional connection and resource overhead. The choice to use TCP or UDP depends on the application.

The additional reliability of TCP means higher implementation overhead of the protocol stack, in terms of code complexity and processing time. This does not mean that using TCP will add complexity and overhead to an application. Depending on what the application is trying to do, using TCP can actually *reduce* application complexity.

Moreover, TCP requires higher networking resources than UDP, since TCP requires an initialization handshake between the sender and receiver, and maintains a connection between the two devices. On the other hand, an application that needs reliable packet delivery will be less complex using TCP than a version building more reliability upon UDP.

If the application requires TCP's guarantees, the programmer will need to do much more work to implement error detection and recovery, fragmentation, and sequencing with UDP.

A significant factor in choosing TCP or UDP would be whether guaranteed delivery is required or not. A closed loop control system, for example, may not need to use TCP since it will monitor the system's behavior from the response. Moreover, if a system has built-in mechanisms for error detection and correction, it can use UDP instead of TCP.

Another factor is the network distance, which includes the number of hops, congestion, and round trip time (RTT) between the source and destination nodes. For example, if the two endpoints are spread far apart and connected through the Internet, UDP is not likely to work well without "adjustments."

Yet another factor to consider is whether the communicating parties need tight synchronization. For example, since TCP ensures that packet will arrive in order, the receiver can act on them in that sequence. On the other hand, packets may arrive out of order with UDP, and we need to consider what effect that will have on the overall system.

While it may first appear that the simplicity of UDP makes it a good choice for embedded systems, choosing TCP versus UDP depends on the needs of the application. For instance, most embedded system use small messages to communicate, which will fit in small datagrams,[9] instead of exchanging many thousands of bytes of data. An application that does not require guaranteed delivery and one where a single datagram can carry all the required data would not need the flow control, buffering, sequencing or reliability offered by TCP.

Let us ponder a bit over the "UDP practical limit of 512 bytes per datagram." The Internet Protocol specification, RFC 791 for IPv4, requires that a host handle an IP datagram size of at least 576 bytes. Therefore, the effective size of the data payload would be: 576 less 20 (IP header without options) less 8 (UDP header), which equals 548 bytes. 512 is likely chosen as a nice binary number.

Another significant concept to consider is that TCP is stream-oriented, while UDP is record-oriented (a record being a packet). If one wants the reliability and simplicity of TCP with a record-oriented interface, it can be built quite easily.

---

[9] UDP has a practical limit of 512 bytes per datagram, but it could be 1472 bytes on local Ethernet. We definitely do not want UDP datagrams larger than the Maximum Transmission Unit (MTU), since that will force fragmentation, and worse, may cause reliability issues.

Broadcast or multicast applications should use UDP. If these applications start creating connections between each client and server, they can overwhelm the network resources. For example, UDP is often used for real-time streaming. Some packets will be dropped, and that's acceptable. On the other hand, if the application needs to record streamed audio at the highest quality, for example, it should use TCP to maintain a good connection.

If the programmer is trying to build a high degree of reliability in communication by implementing flow control, congestion avoidance, error detection, and retransmission, it may be simpler to use TCP and let the lower level routines do the hard work "behind the scenes." Implementing these mechanisms when using UDP will often not be worth it.

Programmers are strongly discouraged from trying to "reinvent TCP" with UDP. The biggest risk is to add congestion to the whole network. In addition, the resultant application will be more complex than the equivalent using the built-in TCP stack. This is primarily because the TCP API is well though out, while if a programmer were to reinvent TCP, they will likely create something clumsy and inefficient, unless this is being done with great expertise.

4.16   Important Dynamic C Library Functions for Socket Programming

We will use a number of Dynamic C library functions in our networking applications. Let us examine these from the perspective of the various phases of the connection. The goal here is not to provide too much detail about each function but to introduce the reader to the available functions. To keep things simple, the parameters to the functions are not shown here.

Parameters to library functions can be examined in Dynamic C by placing the cursor over a function and hitting control-H.

Most of these functions work with both TCP and UDP sockets and the reader has to read about the functions in greater detail or look at sample programs to determine the behavior of each function for use with TCP or UDP communication.

Also note that these functions are current as of Dynamic C 8.61. Future versions may modify these functions or add new ones, and the reader is encouraged to consult the latest Dynamic C documentation available.

4.16.1 Functions Used for Connection Initialization or Termination

- `sock_init`: should be called before using other networking functions. It initializes the necessary low level packet drivers. The return value indicates if the initialization was successful.

  After this function is called, the network interface will take some time to come up, especially if DHCP is being used. The programs in this chapter use a while loop to wait until the interface has completed initialization or has failed to come up.

- `tcp_listen`: tells the low level packet driver that an incoming session for a particular port will be accepted.

- `tcp_open`: actively creates a session with another machine with the given IP address and port number.

- `tcp_extopen`: this is an extended version of tcp_open and also actively creates a session with another machine.

- `sock_close`: closes an open socket. In order to de-allocate resources, it is good practice to close open sockets when the client and server agree to disconnect or when the client or server has established that the other party has gone away for good.

4.16.2 Functions Used to Determine Socket Status

- `sock_established`: helps us determine whether a socket is currently active.

- `sock_bytesready`: indicates whether data is ready to be read from a socket.

- `sock_alive`: indicates one of two states of a socket connection:

  a. reset or fully closed, or

  b. opening, established, listening, or in the process of closing. tcp_tick performs similar checks but involves greater overhead.

- `sock_readable`: indicates whether a socket connection has data available to be read, and how much data is ready to be read.

- `sock_writable`: indicates whether a socket connection can have data written to it, and how much data can be written.

4.16.3   Functions Used to Send or Receive Data

These functions work only for TCP communication, but have their counterparts for exchanging UDP datagrams:

- sock_puts: In binary mode, sends a string; in ASCII mode, appends a carriage return and linefeed to the string.

- sock_gets: It reads a string from a socket and replaces the carriage return or linefeed with a "\0."

- sock_fastwrite: Writes as many bytes possible to the socket and returns the number of bytes it wrote. The equivalent UDP functions are udp_send and udp_sendto.

- sock_fastread: Reads the given number of bytes from a socket, or the number of bytes immediately available. The equivalent UDP functions are udp_recv and udp_recvfrom.

- sock_awrite: Does an "all or none" write to a socket, meaning that it writes the given number of bytes to a socket, and, if that amount of data cannot be written, it writes nothing to the socket.

4.16.4   Blocking versus Nonblocking Functions

Blocking functions [such as sock_write() and sock_read()] can block the program while they wait for an event to occur. Nonblocking functions [such as sock_fastwrite() and sock_fastread()] will either do their task immediately, or indicate to the programmer that they could not (or could only do part of it). Nonblocking functions are more difficult to use, but when an application has other tasks that must be done in a timely fashion, then they must be used (at least in terms of cooperative multitasking). Anything more than a trivial program that uses cooperative multitasking will need to use the nonblocking functions.

> In using these programs, it will be best to configure any firewalls running on the PC to open the ports we need to use. For the sake of consistency, we will use port 8000 for all of our examples on the Rabbit side as well as the PC side, and the firewall may block access to this port on the PC. This will cause some of these programs to not work.

4.17   Project 2: Implementing a Rabbit TCP/IP Server

We will use the RCM3400 prototyping board as a TCP server. A client will establish a connection with the TCP server and will request temperature readings from the server. For each request, the server will send out a sample and a sequence number, followed by a carriage return and a linefeed. The sequence number will ensure that the client will know if it has missed receiving any samples. This precaution is on top of the sequencing mechanism that TCP already implements behind the scenes. This is shown here for illustration purposes only; such precaution, of course, is a waste of bandwidth, unless it is useful to the application.

In order to establish a connection to the server, the client would need to know the server's IP address and port number. While it is easy for us to use dynamic addressing for the server, we will use static addressing so that the client will know for sure the server's IP address. Otherwise, each time the server power cycles, it may come up with a new IP address, and the client may not be able to ever find the server again or that the client code may have to be recompiled again with the server's IP address.

4.17.1   The Server TCP/IP State Machine

"Big loop" implementation for multitasking is a common technique, especially for cooperative multitasking. We will implement TCP and UDP communication using a combination of state machine programming as well as big loops for cooperative multitasking. The TCP Server program implements the TCP state machine shown in Figure 4.10.

The server takes periodic temperature readings, specified by the number of milliseconds between samples in the SAMPLE_DELAY constant. Flashing the sample LED requires 50 ms each time, and this serves as the minimum delay between samples.

Once a connection is established with a client, the server sends the readings and a sequence number to the client. The client application can use the sequence numbers to determine if any samples were not received.

The server sends the number of samples defined by the SAMPLE_LIMIT constant, and then sends a special code, "END." The server also resets the sequence number at this point. The client terminates the connection once it receives the "END" code.

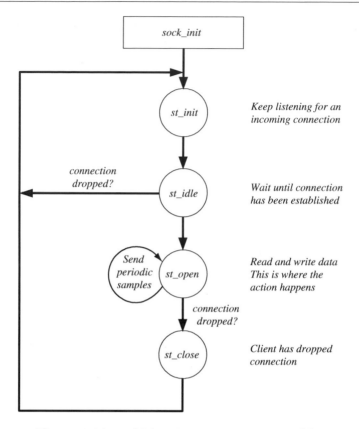

Figure 4.10: Rabbit TCP Server State Machine

The Rabbit TCP server uses the DS1 and DS2 LEDs on the prototyping board to indicate the following:

- The DS1 LED remains on while the client and server are connected.

- The DS2 LED flashes each time a sample gets sent.

The Rabbit server uses two costatements: one to implement the TCP state machine and the other one to take periodic temperature samples. The costate that takes temperature samples keeps an eye on a connection being established, and, once that happens, it starts transmitting the temperature readings and sequence numbers to the client.

Program 4.8 shows a code fragment for the Rabbit Server's TCP state machine:

```
// ADCservTCP.c

<code removed for brevity>

tcp_state = st_init;
sequence = 0;

for(;;)
{
    tcp_tick(NULL);

    costate
    {
    switch (tcp_state)
    {
    case st_init:
        if (tcp_listen(&serv, 8000, 0, 0, NULL, 0))
            tcp_state = st_idle;
        else
        {
            printf ("\nERROR - Cannot initialize TCP Socket\n");
            exit (99);
        }
        break;

    case st_idle:
        if (sock_established(&serv) || sock_bytesready (&serv) != -1)
        {
            tcp_state = st_open;
            connectled(LEDON);
        }
        break;

    case st_open:
        if (!sock_readable (&serv))
        {
            tcp_state = st_close;
            sock_close(&serv);
        }
        break;
```

```
     case st_close:
        // not much to do here; the socket is closed elsewhere

     default:
     } // switch

     if (!sock_alive (&serv))
     {
        tcp_state = st_init;
        connectled(LEDOFF);
     }
   } // costate

 } // for infinite loop
```

Program 4.8: Rabbit-based TCP Server

The costatement that determines whether data needs to be sent to the client is not shown here.

4.17.2 Working with General-Purpose TCP Utilities

Before we write any client code on the PC platform, we must make sure that the Rabbit TCP server runs as planned. This will also help us establish a reference point for the future; if commonly available utilities are able to talk to the Rabbit but our Java or C++ code cannot, we will easily know that we need to start debugging the PC code.

A simple way to check the server is to just use a telnet connection. The telnet utility is described in Section 4.17.1 and can be invoked with the IP address and port number of the TCP server:

```
C:>telnet 192.168.1.50 8000
```

A more versatile utility, Netcat, can be invoked as a TCP or UDP client or a server. For this project, we will invoke Netcat to connect with the Rabbit TCP server (Netcat is described in Section 4.20.4). A screen output is shown in Figure 4.11:

```
C:\Netcat>nc 192.168.1.50 8000
86.784568, 0
86.784568, 1
86.784568, 2
86.784568, 3

<output removed for brevity>

86.784568, 197
86.772720, 198
86.784568, 199
END
86.784568, 0
86.784568, 1
86.784568, 2
86.784568, 3
^C
C:\Netcat>
```

Figure 4.11: Netcat Talking to the Rabbit TCP Server

We can verify that the Rabbit server is sending us the samples and sequence numbers, as well as the "END" code after the maximum number of samples hard coded in the program. Now that we are satisfied that the Rabbit TCP server works as expected, we will use Java and C++ based TCP clients to talk to the Rabbit TCP server.

4.17.3   Working with a Java TCP/IP Client

The core of the Java client is shown in Program 4.9:

```
// tcpClient.java

<code removed for brevity>

    socket = new Socket(server, port);

    BufferedReader reader
        = new BufferedReader(
            new InputStreamReader(socket.getInputStream() ));

    while((line = reader.readLine()) != null)
    {
        if(line.equals("END"))
        {
        break;
      }

        System.out.println(line);
    } // while

    socket.close();
```

Program 4.9: Java-based TCP Client

While it is easy to enter command line arguments for the server's IP address and port number, the Java client uses constants for these values.

The code first creates a socket with the `Socket(server, port)` call, and then creates a stream reader with the `socket.InputStreamReader()` call.

When the work is done, it closes the socket with the `socket.close()` call.

4.17.4   Working with a C++ TCP/IP Client

The TCP client built with Microsoft C++.net has similar functionality as the Java client in the previous section. Although this code is compiled with the .net compiler, it has been built as a Windows32 console application. Similar to the Java program, the TCP client uses constants for the server's IP address and port number; it is easy to change the code so that the user can enter these values manually.

A fragment of the C++ connection code is shown in Program 9.5:

```
// TCPClient.cpp

<code removed for brevity>

// Connection Phase
printf("\nLooking for Rabbit TCP Server at %s:%d\n",
        RABBIT_SERVER, htons(RABBIT_PORT));

SOCKET clientSock=Connect(RABBIT_SERVER);

// Send / Receive Phase
Send(clientSock,"\n");

while (!done)
{
recvbytes=recv(clientSock, receivedStr, 100, 0);
receivedStr[recvbytes-1]='\0';
printf("\n%s", receivedStr);
// printf(": %d bytes", recvbytes);

Send(clientSock,"\n");
if (!strnicmp(receivedStr,"END",3)) done = 1;
}

// Closing Phase
printf("\n\nReceived END message from Rabbit Server!!");
printf("\nTerminating TCP connection ...");

closesocket(clientSock);
WSACleanup();
```

Program 4.10: TCP Client Built in C++

The client uses the *Winsock* library to implement various phases of the communication, such as *connect*, *send/receive*, and *close*. If the C++.net application was built instead of the Windows32 console application, we could have used the various goodies that Windows provides, such as text boxes and scroll bars. Moreover, C++.net would have allowed us to use the try/catch exception handling mechanism that we will use with Java.

The client displays a couple of status messages in the console window, and then starts to display the temperature data (temperature and sequence number). Once the Rabbit server has reached the sample count, the server sends an "END" string to the client, at which point the client terminates the connection.

Figure 4.12 shows an output from the C++ client. To reduce ambiguity and aid in debugging, the client displays the IP address and port number of the server it is seeking.

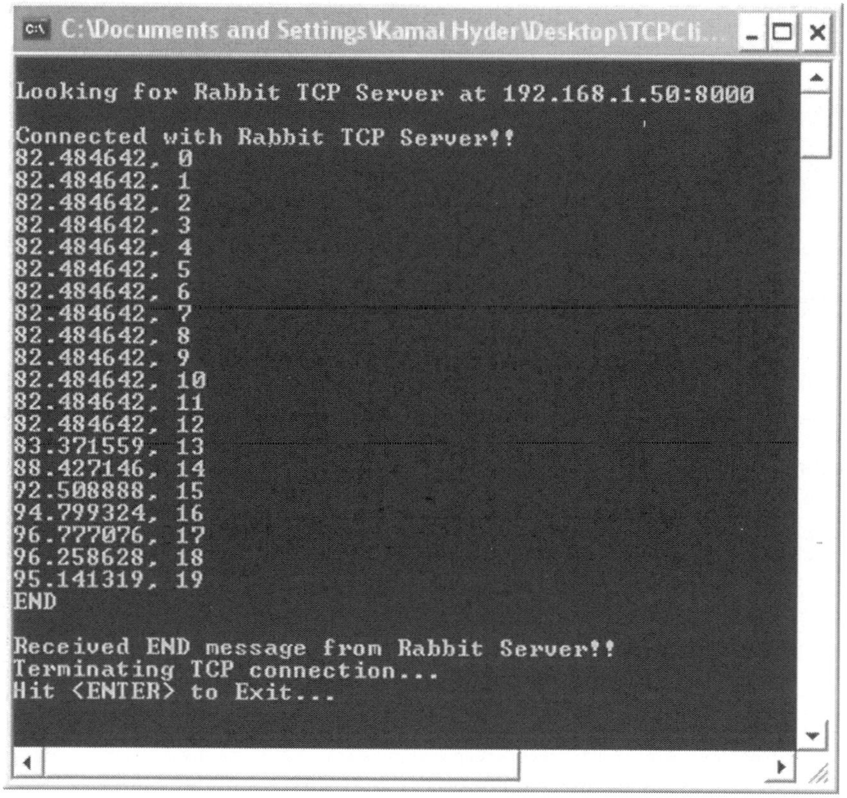

Figure 4.12: Output of the TCP Client

While the code was running, we held the thermistor that made the temperature rise. Then we let go of the thermistor, at which point the temperature started falling. The Rabbit server was programmed to send the "END" terminator after twenty samples, and the client program ended as expected.

4.18 Project 3: Implementing a Rabbit TCP/IP Client

In this project, we will program the RCM3400 prototyping board as a TCP client that will read a channel from the analog-to-digital converter (ADC) and send a reading to the server. When connecting to a server, a TCP client typically does an active open, which requires us to specify the IP address and port number of the server. Once connected, the client carries out two-way communication.

In any two-way communication, it is important to define the highest level protocol. Although TCP takes care of various lower level details for this project, we still need to establish how the client will signal start of communication, how the server will request a channel number for the ADC, and how the client will return the appropriate ADC reading. Just as significant is knowing how the connection will be terminated. We have a simple scheme here—the server can either send a channel number in the range 0 through 7 (yes, the Rabbit client does range checking) or the server sends a "quit" and the client terminates the connection. The higher level connection sequence is shown in Figure 4.13a:

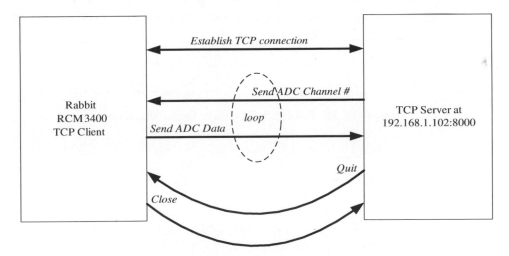

Figure 4.13a: Two-way Communication between Rabbit Client and Server

The state machine for the Rabbit TCP client is shown in Figure 4.13b. This is similar to the state machine for the Rabbit TCP server, except for a couple of states that wait for connection to be established and send data to the server.

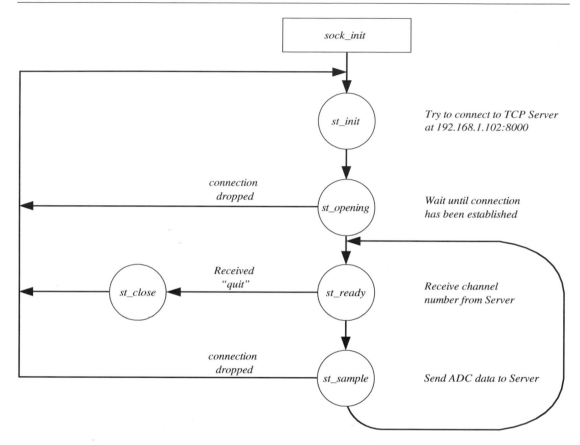

Figure 4.13b: Rabbit TCP Client State Machine

Either party can drop the connection at any time, and we need to account for this possibility.

The Rabbit client comes up with DHCP and displays its IP address in the stdio window. It then tries to connect with a TCP server at IP address 192.168.1.102 port 8000. Program 4.11 shows the Rabbit Client TCP state machine:

```
// ADCclientTCP.c

<code removed for brevity>

switch (tcp_state)
{

case st_init:
    // printf("In Init State");
    connectled(LEDOFF);
    if (tcp_open(&serv, 0, resolve (SERVER), PORT, NULL))
            tcp_state = st_opening;
    else
            exit (EXIT_ON_ERROR);
    break;
case st_opening:
    if (sock_established(&serv) || sock_bytesready (&serv) != -1)
    {
    tcp_state = st_ready;
        connectled(LEDON);
    }
    break;

case st_ready:
    // check for the socket being readable
    // this is important to check in case the client is trying to
    // close a connection

    if (!sock_readable (&serv))
    {
            tcp_state = st_close;
            sock_close(&serv);
            break;
    }

    if (sock_gets(&serv, buff, sizeof(buff))>0)
    {
            if ((!strcmpi(buff, "q")) || (!strcmpi(buff, "quit")))
            {
                    tcp_state = st_close;
                    sock_close(&serv);
            }
            Else
```

```
                {
                    channel = atoi(buff);

                    if ((channel>=0) && (channel <= 7))
                    {
                    printf("\nChannel Number = %d", channel);
                    }
                    else
                    {
                    printf("\nInvalid Channel Number! Data from \
                            channel 7 will be sent");
                    channel = 7;
                    }

                    tcp_state = st_sample;
                } // else
        } // if
        break;

case st_sample:
    // printf("In Sample State");
    // not much to do here; the other costate is doing the sampling

case st_close:
    // printf("In Close State");
    // not much to do here; the socket is closed elsewhere

default:
} // case
```

Program 4.11: Rabbit-based TCP Client

The Rabbit code implements the TCP state machine in a costatement. It tries to initialize the TCP socket in state st_init, and returns an error if it cannot. In st_opening, it proceeds to the ready state if the socket is active or if there are bytes ready to be read. It proceeds to st_sample once it has received a valid channel number from the TCP server.

A different costatement keeps track of whether we are in the st_sample state. If so, the Rabbit code sends a sample and returns to st_ready so that it can receive a new channel number from the server.

In development phase of the project, it is good practice for the server and client to print their IP address on the screen. This will avoid confusion and will aid in debugging. When doing a final build, the developer can simply comment out the printfs or encapsulate them in a

VERBOSE macro. We used a bunch of `printfs` in the code for debugging purpose and then commented them out once we confirmed that the code worked as planned.

4.18.1   Disabling the Windows XP Firewall

At first, we had trouble with this project. While the Linux box connected with the Rabbit client, the Windows PC did not. The machine was running Windows XP Professional, with the Internet Connection Firewall (ICF) enabled. The firewall blocked access to port 8000 for the PC-based servers. We had to choose between disabling the firewall or opening up our ports of choice—we opted to open up port 8000 for both TCP and UDP, and the following procedure can be used to do that on a Windows XP system. Other software firewalls will require different instructions for creating the port 8000 hole.

From the "Network Connections" folder, right click on the appropriate connection. This will most likely be labeled the "local area network." Choose Properties ⇨ Advanced ⇨ Settings.

From the "Advanced Settings" dialog box, click on "Add" to define a new setting. Figure 4.14a shows the values we entered for opening up TCP port 8000. Similarly, we have to add a new setting for UDP port 8000.

Figure 4.14a: Adding a Setting for TCP Port 8000 in Windows XP Professional

After adding another setting for UDP, the "Advanced Settings" dialog would look similar to the one shown in Figure 4.14b. This confirms that we have opened up port 8000 for both TCP and UDP.

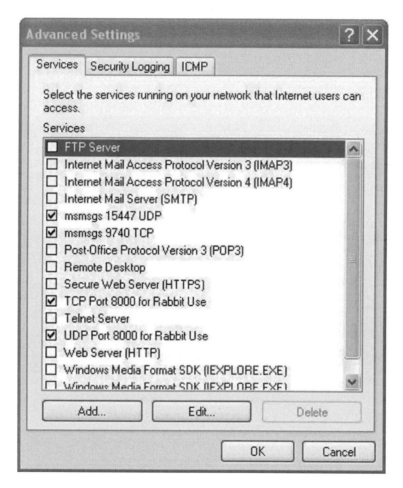

Figure 4.14b: Port 8000 Opened Up for TCP and UDP

If the user is running a software firewall client on the PC, the appropriate steps should be taken to disable the firewall completely or to "punch a port hole" through it.

4.18.2   *Verifying the Client Code*

Simply putting a few `printfs` in the right places in the code can save debugging time. We coded the client to output its IP address in the stdio windows, as well as the IP address and port number of the server it is trying to connect to. Moreover, the client prints the channel number that it receives from the server. Having this information on the screen removes any ambiguity and confusion that may arise when things do not work as expected. The output of the Rabbit client is shown in Figure 9.15a.

```
Welcome to the Rabbit TCP Client!!

My IP address is 192.168.1.103

Trying to connect to TCP Server at 192.168.1.102:8000.


Connected to Server at 192.168.1.102:8000.

The server should enter ADC Channel to read from Client, or QUIT

Channel Number = 4
Channel Number = 7
Bye!
```

Figure 4.15a: Output from Rabbit Client

The output from the Netcat server is shown in Figure 4.15b. The command line arguments shown for Netcat make it listen at 8000 for incoming connections.

```
C:\Netcat>nc -l -p 8000
4
2.000000
7
1070.000000
q

C:\Netcat>
```

Figure 4.15b: Output from Netcat Server

Now that we have verified that the Rabbit TCP client is able to make a connection and that it is working as planned, we can implement a PC-side server in C# and Java.

4.18.3   Working with a Java TCP/IP Server

A fragment of the Java TCP server is shown in Program 4.12:

```java
// TCPClient.java

<code removed for brevity>

try {
    System.out.println("Input details from keyboard" +
    '\n' +"If you want to exit the connection enter 'q' or 'quit':");

    // The Client reads the standard input from keyboard
    BufferedReader inFrmUsr =
    new BufferedReader(new InputStreamReader(System.in));

    // When some client asks for tcpServSocket,There is a connection
    // established between the tcpServSocket and tcpClientSocket
    Socket connSocket = tcpServSocket.accept();

    // This stream provides process input from the socket
    BufferedReader inFrmClient =
    new BufferedReader(new InputStreamReader(connSocket.
    getInputStream()));

    // This stream provides process output to the socket
    DataOutputStream outStream =
        new DataOutputStream(connSocket.getOutputStream());

    while (bflag)
    {
        // Places a line typed by user in the strInput
        strInput = inFrmUsr.readLine() ;

        outStream.writeBytes(strInput + '\n');

        // Places a line from the client
        clientStr = inFrmClient.readLine();

        System.out.println("From Client: " + clientStr);
```

```
        if ((strInput.equalsIgnoreCase("q")) ||
           (strInput.equalsIgnoreCase("quit")))
                   bflag = false;
   }

   // Close the socket
   tcpServSocket.close();
} // end try

catch (IOException exp)
{
    System.out.println("Connection closed between client and server");
} // end catch
```

Program 4.12: Code Fragment that Implements the Java Server

Similar to Project 2 (Section 4.17), the Java code creates `InputStreamReader` instances for the socket and user input. It also creates `DataOutputStream` for output to the Rabbit client and processes input until it receives a "quit" signal.

The try/catch blocks in Java form a useful mechanism for exception handling. The *try* keyword guards a section of the code, and, if a method throws an exception, code execution stops at that point and the *catch* block, the exception handler, is executed. The programmer can insert the appropriate code in the exception handler to inform the user.

The output of the Java server is shown in Figure 4.16:

```
C:\ java>java TCPServer
TCP Server IP Address : desktop-fast-28/192.168.1.102
Server hostname       : desktop-fast-28
Server listening on port 8000

Input details from keyboard
If you want to exit the connection enter 'q' or 'quit':
5
From Client: 2.000000
9
From Client: 1068.000000
7
From Client: 1068.000000
q
From Client: null
Server exiting...

C:\ \java>
```

Figure 4.16: Output from Java Server

If the channel number entered by the server is outside of the range 0 through 7, the Rabbit client outputs the ADC value from channel 7.

4.18.4   Working with a TCP/IP Server in C#

The C# code runs as a server. The life cycle of the server is, to listen to incoming request on a specific port. Once the client establishes the connection, both the client and server enter a loop where the client keeps listening for server request and returns a response based on the channel requested. The server, on the other hand, queries the user for channel numbers, which in turn it sends to the client. A fragment of the C# code is shown in Program 4.13:

```
// TCPServer.cs

<code removed for brevity>

Console.Write("Waiting for a connection on port["+port+"] ... ");

// Perform a blocking call to accept requests.
// We could also user server.AcceptSocket() here.

TcpClient client = server.AcceptTcpClient();
Console.WriteLine("Connected!");

data = null;

// Get a stream object for reading and writing
NetworkStream stream = client.GetStream();
do
{
    // Send the Channel Number from User Input Console.
    WriteLine("Input details from keyboard ['q' or 'quit'  exits]");
    data = Console.ReadLine();
    data =((data.ToLower()=="q")?"quit":data)+"\n";

    // convert data to bytes
    byte[] msg = System.Text.Encoding.ASCII.GetBytes(data);

    // Send the user request.
    stream.Write(msg, 0, msg.Length);

    // make sure the buffered data is written to underlying device
    stream.Flush();

    Console.WriteLine(String.Format("Sent: {0}", data));

    int i;

    // Receive the data sent by the client.
    if((i = stream.Read(bytes, 0, bytes.Length))!=0)
```

```
    {
            // Translate data bytes to a ASCII string.
            data = System.Text.Encoding.ASCII.GetString(bytes, 0, i);
            Console.WriteLine(String.Format("Received: {0}\n", data));
    }
}
while(!data.StartsWith("quit"));
// Shutdown and end connection
Console.WriteLine("Closing connection");
client.Close();
```

Program 4.13: Code Fragment that Implements the C# Server

4.19   *Project 4: Implementing a Rabbit UDP Server*

So far, we have focused on TCP communication, and a critical part of that is connection establishment before data can change hands. With UDP, there is no need to establish such a connection—the UDP server just initializes itself to listen on port 8000 and waits until a client sends a datagram to that port. Unlike TCP, communication happens one datagram at a time instead of on a "per connection" basis.

The UDP server sample presented here does essentially the same thing as the TCP samples presented earlier.

Unless the client sends a "C" to close the connection (akin to a "finish" request), the server keeps sending temperature readings and a sequence number with each reading.

The Rabbit UDP server uses the DS1 and DS2 LEDs on the prototyping board to indicate the following:

- The DS1 LED remains on while the client and server exchange data. Once the server receives a "finish" request, it turns the LED off. The LED will turn on when the server receives the next datagram.

- The DS2 LED flashes each time a sample gets sent.
  The state machine for the UDP server is shown in Figure 4.15.

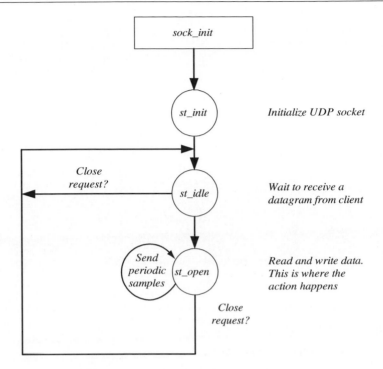

Initialize UDP socket

Wait to receive a datagram from client

Read and write data. This is where the action happens

Figure 4.17: Rabbit-based UDP Server

The code for UDP server's state machine is in Program 4.14. Not shown is the costatement that sends periodic data to the client.

```
// ADCservUDP.c

<code removed for brevity>

for(;;)
{
    // It is important to call tcp_tick periodically otherwise
    // the networking subsystem will not work

    tcp_tick(NULL);

    // UDP state machine
Costate
{
```

```
    switch (udp_state)
    {

    case st_init:
        if (udp_open(&serv, PORT, -1, 0, NULL))
         udp_state = st_idle;
        else
        {
        printf ("\nERROR - Cannot initialize UDP Socket\n");
          exit (EXIT_ON_ERROR);
        }
        break;

    case st_idle:

    // look at data in the buffer for a close request
    buff[0] = 0;

    if ((udp_recvfrom(&serv,buff,sizeof(buff),&cli_ip, &cli_port))
       >= 0)
    {
        if (buff[0] != 'C')
        {
        udp_state = st_open;
        connectled(LEDON);
        }
    }
        break;

    case st_open:
        // nothing to do here; the other costate will check for
        // udp_state
        // to determine whether we are in st_open state

    // keep checking for new packet; if so,
    // update client IP address and port number dynamically

    buff[0] = 0;

    // look at data in the buffer for a close request

    if ((udp_recvfrom(&serv,buff,sizeof(buff),&cli_ip, &cli_port))
       >= 0)
```

```
     {
     if (buff[0] == 'C')

          {
          udp_state = st_idle;
          connectled(LEDOFF);
          }
     }
        break;
     case st_close:

     default:
     } // switch

  } // costate

} // for
```

Program 4.14: Rabbit-based UDP Server

As is the case with most other programs in this chapter, we use two costatements—one to run the UDP server state machine and the other one to send data to the client, once we are in the right state.

4.19.1 Working with a Java UDP Client

To use the program, the user would launch the Java client and would hit <enter> to send a datagram to the UDP server. Each datagram requests a new temperature sample from the Rabbit. The Java client repeats this sequence ten times and then quits.

The Java UDP client is shown in Program 4.15.

```
// udpClient.java

<code removed for brevity>

    for (;;)
    {
        count++;

        String sentence = inFromUser.readLine();
        sendData = sentence.getBytes();

        sendPacket = new DatagramPacket
                    (sendData, sendData.length, remoteAddr, PORT);

        clientSocket.send(sendPacket);


        receivePacket = new DatagramPacket (receiveData,
                          receiveData.length);

        clientSocket.setSoTimeout(2000);
        clientSocket.receive(receivePacket);

        receivedData = new String
            (receivePacket.getData(), 0,
             receivePacket.getLength());

        System.out.println("FROM SERVER: " + receivedData);

        if (count == 10)
        {
            clientSocket.close();
            break;

        }

    } // for
```

Program 4.15: Java UDP Client

Unlike the TCP examples, the Java code creates sendPacket and receivePacket, which are new instances of the DatagramPacket class. It creates a new client socket each time it exchanges data with the UDP server, and then quits once it has counted up to ten responses from the server.

4.19.2   *Working with a C++ UDP Client*

The code functions as a client where the client sends packets containing a channel number to the rabbit server. The client then blocks for response on the same endpoint from the server, displaying it to the user once it receives it. A code fragment of the C++ client is shown in Program 4.16:

```
// cppUDPClient.cpp

<code removed for brevity>

do{
    std::cout << "Input a string :" << std::endl;
    std::cin >> szBuf;
    if(!isalpha(szBuf[0])&& strlen(szBuf)==1 && (szBuf[0]>='0')&&(szBuf[0
]<='7'))
            inValidInput=false;
}
while(inValidInput);
nRet = sendto(theSocket,            // Socket
szBuf,                              // Data buffer
strlen(szBuf),                      // Length of data
0,                                  // Flags
(LPSOCKADDR)&saServer,              // Server address
sizeof(struct sockaddr));           // Length of address

<code removed for brevity>

// get acknowledge from server
nFromLen = sizeof(struct sockaddr);
nret = recvfrom(theSocket,          // Socket
outBuf,                             // Receive buffer
sizeof(outBuf),                     // Length of receive buffer
0,                                  // Flags
(LPSOCKADDR)&saServer,              // Buffer to receive sender's
address&nFromLen);                  // Length of address buffer
```

Program 4.16: Code Fragment for C++ UDP Client

4.20 Some Useful (and Free!) Networking Utilities

We should not take code libraries at face value—things do not sometimes work as planned, and, in the networking world, it is important to be able to independently observe network traffic while debugging. The following tools will be handy in debugging applications.

4.20.1 Ping

This useful utility is found on almost all operating systems, and its simplicity and ubiquity make it a popular application. The key function of Ping is to quickly establish whether a device on the network is reachable. Moreover, because Ping also reports the length of time between the outgoing request and the response, it can help us determine network conditions. Curious programmers should go one step further and explore how Ping uses ICMP messages to perform its function.

Figure 4.18a shows how we used Ping in a Windows XP command window to reach the default gateway in Figure 4.7.

```
C:\>ping 192.168.1.1

Pinging 192.168.1.1 with 32 bytes of data:

Reply from 192.168.1.1: bytes=32 time=1ms TTL=150
Reply from 192.168.1.1: bytes=32 time<1ms TTL=150
Reply from 192.168.1.1: bytes=32 time<1ms TTL=150
Reply from 192.168.1.1: bytes=32 time<1ms TTL=150

Ping statistics for 192.168.1.1:
    Packets: Sent = 4, Received = 4, Lost = 0 (0% loss),
Approximate round trip times in milli-seconds:
    Minimum = 0ms, Maximum = 1ms, Average = 0ms
```

Figure 4.18a: Pinging with Windows XP

Ping also resolves fully qualified domain names with the DNS server to check for connectivity with remote hosts. For example, Figure 4.18b shows how we can try to ping Rabbit Semiconductor's URL.

```
C:\>ping www.rabbitsemiconductor.com

Pinging www.rabbitsemiconductor.com [216.167.101.65] with 32 bytes
of data:

Reply from 216.167.101.65: bytes=32 time=83ms TTL=238
Reply from 216.167.101.65: bytes=32 time=82ms TTL=238
Reply from 216.167.101.65: bytes=32 time=83ms TTL=238
Reply from 216.167.101.65: bytes=32 time=82ms TTL=238

Ping statistics for 216.167.101.65:
    Packets: Sent = 4, Received = 4, Lost = 0 (0% loss),
Approximate round trip times in milli-seconds:
    Minimum = 82ms, Maximum = 83ms, Average = 82ms
```

Figure 4.18b: Pinging a URL

The utility resolves the URL and displays the resulting IP address that it is trying to reach. We can look for packet loss and average response times as well.

4.20.2   Traceroute

While Ping helps us establish whether we can send a packet to a network device, we can use traceroute to find out how to trace the path of the packet and discover how the packet gets to its destination. We can examine each line of the response, which gives us the IP address of a router along the path.

If we discover that a network host can no longer be pinged, we can use traceroute to determine how far we can get to the destination before we lose connectivity.

Figure 4.19 shows the results in a Windows XP environment; we can examine the path taken by a packet to Rabbit Semiconductor's URL.

4.20.3   Ethereal

This open source protocol analyzer is popular with developers and is a useful troubleshooting aid. We can capture live packets as they traverse the network and the utility, with its knowledge of hundreds of protocols, helps us easily decipher what it is seeing on the network. In a Windows environment, it requires us to install the WinPcap packet capture driver.

```
C:\>tracert www.rabbitsemiconductor.com

Tracing route to www.rabbitsemiconductor.com [216.167.101.65]
over a maximum of 30 hops:

  1    10 ms    11 ms    12 ms   10.149.184.1
  2    15 ms    17 ms    11 ms   12.244.101.113
  3    11 ms    22 ms    13 ms   12.244.67.169
  4    13 ms    13 ms    17 ms   12.244.67.14
  5    27 ms    23 ms    27 ms   12.244.72.206
  6    23 ms    22 ms    22 ms   gbr2-p50.sffca.ip.att.net [12.123.13.62]
  7    41 ms    27 ms    23 ms   tbr1-p012702.sffca.ip.att.net
[12.122.11.69]
  8    22 ms    27 ms    23 ms   ggr2-p300.sffca.ip.att.net
[12.123.13.190]
  9    23 ms    24 ms    24 ms   p16-0-1-1.r20.plalca01.us.bb.verio.net
[129.250.9.73]
 10    84 ms    86 ms    99 ms   p16-0-1-3.r21.asbnva01.us.bb.verio.net
[129.250.2.193]
 11    88 ms    95 ms    84 ms   p64-0-0-0.r20.asbnva01.us.bb.verio.net
[129.250.2.34]
 12    91 ms    88 ms    88 ms   p16-5-0-0.r01.mclnva02.us.bb.verio.net
[129.250.2.180]
 13    91 ms   113 ms    84 ms   p4-9-0.a00.alxnva02.us.da.verio.net
[129.250.17.58]
 14    92 ms   103 ms    86 ms   ge-5-0.a01.alxnva02.us.da.verio.net
[129.250.61.10]
 15    85 ms    88 ms    90 ms   ge0031.ed2.wdc.dn.net [216.167.88.124]
 16    83 ms   105 ms    82 ms   rabbitsemiconductor.com [216.167.101.65]

Trace complete.
```

Figure 4.19: Tracing the Route to a Destination

This utility can be downloaded from http://www.ethereal.com. The URL also contains the mirror to the WinPcap packet capture driver.

4.20.4   Netcat

Netcat started out as a useful debugging tool on the UNIX platform and has made its way to the Windows environment. For our purposes, this open source tool serves as an invaluable "reference" TCP or UDP server or client that handles inbound or outbound connections. If the Rabbit code that we write connects with Netcat, we know it works.

Netcat for Windows can be downloaded from http://www.atstake.com/research/tools/network_utilities/. A version of netcat for the Cygwin environment is also available.

4.20.5   Online Tools

A number of tools are available online as well for searching through the Internet. These tools can help us Ping destinations on the Internet, explore DNS, and registration records. Some of the tools are available at:

http://network-tools.com/
http://samspade.org/t/.

4.21   Final Thought

In this chapter, we have merely scratched the surface of what is possible with a networked embedded system. Any networked embedded system will likely not work in isolation. The designer has to think from a *systems* perspective and which networked element can provide what services. For example, in a given industry vertical, an embedded system may serve a monitoring, diagnostic, or billing function, but it will have to report to a system at a higher layer. In such an environment, the designer has to consider systems-wide issues involving access, performance, reliability, availability, and security.

As our world becomes more connected, people will come to expect the devices in their daily lives to be both user friendly and interconnected. The full featured TCP/IP stack from Rabbit along with other more proprietary tools, such as RabbitWeb, allow a system designer to web enable a design with a minimum investment in time and money.

As with any technology, tools used to mold the technology into products are as important as the technology. Rabbit's highly integrated embedded processor is a remarkable technology. Ethernet along with the various protocols we have discussed is also a very useful technology. C-compilers, assemblers and debuggers, and a full-featured TCP/IP stack are the tools used to marry these technologies and create products.

Dynamic C has been used throughout this book for developing the Rabbit side code. As we have also used Java, C#, and C++ under both Windows and Linux for the PC side, engineers have tool choices on the Rabbit side as well.

For all good processors, third party tools exist. Rabbit is no exception. Softools has created an ANSI C compiler, an assembler, and linker bundled under an IDE. The tools offer a traditional development environment with multiple source files.

CHAPTER 5

Error Handling and Debugging

Jack Ganssle
Kamal Hyder
Bob Perrin

In this chapter, we look at error handling, error management, managing changes to memory, and debugging techniques.

You have to think through error handling from the outset of designing a product. Can the system recover from a detected error? How? Will you permit a system crash or at least leave some debugging breadcrumbs behind? How? Does a long jump back to main() suitably reset the system after a catastrophic error is detected? Are you sure the peripherals will reset properly? Maybe it makes sense to halt and let the watchdog time out. Above all, error handling is fundamental to building a reliable embedded system.

The following sections discuss various techniques to avoid, find, and handle errors, and offer numerous tips for effective debugging and troubleshooting, including an important section on managing changes to memory. (In 2003 a woman almost died when her rice cooker reprogrammed her pacemaker's Flash memory, and a man collapsed when a retail store's security device did the same to his!)

5.1 The Zen of Embedded Systems Development and Troubleshooting

Troubleshooting is puzzle solving. When it comes to troubleshooting, a software engineer's mind-set is 80% of the game. The techniques and tools in the toolbox are the remaining 20%.

Before discussing the nitty-gritty technical details of debugging in Dynamic C, we will touch on some concepts to help the reader develop a state of mind conducive to embedded systems development and troubleshooting.

While developing an embedded system, an engineer wears three distinct hats. They are inscribed: DEVELOPER, FINDER, and FIXER.

Wearing the DEVELOPER hat implies a responsibility to find or create a cost-effective solution to the control problem at hand.

An engineer dons the FINDER hat when a malfunction or bug is observed. The responsibility of the FINDER is to delve deeply into the bug and determine the root cause of the malfunction.

Once a bug is identified, the engineer slips on the FIXER hat. The FIXER, much like the DEVELOPER, must find a cost effective solution to a problem. However, unlike a DEVELOPER, a FIXER is usually constrained to an existing design and seldom has the leeway a DEVELOPER does.

5.1.1　The DEVELOPER Hat

Embedded systems developers have many details with which to contend. An enormously helpful philosophy is that of "baby steps for the engineer." This philosophy derives from the fact that diagnosing runtime errors in embedded systems is often significantly more difficult than diagnosing problems in more controlled environments

Software engineers writing middleware code for SAP, C++, or Perl for web servers will often knock out a few hundred lines of code in a module before testing it. In embedded systems, a good rule of thumb is to try to test the code every 20 to 50 lines. Start simple. Progress in small steps. Be sure that each function, each module, and each library operates correctly both with "normal data" as well as with data outside the boundary conditions.

5.1.2　Regression Testing—Test Early and Test Often

Another bit of philosophy admonishes, "Test early and test often."

When testing embedded systems, it is useful to develop a suite of tests that can be run repeatedly throughout the code development. These tests might be as simple as a sequence of button presses on the MMI, or as complex as replacing hardware drivers with functions (often by substituting a library file) that, instead of acquiring live sensor data, report simulated sensor data.

If the test suite runs successfully on a previous code build, but not on the current code build, and the programmer has only made incremental changes between the two code builds, then we have a good idea where to start looking for defective code.

This test suite is also useful when upgrading a compiler. If the code compiles and successfully completes the tests with a previous version of the compiler, but not with the current version, then we have both an indication that something is wrong, and that changing compiler versions caused the bug.

This technique of running the same old tests on new pieces of code is called regression testing.

5.1.3 Case Study—Big Bang Integration and No Regression Test Suite

Not every system will have difficult-to-find bugs, but when one does crop up, the time it takes to reproduce can be staggering. Here is one company's experience with an obscure bug:

The company made equipment that printed numbers in magnetic ink on the bottom of checks. The engineering group was long overdue releasing the new high-speed document printer. Marketing had already announced the new product and sales of the older products had subsequently tanked. Customers were holding out for the announced, but unreleased, high capacity, high-speed printer.

The new printer had a hopper that could hold 5000 documents and could process the documents as fast as the fastest human operator could enter the information to be printed.

The document handler generally operated as expected. With the exception that every 16,000 to 47,000 documents, the code would "lock-up." After a few seconds, the onboard watchdog would reset the system. The log files stored in RAM would invariably be corrupted.

The software team had opted for a big-bang integration of software modules written by five talented engineers for the system's microprocessors. Each engineer had his or her own style and method of "testing" their code. No collaborative regression tests existed.

The hardware design team was confident in the design. Simple bits of code were used to verify that the motors and solenoids were properly under software control. Digital storage oscilloscopes were used to verify noise levels and transients were well within design tolerances.

The software engineers were sure it was an issue associated with the big bang integration. The software team burned many a gallons of midnight oil looking for solutions and spent months developing simulators and regression tests to try and find the bug.

Through robust testing, they found and fixed many bugs that would have eventually caused customers grief. However, through software testing alone, they were unable to reproduce or identify the cause of the "it just crashes every now and again" bug.

After several months, several hundred thousand dollars, and a 30% companywide reduction in force due to lack of sales of existing product, the problem turned out to be ESD (electrostatic discharge). As the checks were pulled along by little rubber rollers, ESD built up on the documents and was accumulated on a plastic photo-sensor. If conditions were dry enough, eventually the charge would discharge into a nearby PCB trace and would addle the CPU.

The solution was to add a couple of grounded little tinsel brushes to wipe off the static build up from the documents.

This simple little problem almost killed the company. It did cost forty people their jobs. What could have been done?

The hardware group had proceeded with their design in incremental steps. Each piece of hardware was tested. The integrated units were tested. The careful testing and development gave the hardware group, and the engineering management, confidence in the hardware design.

The software group had been much more cavalier. Bright people had coded whole modules quickly—sometimes overnight. The rapid development, coupled with the lack of formal regression testing or integration testing, inspired a lack of confidence in the final code base.

Not only did management feel the issue was a software problem, so did the software engineers. They just *knew* that some buffer was overflowing or some pointer was errant.

The fact that as they developed tests for each module, and for the integration of modules they found bugs, further enforced the belief that the code base was unstable.

It wasn't until months had passed and the software group had found and fixed many bugs in the code base that the company developed enough confidence in the code base to begin to seriously look at potential hardware issues again. Which is where the "show stopper" bug was found.

The lesson to be learned here is that not all problems that appear to be software actually are. Additionally, if one is not confident with the code base, a lot of time and money may be spent looking in the wrong place.

5.1.4   The FINDER Hat

Troubleshooting requires a very methodical mind-set. Never assume anything. Never assume that the tools are working flawlessly. Never assume that the hardware is good. Never assume that the code is without bugs.

When troubleshooting, don't look at anything on the bench as being black or white. Think in terms of gray. Each piece of code or hardware or test gear should be assigned a degree of confidence.

For example, consider a digital multimeter (DMM) that is telling us that we have a low supply voltage. Don't assume that the DMM is correct. Crosscheck it.

We could measure a fresh 9-volt battery. If we have confidence in the freshness of the 9-volt battery, then our confidence has increased in the DMM's accuracy (assuming the DMM displayed about 9 volts when connected to the battery).

We could crosscheck the DMM's measurement of the "low" supply voltage with another DMM or better yet, an analog meter. If we get the same measurement from both instruments, the confidence level improves in both the test instruments (reading similarly) and that the measured supply rail might be low.

However, we're not done yet. We can hook up an oscilloscope to the supply rail in question. Even a crummy old slow scope will do in most situations. What we want to eliminate is the possibility that the DMM is giving us a false measurement due to alternating current (AC) noise on the supply rail—which would still be a problem, but a different one than a "low voltage" rail.

Each of the above steps gives us a clue. Each step helps us build confidence in our understanding of the situation. Never assume—always double check.

When debugging a system, there are two distinct hats that must be worn. The first hat is inscribed FINDER, the second FIXER. We should always be cognizant of which hat we are wearing.

The FINDER hat is worn during the diagnostic phase. Before we can properly "fix" a problem, we must understand it.

Some engineers sometimes approach a malfunctioning system with shotgun solutions like these:

- Just make the stack bigger
- Just add an extra time delay here or there

- Just put a few capacitors on the power supply or in the feedback loop

- Just ground the cable shields

- Just disable interrupts

- Just disable the watchdog

- Just make everything a global variable

People that don't necessarily understand the distinction between FINDER and FIXER take this sort of shotgun approach.

Shotgun solutions don't always address the root cause of the problem. They may mask a problem in a particular unit or prototype, or under a given set of conditions, but the problem may reassert itself later in the field.

For example, an engineer may see a problem in a function's behavior, and determine that "all this passing parameters by address is overly difficult—I'll just make this function's variables global." This engineer may have fixed this problem for the function in question, but nothing was done to address the root cause of the problem. The engineer that wrote the code may not have understood how to pass parameters into and out of functions. Quite likely another place in the code has similar problems. These problems may just not be asserting themselves at the present time.

A much better solution would have been for the engineer to recognize the deficiency, correct the function that was exhibiting the odd behavior, and then carefully comb through the code base looking for similar problems.

The FINDER must carefully amass clues, generate a hypothesis for the problem, and then build confidence in the hypothesis through experimentation.

Part of the process of amassing clues involves **never changing more than one thing at a time**. If two or three tweaks are made to a system as part of the exploration of a problem, there is no way to tell which change affected the behavior of the problem.

If we make a change to a system and the change doesn't seem to change the system's behavior, in most circumstances, the next move should be to restore the system to the original configuration (undo the change) and verify the problem still exists.

In many situations, we may make a change that we feel should be incorporated into the final project, even if it didn't affect the problem at hand. If that occurs, be disciplined, undo the change, and proceed with the FINDER's duty. Once the root cause of the problem is found, we can always come back (wearing the DEVELOPER's hat) and make additional changes that we might consider good engineering practice.

For example, consider a target system that is exhibiting difficulty communicating with another system. We notice that the communications cable's shield is ungrounded. We ground the shield, but the communications problem still exists. Even though we might consider it a good engineering practice to ground the shield, the system should still be placed back into the original state until we get to the bottom of the communications problem. Always make just one change at a time. Always go back to the initial configuration that caused the problem.

After making a change to a system and observing the behavior, a useful practice is to write a few notes about what change was made and what behavior was observed. The reasons for this are simple. Humans forget. Humans get confused.

An experiment run a couple of minutes ago might be clear in one's mind, but after another four hours of tracking down a difficult bug, the experiment and results will be difficult to recall with confidence. Time is the enemy of recollection. Take notes. All the best detectives have a little notepad.

Bob Pease is an internationally revered engineer and author. In his book *Troubleshooting Analog Circuits* Pease introduces Milligan's Law—"When you are taking data, if you see something funny, Record Amount of Funny." This is as important in software troubleshooting or system level troubleshooting as it is in analog troubleshooting. Take copious and clear notes.

5.1.5 The FIXER Hat

Once the root cause of a problem is understood, the engineer wearing the FIXER hat can devise a solution. The duty of the FIXER is the same as that of any design engineer—balance the cost and timeliness of a solution with the effectiveness of the solution.

Beyond repairing the bug, the FIXER has an institutional responsibility to provide feedback to the engineer or engineering group that introduced the bug. Without this feedback, the same bug may be introduced into future products.

Depending on the production status of the defective product, the FIXER may have the additional burden of devising material dispositions for existing stock, work in progress, and systems deployed at customer sites. In some situations, the FIXER may be called upon to do as much diplomacy as engineering.

5.2   Avoid Debugging Altogether—Code Smart

These are some guidelines that are useful to both the DEVELOPER and the FIXER. As with any guideline, following these to the extreme is probably not going to be either possible or desirable for embedded systems development. An engineer should keep the spirit of these guidelines in mind.

For example, a guideline might say a NULL pointer check before every pointer access is a good idea. On a PC application this might be acceptable, but a NULL pointer check and special handling of NULL pointer cases can bloat and slow down code too much for an 8-bit system. The programmer must take extra care to make sure the situation doesn't happen in the first place. The spirit of the guideline is clearly "be careful that pointers are initialized correctly."

5.2.1   Guideline #1: Use Small Functions

Keep functions to a page or less of code when possible. Minimize their side effects; for example, don't modify global variables whenever possible. Test functions by themselves to make sure they work as expected for various input values, especially boundary conditions. Use and check return values where invalid input is a possibility. Remember:

- Baby Steps

- Test early. Test often.

5.2.2   Guideline #2: Use Pointers with Great Care

An uninitialized or badly initialized pointer can point to anywhere in memory and therefore corrupt any location in RAM during a write. Be careful that pointers are initialized correctly.

5.2.3   Guideline #3: Comment Code Well

Write good descriptions for functions that include what inputs are used for and what output is to be expected. Comment code lines where the purpose of the code isn't obvious. The code that the programmer writes today may not be so familiar in six months when a bug needs to be fixed.

5.2.4   Guideline #4: Avoid "Magic Numbers"

Use a single macro to define a constant value that is or may be reused so that we only have to change it in one place. For example, the following code uses a macro to define BIGARRAYSIZE, which is then used in more than one place:

```
#define BIGARRAYSIZE 500
char bigarray[BIGARRAYSIZE];
...
memset (bigarray, 0, BIGARRAYSIZE);
```

The following code segment is an example of how NOT to define arrays. Use of magic numbers (like 500) often leads to confusion—especially during future code maintenance or debugging.

```
char bigarray[500];
...
memset (bigarray, 0, 500)
```

5.3   Proactive Debugging

Academics who study software engineering have accumulated an impressive number of statistics about where bugs come from and how many a typical program will have once coding stops but before debugging starts. Amazingly, debugging burns about half the total engineering time for most products. That suggests minimizing debugging is the first place to focus our schedule optimization efforts.

Defect reduction is not rocket science. Start with a well-defined specifications/requirements document, use a formal process of inspections on all work products (from the specifications to the code itself), and give the developers powerful and reliable tools. Skip any of these and bugs will consume too much time and sap our spirits.

However, no process, no matter how well defined, will eliminate all defects. We make mistakes!

Typically 5–10% of the source code will be wrong. Inspections catch 70–80% of those errors. A little 10,000-line program will still, after employing the very best software engineering practices, contain hundreds of bugs we've got to chase down before shipping. Use poor practices and the number (and development time) skyrockets.

Debugging is hard, slow, and frustrating. Since we know we'll have bugs, and we know some will be god-awful hard to find, let's look for things we can do to our code to catch problems automatically or more easily. I call this *proactive debugging*, which is the art of anticipating problems and instrumenting the code accordingly.

5.4   Stacks and Heaps

Do you use the standard, well-known method for computing stack size? After all, undersize the stack and your program will crash in a terribly-hard-to-find manner. Allocate too much and you're throwing money away. High-speed RAM is expensive.

The standard stack-sizing methodology is to take a wild guess and hope. There is no scientific approach, nor even a rule of thumb. This isn't too awful in a single-task environment, since we can just stick the stack at the end of RAM and let it grow downwards, all the time hoping, of course, that it doesn't bang into memory already used. Toss in an RTOS, though, and such casual and sanguine allocation fails, since every task needs its own stack, each of which we'll allocate using the "take a guess and hope" methodology.

Take a wild guess and hope. Clearly this means we'll be wrong from time to time; perhaps even usually wrong. *If we're doing something that will likely be wrong, proactively take some action to catch the likely bug.*

Some RTOSes, for instance, include monitors that take an exception when any individual stack grows dangerously small. Though there's a performance penalty, consider turning these on for initial debug.

In a single task system, when debugging with an emulator or other tool with lots of hardware breakpoints, configure the tool to automatically (every time you fire it up) set a memory-write breakpoint near the end of the stack.

As soon as I allocate a stack I habitually fill each one with a pattern, like 0x55aa, even when using more sophisticated stack monitoring tools. After running the system for a minute or a week—whatever's representative for that application—I'll stop execution with the debugger and examine the stack(s). Thousands of words loaded with 0x55aa means I'm wasting RAM, and few or none of these words means the end is near, the stack is too small, and a crash is imminent.

Heaps are even more problematic. Malloc() is a nightmare for embedded systems. As with stacks, figuring the heap size is tough at best, a problem massively exacerbated by multitasking. `Malloc()` leads to heap fragmentation—though it may contain vast amounts of free memory, the heap may be so broken into small, unusable chunks that `malloc()` fails.

In simpler systems it's probably wise to avoid `malloc()` altogether. When there's enough RAM, allocating all variables and structures statically yields the fastest and most deterministic behavior, though at the cost of using more memory.

When dynamic allocation is unavoidable, by all means remember that `malloc()` has a return value! I look at a tremendous amount of firmware yet rarely see this function tested. It must be a guy thing. Testosterone. We're gonna malloc that puppy, by gawd, and that's that! Fact is, it may fail, which will cause our program to crash horribly. If we're smart enough—proactive enough—to test every `malloc()` then an allocation error will still cause the program to crash horribly, but at least we can set a debug trap, greatly simplifying the task of finding the problem.

An interesting alternative to `malloc()` is to use multiple heaps. Perhaps a heap for 100-byte allocations, one for 1000 bytes, and another for 5000. Code a replacement `malloc()` that takes the heap identifier as its argument. Need 643 bytes? Allocate a 1000-byte block from the 1000-byte heap. Memory fragmentation becomes extinct, your code runs faster, though some RAM will be wasted for the duration of the allocation. A few commercial RTOSes do provide this sort of replacement `malloc()`.

Finally, if you do decide to use the conventional `malloc()`, at least for debugging purposes link in code to check the success of each allocation.

Walter Bright's memory allocation test code, put into public domain many years ago can be found at www.snippets.org/MEM.TXT (the link is case-sensitive) along with companion files. MEM is a few hundred lines of C that replaces the library's standard memory functions with versions that diagnose common problems.

MEM looks for out-of-memory conditions, so if you've inherited a lot of poorly written code that doesn't properly check `malloc()`'s return value, use MEM to pick up errors. It verifies that frees match allocations. Before returning a block it sets the memory to a nonzero state to increase the likelihood that code expecting an initialized data set fails.

An interesting feature is that it detects pointer over- and underruns. By allocating a bit more memory than you ask for, and writing a signature pattern into your pre- and post-buffer memory, when the buffer is freed MEM can check to see if a pointer wandered beyond the buffer's limits.

Geodesic Systems (www.geodesic.com) builds a commercial and much more sophisticated memory allocation monitor targeted at desktop systems. They claim that 99% of all PC programs suffer from memory leaks (mostly due to memory that is allocated but never freed). I have no idea how true this statement really is, but the performance of my PC sure seems to support their proposition. On a PC a memory leak isn't a huge problem, since the programs are either closed regularly or crash sufficiently often to let the OS reclaim that leaked resource.

Firmware, though, must run for weeks, months, even years without crashing. If 99% of PC applications suffer from leaks, I'd imagine a large number of embedded projects share similar problems. One of MEM's critical features is that it finds these leaks, generally before the system crashes and burns.

MEM is a freebie and requires only a small amount of extra code space, yet will find many classes of very common problems. The wise developer will link it, or other similar tools, into every project proactively, before the problems surface.

5.5   Seeding Memory

Bugs lead to program crashes. A "crash," though, can be awfully hard to find. First we notice a symptom—the system stops responding. If the debugger is connected, stopping execution shows that the program has run amok. But why? Hundreds of millions of instructions might elapse between ours seeing the problem and starting troubleshooting. No trace buffer is that large. So we're forced to recreate the problem—if we can—and use various strategies to capture the instant when things fall apart. Yet "crash" often means that the code branches to an area of memory where it simply should not be. Sometimes this is within the body of code itself; often it's in an address range where there's neither code nor data.

Why do we continue to leave our unused ROM space initialized to some default value that's a function of the ROM technology and not what makes sense for us? Why don't we make a practice of setting all unused memory, both ROM and RAM, to a software interrupt instruction that immediately vectors execution to an exception handler?

Most CPUs have single byte or single word opcodes for a software interrupt. The Z80's RST7 was one of the most convenient, as its 0xff, which is the defaults state of unprogrammed EPROM. x86 processors, all support the single byte INT3 software interrupt. Motorola's 68k family, and other processors, have an illegal instruction word.

Set all unused memory to the appropriate instruction, and write a handler that captures program flow if the software interrupt occurs. The stack often contains a wealth of clues about where things were and what was going on when the system crashed, so copy it to a debug area. In a multitasking application the OS's task control block and other data structures will have valuable hints. Preserve this critical tidbits of information.

Make sure the exception handler stops program flow; lock up in an infinite loop or something similar, ensure all interrupts and DMA are off, to stop the program from wandering away.

There's no guarantee that seeding memory will capture all crashes, but if it helps in even a third of the cases you've got a valuable bit of additional information to help diagnose problems.

But there's more to initializing memory than just seeding software interrupts. Other kinds of crashes require different proactive debug strategies. For example, a modern microprocessor might support literally hundreds of interrupt sources, with a vector table that dispatches ISRs for each. Yet the average embedded system might use a few, or perhaps a dozen, interrupts. What do we do with the unused vectors in the table?

Fill them, of course, with a vector aimed at an error handler! It's ridiculous to leave the unused vectors aimed at random memory locations. Sometime, for sure, you'll get a spurious interrupt, something awfully hard to track down. These come from a variety of sources, such as glitchy hardware (you're probably working on a barely functional hardware prototype, after all).

More likely is a mistake made in programming the vectors into the interrupting hardware. Peripherals have gotten so flexible that they're often impossible to manage. I've used parts with hundreds of internal registers, each of which has to be set just right to make the device function properly. Motorola's TPU, which is just a lousy timer, has a 142-page data book that documents some 36 internal registers. For a timer. I'm not smart enough to set them correctly first try, every time. Misprogramming any of these complex peripherals can easily lead to spurious interrupts.

The error handler can be nothing more than an infinite loop. Be sure to set up your debug tool so that every time you load the debugger it automatically sets a breakpoint on the handler. Again, this is nothing more than anticipating a tough problem, writing a tiny bit of code to capture the bug, and then configuring the tools to stop when and if it occurs.

5.6   *Wandering Code*

Embedded code written in any language seems determined to exit the required program flow and miraculously start running from data space or some other address range a very long way from code store. Sometimes keeping the code executing from ROM addresses feels like herding a flock of sheep, each of whom is determined to head off in its own direction.

In assembly a simple typographical error can lead to a jump to a data item; C, with support for function pointers, means state machines not perfectly coded might execute all over the CPU's address space. Hardware issues—like interrupt service routines with improperly initialized vectors and controllers—also lead to sudden and bizarre changes in program context.

Over the course of a few years I checked a couple of dozen embedded systems sent into my lab. The logic analyzer showed writes to ROM (surely an exercise in futility and a symptom of a bug) in more than half of the products.

Though there's no sharp distinction between wandering code and wandering pointers (as both often come from the same sorts of problems), diagnosing the problems requires different strategies and tools.

Quite a few companies sell products designed to find wandering code, or that can easily be adapted to this use. Some emulators, for instance, let you set up rules for the CPU's address space: a region might be enabled as execute-only, another for data read-writes but no executions, and a third tagged as no accesses allowed. When the code violates a rule the emulator stops, immediately signaling a serious bug. If your emulator includes this sort of feature, use it!

One of the most frustrating parts of being a tool vendor is that most developers use 10% of a tool's capability. We see engineers fighting difficult problems for hours, when a simple built-in feature might turn up the problem in seconds. I found that less than 1% of people I've worked with use these execution monitors, yet probably 100% run into crashes stemming from code flaws that the tools would pick up instantly.

Developers fall into four camps when using an execution-monitoring device: the first bunch don't have the tool. Another group has one but never uses it, perhaps because they have simply not learned its fundamentals. To have unused debugging power seems a great pity to me. A third segment sets up and arms the monitoring tool only when it's obvious the code indeed wanders off somewhere, somehow.

The fourth, and sadly tiny, group builds a configuration file loaded by their ICE or debugger on every startup, that profiles what memory is where. These, in my mind, are the professional developers, the ones who prepare for disaster long before it inevitably strikes. Just like with make files, building configuration files takes tens of minutes so is too often neglected.

If your debugger or ICE doesn't come with this sort of feature, then adapt something else! A simple trick is to monitor the address bus with a logic analyzer programmed to look for illegal memory references. Set it to trigger on accesses to unused memory (most embedded systems use far less than the entire CPU address space; any access to an unused area indicates something is terribly wrong), or data-area executes, and so on.

A couple of years ago I heard an interesting tale: an engineer, searching out errant writes, connected an analyzer to his system and quickly found the bug. Most of us would stop there, disconnect the tool, and continue traditional debugging. Instead, he left the analyzer connected for a number of weeks till completing debug. In that time he caught seven—count 'em—similar problems that may very well have gone undetected. These were weird writes to code or unused address spaces, bugs with no immediately apparent symptoms.

What brilliant engineering! He identified a problem, then developed a continuous process to always find new instances of the issue. Without this approach the unit would have shipped with these bugs undetected.

I believe that one of the most insidious problems facing firmware engineers are these lurking bugs. *Code that does things it shouldn't is flawed, even if the effects seem benign.*

Various studies suggest that *up to half the code in a typical system never gets tested.* Deeply nested ifs, odd exception/error conditions, and similar ills defy even well-designed test plans. A rule of thumb predicts that (uninspected) code has about a 5% error rate. This suggests that a 10,000 line project (not big by any measure) likely has 250 bugs poised to strike at the worst possible moment.

5.7 Special Decoders

Another option is to build a special PAL or PLD, connected to address and status lines that flags errant bus transactions. Trigger an interrupt or a wait state when, say, the code writes to ROM.

If your system already uses a PAL, PLD or FPGA to decode memory chip selects, why not add an output that flags an error? The logic device—required anyway to enable memory banks—can also flash an LED or interrupt the CPU when it finds an error. A single added bit creates a complete self-monitoring system. More than just a great debug aid, it's a very intriguing adjunct to high-reliability systems. Cheap, too.

A very long time ago—way back in the 20th century, actually—virtually all microprocessor designs used an external programmable device to decode addresses. These designs were thus all ripe for this simple yet powerful addition. Things have changed; now CPUs sport sometimes dozens of chip select outputs. With so many selects the day of the external decoder is fading away.

Few modern designs actually use all of those chip selects, though, opening up some interesting possibilities. Why not program one to detect accesses to unused memory? Fire off an interrupt or generate a wait state when such a condition occurs. The cost: zero. Effort needed: a matter of minutes. Considering the potential power of this oh-so-simple tool, the cost/benefit ratio is pretty stunning.

If you're willing to add a trivial amount of hardware, then exploit more of the unused chip selects. Have one monitor the bottom of the stack; AND it with `write` to detect stack overflows.

Be sure to AND the ROM chip select with `write` to find those silly code overwrites as well.

5.8   MMUs

If your system has an integrated memory management unit, more options exist. Many MMUs both translate addresses and provide memory protection features. Though MMUs are common only on higher end CPUs, where typically people develop pretty big applications, most of the systems I see pretty much bypass the MMU.

The classic example of this is protected mode in the x86 architecture. Intel's 16 bit x86 processors are all limited to a 1 Mb address range, using what's called "real mode." It's a way of getting direct access to 64 k chunks of memory; with a little segment register, fiddling the entire 1 Mb space is available.

But segmenting memory into 64 k blocks gets awfully awkward as memory size increases. "Protected mode," Intel's answer to the problem, essentially lets you partition 4 Gb of memory into thousands of variable size segments.

Protected mode resulted from a very reasonable design decision, that of preserving as much software compatibility as possible with the old real mode code. The ugly feelings about segments persisted, though, so even now most of the x86 protected mode embedded applications I see use map memory into a single, huge, 4 Gb segment. Doing this—a totally valid way of using the device's MMU—indeed gives the designer the nice flat address space we've all admired in the 68 k family. But it's a dumb idea.

The idea of segments, especially as implemented in protected mode, offers some stunning benefits. Break tasks into their own address spaces, with access rights for each. For the x86 MMU associates access rules for each segment. A cleverly written program can ensure that no one touches routines or data unless entitled to do so. The MMU checks each and every memory reference automatically, creating an easily-trapped exception when a violation occurs.

Most RTOSes targeted at big x86 applications explicitly support segmentation for just this reason. Me, I'd never build a 386 or bigger system without a commercial RTOS.

Beyond the benefits accrued from using protected mode to ensure data and code doesn't fall victim to cruddy code, why not use this very powerful feature for debugging?

First, do indeed plop code into segments whose attributes prohibit write accesses. Write an exception handler that traps these errors, and that logs the source of the error.

Second, map every unused area of memory into its own segment, which has access rights guaranteeing that any attempt to read or write from the region results in an exception.

It's a zero cost way to both increase system reliability in safety-critical applications, and to find those pesky bugs that might not manifest themselves as symptoms for a very long time.

5.9   Conclusion

With the exception of the logic analyzer—though a tool virtually all labs have anyway—all of the suggestions above require no tools. They are nearly zero-cost ways to detect a very common sort of problem. We know we'll have these sorts of problems; it's simply amateurish to not instrument your system *before* the problems manifest a symptom.

Solving problems is a high-visibility process; preventing problems is low-visibility. This is illustrated by an old parable: In ancient China there was a family of healers, one of whom was known throughout the land and employed as a physician to a great lord. The physician was asked which of his family was the most skillful healer. He replied, "I tend to the sick and dying with drastic and dramatic treatments, and on occasion someone is cured and my name gets out among the lords."

"My elder brother cures sickness when it just begins to take root, and his skills are known among the local peasants and neighbors."

"My eldest brother is able to sense the spirit of sickness and eradicate it before it takes form. His name is unknown outside our home."

Great developers recognize that their code will be flawed, so instrument their code, and create tool chains designed to sniff out problems before a symptom even exists.

5.10   Implementing Downloadable Firmware with Flash Memory

The problem with any approach to *in-situ* firmware updates is that when such a feature contains a flaw, the target system may become an expensive doorstop—and perhaps injure the user in the process. Many of the potential pitfalls are obvious and straightforward to correct, but other, insidious defects may not appear until after a product has been deployed in its application environment.

Users are unequaled in their abilities to expose and exploit product defects, and to make matters worse, users also generally fail to heed warnings like "system damage will occur if power is interrupted while programming is underway." They will happily attempt to reboot an otherwise functional system in the middle of the update process, and then file a warranty claim for the now "defective" product.

Any well-designed, user-ready embedded system must include the ability to recover from user errors and other catastrophic events to the fullest extent possible. The best way to accomplish this is to implement a fundamentally sound software update strategy that avoids these problems entirely. This chapter presents one such design.

5.11   The Microprogrammer

The following sections describe a *microprogrammer*-based approach to the implementation of a downloadable firmware feature for embedded systems. This approach is suitable for direct implementation as described, but can also be modified for use in situations where some of its features are not needed, or must be avoided.

The definition of a *microprogrammer* is a system-level description of how the embedded system behaves before, during, and after the firmware update process. This behavior is carefully defined to help avoid many of the problems associated with other approaches to downloadable firmware. A careful implementation of this behavior eliminates the remaining concerns.

The first step in a microprogrammer-based firmware update process is to place the embedded system into a state where it is expecting the download process to occur. The transition to this state could be caused by the user pressing a button marked **UPGRADE** on the device's interface panel, by the system detecting the start or end of a file transfer, or by some other means. In any case, the target system now realizes that its firmware will soon be updated, and brings any controlled processes to a safe and stable halt configuration.

Next, the target system is sent a small application called a *microprogrammer*. The microprogrammer assumes control of the system, and begins receiving the new application firmware and programming it into flash. At the conclusion of this process, the target system begins running the new firmware.

5.12 Advantages of Microprogrammers

One of the biggest selling points of a microprogrammer-based approach to downloadable firmware is its flexibility. The microprogrammer's implementation can be changed even in the final moments before the firmware update process begins, which allows bug fixes and enhancements to be applied retroactively to deployed systems.

A microprogrammer does not consume resources in the target system except when programming is actually underway. Furthermore, since an effective microprogrammer can be as small as 10 K of code or less, the target system does not require a bulky, sophisticated communications protocol in order to receive the microprogrammer at the start of the process—a simple text file transfer is often reliable enough.

The safety of a microprogrammer-based firmware update process is sometimes its most compelling advantage. When the target system's flash chip is not used for any purpose other than firmware storage, the code needed to erase and program the flash chip does not exist in the system until it is actually needed. As such, the system is highly unlikely to accidentally erase its own firmware, even in severe cases of program runaway.

5.13 Disadvantages of Microprogrammers

One of the drawbacks of microprogrammers is that the microprogrammer itself is usually implemented as a stand-alone program, which means that its code is managed separately from both the application that downloads it to the target system, and the application that the microprogrammer delivers to the target system. This management effort requires additional developer resources, the quantity of which is highly dependent on how closely coupled these applications are to each other. Careful attention to program modularity helps to minimize the workload.

A microprogrammer is generally downloaded to and run from RAM, which means that the target system needs some quantity of memory available to hold the microprogrammer. This memory can be shared with the preexisting target application when necessary, but an embedded system that only has a few hundred bytes of RAM in total will probably need a different strategy.

And finally, in a single-flash system the microprogrammer approach requires the target system to be able to run code from RAM. This simply isn't possible for certain microprocessor architectures, in particular the venerable 8051 and family. Hardware-oriented workarounds exist for these cases, but the limitations they impose often outweigh their benefits.

5.14   Receiving a Microprogrammer

The code in Listing 5.1 illustrates the functionality needed by a target system to download and run a microprogrammer. In the example, the target is sent a plain text Motorola S Record file from some I/O channel (perhaps a serial port), which is decoded and written to RAM. The target then activates the microprogrammer by jumping into the downloaded code at the end of the transfer.

Notice that the `programmer_buf[]` memory space is allocated as an automatic variable, which means that it has no fixed location in the target system's memory image. This implies both that the addresses in the incoming S Records are relative rather than absolute, and that the incoming code is position-independent. If your compiler cannot produce position-independent code, then `programmer_buf[]` must be assigned to a fixed location in memory and the addresses in the incoming S Records must be located within that memory space.

The incoming microprogrammer image can be placed on top of other data if the target system does not have the resources to permanently allocate `programmer_buf[]`. At this point the embedded system has suspended normal operations anyway, making the bulk of its total RAM space available for use by the microprogrammer.

Listing 5.1: Code that Downloads and Runs a Microprogrammer

```c
enum srec_type_t {
    SREC_TYPE_S1, SREC_TYPE_S2, SREC_TYPE_S3,
    SREC_TYPE_S7, SREC_TYPE_S9
};
typedef void (*entrypoint_t)(void);

void microprogrammer()
{
    char programmer_buf[8192];
    int len;
    char sbuf[256];
    unsigned long addr;
    enum srec_type_t type;
    entrypoint_t entrypoint;

    while (1) {
        if (read_srecord(&type, &len, &addr, sbuf)) {
            switch (type) {
                case SREC_TYPE_S1:
                case SREC_TYPE_S2:
                case SREC_TYPE_S3:
                    /* record contains data (code) */
                    memcpy(programmer_buf + addr, sbuf, len);
                    break;

                case SREC_TYPE_S7:
                    /* record contains address of downloaded main() */
                    entrypoint = (entrypoint_t)(programmer_buf + addr);
                    break;

                case SREC_TYPE_S9:
                    /* record indicates end of data (code)- run it */
                    entrypoint();
                    break;

            }
        }
    }
}
```

5.15   A Basic Microprogrammer

The top-level code for a microprogrammer is shown in Listing 5.2. For consistency with the previous example, this code also receives an S Record file from some source and decodes it. The microprogrammer writes the incoming data to flash, and the system is rebooted at the end of the file transfer. Although overly simplistic (a plain text file transfer is probably not reliable enough for large programs), this code illustrates all the important features of a microprogrammer.

Listing 5.2: A Basic Microprogrammer

```
void programmer()
{
    int len;
    char buf[256];
    unsigned long addr;
    enum srec_type_t type;

    while (1) {
        if (read_srecord(&type, &len, &addr, buf)) {
            switch (type) {
                case SREC_TYPE_S1:
                case SREC_TYPE_S2:
                case SREC_TYPE_S3:
                    /* record contains data or code- program it */
                    if (!is_section_erased(addr, len))
                        erase_flash(addr, len);
                    write_flash(addr, len, buf);
                    break;


                case SREC_TYPE_S9:
                    /* this record indicates end of data-
                        execute system reset to run new application */
                    reset();
                    break;

            }
        }
    }
}
```

In addition to actually erasing flash, the function `erase_flash()` also manages a simple data structure that keeps track of which sections of the flash chip need to be erased, and which ones have already been erased. This data structure is checked by the `is_section_erased()` function, which prevents multiple erasures of flash sections when data arrives out of order—which is a common occurrence in an S Record file.

5.16   Common Problems and Their Solutions

Regardless of how you modify the microprogrammer-based system description to fit your own requirements, you will encounter some common problems in the final implementation. These problems, along with their solutions, are the subject of this section.

5.16.1   Debugger Doesn't Like Writeable Code Space

Some debuggers, emulators in particular, struggle with the idea of code space that can be written to by the target microprocessor. Most debugging tools treat code space as read-only, and some will generate error messages or simply disallow a writing operation when they detect one. In general, the debugger's efforts to protect code space are well-intentioned. A program writing to its code space usually signals a serious programming error, *except* when a firmware update is in progress. The debugger can't tell the difference, of course. The remedy is to implement a *memory alias* for the flash chip, in a different memory space that is not considered read-only by the debugger.

Consider a 512 KB flash chip that starts at address 0 in the target's memory space. By utilizing a chip select line that responds to an address in the range of 0–1024 KB, you can access the first byte in the flash chip by writing to address 0x80000 instead of address 0, and simultaneously avoid any intrusion from the debugger or emulator.

The memory region between 512 KB and 1024 KB is sometimes called an *alias* of the region 0–512 KB, because the underlying physical hardware cannot distinguish between the two addresses and thus maps them to the same physical location in the flash chip. The debugger can distinguish between the two address ranges, however, and can therefore be configured to ignore (and thereby permit) write accesses in the alias region.

The typical implementation of a memory alias is straightforward: simply double the size of the chip select used to activate the device, and apply an offset to any address that refers to a location in the region designated as the "physical" address region, to move it to the "alias" region. The best place to apply this offset is usually in the flash chip driver itself, as shown in the hypothetical `write_flash()` function in Listing 5.3.

Listing 5.3: Implementing Memory Aliasing in a write_flash() Function

```
#define PHYS_BASE_ADDRESS 0 // physical base address
#define ALIAS_BASE_ADDRESS 0x80000 // alias base address
void write_flash (long addr, unsigned char* data, int len)
{
    addr = (addr - PHYS_BASE_ADDRESS + ALIAS_BASE_ADDRESS);
    while (length) {
        ...
    }
}
```

5.16.2   Debugger Doesn't Like Self-relocating Code

One variation on the microprogrammer approach is to build the microprogrammer's functionality into the target system—a so-called *integral programmer*—instead of downloading it to RAM as the first step of the firmware update process. This strategy has its advantages, but it creates the need to copy the programmer's code from flash to RAM before the flash chip is erased and reprogrammed. In other words, the code must *self-relocate* to RAM at some point during its operation, so that it can continue to run after the flash chip is erased. This approach also requires that the code involved be position-independent.

The code in Listing 5.4 illustrates how to copy the integral programmer's code into RAM, and how to find the RAM version of the function `programmer()`. The symbols `RAM_PROG_START`, `PROG_LEN` and `ROM_PROG_START` mark the regions in RAM and ROM where the programmer's code (which may be a subset of the total application) is located, and can often be computed automatically by the program's linker. The complicated-looking casting in the entry point calculation forces the compiler to do byte-sized address calculations when computing the value of `entrypoint`.

Listing 5.4: Copying Code into RAM

```c
typedef int(*entrypoint_t)(void);
entrypoint_t relocate_programmer()
{
    entrypoint_t entrypoint;

    /* relocate the code */
    memcpy(RAM_PROG_START, ROM_PROG_START, PROG_LEN);

    /* find programmer() in ram: its location is the same
       offset from RAM_PROG_START as the rom version is
       from ROM_PROG_START */
    entrypoint = (entrypoint_t)((char*)programmer
        -(char*)ROM_PROG_START + (char*)RAM_PROG_START);

    return entrypoint;
}
```

When the caller invokes the function at the address returned by
`relocate_programmer()`, control passes to the RAM copy of the microprogrammer
code—and your debugger, if in use, stops showing any symbolic information related to the
`programmer()` function. Why? Because `programmer()` is now running from an
address that is different from the address it was originally located at, so the symbol
information provided to the debugger by the linker is now meaningless.

One solution to this problem is to relink the application with `programmer()` at the RAM
address, and then import this symbol information into the debugger. This would be a
convenient fix, except that not all debuggers support incremental additions to their symbol
table. Another option is to simply suffer until the debugging of `programmer()` is
complete, at which point you don't need to look at the code any more, in theory at least.

If the development environment is based on a hardware emulator rather than a self-hosted
debugging agent, then you can completely avoid the hassles of code relocation by simply
not relocating `programmer()` when an emulator is in use. When an emulator is present
such relocation is, in fact, unnecessary: the opcodes associated with `programmer()` are
actually located in the emulator's memory, rather than flash, so there is no worry that these
instructions will disappear when the flash chip is erased and reprogrammed. This may also
be the case for a self-hosted debugger setup, if the code being debugged and the debugging
agent itself are both running from RAM.

Listing 5.5 illustrates an enhanced `relocate_programmer()` that does not copy code to RAM when an emulator is in use. Instead of using a #if compilation block to selectively enable or disable code copying, the function checks for an emulator at runtime, and skips the code relocation steps if it finds one.

Listing 5.5: A Smarter Code Relocation Strategy, Which Does Not Move Code Except When Necessary

```
typedef int(*entrypoint_t)(void);
entrypoint_t relocate_programmer()
{
    entrypoint_t entrypoint;

    /* test for an emulator, and only relocate code if necessary */
    if (memcmp(FLASH_START, FLASH_START + FLASH_SIZE, FLASH_SIZE))
        entrypoint = programmer;
    else {
        /* no emulator; copy programmer's memory section to ram */
        memcpy(RAM_PROG_START, ROM_PROG_START, PROG_LEN);
        entrypoint = (entrypoint_t)((char*)programmer
            -(char*)ROM_PROG_START + (char*)RAM_PROG_START);
    }
    return entrypoint;
}
```

The test for the presence of an emulator exploits the nature of the *memory alias* strategy discussed in the previous section. Without an emulator attached, the contents of memory in the region `FLASH_START` to (`FLASH_START+FLASH_SIZE`) must be identical to the memory in the region's memory alias, in this case assumed to start at (`FLASH_START+FLASH_SIZE`), because the two address regions actually resolve to the same physical addresses in the flash chip.

With an emulator attached, however, a portion of flash memory is remapped to the emulator's internal memory space, so differences in comparisons can and do occur. These differences cause the `memcmp()` call to return nonzero, disclosing the presence of the emulator to `relocate_programmer()`. To improve performance, the region used for comparison can be reduced to just a handful of bytes that are known to change during compilation (a text string containing the time and date of the compilation, for example), or a known blank location in the flash chip that is preinitialized in the emulator's memory space to a value other than 0xff (the value of an erased flash memory cell).

Testing a blank flash memory cell also discloses the presence of an emulator when a memory alias of the flash chip is not available, which is useful in cases where the flash chip is so large that it occupies more than half the target processor's total address space—thereby making a complete memory alias impossible.

5.16.3 Can't Generate Position-independent Code

Not all embedded development tool chains can produce code that is relocatable to arbitrary locations at runtime. Such position-*dependent* code must run from the address it was placed at by the linker, or the program will crash.

When a position-dependent microprogrammer is all that is available, then it obviously must be downloaded to the location in RAM that the linker intended it to run from. This implies that the memory reserved to hold the microprogrammer must be allocated at a known location.

With an integral programmer approach, there are two options. The first option is to compile the programmer code as a stand-alone program, located at its destination address in RAM. This code image is then included in the application image (perhaps by translating its binary image into a constant character array), and copied to RAM at the start of the firmware update process. This is a lot like a microprogrammer implementation, with the microprogrammer "downloaded" from the target's onboard memory instead of from a serial port.

The second option is to handle the integral programmer code as initialized data, and use the runtime environment's normal initialization procedures to copy the code into RAM. The GNU compiler supports this using its __attribute__ language extension, and several commercial compilers provide this capability as well. The only limitation of this strategy is that it requires enough RAM space to hold the integral programmer code *plus* the program's other data.

5.16.4 No Firmware at Boot Time

Even in the most carefully designed downloadable firmware feature, the possibility still exists that the target hardware could attempt a startup using an accidentally blank flash chip. The outcome to avoid in this case is the sudden activation of the application's rotating machinery, which subsequently gobbles up an unprepared user.

The prescription for this case—which is best applied before a user is actually eaten—involves a careful study of the target processor's reaction to the illegal instructions and/or data represented by the $0xff$'s of an unprogrammed section of flash memory. Many embedded processors eventually halt processing, and tri-state their control signals to let them float to whatever value is dictated by external hardware. Without pull-up resistors or other precautions in place to force these uncontrolled signals to safe states, unpredictable and potentially lethal results are likely.

5.16.5   Persistent Watchdog Time-out

In systems that support downloadable firmware, an unavoidable application defect that forces a watchdog time-out and system reset can lock the microprogrammer out of the embedded system. The extreme case is an accidental `while(1);` statement in a program's `main()` function: the loop-within-a-loop that results (the infinite program loop, wrapped by the infinite watchdog time-out and system reset loop) keeps the target from responding to the **UPGRADE** button, because the system restarts before the button is even checked.

Systems that support downloadable firmware must carefully examine all available status circuitry to determine the reason the system is starting up, and force a transition to **UPGRADE** mode in the event that an excessive number of watchdog or other resets are detected. Many embedded systems do not interrupt power to RAM when a watchdog timeout occurs, so it is safe to store a "magic number" and count of the number of resets there; the counter is incremented on each reset, and once a certain number is reached the system halts the application in an effort to avoid the code that forces the reset.

5.16.6   Unexpected Power Interruption

If power is lost unexpectedly during flash reprogramming, then the target's flash chip is left in a potentially inconsistent state when power is restored: maybe the programming operation finished and everything is fine, but probably not. The best case is when the system's boot code is intact, but portions of application code are missing; the worst case is where the flash chip is completely blank.

In the first case, a checksum can be used to detect the problem, and the system can force a transition to **UPGRADE** mode whether the user is requesting one or not. The only solution to the second case is to avoid ever having it happen.

One way to avoid producing a completely blank flash chip is to never erase the section of flash that contains the system's boot and programmer firmware. By leaving this code intact at all times, a hopefully-adequate environment is maintained even if power is interrupted. This approach may not always be an option, however: the flash chip may be a single-sector part that can only do an all-or-nothing erase, or the "boot sector" of the flash chip may be larger than the boot code it contains, and the wasted space is needed for application code.

A careful strategy for erasing and reprogramming the boot region of a flash chip can minimize the risk of damage to a system from unexpected power interruption, and in some cases eliminate it entirely.

The strategy works as follows: When a request to reprogram the section of flash containing boot code is detected, the programmer first creates a copy of the position-independent code in that section to another section in flash, or assumes that a prepositioned clone of the system's boot code exists. The boot code is then erased, and the target's reset vector is quickly restored to point to the temporary copy of the boot code. Once the boot sector is programmed, the reset vector is rewritten to point to the new startup code.

One of the two keys to the success of this strategy hinges on careful selection of the addresses used for the temporary and permanent copies of the boot code. If the permanent address is lower than the temporary address by a power of two, then the switch from the temporary reset vector to the permanent one can take place by only changing bits from one to zero, which makes a new erase of the flash sector unnecessary. This makes the switch from the temporary reset vector to the permanent one an atomic operation: no matter when power is interrupted, the vector still points to either one location or the other.

Obviously, the other critical element of this strategy is to eliminate the risk of power interruption in the moment between when the boot sector is erased and when the temporary reset vector is written. The amount of energy required to complete this operation can be computed by looking at the power consumption and timing specifications in the datasheet for the flash chip and microprocessor, and is usually in the range where additional capacitance in the system's power supply circuitry can bridge the gap if a power loss occurs. By checking that power has not been already lost at the start of the operation, and running without interruption until the sector is erased and the temporary reset vector is written, the remaining opportunity for damage is eliminated.

The limitation of this strategy is that it depends on a microprocessor startup process that reads a reset vector to get the value of the initial program counter. In processors that simply start running code from a fixed address after a reset, it may be possible to modify this strategy to accomplish the same thing with clever combinations of jmp and similar opcodes.

5.17   Hardware Alternatives

The microprogrammer-based firmware update strategy and its variations are all firmware-based, because they require that some code exist in the target system before that code can be reprogrammed. This creates a sort of chicken-and-egg problem: if you need code in the embedded system to put new code into the embedded system, how do you get the initial code there in the first place?

At least two existing hardware-based methods can be used to jump-start this process: *BDM* and *JTAG*.

A processor with a BDM port provides what is in essence a serial port tied directly to the guts of the microprocessor itself. By sending the right commands and data through this port, you can push a copy of your microprogrammer into the target's RAM space and hand over control to it. The BDM port can also be used to stimulate the I/O lines of the flash chip, thus programming it directly. Many BDM-based development systems include scripts and programs that implement this functionality, but it can also be implemented by hand with a few chips, a PC's printer port, and a careful study of the processor's datasheets.

JTAG is a fundamentally different technology designed to facilitate reading and writing of a chip's I/O lines, usually while the chip's microprocessor (if it has one) is held in reset. Like BDM, however, this capability can be used to stimulate a RAM or flash chip to push a microprogrammer application into it. And also like BDM, a JTAG interface can be built with just a few components and some persistent detective work in the target processor's manual.

A JTAG bus transceiver chip, versions of which are available from several vendors, can be added to systems that lack JTAG support.

5.17.1   Separating Code and Data

The ultimate goal of any downloadable firmware implementation effort is exactly that: a working downloadable firmware feature. Once this capability is safely in place, however, it is important to consider some other capabilities that the system can suddenly offer.

By separating an application's code and data into separate flash sectors, the possibility exists to update the two independently. This is useful if an application uses custom data like tuned parameters or accumulated measurements that cannot be easily replaced if lost. Such data tables must contain version information, however, so that later versions of an application can read old table formats—and so that old applications are not confused by new table formats.

The code in Listing 5.6 demonstrates one way to define a data table containing version information, and how to select from one of several data table formats at runtime.

5.17.2 Flexible and Safe

A microprogrammer-based downloadable firmware feature, when properly implemented, can safely add considerable flexibility to an embedded system that uses flash memory. The techniques described here will help you avoid common mistakes and reap the uncommon benefits that microprogrammers and flash memory can offer.

Listing 5.6: Supporting Multiple Data Table Format

```
/* the original table format,
   a.k.a. version 1 */
typedef struct {
  int x;
  int y;
} S_data_ver_1;

/* version 2 of the table format,
   which adds a 'z' field */
typedef struct {
  int x;
  int y;
  int z;
} S_data_ver_2;

/* the data, which always starts
   with a version identifier */
typedef struct{
  const int ver;
  union {
    S_data_ver_1 olddata[N];
    S_data_ver_2 newdata[N];
  };
} data_table;

void foo ( data_table* dt )
{
  int x, y, z, wdata;
  S_data_ver_1* dv1;
  S_data_ver_2* dv2;
```

```
  switch(dt->ver) {
  case 1:
    for( wdata = 0,
         dv1 = dt->olddata;
         wdata < N; wdata++, dv1++ ) {
      x = dv1->x;
      y = dv1->y;
       /* old data format did not include 'z',
          impose a default value */
      z = 0;
    }
    break;
  case 2:
    for( wdata = 0,
         dv2 = dt->newdata;
         wdata < N; wdata++, dv2++ ) {
      x = dv2->x;
      y = dv2->y;
      z = dv2->z;

    }
    break;
  default:
    /* unsupported format,
       select reasonable defaults */
    x = y = z = 0;
  }
}
```

5.18   Memory Diagnostics

In "A Day in the Life" John Lennon wrote, "He blew his mind out in a car; he didn't notice that the lights had changed." As a technologist this always struck me as a profound statement about the complexity of modern life. Survival in the big city simply doesn't permit even a very human bit of daydreaming. Twentieth-century life means keeping a level of awareness and even paranoia that our ancestors would have found inconceivable.

Since this song's release in 1967, survival has become predicated on much more than the threat of a couple of tons of steel hurtling though a red light. Software has been implicated in many deaths, for example, plane crashes, radiation overexposures, and pacemaker misfires.

Perhaps a single bit, something so ethereal that it is nothing more than the charge held in an impossibly small well, is incorrect—that's all it takes to crash a system. Today's version of the Beatles song might include the refrain "He didn't notice that the bit had flipped."

Beyond software errors lurks the specter of a hardware failure that causes our correct code to die. Many of us write diagnostic code to help contain the problem.

5.19 ROM Tests

It doesn't take much to make at least the kernel of an embedded system run. With a working CPU chip, memories that do their thing, perhaps a dash of decoder logic, you can count on the code starting off . . . perhaps not crashing until running into a problem with I/O.

Though the kernel may be relatively simple, with the exception of the system's power supply it's by far the most intolerant portion of an embedded system to any sort of failure. The tiniest glitch, a single bit failure in a huge memory array, or any problem with the processor pretty much guarantees that nothing in the system stands a change of running.

Nonkernel failures may not be so devastating. Some I/O troubles will cause just part of the system to degrade, leaving much of the rest up. My car's black box seems to have forgotten how to run the cruise control, yet it still keeps the fuel injection and other systems running.

In the minicomputer era, most booted with a CPU test that checked each instruction. That level of paranoia is no longer appropriate, as a highly integrated CPU will generally fail disastrously. If the processor can execute any sort of a self test, it's pretty much guaranteed to be intact.

Dead decoder logic is just as catastrophic. No code will execute if the ROMs can't be selected.

If your boot ROM is totally misprogrammed or otherwise nonfunctional, then there's no way a ROM test will do anything other than crash. The value of a ROM test is limited to dealing with partially programmed devices (due, perhaps, to incomplete erasure, or inadvertently removing the device before completion of programming).

There's a small chance that ROM tests will pick up an addressing problem, if you're lucky enough to have a failure that leaves the boot and ROM test working. The odds are against it, and somehow Mother Nature tends to be very perverse.

Some developers feel that a ROM checksum makes sense to insure the correct device is inserted. This works best only if the checksum is stored outside of the ROM under test.

Otherwise, inserting a device with the wrong code version will not show an error, as presumably the code will match the (also obsolete) checksum.

In multiple-ROM systems a checksum test can indeed detect misprogrammed devices, assuming the test code lives in the boot ROM. If this one device functions, and you write the code so that it runs without relying on any other ROM, then the test will pick up many errors.

Checksums, though, are passé. It's pretty easy for a couple of errors to cancel each other out. Compute a CRC (Cyclic Redundancy Check), a polynomial with terms fed back at various stages. CRCs are notoriously misunderstood but are really quite easy to implement. The best reference I have seen to date is "A Painless Guide to CRC Error Detection Algorithms," by Ross Williams. It's available via anonymous FTP from ftp.adelaide.edu.au/pub/rocksoft/crc_v3.txt.

The following code computes the 16 bit CRC of a ROM area (pointed to by `rom`, of size `length`) using the $x^{16} + x^{12} + x^5 + 1$ CRC:

```
#define CRC_P 0x8408
WORD rom_crc(char *rom, WORD length)
{
  unsigned char i;
  unsigned int value;
  unsigned int crc = 0xffff;

  do
  {
    for (i=0, value=(unsigned int)0xff & *rom++;
        i < 8;
        i++, value >>= 1)
    {
      if ((crc & 0x0001) ^ (value & 0x0001))
          crc = (crc >> 1) ^ CRC_P;
      else crc >>= 1;
    }
  } while (-length);
        crc = ~crc;
        value = crc;
        crc = (crc << 8) | ((value >> 8) & 0xff);

        return (crc);
    }
```

It's not a bad idea to add death traps to your ROM. On a Z80 0xff is a call to location 38. Conveniently, unprogrammed areas of ROMs are usually just this value. Tell your linker to set all unused areas to 0xff; then, if an address problem shows up, the system will generate lots of spurious calls. Sure, it'll trash the stack, but since the system is seriously dead anyway, who cares? Technicians can see the characteristic double write from the call, and can infer pretty quickly that the ROM is not working.

Other CPUs have similar instructions. Browse the op code list with a creative mind.

5.20   RAM Tests

Developers often adhere to beliefs about the right way to test RAM that are as polarized as disparate feelings about politics and religion. I'm no exception, and happily have this forum for blasting my own thoughts far and wide . . . so will I shamelessly do so.

Obviously, a RAM problem will destroy most embedded systems. Errors reading from the stack will surely crash the code. Problems, especially intermittent ones, in the data areas may manifest bugs in subtle ways. Often you'd rather have a system that just doesn't boot, rather than one that occasionally returns incorrect answers.

Some embedded systems are pretty tolerant of memory problems. We hear of NASA spacecraft from time to time whose core or RAM develops a few bad bits, yet somehow the engineers patch their code to operate around the faulty areas, uploading the corrections over the distances of billions of miles.

Most of us work on systems with far less human intervention. There are no teams of highly trained personnel anxiously monitoring the health of each part of our products. It's our responsibility to build a system that works properly when the hardware is functional.

In some applications, though, a certain amount of self-diagnosis either makes sense or is required; critical life support applications should use every diagnostic concept possible to avoid disaster due to a submicron RAM imperfection.

So, my first belief about diagnostics in general, and RAM tests in particular, is to define your goals clearly. Why run the test? What will the result be? Who will be the unlucky recipient of the bad news in the event an error is found, and what do you expect that person to do?

Will a RAM problem kill someone? If so, a very comprehensive test, run regularly, is mandatory.

Is such a failure merely a nuisance? For instance, if it keeps a cell phone from booting, if there's nothing the customer can do about the failure anyway, then perhaps there's no reason for doing a test. As a consumer I could care less why the damn phone stopped working, if it's dead I'll take it in for repair or replacement.

Is production test—or even engineering test—the real motivation for writing diagnostic code? If so, then define exactly what problems you're looking for and write code that will find those sorts of troubles.

Next, inject a dose of reality into your evaluation. Remember that today's hardware is often very highly integrated. In the case of a microcontroller with onboard RAM the chances of a memory failure that doesn't also kill the CPU is small. Again, if the system is a critical life support application it may indeed make sense to run a test as even a minuscule probability of a fault may spell disaster.

Does it make sense to ignore RAM failures? If your CPU has an illegal instruction trap, there's a pretty good chance that memory problems will cause a code crash you can capture and process. If the chip includes protection mechanisms (like the x86 protected mode), count on bad stack reads immediately causing protection faults your handlers can process. Perhaps RAM tests are simply not required given these extra resources.

Too many of us use the simplest of tests—writing alternating 0x55 and 0xAA values to the entire memory array, and then reading the data to ensure it remains accessible. It's a seductively easy approach that will find an occasional problem (like, someone forgot to load all of the RAM chips), but that detects few real world errors.

Remember that RAM is an array divided into columns and rows. Accesses require proper chip selects and addresses sent to the array—and not a lot more. The 0x55/0xAA symmetrical pattern repeats massively all over the array; accessing problems (often more common than defective bits in the chips themselves) will create references to incorrect locations, yet almost certainly will return what appears to be correct data.

Consider the physical implementation of memory in your embedded system. The processor drives address and data lines to RAM—in a 16 bit system there will surely be at least 32 of these. Any short or open on this huge bus will create bad RAM accesses. Problems with the PC board are far more common than internal chip defects, yet the 0x55/0xAA test is singularly poor at picking up these, the most likely failures.

Yet, the simplicity of this test and its very rapid execution have made it an old standby used much too often. Isn't there an equally simple approach that will pick up more problems?

If your goal is to detect the most common faults (PCB wiring errors and chip failures more substantial than a few bad bits here or there), then indeed there is. Create a short string of almost random bytes that you repeatedly send to the array until all of memory is written. Then, read the array and compare against the original string.

I use the phrase "almost random" facetiously, but in fact it little matters what the string is, as long as it contains a variety of values. It's best to include the pathological cases, like 00, 0xaa, ox55, and 0xff. The string is something you pick when writing the code, so it is truly not random, but other than these four specific values you fill the rest of it with nearly any set of values, since we're just checking basic write/read functions (remember: memory tends to fail in fairly dramatic ways). I like to use very orthogonal values—those with lots of bits changing between successive string members—to create big noise spikes on the data lines.

To make sure this test picks up addressing problems, ensure the string's length is not a factor of the length of the memory array. In other words, you don't want the string to be aligned on the same low-order addresses, which might cause an address error to go undetected. Since the string is much shorter than the length of the RAM array, you ensure it repeats at a rate that is not related to the row/column configuration of the chips.

For 64 k of RAM, a string 257 bytes long is perfect. 257 is prime, and its square is greater than the size of the RAM array. Each instance of the string will start on a different low order address. 257 has another special magic: you can include every byte value (00 to 0xff) in the string without effort. Instead of manually creating a string in your code, build it in real time by incrementing a counter that overflows at 8 bits.

Critical to this, and every other RAM test algorithm, is that you write the pattern to all of RAM before doing the read test. Some people like to do nondestructive RAM tests by testing one location at a time, then restoring that location's value, before moving onto the next one. Do this and you'll be unable to detect even the most trivial addressing problem.

This algorithm writes and reads every RAM location once, so is quite fast. Improve the speed even more by skipping bytes, perhaps writing and reading every third or fifth entry. The test will be a bit less robust yet will still find most PCB and many RAM failures.

Some folks like to run a test that exercises each and every bit in their RAM array. Though I remain skeptical of the need since most semiconductor RAM problems are rather catastrophic, if you do feel compelled to run such a test, consider adding another iteration of the algorithm just described, with all of the data bits inverted.

Sometimes, though, you'll want a more thorough test, something that looks for difficult hardware problems at the expense of speed.

When I speak to groups I'll often ask "What makes you think the hardware really works?" The response is usually a shrug of the shoulders, or an off-the-cuff remark about everything seeming to function properly, more or less, most of the time.

These qualitative responses are simply not adequate for today's complex systems. All too often, a prototype that seems perfect harbors hidden design faults that may only surface after you've built a thousand production units. Recalling products due to design bugs is unfair to the customer and possibly a disaster to your company.

Assume the design is absolutely ridden with problems. Use reasonable methodologies to find the bugs before building the first prototype, but then use that first unit as a test bed to find the rest of the latent troubles.

Large arrays of RAM memory are a constant source of reliability problems. It's indeed quite difficult to design the perfect RAM system, especially with the minimal margins and high speeds of today's 16 and 32 bit systems. If your system uses more than a couple of RAM parts, count on spending some time qualifying its reliability via the normal hardware diagnostic procedures. Create software RAM tests that hammer the array mercilessly.

Probably one of the most common forms of reliability problems with RAM arrays is pattern sensitivity. Now, this is not the famous pattern problems of yore, where the chips (particularly DRAMs) were sensitive to the groupings of ones and zeroes. Today the chips are just about perfect in this regard. No, today pattern problems come from poor electrical characteristics of the PC board, decoupling problems, electrical noise, and inadequate drive electronics.

PC boards were once nothing more than wiring platforms, slabs of tracks that propagated signals with near perfect fidelity. With very high speed signals, and edge rates (the time it takes a signal to go from a zero to a one or back) under a nanosecond, the PCB itself assumes all of the characteristics of an electronic component—one whose virtues are almost all problematic. It's a big subject (refer to *High Speed Digital Design—a Handbook of Black Magic* by Howard Johnson and Martin Graham [1993 PTR Prentice Hall, NJ] for the canonical words of wisdom on this subject), but suffice to say a poorly designed PCB will create RAM reliability problems.

Equally important are the decoupling capacitors chosen, as well as their placement. Inadequate decoupling will create reliability problems as well.

Modern DRAM arrays are massively capacitive. Each address line might drive dozens of chips, with 5 to 10 pf of loading per chip. At high speeds the drive electronics must somehow drag all of these pseudo-capacitors up and down with little signal degradation. Not an easy job! Again, poorly designed drivers will make your system unreliable.

Electrical noise is another reliability culprit, sometimes in unexpected ways. For instance, CPUs with multiplexed address/data buses use external address latches to demux the bus. A signal, usually named ALE (Address Latch Enable) or AS (Address Strobe) drives the clock to these latches. The tiniest, most miserable amount of noise on ALE/AS will surely, at the time of maximum inconvenience, latch the data part of the cycle instead of the address. Other signals are also vulnerable to small noise spikes.

Many run-of-the-mill RAM tests, run for several hours, as you cycle the product through it's design environment (temperature and so forth) will show intermittent RAM problems. These are symptoms of the design faults I've described, and always show a need for more work on the product's engineering.

Unhappily, all too often the RAM tests show no problem when hidden demons are indeed lurking. The algorithm I've described, as well as most of the others commonly used, trade-off speed versus comprehensiveness. They don't pound on the hardware in a way designed to find noise and timing problems.

Digital systems are most susceptible to noise when large numbers of bits change all at once. This fact was exploited for data communications long ago with the invention of the Gray Code, a variant of binary counting, where no more than one bit changes between codes. Your worst nightmares of RAM reliability occur when all of the address and/or data bits change suddenly from zeroes to ones.

For the sake of engineering testing, write RAM test code that exploits this known vulnerability. Write 0xffff to 0x0000 and then to 0xffff, and do a read-back test. Then write zeroes. Repeat as fast as your loop will let you go.

Depending on your CPU, the worst locations might be at 0x00ff and 0x0100, especially on 8 bit processors that multiplex just the lower 8 address lines. Hit these combinations, hard, as well.

Other addresses often exhibit similar pathological behavior. Try 0x5555 and 0xaaaa, which also have complementary bit patterns.

The trick is to write these patterns back-to-back. Don't test all of RAM, with the understanding that both 0x0000 and 0xffff will show up in the test. You'll stress the system most effectively by driving the bus massively up and down all at once.

Don't even think about writing this sort of code in C. Any high level language will inject too many instructions between those that move the bits up and down. Even in assembly the processor will have to do fetch cycles from wherever the code happens to be, which will slow down the pounding and make it a bit less effective.

There are some tricks, though. On a CPU with a prefetcher (all x86, 68 k, and so on) try to fill the execution pipeline with code, so the processor does back-to-back writes or reads at the addresses you're trying to hit. And, use memory-to-memory transfers when possible. For example:

```
mov si, 0xaaaa
mov di, 0x5555
mov [si], 0xff
mov [di],[si]
```

5.21   Nonvolatile Memory

Many of the embedded systems that run our lives try to remember a little bit about us, or about their application domain, despite cycling power, brownouts, and all of the other perils of fixed and mobile operation. In the bad old days before microprocessors we had core memory, a magnetic medium that preserved its data when powered or otherwise.

Today we face a wide range of choices. Sometimes Flash or EEPROM is the natural choice for nonvolatile applications. Always remember, though, that these devices have limited numbers of write cycles. Worse, in some cases writes can be very slow.

Battery-backed up RAMs still account for a large percentage of nonvolatile systems. With robust hardware and software support they'll satisfy the most demanding of reliability fanatics; a little less design care is sure to result in occasional lost data.

5.22   Supervisory Circuits

In the early embedded days we were mostly blissfully unaware of the perils of losing power. Virtually all reset circuits were nothing more than a resistor/capacitor time constant. As Vcc ramped from 0 to 5 volts, the time constant held the CPU's reset input low—or lowish—long enough for the system's power supply to stabilize at 5 volts.

Though an elegantly simple design, RC time constants were flawed on the back end, when power goes away. Turn the wall switch off, and the 5 volt supply quickly decays to zero. Quickly only in human terms, of course, as many milliseconds went by while the CPU was powered by something between 0 and 5. The RC circuit is, of course, at this point at a logic one (not-reset), so it allows the processor to run.

And run they do! With Vcc down to 3 or 4 volts most processors execute instructions like mad. Just not the ones you'd like to see. Run a CPU with out-of-spec power and expect random operation. There's a good chance the machine is going wild, maybe pushing and calling and writing and generally destroying the contents of your battery backed up RAM.

Worse, brown-outs, the plague of summer air conditioning, often cause small dips in voltage. If the AC mains decline to 80 volts for a few seconds a power supply might still crank out a few volts. When AC returns to full rated values the CPU is still running, back at 5 volts, but now horribly confused. The RC circuit never notices the dip from 5 to 3 or so volts, so the poor CPU continues running in its mentally unbalanced state. Again, your RAM is at risk.

Motorola, Maxim, and others developed many ICs designed specifically to combat these problems. Though features and specs vary, these supervisory circuits typically manage the processor's reset line, battery power to the RAM, and the RAM's chip selects.

Given that no processor will run reliably outside of its rated Vcc range, the first function of these chips is to assert reset whenever Vcc falls below about 4.7 volts (on 5 volt logic). Unlike an RC circuit that limply drools down as power fails, supervisory devices provide a snappy switch between a logic zero and one, bringing the processor to a sure, safe stopped condition.

They also manage the RAM's power, a tricky problem since it's provided from the system's Vcc when power is available, and from a small battery during quiescent periods. The switchover is instantaneous to keep data intact.

With RAM safely provided with backup power and the CPU driven into a reset state, a decent supervisory IC will also disable all chip selects to the RAM. The reason? At some point after Vcc collapses you can't even be sure the processor, and your decoding logic, will not create rogue RAM chip selects. Supervisory ICs are analog beasts, conceived outside of the domain of discrete ones and zeroes, and will maintain safe reset and chip select outputs even when Vcc is gone.

But check the specs on the IC. Some disable chip selects at exactly the same time they assert reset, asynchronously to what the processor is actually doing. If the processor initiates a write to RAM, and a nanosecond later the supervisory chip asserts reset and disables chip select, which write cycle will be one nanosecond long. *You cannot play with write timing and expect predictable results.* Allow any write in progress to complete before doing something as catastrophic as a reset.

Some of these chips also assert an NMI output when power starts going down. Use this to invoke your "oh_my_god_we're_dying" routine.

Since processors usually offer but a single NMI input, when using a supervisory circuit never have any other NMI source. You'll need to combine the two signals somehow; doing so with logic is a disaster, since the gates will surely go brain dead due to Vcc starvation.

Check the specifications on the parts, though, to ensure that NMI occurs *before* the reset clamp fires. Give the processor a handful of microseconds to respond to the interrupt before it enters the idle state.

There's a subtle reason why it makes sense to have an NMI power-loss handler: you want to get the CPU away from RAM. Stop it from doing RAM writes *before* reset occurs. If reset happens in the middle of a write cycle, there's no telling what will happen to your carefully protected RAM array. Hitting NMI first causes the CPU to take an interrupt exception, first finishing the current write cycle if any. This also, of course, eliminates troubles caused by chip selects that disappear synchronously to reset.

Every battery-backed up system should use a decent supervisory circuit; you just cannot expect reliable data retention otherwise. Yet, these parts are no panacea. The firmware itself is almost certainly doing things destined to defeat any bit of external logic.

5.23   Multibyte Writes

There's another subtle failure mode that afflicts all too many battery-backed up systems. He observed that in a kinder, gentler world than the one we inhabit all memory transactions would require exactly one machine cycle, but here on Earth 8 and 16 bit machines constantly manipulate large data items. Floating point variables are typically 32 bits, so any store operation requires two or four distinct memory writes. Ditto for long integers.

The use of high-level languages accentuates the size of memory stores. Setting a character array, or defining a big structure, means that the simple act of assignment might require tens or hundreds of writes.

Consider the simple statement:

```
a=0x12345678;
```

An x86 compiler will typically generate code like:

```
mov[bx], 5678
```

```
mov[bx+2], 1234
```

which is perfectly reasonable and seemingly robust.

In a system with a heavy interrupt burden it's likely that sooner or later an interrupt will switch CPU contexts between the two instructions, leaving the variable "a" half-changed, in what is possibly an illegal state. This serious problem is easily defeated by avoiding global variables—as long as "a" is a local, no other task will ever try to use it in the half-changed state.

Power-down concerns twist the problem in a more intractable manner. As Vcc dies off a seemingly well-designed system will generate NMI while the processor can still think clearly. If that interrupt occurs during one of these multibyte writes—as it eventually surely will, given the perversity of nature—your device will enter the power-shutdown code with data now corrupt. It's quite likely (especially if the data is transferred via CPU registers to RAM) that there's no reasonable way to reconstruct the lost data.

The simple expedient of eliminating global variables has no benefit to the power-down scenario.

Can you imagine the difficulty of *finding* a problem of this nature? One that occurs maybe once every several thousand power cycles, or less? In many systems it may be entirely reasonable to conclude that the frequency of failure is so low the problem might be safely ignored. This assumes you're not working on a safety-critical device, or one with mandated minimal MTBF numbers.

Before succumbing to the temptation to let things slide, though, consider implications of such a failure. Surely once in a while a critical data item will go bonkers. Does this mean your instrument might then exhibit an accuracy problem (for example, when the numbers are calibration coefficients)? Is there any chance things might go to an unsafe state? Does the loss of a critical communication parameter mean the device is dead until the user takes some presumably drastic action?

If the only downside is that the user's TV set occasionally—and rarely—forgets the last channel selected, perhaps there's no reason to worry much about losing multibyte data. Other systems are not so forgiving.

It was suggested to implement a data integrity check on power-up, to insure that no partial writes left big structures partially changed. I see two different directions this approach might take.

The first is a simple power-up check of RAM to make sure all data is intact. Every time a truly critical bit of data changes, update the CRC, so the boot-up check can see if data is intact. If not, at least let the user know that the unit is sick, data was lost, and some action might be required.

A second, and more robust, approach is to complete every data item write with a checksum or CRC of just that variable. Power-up checks of each item's CRC then reveals which variable was destroyed. Recovery software might, depending on the application, be able to fix the data, or at least force it to a reasonable value while warning the user that, while all is not well, the system has indeed made a recovery.

Though CRCs are an intriguing and seductive solution I'm not so sanguine about their usefulness. Philosophically it *is* important to warn the user rather than to crash or use bad data. But it's much better to never crash at all.

We can learn from the OOP community and change the way we write data to RAM (or, at least the critical items for which battery back-up is so important).

First, hide critical data items behind drivers. The best part of the OOP triptych mantra "encapsulation, inheritance, polymorphism" is "encapsulation." Bind the data items with the code that uses them. Avoid globals; change data by invoking a routine, a method that does the actual work. Debugging the code becomes much easier, and reentrancy problems diminish.

Second, add a "flush_writes" routine to every device driver that handles a critical variable. "Flush_writes" finishes any interrupted write transaction. Flush_writes relies on the fact that only one routine—the driver—ever sets the variable.

Next, enhance the NMI power-down code to invoke all of the flush_write routines. Part of the power-down sequence then finishes all pending transactions, so the system's state will be intact when power comes back.

The downside to this approach is that you'll need a reasonable amount of time between detecting that power is going away, and when Vcc is no longer stable enough to support reliable processor operation. Depending on the number of variables needed flushing this might mean hundreds of microseconds.

Firmware people are often treated as the scum of the earth, as they inevitably get the hardware (late) and are still required to get the product to market on time. Worse, too many hardware groups don't listen to, or even solicit, requirements from the coding folks before cranking out PCBs. This, though, is a case where the firmware requirements clearly drive the hardware design. If the two groups don't speak, problems will result.

Some supervisory chips do provide advanced warning of imminent power-down. Maxim's (www.maxim-ic.com) MAX691, for example, detects Vcc failing below some value before shutting down RAM chip selects and slamming the system into a reset state. It also includes a separate voltage threshold detector designed to drive the CPU's NMI input when Vcc falls below some value you select (typically by selecting resistors). It's important to set this threshold above the point where the part goes into reset. Just as critical is understanding how power fails in your system. The capacitors, inductors, and other power supply components determine how much "alive" time your NMI routine will have before reset occurs. Make sure it's enough.

I mentioned the problem of power failure corrupting variables to Scott Rosenthal, one of the smartest embedded guys I know. His casual "yeah, sure, I see that all the time" got me interested. It seems that one of his projects, an FDA-approved medical device, uses hundreds of calibration variables stored in RAM. Losing any one means the instrument has to go back for readjustment. Power problems are just not acceptable.

His solution is a hybrid between the two approaches just described. The firmware maintains two separate RAM areas, with critical variables duplicated in each. Each variable has its own driver.

When it's time to change a variable, the driver sets a bit that indicates "change in process." It's updated, and a CRC is computed for that data item and stored with the item. The driver unasserts the bit, and then performs the exact same function on the variable stored in the duplicate RAM area.

On power-up the code checks to insure that the CRCs are intact. If not, that indicates the variable was in the process of being changed, and is not correct, so data from the mirrored address is used. If both CRCs are OK, but the "being changed" bit is asserted, then the data protected by that bit is invalid, and correct information is extracted from the mirror site.

The result? With thousands of instruments in the field, over many years, not one has ever lost RAM.

5.24 Testing

Good hardware and firmware design leads to reliable systems. You won't know for sure, though, if your device really meets design goals without an extensive test program. Modern embedded systems are just too complex, with too much hard-to-model hardware/firmware interaction, to expect reliability without realistic testing.

This means you've got to pound on the product, and look for every possible failure mode. If you've written code to preserve variables around brown-outs and loss of Vcc, and don't conduct a meaningful test of that code, you'll probably ship a subtly broken product.

In the past I've hired teenagers to mindlessly and endlessly flip the power switch on and off, logging the number of cycles and the number of times the system properly comes to life. Though I do believe in bringing youngsters into the engineering labs to expose them to the cool parts of our profession, sentencing them to mindless work is a sure way to convince them to become lawyers rather than techies.

Better, automate the tests. The Poc-It, from Microtools (www.microtoolsinc.com/products.htm) is an indispensable $250 device for testing power-fail circuits and code. It's also a pretty fine way to find uninitialized variables, as well as isolating those awfully hard to initialize hardware devices like some FPGAs.

The Poc-It brainlessly turns your system on and off, counting the number of cycles. Another counter logs the number of times a logic signal asserts after power comes on. So, add a bit of test code to your firmware to drive a bit up when (and if) the system properly comes to life. Set the Poc-It up to run for a day or a month; come back and see if the number of power cycles is exactly equal to the number of successful assertions of the logic bit. Anything other than equality means something is dreadfully wrong.

5.25 Conclusion

When embedded processing was relatively rare, the occasional weird failure meant little. Hit the reset button and start over. That's less of a viable option now. We're surrounded by hundreds of CPUs, each doing its thing, each affecting our lives in different ways. Reliability will probably be the watchword of the next decade as our customers refuse to put up with the quirks that are all too common now.

The current drive is to add the maximum number of features possible to each product. I see cell phones that include games. Features are swell . . . if they work, if the product always fulfills its intended use. Cheat the customer out of reliability and your company is going to lose. Power cycling is something every product does, and is too important to ignore.

5.26   Building a Great Watchdog

Launched in January 1994, the Clementine spacecraft spent two very successful months mapping the moon before leaving lunar orbit to head toward near-Earth asteroid Geographos.

A dual-processor Honeywell 1750 system handled telemetry and various spacecraft functions. Though the 1750 could control Clementine's thrusters, it did so only in emergency situations; all routine thruster operations were under ground control.

On May 7 the 1750 experienced a floating point exception. This wasn't unusual; some 3000 prior exceptions had been detected and handled properly. But immediately after the May 7 event downlinked data started varying wildly and nonsensically. Then the data froze. Controllers spent 20 minutes trying to bring the system back to life by sending software resets to the 1750; all were ignored. A hardware reset command finally brought Clementine back online.

Alive, yes, even communicating with the ground, but with virtually no fuel left.

The evidence suggests that the 1750 locked up, probably due to a software crash. While hung the processor turned on one or more thrusters, dumping fuel and setting the spacecraft spinning at 80 RPM. In other words, it appears the code ran wild, firing thrusters it should never have enabled; they kept firing till the tanks ran nearly dry and the hardware reset closed the valves. The mission to Geographos had to be abandoned.

Designers had worried about this sort of problem and implemented a software thruster time-out. That, of course, failed when the firmware hung.

The 1750's built-in watchdog timer hardware was not used, over the objections of the lead software designer. With no automatic "reset" button, success of the mission rested in the abilities of the controllers on Earth to detect problems quickly and send a hardware reset. For the lack of a few lines of watchdog code the mission was lost.

Though such a fuel dump had never occurred on Clementine before, roughly 16 times before the May 7 event hardware resets from the ground had been required to bring the spacecraft's firmware back to life. One might also wonder why some 3000 previous floating point exceptions were part of the mission's normal firmware profile.

Not surprisingly, the software team wished they had indeed used the watchdog, and had not implemented the thruster time-out in firmware. They also noted, though, that a normal, simple, watchdog may not have been robust enough to catch the failure mode.

Contrast this with Pathfinder, a mission whose software also famously hung, but which was saved by a reliable watchdog. The software team found and fixed the bug, uploading new code to a target system 40 million miles away, enabling an amazing roving scientific mission on Mars.

Watchdog timers (WDTs) are our fail-safe, our last line of defense, an option taken only when all else fails—right? These missions (Clementine had been reset 16 times prior to the failure) and so many others suggest to me that WDTs are not emergency outs, but integral parts of our systems. The WDT is as important as `main()` or the runtime library; it's an asset that is likely to be used, and maybe used a lot.

Outer space is a hostile environment, of course, with high intensity radiation fields, thermal extremes, and vibrations we'd never see on Earth. Do we have these worries when designing Earth-bound systems?

Maybe so. Intel revealed that the McKinley processor's ultra fine design rules and huge transistor budget means cosmic rays may flip on-chip bits. The Itanium 2 processor, also sporting an astronomical transistor budget and small geometry, includes an onboard system management unit to handle transient hardware failures. The hardware ain't what it used to be—even if our software were perfect.

But too much (all?) firmware is not perfect. Consider this unfortunately true story from Ed VanderPloeg:

> The world has reached a new embedded software milestone: I had to reboot my hood fan. That's right, the range exhaust fan in the kitchen. It's a simple model from a popular North American company. It has six buttons on the front: 3 for low, medium, and high fan speeds and 3 more for low, medium, and high light levels. Press a button once and the hood fan does what the button says. Press the same button again and the fan or lights turn off. That's it. Nothing fancy. And it needed rebooting via the breaker panel.
>
> Apparently the thing has a micro to control the light levels and fan speeds, and it also has a temperature sensor to automatically switch the fan to high speed if the temperature exceeds some fixed threshold. Well, one day we were cooking dinner as usual, steaming a pot of potatoes, and suddenly the fan kicks into high speed and the lights start flashing. "Hmm, flaky sensor or buggy sensor software," I think to myself.

The food happened to be done so I turned off the stove and tried to turn off the fan, but I suppose it wanted things to cool off first. Fine. So after ten minutes or so the fan and lights turned off on their own. I then went to turn on the lights, but instead they flashed continuously, with the flash rate depending on the brightness level I selected.

So just for fun I tried turning on the fan, but any of the three fan speed buttons produced only high speed. "What 'smart' feature is this?," I wondered to myself. Maybe it needed to rest a while. So I turned off the fan and lights and went back to finish my dinner. For the rest of the evening the fan and lights would turn on and off at random intervals and random levels, so I gave up on the idea that it would self-correct. So with a heavy heart I went over to the breaker panel, flipped the hood fan breaker to and fro, and the hood fan was once again well-behaved.

For the next few days, my wife said that I was moping around as if someone had died. I would tell everyone I met, even complete strangers, about what happened: "Hey, know what? I had to reboot my hood fan the other night!" The responses were varied, ranging from "Freak!" to "Sounds like what happened to my toaster . . . " Fellow programmers would either chuckle or stare in common disbelief.

What's the embedded world coming to? Will programmers and companies everywhere realize the cost of their mistakes and clean up their act? Or will the entire world become accustomed to occasionally rebooting everything they own? Would the expensive embedded devices then come with a "reset" button, advertised as a feature? Or will programmer jokes become as common and ruthless as lawyer jokes? I wish I knew the answer. I can only hope for the best, but I fear the worst.

One developer admitted to me that his consumer products company could care less about the correctness of firmware. Reboot—who cares? Customers are used to this, trained by decades of desktop computer disappointments. Hit the reset switch, cycle power, remove the batteries for 15 minutes, even preteens know the tricks of coping with legions of embedded devices.

Crummy firmware is the norm, but in my opinion is totally unacceptable. Shipping a defective product in any other field is like opening the door to torts. So far the embedded world has been mostly immune from predatory lawyers, but that Brigadoon-like isolation is unlikely to continue. Besides, it's simply unethical to produce junk.

But it's hard, even impossible, to produce perfect firmware. We must strive to make the code correct, but also design our systems to cleanly handle failures. In other words, a healthy dose of paranoia leads to better systems.

A Watchdog Timer is an important line of defense in making reliable products. Well-designed watchdog timers fire off a lot, daily and quietly saving systems and lives without the esteem offered to other, human, heroes. Perhaps the developers producing such reliable WDTs deserve a parade. Poorly-designed WDTs fire off a lot, too, sometimes saving things, sometimes making them worse. A simple-minded watchdog implemented in a nonsafety critical system won't threaten health or lives, but can result in systems that hang and do strange things that tick off our customers. No business can tolerate unhappy customers, so unless your code is perfect (whose is?) it's best in all but the most cost-sensitive applications to build a really great WDT.

An effective WDT is far more than a timer that drives reset. Such simplicity might have saved Clementine, but would it fire when the code tumbles into a really weird mode like that experienced by Ed's hood fan?

5.27 Internal WDTs

Internal watchdogs are those that are built into the processor chip. Virtually all highly integrated embedded processors include a wealth of peripherals, often with some sort of watchdog. Most are brain-dead WDTs suitable for only the lowest-end applications.

Let's look at a few. Toshiba's TMP96141AF is part of their TLCS-900 family of quite nice microprocessors, which offers a wide range of extremely versatile onboard peripherals. All have pretty much the same watchdog circuit. As the data sheet says, "The TMP96141AF is containing watchdog timer of Runaway detecting."

Ahem. And I thought the days of Jinglish were over. Anyway, the part generates a nonmaskable interrupt when the watchdog times out, which is either a very, very bad idea or a wonderfully clever one. It's clever only if the system produces an NMI, waits a while, and only then asserts reset, which the Toshiba part unhappily cannot do. Reset and NMI are synchronous.

A nice feature is that it takes two different I/O operations to disable the WDT, so there are slim chances of a runaway program turning off this protective feature.

Motorola's widely-used 68332 variant of their CPU32 family (like most of these 68 k embedded parts) also includes a watchdog. It's a simple-minded thing meant for low-reliability applications only. Unlike a lot of WDTs, user code must write two different values (0x55 and 0xaa) to the WDT control register to ensure the device does not time out. This is a very good thing—it limits the chances of rogue software accidentally issuing the command needed to appease the watchdog. I'm not thrilled with the fact that any amount of time may elapse between the two writes (up to the time-out period). Two back-to-back writes would further reduce the chances of random watchdog tickles, though once would have to ensure no interrupt could preempt the paired writes. And the 0x55/0xaa twosome is often used in RAM tests; since the 68 k I/O registers are memory mapped, a runaway RAM test could keep the device from resetting.

The 68332's WDT drives reset, not some exception handling interrupt or NMI. This makes a lot of sense, since any software failure that causes the stack pointer to go odd will crash the code, and a further exception-handling interrupt of any sort would drive the part into a "double bus fault." The hardware is such that it takes a reset to exit this condition.

Motorola's popular Coldfire parts are similar. The MCF5204, for instance, will let the code write to the WDT control registers only once. Cool! Crashing code, which might do all sorts of silly things, cannot reprogram the protective mechanism. However, it's possible to change the reset interrupt vector at any time, pretty much invalidating the clever write-once design.

Like the CPU32 parts, a 0x55/0xaa sequence keeps the WDT from timing out, and back-to-back writes aren't required. The Coldfire datasheet touts this as an advantage since it can handle interrupts between the two tickle instructions, but I'd prefer less of a window. The Coldfire has a fault-on-fault condition much like the CPU32's double bus fault, so reset is also the only option when WDT fires—which is a good thing.

There's no external indication that the WDT timed out, perhaps to save pins. That means your hardware/software must be designed so at a warm boot the code can issue a from-the-ground-up reset to every peripheral to clear weird modes that may accompany a WDT time-out.

Philip's XA processors require two sequential writes of 0xa5 and 0x5a to the WDT. But like the Coldfire there's no external indication of a time-out, and it appears the watchdog reset isn't even a complete CPU restart—the docs suggest it's just a reload of the program counter. Yikes—what if the processor's internal states were in disarray from code running amok or a hardware glitch?

Dallas Semiconductor's DS80C320, an 8051 variant, has a very powerful WDT circuit that generates a special watchdog interrupt 128 cycles before automatically—and irrevocably—performing a hardware reset. This gives your code a chance to safe the system, and leave debugging breadcrumbs behind before a complete system restart begins. Pretty cool.

Summary: What's Wrong with Many Internal WDTs:

- A watchdog time-out must assert a hardware reset to guarantee the processor comes back to life. Reloading the program counter may not properly reinitialize the CPU's internals.

- WDTs that issue NMI without a reset may not properly reset a crashed system.

- A WDT that takes a simple toggle of an I/O line isn't very safe.

- When a pair of tickles uses common values like 0x55 and 0xaa, other routines—like a RAM test—may accidentally service the WDT.

- Watch out for WDTs whose control registers can be reprogrammed as the system runs; crashed code could disable the watchdog.

- If a WDT time-out does not assert a pin on the processor, you'll have to add hardware to reset every peripheral after a time-out. Otherwise, though the CPU is back to normal, a confused I/O device may keep the system from running properly.

5.28 External WDTs

Many of the supervisory chips we buy to manage a processor's reset line include built-in WDTs.

TI's UCC3946 is one of many nice power supervisor parts that does an excellent job of driving reset only when Vcc is legal. In a nice small 8 pin SMT package it eats practically no PCB real estate. It's not connected to the CPU's clock, so the WDT will output a reset to the hardware safeing mechanisms even if there's a crystal failure. But it's too darn simple: to avoid a time-out just wiggle the input bit once in a while. Crashed code could do this in any of a million ways.

TI isn't the only purveyor of simplistic WDTs. Maxim's MAX823 and many other versions are similar. The catalogs of a dozen other vendors list equally dull and ineffective watchdogs.

But both TI and Maxim do offer more sophisticated devices. Consider TI's TPS3813 and Maxim's MAX6323. Both are "Window Watchdogs." Unlike the internal versions described above that avoid time-outs using two different data writes (like a 0x55 and then 0xaa), these require tickling within certain time bands. Toggle the WDT input too slowly, too fast, or not at all, and a time-out will occur. That greatly reduces the chances that a program run amok will create the precise timing needed to satisfy the watchdog. Since a crashed program will likely speed up or bog down if it does anything at all, errant strobing of the tickle bit will almost certainly be outside the time band required.

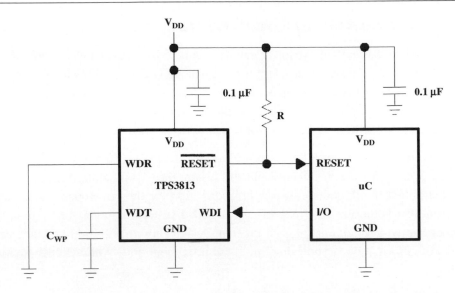

Figure 5.1: TI's TPS3813 Is Easy to Use and Offers a Nice Windowing WDT Feature

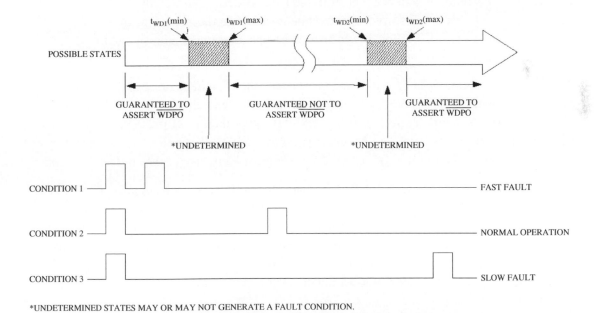

*UNDETERMINED STATES MAY OR MAY NOT GENERATE A FAULT CONDITION.

Figure 5.2: Window Timing of Maxim's Equally Cool MAX6323

5.29   *Characteristics of Great WDTs*

What's the rationale behind an awesome watchdog timer? The perfect WDT should detect all erratic and insane software modes. It must not make any assumptions about the condition of the software or the hardware; in the real world anything that can go wrong will. It must bring the system back to normal operation no matter what went wrong, whether from a software defect, RAM glitch, or bit flip from cosmic rays.

It's impossible to recover from a hardware failure that keeps the computer from running properly, but at the least the WDT must put the system into a safe state. Finally, it should leave breadcrumbs behind, generating debug information for the developers. After all, a watchdog time-out is the yin and yang of an embedded system. It saves the system, keeping the customer happy, yet demonstrates an inherent design flaw that should be addressed. Without debug information, troubleshooting these infrequent and erratic events is close to impossible.

What does this mean in practice?

An effective watchdog is independent from the main system. Though all WDTs are a blend of interacting hardware and software, something external to the processor must always be poised, like the sword of Damocles, ready to intervene as soon as a crash occurs. Pure software implementations are simply not reliable.

There's only one kind of intervention that's effective: an immediate reset to the processor and all connected peripherals. Many embedded systems have a watchdog that initiates a nonmaskable interrupt. Designers figure that firing off NMI rather than reset preserves some of the system's context. It's easy to seed debugging assets in the NMI handler (like a stack capture) to aid in resolving the crash's root cause. That's a great idea, except that it does not work.

All we really know when the WDT fires is that something truly awful happened. Software bug? Perhaps. Hardware glitch? Also possible. Can you ensure that the error wasn't something that totally scrambled the processor's internal logic states? I worked with one system where a motor in another room induced so much EMF that our instrument sometimes went bonkers. We tracked this down to a subnanosecond glitch on one CPU input, a glitch so short that the processor went into an undocumented weird mode. Only a reset brought it back to life.

Some CPUs, notably the 68 k and ColdFire, will throw an exception if a software crash causes the stack pointer to go odd. That's not bad, except that any watchdog circuit that then drives the CPU's nonmaskable interrupt will unavoidably invoke code that pushes the system's context, creating a second stack fault. The CPU halts, staying halted till a reset, and only a reset, comes along.

Drive reset; it's the only reliable way to bring a confused microprocessor back to lucidity. Some clever designers, though, build circuits that drive NMI first, and then after a short delay pound on reset. If the NMI works then its exception handler can log debug information and then halt. It may also signal other connected devices that this unit is going offline for a while. The pending reset guarantees an utterly clean restart of the code. Don't be tempted to use the NMI handler to safe dangerous hardware; that task always, in every system, belongs to a circuit external to the possibly confused CPU.

Don't forget to reset the whole computer system; a simple CPU restart may not be enough. Are the peripherals absolutely, positively, in a sane mode? Maybe not. Runaway code may have issued all sorts of I/O instructions that placed complex devices in insane modes. Give every peripheral a *hardware* reset; software resets may get lost in all of the I/O chatter.

Consider what the system must do to be totally safe after a failure. Maybe a pacemaker needs to reboot in a heartbeat (so to speak), or maybe backup hardware should issue a few ticks if reboots are slow.

One thickness gauge that beams high energy gamma rays through 4 inches of hot steel failed in a spectacular way. Defective hardware crashed the code. The WDT properly closed the protective lead shutter, blocking off the 5 curie cesium source. I was present, and watched incredulously as the engineering VP put his head in path of the beam; the crashed code, still executing something, tricked the watchdog into opening the shutter, beaming high intensity radiation through the veep's forehead. I wonder to this day what eventually became of the man.

A really effective watchdog cannot use the CPU's clock, which may fail. A bad solder joint on the crystal, poor design that doesn't work well over temperature extremes, or numerous other problems can shut down the oscillator. This suggests that no WDT internal to the CPU is really safe. All (that I know of) share the processor's clock.

Under no circumstances should the software be able to reprogram the WDT or any of its necessary components (like reset vectors, I/O pins used by the watchdog, and so on). Assume runaway code runs under the guidance of a malevolent deity.

Build a watchdog that monitors the *entire system*'s operation. Don't assume that things are fine just because some loop or ISR runs often enough to tickle the WDT. A software-only watchdog should look at a variety of parameters to insure the product is healthy, kicking the dog only if everything is OK. What is a software crash, after all? Occasionally the system executes a HALT and stops, but more often the code vectors off to a random location, continuing to run instructions. Maybe only one task crashed. Perhaps only one is still alive—no doubt that which kicks the dog.

Think about what can go wrong in your system. Take corrective action when that's possible, but initiate a reset when it's not. For instance, can your system recover from exceptions like floating point overflows or divides by zero? If not, these conditions may well signal the early stages of a crash. Either handle these competently or initiate a WDT time-out. For the cost of a handful of lines of code you may keep a *60 Minutes* camera crew from appearing at your door.

It's a good idea to flash an LED or otherwise indicate that the WDT kicked. A lot of devices automatically recover from time-outs; they quickly come back to life with the customer totally unaware a crash occurred. Unless you have a debug LED, how do you know if your precious creation is working properly, or occasionally invisibly resetting? One outfit complained that over time, and with several thousand units in the field, their product's response time to user inputs degraded noticeably. A bit of research showed that their system's watchdog properly drove the CPU's reset signal, and the code then recognized a warm boot, going directly to the application with no indication to the users that the time-out had occurred. We tracked the problem down to a floating input on the CPU, that caused the software to crash—up to several thousand times per second. The processor was spending most of its time resetting, leading to apparently slow user response. An LED would have shown the problem during debug, long before customers started yelling.

Everyone knows we should include a jumper to disable the WDT during debugging. But few folks think this through. The jumper should be *inserted to enable debugging*, and removed for normal operation. Otherwise if manufacturing forgets to install the jumper, or if it falls out during shipment, the WDT won't function. And there's no production test to check the watchdog's operation.

Design the logic so the jumper disconnects the WDT from the reset line (possibly though an inverter so an inserted jumper sets debug mode). Then the watchdog continues to function even while debugging the system. It won't reset the processor but will flash the LED. The light will blink a lot when break pointing and single stepping, but should never come on during full-speed testing.

Characteristics of Great WDTs:

- Make no assumptions about the state of the system after a WDT reset; hardware and software may be confused.

- Have *hardware* put the system into a safe state.

- Issue a hardware reset on time-out.

- Reset the peripherals as well.

- Ensure a rogue program cannot reprogram WDT control registers.

- Leave debugging breadcrumbs behind.

- *Insert* a jumper to disable the WDT for debugging; *remove* it for production units.

5.30 Using an Internal WDT

Most embedded processors that include high integration peripherals have some sort of built-in WDT. Avoid these except in the most cost-sensitive or benign systems. Internal units offer minimal protection from rogue code. Runaway software may reprogram the WDT controller, many internal watchdogs will not generate a proper reset, and any failure of the processor will make it impossible to put the hardware into a safe state. A great WDT must be independent of the CPU it's trying to protect.

However, in systems that really must use the internal versions, there's plenty we can do to make them more reliable. The conventional model of kicking a simple timer at erratic intervals is too easily spoofed by runaway code.

A pair of design rules leads to decent WDTs: kick the dog only after your code has done *several unrelated* good things, and make sure that erratic execution streams that wander into your watchdog routine won't issue incorrect tickles.

This is a great place to use a simple state machine. Suppose we define a global variable named "state." At the beginning of the main loop set state to 0x5555. Call watchdog routine A, which adds an offset—say 0x1111—to state and then ensures the variable is now 0x66bb. Return if the compare matches; otherwise halt or take other action that will cause the WDT to fire.

Later, maybe at the end of the main loop, add another offset to state, say 0x2222. Call watchdog routine B, which makes sure state is now 0x8888. Set state to zero. Kick the dog if the compare worked. Return. Halt otherwise.

This is a trivial bit of code, but now runaway code that stumbles into any of the tickling routines cannot errantly kick the dog. Further, no tickles will occur unless the entire main loop executes in the proper sequence. If the code just calls routine B repeatedly, no tickles will occur because it sets state to zero before exiting.

Add additional intermediate states as your paranoia or fear of litigation dictates.

Normally I detest global variables, but this is a perfect application. Cruddy code that mucks with the variable, errant tasks doing strange things, or any error that steps on the global will make the WDT time-out.

Do put these actions in the program's main loop, not inside an ISR. It's fun to watch a multitasking product crash—the entire system might be hung, but one task still responds to interrupts. If your watchdog tickler stays alive as the world collapses around the rest of the code, then the watchdog serves no useful purpose.

If the WDT doesn't generate an external reset pulse (some processors handle the restart internally) make sure the code issues a hardware reset to all peripherals immediately after start-up. That may mean working with the EEs so an output bit resets every resetable peripheral.

If you must take action to safe dangerous hardware, well, since there's no way to guarantee the code will come back to life, stay away from internal watchdogs. Broken hardware will obviously cause this—but so can lousy code. A digital camera was recalled recently when users found that turning the device off when in a certain mode meant it could never be turned on again. The code wrote faulty information to flash memory that created a permanent crash.

```
main(){                          wdt_a(){
    state=0x5555;                    if (state!= 0x5555) halt;
    wdt_a();                         state+=0x1111;
    .                            }
    .
    .
                                 wdt_b(){
    state+=0x2222;                   if (state!= 0x8888) halt;
    wdt_b();                         state=0;
}                                    kick dog;
                                 }
```

Figure 5.3: Pseudo Code for Handling an Internal WDT

5.31   An External WDT

The best watchdog is one that doesn't rely on the processor or its software. It's external to the CPU, shares no resources, and is utterly simple, thus devoid of latent defects.

Use a PIC, a Z8, or other similar dirt-cheap processor as a system health monitor. These parts have an independent clock, onboard memory, and the built-in timers we need to build a truly great WDT. Being external, you can connect an output to hardware interlocks that put dangerous machinery into safe states.

But when selecting a watchdog CPU check the part's specifications carefully. Tying the tickle to the watchdog CPU's interrupt input, for instance, may not work reliably. A slow part—like most PICs—may not respond to a tickle of short duration. Consider TI's MSP430 family or processors. They're a very inexpensive (half a buck or so) series of 16 bit processors that use virtually no power and no PCB real estate.

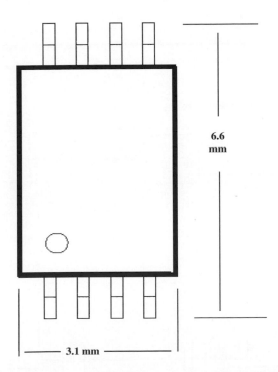

6.6 mm

3.1 mm

Figure 5.4: The MSP430—a 16 Bit Processor that Uses No PCB Real Estate. For Metrically-Challenged Readers, This Is about 1/4" x 1/8"

Tickle it using the same sort of state-machine described above. Like the windowed watchdogs (TI's TPS3813 and Maxim's MAX6323), define min *and* max tickle intervals, to further limit the chances that a runaway program deludes the WDT into avoiding a reset.

Perhaps it seems extreme to add an entire computer just for the sake of a decent watchdog. We'd be fools to add extra hardware to a highly cost-constrained product. Most of us, though, build lower volume higher margin systems. A fifty cent part that prevents the loss of an expensive mission, or that even saves the cost of one customer support call, might make a lot of sense.

In a multiprocessor system it's easy to turn all of the processors into watchdogs. Have them exchange "I'm OK" messages periodically. The receiver resets the transmitter if it stops speaking. This approach checks a lot of hardware and software, and requires little circuitry.

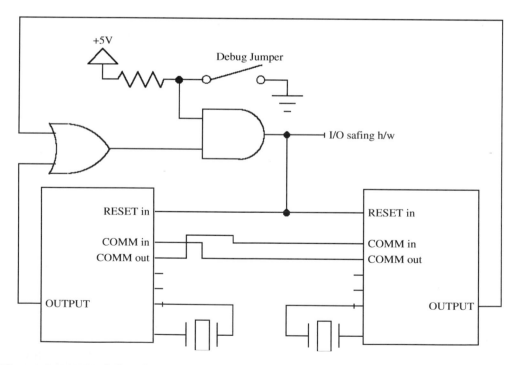

Figure 5.5: Watchdog for a Dual-Processor System—Each CPU Watches the Other

5.32   WDTs for Multitasking

Tasking turns a linear bit of software into a multidimensional mix of tasks competing for processor time. Each runs more or less independently of the others, which means each can crash on its own, without bringing the entire system to its knees.

You can learn a lot about a system's design just by observing its operation. Consider a simple instrument with a display and various buttons. Press a button and hold it down; if the display continues to update, odds are the system multitasks.

Yet in the same system a software crash might go undetected by conventional watchdog strategies. If the display or keyboard tasks die, the main line code or a WDT task may continue to run.

Any system that uses an ISR or a special task to tickle the watchdog, but that does not examine the health of all other tasks, is not robust. Success lies in weaving the watchdog into the fabric of all of the system's tasks, which is happily much easier than it sounds.

First, build a watchdog task. It's the only part of the software allowed to tickle the WDT. If your system has an MMU, mask off all I/O accesses to the WDT except those from this task, so rogue code traps on an errant attempt to output to the watchdog.

Next, create a data structure that has one entry per task, with each entry being just an integer.

When a task starts it increments its entry in the structure. Tasks that only start once and stay active forever can increment the appropriate value each time through their main loops.

Increment the data *atomically*—in a way that cannot be interrupted with the data half-changed. ++TASKi (if TASK is an integer array) on an 8 bit CPU might not be atomic, though it's almost certainly OK on a 16 or 32 bitter. The safest way to both encapsulate and ensure atomic access to the data structure is to hide it behind another task. Use a semaphore to eliminate concurrent shared accesses. Send increment messages to the task, using the RTOS's messaging resources.

As the program runs the number of counts for each task advances. Infrequently but at regular intervals the watchdog task runs. Perhaps once a second, or maybe once a msec—it's all a function of your paranoia and the implications of a failure.

The watchdog task scans the structure, checking that the count stored for each task is reasonable. One that runs often should have a high count; another which executes infrequently will produce a smaller value. Part of the trick is determining what's reasonable for each task; stick with me—we'll look at that shortly.

If the counts are unreasonable, halt and let the watchdog time-out. If everything is OK, set all of the counts to zero and exit.

Why is this robust? Obviously, the watchdog monitors every task in the system. But it's also impossible for code that's running amok to stumble into the WDT task and errantly tickle the dog; by zeroing the array we guarantee it's in a "bad" state.

I skipped over a critical step—how do we decide what's a reasonable count for each task? It might be possible to determine this analytically. If the WDT task runs once a second, and one of the monitored tasks starts every 50 msec, then surely a count of around 20 is reasonable.

Other activities are much harder to ascertain. What about a task that responds to asynchronous inputs from other computers, say data packets that come at irregular intervals? Even in cases of periodic events, if these drive a low-priority task they may be suspended for rather long intervals by higher-priority problems.

The solution is to broaden the data structure that maintains count information. Add minimum (min) and maximum (max) fields to each entry. Each task must run at least min, but no more than max times.

Now redesign the watchdog task to run in one of two modes. The first is the one already described, and is used during normal system operation.

The second mode is a debug environment enabled by a compile-time switch that collects min and max data. Each time the WDT task runs it looks at the incremented counts and sets new min and max values as needed. It tickles the watchdog each time it executes.

Run the product's full test suite with this mode enabled. Maybe the system needs to operate for a day or a week to get a decent profile of the min/max values. When you're satisfied that the tests are representative of the system's real operation, manually examine the collected data and adjust the parameters as seems necessary to give adequate margins to the data.

What a pain! But by taking this step you'll get a great watchdog—and a deep look into your system's timing. I've observed that few developers have much sense of how their creations perform in the time domain. "It seems to work" tells us little. Looking at the data acquired by this profiling, though might tell a lot. Is it a surprise that task A runs 400 times a second? That might explain a previously-unknown performance bottleneck.

In a real time system we must manage and measure time; it's every bit as important as procedural issues, yet is oft ignored until a nagging problem turns into an unacceptable symptom. This watchdog scheme forces you to think in the time domain, and by its nature profiles—admittedly with coarse granularity—the time-operation of your system.

There's yet one more kink, though. Some tasks run so infrequently or erratically that any sort of automated profiling will fail. A watchdog that runs once a second will miss tasks that start only hourly. It's not unreasonable to exclude these from watchdog monitoring. Or, we can add a bit of complexity to the code to initiate a watchdog time-out if, say, the slow tasks don't start even after a number of hours elapse.

5.33 Summary and Other Thoughts

I remain troubled by the fan failure described earlier. It's easy to dismiss this as a glitch, an unexplained failure caused by a hardware or software bug, cosmic rays, or meddling by aliens. But others have written about identical situations with their vent fans, all apparently made by the same vendor.

When we blow off a failure, calling it a "glitch" as if that name explains something, we're basically professing our ignorance. There are no glitches in our macroscopically deterministic world. Things happen for a reason.

The fan failures didn't make the evening news and hurt no one. So why worry? Surely the customers were irritated, and the possible future sales of that company at least somewhat diminished. The company escalated the general rudeness level of the world, and thus the sum total incipient anger level, by treating their customers with contempt. Maybe a couple more Valiums were popped, a few spouses yelled at, some kids cowered until dad calmed down. In the grand scheme of things perhaps these are insignificant blips. Yet we must remember the purpose of embedded control is to help people, to improve lives, not to help therapists garner new patients.

What concerns me is that if we cannot even build reliable fan controllers, what hope is there for more mission-critical applications?

I don't know what went wrong with those fan controllers, and I have no idea if a WDT—well designed or not—is part of the system. I do know, though, that the failures are unacceptable and avoidable. But maybe not avoidable by the use of a conventional watchdog. A WDT tells us the code is running. A windowing WDT tells us it's running with pretty much the right timing. No watchdog, though, flags software executing with corrupt data structures, unless the data is so bad it grossly affects the execution stream.

Why would a data structure become corrupt? Bugs, surely. Strange conditions the designers never anticipated will also create problems, like the never-ending flood of buffer overflow conditions that plague the net, or unexpected user inputs ("We never thought the user would press all 4 buttons at the same time!").

Is another layer of self-defense, beyond watchdogs, wise? Safety critical applications, where the cost of a failure is frighteningly high, should definitely include integrity checks on the data. Low threat equipment—like this oven fan—can and should have at least a minimal amount of code for trapping possible failure conditions.

Some might argue it makes no sense to "waste" time writing defensive code for a dumb fan application. Yet the simpler the system, the easier and quicker it is to plug in a bit of code to look for program and data errors.

Very simple systems tend to translate inputs to outputs. Their primary data structures are the I/O ports. Often several unrelated output bits get multiplexed to a single port. To change one bit means either reading the port's current status, or maintaining a copy of the port in RAM. Both approaches are problematic.

Computers are deterministic, so it's reasonable to expect that, in the absence of bugs, they'll produce correct results all the time. So it's apparently safe to read a port's current status, AND off the unwanted bits, OR in new ones, and output the result. This is a state machine; the outputs evolve over time to deal with changing inputs. But the process works only if the state machine never incorrectly flips a bit. Unfortunately, output ports are connected to the hostile environment of the real world. It's entirely possible that a bit of energy from starting the fan's highly inductive motor will alter the port's setting. I've seen this happen many times.

So maybe it's more reliable to maintain a memory image of the port. The downside is that a program bug might corrupt the image. Most of the time these are stored as global variables, so any bit of sloppy code can accidentally trash the location. Encapsulation solves that problem, but not the one of a wandering pointer walking over the data, or of a latent reentrancy issue corrupting things. You might argue that writing correct code means we shouldn't worry about a location changing, but we added a WDT to, in part, deal with bugs. Similar concerns about our data are warranted.

In a simple system look for a design that resets data structures from time to time. In the case of the oven fan, whenever the user selects a fan speed reset all I/O ports and data structures. It's that simple.

In a more complicated system the best approach is the oldest trick in software engineering: check the parameters passed to functions for reasonableness. In the embedded world we chose not to do this for three reasons: speed, memory costs, and laziness. Of these, the third reason is the real culprit most of the time.

Cycling power is the oldest fix in the book; it usually means there's a lurking bug and a poor WDT implementation. Embedded developer Peter Putnam wrote:

Last November, I was sitting in one of a major airline's newer 737-900 aircraft on the ramp in Cancun, Mexico, waiting for departure when the pilot announced there would be a delay due to a computer problem. About twenty minutes later a group of maintenance personnel arrived. They poked around for a bit, apparently to no avail, as the captain made another announcement. "Ladies and Gentlemen," he said, "we're unable to solve the problem, so we're going to try turning off all aircraft power for thirty seconds and see if that fixes it."

Sure enough, after rebooting the Boeing 737, the captain announced that "All systems are up and running properly."

Nobody saw fit to leave the aircraft at that point, but I certainly considered it

---

CHAPTER 6

Hardware/Software Co-Verification

Jason Andrews

6.1   Embedded System Design Process

The process of embedded system design generally starts with a set of requirements for what the product must do and ends with a working product that meets all of the requirements. Following is a list of the steps in the process and a short summary of what happens at each state of the design. The steps are shown in Figure 6.1.

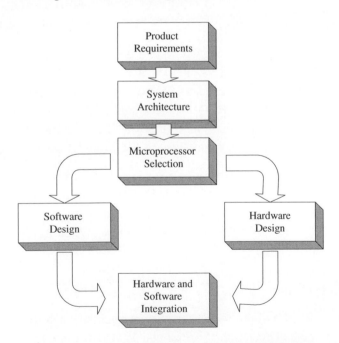

Figure 6.1: Embedded System Design Process

6.1.1   Requirements

The requirements and product specification phase documents and defines the required features and functionality of the product. Marketing, sales, engineering, or any other individuals who are experts in the field and understand what customers need and will buy to solve a specific problem, can document product requirements. Capturing the correct requirements gets the project off to a good start, minimizes the chances of future product modifications, and ensures there is a market for the product if it is designed and built. Good products solve real needs, have tangible benefits, and are easy to use.

6.1.2   System Architecture

System architecture defines the major blocks and functions of the system. Interfaces, bus structure, hardware functionality, and software functionality are determined. System designers use simulation tools, software models, and spreadsheets to determine the architecture that best meets the system requirements. System architects provide answers to questions such as, "How many packets/sec can this router design handle?" or "What is the memory bandwidth required to support two simultaneous MPEG streams?"

6.1.3   Microprocessor Selection

One of the most difficult steps in embedded system design can be the choice of the microprocessor. There are an endless number of ways to compare microprocessors, both technical and nontechnical. Important factors include performance, cost, power, software development tools, legacy software, RTOS choices, and available simulation models. Benchmark data is generally available, though apples-to-apples comparisons are often difficult to obtain. Creating a feature matrix is a good way to sift through the data to make comparisons.

Software investment is a major consideration for switching the processor. Embedded guru Jack Ganssle says the rule of thumb is to decide if 70% of the software can be reused; if so, don't change the processor. Most companies will not change processors unless there is something seriously deficient with the current architecture. When in doubt, the best practice is to stick with the current architecture.

6.1.4   Hardware Design

Once the architecture is set and the processor(s) have been selected, the next step is hardware design, component selection, Verilog and VHDL coding, synthesis, timing analysis, and physical design of chips and boards.

The hardware design team will generate some important data for the software team such as the CPU address map(s) and the register definitions for all software programmable registers. As we will see, the accuracy of this information is crucial to the success of the entire project.

6.1.5   Software Design

Once the memory map is defined and the hardware registers are documented, work begins to develop many different kinds of software. Examples include boot code to start up the CPU and initialize the system, hardware diagnostics, real-time operating system (RTOS), device drivers, and application software.

During this phase, tools for compilation and debugging are selected and coding is done.

6.1.6   Hardware and Software Integration

The most crucial step in embedded system design is the integration of hardware and software. Somewhere during the project, the newly coded software meets the newly designed hardware. How and when hardware and software will meet for the first time to resolve bugs should be decided early in the project. There are numerous ways to perform this integration. Doing it sooner is better than later, though it must be done smartly to avoid wasted time debugging good software on broken hardware or debugging good hardware running broken software.

6.2   Verification and Validation

Two important concepts of integrating hardware and software are verification and validation. These are the final steps to ensure that a working system meets the design requirements.

6.2.1   Verification: Does It Work?

Embedded system verification refers to the tools and techniques used to verify that a system does not have hardware or software bugs. Software verification aims to execute the software and observe its behavior, while hardware verification involves making sure the hardware performs correctly in response to outside stimuli and the executing software. The oldest form of embedded system verification is to build the system, run the software, and hope for the best. If by chance it does not work, try to do what you can to modify the software and hardware to get the system to work. This practice is called testing and it is not as comprehensive as verification. Unfortunately, finding out what is not working while the system is running is not always easy. Controlling and observing the system while it is running may not even be possible. To cope with the difficulties of debugging the embedded system many tools and techniques have been introduced to help engineers get embedded systems working sooner and in a more systematic way. Ideally, all of this verification is done before the hardware is built. The earlier in the process problems are discovered, the easier and cheaper they are to correct. Verification answers the question, "Does the thing we built work?"

6.2.2   Validation: Did We Build the Right Thing?

Embedded system validation refers to the tools and techniques used to validate that the system meets or exceeds the requirements. Validation aims to confirm that the requirements in areas such as functionality, performance, and power are satisfied. It answers the question, "Did we build the right thing?" Validation confirms that the architecture is correct and the system is performing optimally.

I once worked with an embedded project that used a common MIPS processor and a real-time operating system (RTOS) for system software. For various reasons it was decided to change the RTOS for the next release of the product. The new RTOS was well suited for the hardware platform and the engineers were able to bring it up without much difficulty. All application tests appeared to function properly and everything looked positive for an on-schedule delivery of the new release. Just before the product was ready to ship, it was discovered that the applications were running about 10 times slower than with the previous RTOS. Suddenly, panic set in and the project schedule was in danger. Software engineers who wrote the application software struggled to figure out why the performance was so much lower since not much had changed in the application code. Hardware engineers tried to study the hardware behavior, but using logic analyzers that are better suited for triggering on errors than providing wide visibility over a long range of time, it was difficult to even decide where to look. The RTOS vendor provided most of the system software and so there was little source code to study. Finally, one of the engineers had a hunch that the cache of the MIPS processor was not being properly enabled. This indeed turned out to be the case and after the problem was corrected, system performance was confirmed. This example demonstrates the importance of validation. Like verification, it is best to do this before the hardware is built. Tools that provide good visibility make validation easier.

6.3   Human Interaction

Embedded system design is more than a robotic process of executing steps in an algorithm to define requirements, implement hardware, implement software, and verify that it works. There are numerous human aspects to a project that play an important role in the success or failure of a project.

The first place to look is the organizational structure of the project teams. There are two commonly used structures. Figure 6.2 shows a structure with separate hardware and software teams, whereas Figure 6.3 shows a structure with one group of combined hardware and software engineers that share a common management team.

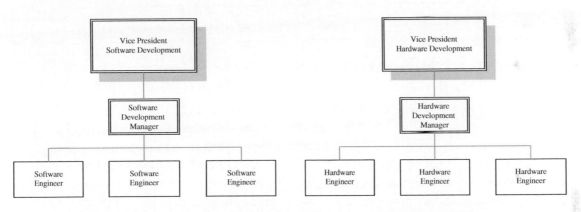

Figure 6.2: Management Structure with Separate Engineering Teams

Separate project teams make sense in markets where time-to-market is less critical. Staggering the project teams so that the software team is always one project behind the hardware team can be used to increase efficiency. This way, the software team always has available hardware before they start any software integration phase. Once the hardware is passed to the software engineers, the hardware engineers can go on to the next project. This structure avoids having the software engineers sitting around waiting for hardware.

A combined project team is most efficient for addressing time-to-market constraints. The best situation to work under is a common management structure that is responsible for project success, not just one area such as hardware engineers or software engineers. Companies that are running most efficiently have removed structural barriers and work together to get the project done. In the end, the success of the project is based on the entire product working well, not just the hardware or software.

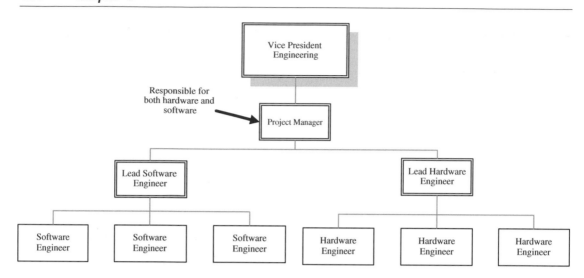

Figure 6.3: Management Structure with Combined Engineering Teams

I once worked in a company that totally separated hardware and software engineers. There was no shared management. When the prototypes were delivered and brought up in the lab, the manager of each group would pace back and forth trying to determine what worked and what was broken. What usually ended up happening was that the hardware engineer would tell his manager that there was something wrong with the software just to get the manager to go away. Most engineers prefer to be left alone during these critical project phases. There is nothing worse than a status meeting to report that your design is not working when you could be working to fix the problems instead of explaining them. I do not know what the software team was communicating to its management, but I also envisioned something about the hardware not working or the inability to get time to use the hardware. At the end of the day, the two managers probably went to the CEO to report the other group was still working to fix its bugs.

Everybody has a role to play on the project team. Understanding the roles and skills of each person as well as the personalities makes for a successful project as well as an enjoyable work environment. Engineers like challenging technical work. I have no data to confirm it, but I think more engineers seek new employment because of difficulties with the people they work with or the morale of the group than because they are seeking new technical challenges.

A recent survey into embedded systems projects found that more than 50% of designs are not completed on time. Typically, those designs are 3 to 4 months off the pace, project cancellations average 11–12%, and average time to cancellation is 4-and-a-half months (Jerry Krasner of Electronics Market Forecasters June 2001).

> Hardware/software co-verification aims to verify embedded system software executes correctly on a representation of the hardware design. It performs early integration of software with hardware, before any chips or boards are available.

The primary focus of this chapter is on system-on-a-chip (SoC) verification techniques. Although all embedded systems with custom hardware can benefit from co-verification, the area of SoC verification is most important because it involves the most risk and is positioned to reap the most benefit. The ARM architecture is the most common microprocessor used in SoC design and serves as a reference to teach many of the concepts presented in the book.

If any of the following statements are true for you, this chapter will provide valuable information:

1. You are a software engineer developing code that interacts directly with hardware.

2. You are curious about the relationship between hardware and software.

3. You would like to learn more about debugging hardware and software interaction problems.

4. You desire to learn more about either the hardware or software design processes for SoC projects.

5. You are an application engineer in a company selling co-verification products.

6. You want to get your projects done sooner and be the hero at your company.

7. You are getting tired of the manager bugging you in the lab asking, "Does it work yet?"

8. You are a manager and you are tired of bugging the engineers asking, "Does it work yet?" and would like to pester the engineers in a more meaningful way.

9. You have no clue what this stuff is all about and want to learn something to at least sound intelligent about the topic at your next interview.

6.4 Co-Verification

Although hardware/software co-verification has been around for many years, over the last few years, it has taken on increased importance and has become a verification technique used by more and more engineers. The trend toward greater system integration, such as the demand for low-cost, high-volume consumer products, has led to the development of the system-on-a-chip (SoC). The SoC was defined as a single chip that includes one or more microprocessors, application specific custom logic functions, and embedded system software. Including microprocessors and DSPs inside a chip has forced engineers to consider software as part of the chip's verification process in order to ensure correct operation. The techniques and methodologies of hardware/software co-verification allow projects to be completed in a shorter time and with greater confidence in the hardware and software. In the *EE Times* "2003 Salary Opinion Survey," a good number of engineers reported spending more than one-third of their day on software tasks, especially integrating software with new hardware. This statistic reveals that the days of throwing the hardware over the cubicle wall to the software engineers are gone. In the future, hardware engineers will continue to spend more and more time on software related issues. This chapter presents an introduction to commonly used co-verification techniques.

6.4.1   History of Hardware/Software Co-Verification

Co-verification addresses one of the most critical steps in the embedded system design process, the integration of hardware and software. The alternative to co-verification has always been to simply build the hardware and software independently, try them out in the lab, and see what happens. When the PCI bus began supporting automatic configuration of peripherals without the need for hardware jumpers, the term plug-and-play became popular. About the same time I was working on projects that simply built hardware and software independently and differences were resolved in the lab. This technique became known as plug-and-debug. It is an expensive and very time-consuming effort. For hardware designs putting off-the-shelf components on a board it may be possible to do some rework on the board or change some programmable logic if problems with the interaction of hardware and software are found. Of course, there is always the "software workaround" to avoid aggravating hardware problems. As integration continued to increase, something more was needed to perform integration earlier in the design process. The solution is co-verification.

Co-verification has its roots in logic simulation. The HDL logic simulator has been used since the early 1990s as the standard way to execute the representation of the hardware before any chips or boards are fabricated. As design sizes have increased and logic simulation has not provided the necessary performance, other methods have evolved that involve some form of hardware to execute the hardware design description. Examples of hardware methods include simulation acceleration, emulation, and prototyping. In this chapter, we will examine each of these basic execution engines as a method for co-verification.

Co-verification borrows from the history of microprocessor design and verification. In fact, logic simulation history is much older than the products we think of as commercial logic simulators today. The microprocessor verification application is not exactly co-verification since we normally think of the microprocessor as a known good component that is put into an embedded system design, but nevertheless, microprocessor verification requires a large amount of software testing for the CPU to be successfully verified. Microprocessor design companies have done this level of verification for many years. Companies designing microprocessors cannot commit to a design without first running many sequences of instructions ranging from small tests of random instruction sequences to booting an operating system like Windows or UNIX. This level of verification requires the ability to simulate the hardware design and have methods available to debug the software sequences when problems occur. As we will see, this is a kind of co-verification.

I became interested in co-verification after spending many hours in a lab trying to integrate hardware and software. I think it was just too many days of logic analyzer probes falling off, failed trigger conditions, making educated guesses about what might be happening, and sometimes just plain trial-and-error. I decided there must be a better way to sit in a quiet, air-conditioned cubicle and figure out what was happening. Fortunately for me, there *were* better ways and I was fortunate enough to get jobs working on some of them.

6.4.1.1 *Commercial Co-Verification Tools Appear*

The first two commercial co-verification tools specifically targeted at solving the hardware/software integration problem for embedded systems were Eaglei from Eagle Design Automation and Seamless CVE from Mentor Graphics. These products appeared on the market within six months of each other in the 1995–1996 time frame and both were created in Oregon. Eagle Design Automation Inc. was founded in 1994 and located in Beaverton. The Eagle product was later acquired by Synopsys, became part of Viewlogic, and was finally killed by Synopsys in 2001 due to lack of sales. In contrast, Mentor Seamless produced consistent growth and established itself as the leading co-verification product. Others followed that were based on similar principles, but Seamless has been the most successful of the commercial co-verification tools. Today, Seamless is the only product listed in market share studies for hardware/software co-verification by analysts such as Dataquest.

The first published article about Seamless was in 1996, at the 7th IEEE International Workshop on Rapid System Prototyping (RSP '96). The title of the paper was: "Miami: A Hardware Software Co-simulation Environment." In this paper, Russ Klein documented the use of an instruction set simulator (ISS) co-simulating with an event-driven logic simulator. As we will see in this chapter, the paper also detailed an interesting technique of dynamically partitioning the memory data between the ISS and logic simulator to improve performance.

I was fortunate to meet Russ a few years later in the Minneapolis airport and hear the story of how Seamless (or maybe it's Miami) was originally prototyped. When he first got the idea for a product that combined the ISS (a familiar tool for software engineers) with the logic simulator (a familiar tool for hardware engineers) and used optimization techniques to increase performance from the view of the software, the value of such an idea wasn't immediately obvious. To investigate the idea in more detail he decided to create a prototype to see how it worked. Testing the prototype required an instruction set simulator for a microprocessor, a logic simulation of a hardware design, and software to run on the system. He decided to create the prototype based on his old CP/M personal computer he used back in college. CP/M was the operating system that later evolved into DOS back around 1980. The machine used the Z80 microprocessor and software located in ROM to start execution and would later move to a floppy disk to boot the operating system (much like today's PC BIOS). Of course, none of the source code for the software was available, but Russ was able to extract the data from the ROM and the first couple of tracks of the boot floppy using programs he wrote. From there he was able to get it into a format that could be loaded into the logic simulator. Working on this home-brew simulation, he performed various experiments to simulate the operation of the PC, and in the end concluded that this was a valid co-simulation technique for testing embedded software running on simulated hardware. Eventually the simulation was able to boot CP/M and used a model of the keyboard and screen to run a Microsoft Basic interpreter that could load Basic programs and execute them. In certain modes of operation, the simulation ran faster than the actual computer!

Russ turned his work into an internal Mentor project that would eventually become a commercial EDA product. In parallel, Eagle produced a prototype of a similar tool. While Seamless started with the premise of using the ISS to simulate the microprocessor internals, Eagle started using native-compiled C programs with special function calls inserted for memory accesses into the hardware simulation environment. At the time, this strategy was thought to be good enough for software development and easier to proliferate since it did not require a full instruction set simulator for each CPU, only a bus functional model. The founders of Eagle, Gordon Hoffman and Geoff Bunza, were interested in looking for larger

EDA companies to market and sell Eaglei (and possibly buy their startup company). After they pitched the product to Mentor Graphics, Mentor was faced with a build versus buy decision. Should they continue with the internal development of Seamless or should they stop development and partner or acquire the Eagle product? According to Russ, the decision was not an easy one and went all the way to Mentor CEO Wally Rhines before Mentor finally decided to keep the internal project alive. The other difficult decision was to decide whether to continue the use of instruction set simulation or follow Eagle into host-code execution when Eagle already had a lead in product development. In the end, Mentor decided to allow Eagle to introduce the first product into the market and confirmed their commitment to instruction set simulation with the purchase of Microtec Research Inc., an embedded software company known for its VRTX RTOS, in 1996. The decision meant Seamless was introduced six months after Eagle, but Mentor bet that the use of the ISS would be a differentiator that would enable them to win in the marketplace.

Another commercial co-verification tool that took a different road to market was V-CPU. V-CPU was developed inside Cisco Systems about the same time as Seamless. It was engineered by Benny Schnaider, who was working for Cisco as a consultant in design verification, for the purpose of early integration of software running with a simulation of a Cisco router. Details of V-CPU were first published at the 1996 Design Automation Conference in a paper titled "Software Development in a Hardware Simulation Environment."

As V-CPU was being adopted by more and more engineers at Cisco, the company was starting to worry about having a consultant as the single point of failure on a piece of software that was becoming critical to the design verification environment. Cisco decided to search the marketplace in hope of finding a commercial product that could do the job and be supported by an EDA vendor. At the time there were two possibilities, Mentor Seamless and Eaglei. After some evaluation, Cisco decided that neither was really suitable since Seamless relied on the use of instruction set simulators and Eaglei required software engineers to put special C calls into the code when they wanted to access the hardware simulation. In contrast, V-CPU used a technique that automatically captured the software accesses to the hardware design and required little or no change to the software. In the end, Cisco decided to partner with a small EDA company in St. Paul, MN, named Simulation Technologies (Simtech) and gave them the rights to the software in exchange for discounts and commercial support. Dave Von Bank and I were the two engineers that worked for Simtech and worked with Cisco to receive the internal tool and make it into a commercial co-verification tool that was launched in 1997 at the International Verilog Conference (IVC) in Santa Clara. V-CPU is still in use today at Cisco. Over the years the software has changed hands many times and is now owned by Summit Design.

6.4.2   Co-Verification Defined

6.4.2.1   Definition

At the most basic level HW/SW co-verification means verifying embedded system software executes correctly on embedded system hardware. It means running the software on the hardware to make sure there are no hardware bugs before the design is committed to fabrication. As we will see in this chapter, the goal can be achieved using many different ways that are differentiated primarily by the representation of the hardware, the execution engine used, and how the microprocessor is modeled. But more than this, a true co-verification tool also provides control and visibility for both software and hardware engineers and uses the types of tools they are familiar with, at the level of abstraction they are familiar with. A working definition is given in Figure 6.4. This means that for a technique to be considered a co-verification product it must provide at least software debugging using a source code debugger and hardware debugging using waveforms as shown in Figure 6.5. This chapter describes many different methods that meet these criteria.

> HW/SW Co-Verification is the process of verifying embedded system software runs correctly on the hardware design before the design is committed for fabrication.

Figure 6.4: Definition of Co-Verification

Co-verification is often called virtual prototyping since the simulation of the hardware design behaves like the real hardware, but is often executed as a software program on a workstation. Using the definition given above, running software on any representation of the hardware that is not the final board, chip, or system qualifies as co-verification. This broad

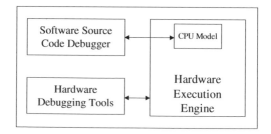

Figure 6.5: Co-Verification Is about Debugging Hardware and Software

definition includes physical prototyping as co-verification as long as the prototype is not the final fabrication of the system and is available earlier in the design process.

A narrower definition of co-verification limits the hardware execution to the context of the logic simulator, but as we will see, there are many techniques that do not involve logic simulation and should be considered co-verification.

6.4.2.2 Benefits of Co-Verification

Co-verification provides two primary benefits. It allows software that is dependent on hardware to be tested and debugged before a prototype is available. It also provides an additional test stimulus for the hardware design. This additional stimulus is useful to augment test benches developed by hardware engineers since it is the true stimulus that will occur in the final product. In most cases, both hardware and software teams benefit from co-verification. These co-verification benefits address the hardware and software integration problem and translate into a shorter project schedule, a lower cost project, and a higher quality product.

The primary benefits of co-verification are:

- Early access to the hardware design for software engineers

- Additional stimulus for the hardware engineers

6.4.2.3 Project Schedule Savings

For project managers, the primary benefit of co-verification is a shorter project schedule. Traditionally, software engineers suffer because they have no way to execute the software they are developing if it interacts closely with the hardware design. They develop the software, but cannot run it so they just sit and wait for the hardware to become available. After a long delay, the hardware is finally ready, and management is excited because the project will soon be working, only to find out there are many bugs in the software since it is brand new and this is the first time is has been executed. Co-verification addresses the problem of software waiting for hardware by allowing software engineers to start testing code much sooner. By getting all the trivial bugs out, the project schedule improves because the amount of time spent in the lab debugging software is much less. Figure 6.6 shows the project schedule without co-verification and Figure 6.7 shows the new schedule with co-verification and early access to the hardware design.

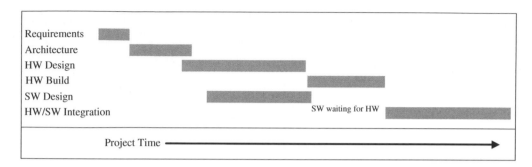

Figure 6.6: Project Schedule without Co-Verification

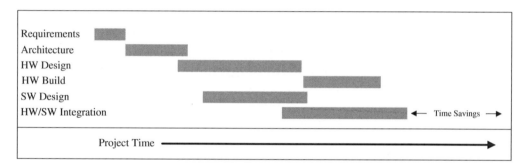

Figure 6.7: Project Schedule with Co-Verification

6.4.2.4   *Co-Verification Enables Learning by Providing Visibility*

Another greatly overlooked benefit of co-verification is visibility. There is no substitute for being able to run software in a simulated world and see exactly the correlation between hardware and software. We see what is really happening inside the microprocessor in a nonintrusive way and see what the hardware design is doing. Not only is this useful for debugging, but it can be even more useful in providing a way to understand how the microprocessor and the hardware work. We will see in future examples that co-verification is an ideal way to really learn how an embedded system works. Co-verification provides information that can be used to identify such things as bottlenecks in performance using information about bus activity or cache hit rates. It is also a great way to confirm the hardware is programmed correctly and operations are working as expected. When software engineers get into a lab setting and run code, there is really no way for them to see how the hardware is acting. They usually rely on some print statements to follow execution and assume if the system does not crash it must be working.

6.4.2.5   Co-Verification Improves Communication

For some projects, the real benefit of co-verification has nothing to do with early access to hardware, improved hardware stimulus, or even a shorter schedule. Sometimes the real benefit of co-verification is improved communication between hardware and software teams. Many companies separate hardware and software teams to the extent that each does not really care about what the other one is doing, a kind of "not my problem" attitude. This results in negative attitudes and finger pointing. It may sound a bit far fetched, but sometimes the introduction of co-verification enables these teams to work together in a positive way and make a positive improvement in company culture. Figure 6.8 shows what Brian Bailey, one of the early engineers on Seamless, had to say about communication:

```
"Software engineering for electronic systems is a very different
culture, they have very different ways of doing things. We're just
beginning to find ways that the two groups can communicate. It's
getting to be a cliché now. When we first started going out and
telling people about Seamless, we would insist on companies that
we talked to having both hardware and software engineers there for
our meeting. In many of those meetings, the hardware and software
guys (from within the same potential customer) literally met for
the first time and exchanged business cards."
"There is still a big divide. We find there is no common boss until
perhaps the vice-president level. And we are not seeing that change
quickly."
Brian Bailey, chief technologist, Mentor Graphics, December 2000
```

Figure 6.8: Brian Bailey on Communication

6.4.2.6   Co-Verification versus Co-Simulation

A similar term to co-verification is co-simulation. In fact, the first paper published about Seamless used this term in the title. Co-simulation is defined as two or more heterogeneous simulators working together to produce a complete simulation result. This could be an ISS working with a logic simulator, a Verilog simulator working with a VHDL simulator, or a digital logic simulator working with an analog simulator. Some co-verification techniques involve co-simulation and some do not.

6.4.2.7   Co-Verification versus Codesign

Often co-verification is lumped together with codesign, but they are really two different things. Earlier, verification was defined as the process of determining something works as intended. Design is the process of deciding how to implement a required function of a system. In the context of embedded systems, design might involve deciding if a function should be implemented in hardware or software. For software, design may involve deciding on a set of software layers to form the software architecture. For hardware, design may involve deciding how to implement a DMA controller on the bus and what programmable registers are needed to configure a DMA channel from software. Design is deciding what to create and how to implement it. Verification is deciding if the thing that was implemented is working correctly. Some co-verification tools provide profiling and other feedback to the user about hardware and software execution, but this alone does not make them codesign tools since they can do this only after hardware and software have been partitioned.

6.4.2.8 Is Co-Verification Really Necessary?

After learning the definition of co-verification and its benefits, the next logical question asks if co-verification is really necessary. Theoretically, if the hardware design has no bugs and is perfect according to the requirements and specifications then it really does not matter what the software does. For this situation, from the hardware engineer's point of view, there is no reason to execute the software before fabricating the design.

Similarly, software engineers may think that early access to hardware is a pain, not a benefit, since it will require extra work to execute software with co-verification. For some software engineers, no hardware equals no work to do. In addition, at these early stages the hardware may be still evolving and have bugs. There is nothing worse for software engineers than to try to run software on buggy hardware since it makes isolating problems more difficult.

The point is that while individual engineers may think co-verification is not for them, almost every project with custom hardware and software will benefit from co-verification in some way. Most embedded projects do not get the publicity of an Intel microprocessor, but most of us remember the famous (or infamous) Pentium FDIV bug where the CPU did not divide correctly. Hardware always has bugs, software always has bugs, and getting rid of them is good.

6.4.3   Co-Verification Methods

Most co-verification methods can be classified based on the execution engine used to run the hardware design. A secondary classification exists based on the method used to model the embedded system microprocessor. Before discussing specific co-verification methods, a quick review of some of the key ingredients in co-verification is useful.

6.4.3.1   Native Compiling Software

Many software engineers prefer to work as much as possible in the host environment (on a PC or workstation) before moving to the embedded system in a lab setting. There are two ways to do software development and software simulation in the host environment. The first is to use workstation tools to compile the embedded system software for the host processor (instead of the embedded processor) and execute it on the workstation. If the embedded system software is written in C or C++, host compiled simulation works very well for functional testing. The embedded system software now becomes a program that runs on a PC or workstation and uses all of the compilers, debuggers, profilers, and other analysis tools available for writing workstation software. Workstation tools are more plentiful and higher quality since more programmers are making use of them (remember, the embedded system space is extremely fragmented). Errors like memory leaks and bad pointers are a joy to fix on the workstation when compared to the tools available on the target system in the lab.

6.4.3.2   Instruction Set Simulation

The second method to work in the host environment is to compile the embedded system software for the target processor using a cross compiler and simulate the software using an application called an instruction set simulator. The ISS is a model of the target microprocessor at the instruction level. It has the ability to load programs compiled for the target instruction set, it contains a model of the registers, and it can decode and model all of the processor's instruction set. Typically, this type of tool is accurate at the instruction level. It runs the given program in a sequential manner and does not model the instruction pipeline, superscalar execution, or any timing of the microprocessor at the hardware level in terms of a clock or digital logic. For this reason a good, fast, functional simulation is provided, but detailed timing and performance estimation is not available. Most instruction set simulators come with an interface to one or more software debuggers. The same embedded software tool companies that provide debuggers and cross-compilers may also provide the instruction set simulators. The ISS is also useful for testing compilers and debuggers without requiring a real processor on a working board. When a new processor is developed, compilers must be developed in parallel with silicon, and the ISS enables a compiler to be ready when the silicon is ready so software can be run immediately upon silicon availability.

6.4.3.3   Hardware Stubs

The major drawback of working on the host with native compiled code or the ISS is the lack of a model of the rest of the embedded system hardware. Much of the embedded system software is dependent on the hardware. Software such as diagnostics and device drivers cannot be tested without a model of how the hardware will react. This hardware dependent software is usually the most important software during the crucial hardware and software integration phase of the project. To combat this limitation, software engineers started using C code to implement simple behavioral models, or stubs, of how the target hardware is expected to behave. These stubs can provide the expected results for system peripherals and other system interfaces. Some instruction set simulators also started to incorporate hardware stubs that could be included in the simulation by providing a C interface to the memory model of the ISS. Peripherals such as timers, UARTs, and even Ethernet controllers can be included in the simulation. The number of hardware models needed to make the ISS useful will determine whether it is worth investing in creating C models of the hardware. For a large system, it can be more work to create the stubs than creating the embedded system software itself. Figure 6.9 shows a diagram of an ISS with a memory model interface that allows the user to add C code to take care of the memory accesses. Figure 6.10 shows a fragment of a simple stub model that returns the ID register of a CPU so the executing software does not get an error when it reads an expected ID code.

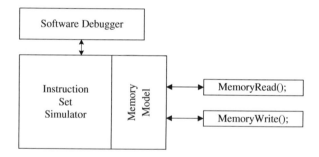

Figure 6.9: ISS with Memory Model Interface

```
static void Access(int nRW, unsigned long addr, unsigned long *data)
    {
    if (!nRW)   /* read */
    {
        if (addr == ID_REGISTER)
        {
            *data = 0x7926F; /* return ID value */
        }
    }
}
```

Figure 6.10: Code for a Simple Stub

6.4.3.4 Real-Time Operating System (RTOS) Simulator

For projects that use real time operating systems, it is possible to use a host-compiled version of the RTOS. Some commercial operating system vendors provide the host-compiled version that can be run on a workstation. For custom or proprietary operating systems, the RTOS code can usually be "ported" to the host. The RTOS simulator is fast and most useful for higher levels of software. It can be used to test the calls to RTOS libraries for tasking, mailboxes, semaphores, and so forth. The RTOS simulator is more abstract then the ISS, and usually runs at a higher speed. Since the software is compiled for the host machine, it does not allow the use of any assembly language. Again, it suffers from the same limitation of the ISS since the custom hardware is not available.

An example of an RTOS simulator is VxSim, a simulation of the popular RTOS VxWorks from Wind River. VxSim allows device drivers and applications to be tested in the host environment before moving to the embedded system. Drivers usually require hardware stubs to provide simulated responses.

6.4.3.5 Microprocessor Evaluation Board

Among software engineers, the most popular tool used for learning a processor and testing code before the target system is ready is the microprocessor evaluation board. This is a board with the target microprocessor and some memory that typically uses a network connection or a serial port to communicate with the host. It allows initial code to be developed, downloaded, and tested. Target tools are used to debug and verify the code. Many software engineers prefer to use the evaluation board since the tools are the same as those that will be used when the system is ready and it is most like working with the true product being developed. Every microprocessor vendor has an evaluation board for sale soon after the processor is available, usually at a very reasonable price. Vendors also provide sample code and even hardware schematics for the board. Some embedded system designs even go so far as to copy the evaluation board and just add a small amount of custom hardware or even buy and use the evaluation board in a product without modification. This is very tempting to get a hardware design quickly, but the boards are not usually designed for higher production volume products. Check the cost and the reliability of the design before directly using an evaluation board as part of a product.

If the embedded system contains a fair amount of custom hardware, the evaluation board is less useful. Depending on the amount and nature of the custom hardware, it may be possible to modify the evaluation board by including extra programmable logic or other semiconductor devices to make it look and act more like the target system design.

6.4.3.6 Waveforms, Log Files, and Disassembly

For SoC designs, many software engineers are forced to do early software verification with full-functional logic simulation models and waveforms in a hardware design environment. Engineers who are skilled in both software development and hardware design may be able to debug this way, but it is not the most comfortable debugging environment for most software engineers. A source level debugger with C code is preferred to bus waveforms and large log files from a Verilog or VHDL simulator.

I once introduced co-verification to a project team working on a complex video chip with four ARM CPU cores. After preaching the benefits of co-verification and the ability to debug software using a source level debugger the software engineers shook their heads and seemed to understand. Their current setup involved the use of the RTL code for the ARM cores running in a logic simulator. As part of this environment, they included a model that monitored the execution of the ARM cores and output a log file with the disassembly of the executing software as a way to track software progress. Since the tests ran very slow, they would wait patiently for simulation to complete and then get this log file and try to correlate it with the source code to see what happened. When they went to start co-verification they immediately asked if the co-verification tool could output the same kind of log file so they could track execution after the test finished. Of course, it could, but this type of debugging does not really improve their situation. After some coaxing, they agreed to try interactive software debugging with a source-level debugger and were pleased to discover this type of debugging was possible.

6.4.4 A Sample of Co-Verification Methods

This section introduces some of the commonly used co-verification methods and architectures used to verify embedded software running on the hardware design. All of these have some pros and cons. That is why there is so many of them and it can be difficult to sort out the choices.

6.4.4.1 Host-Code Mode with Logic Simulation

Host-code mode is a technique to compile the embedded system software, not for the embedded processor in the hardware design, but instead for the host workstation. This is also referred to as native compile. To perform co-verification the resulting executable is run on the host machine, and it connects to a logic simulator that executes the hardware design. Some type of inter-process communication (IPC) is required to exchange information between the host-compiled embedded software and the logic simulator. The IPC implementation could be a socket that allows each of the two processes to be on different machines on the network or shared memory that runs both processes on the same machine.

Host-code mode is not limited to using a logic simulator as the hardware execution engine. Any hardware execution engine can be used. Some others that have been used with host-code mode are an accelerator/emulator and a prototyping platform.

With host-code mode, a bus functional model is used in the hardware execution engine to create bus transactions for the bus interface of the microprocessor. The combination of the host-compiled program plus the bus functional model serves as a microprocessor model.

Host-code mode provides an attractive environment for both software and hardware engineers. Software engineers can continue to use the software tools they are already using, including source code debuggers and other development and debug tools on the host. Hardware engineers can also use the tools they are already using as part of the design process; a Verilog or VHDL logic simulator and associated debug tools. This requires a minimal methodology change for both groups of engineers and can benefit both software and hardware verification. The ability to do pre-silicon co-verification is a great benefit when the processor does not yet exist. Figure 6.11 shows the basic architecture.

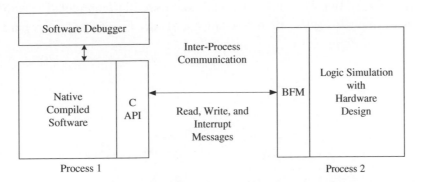

Figure 6.11: Host-Code Execution with Logic Simulation

Host-code mode can also be used when the software does not access the hardware design via a microprocessor bus, but instead via a generic bus interface like PCI. Many chips do not have an embedded microprocessor, but are designed with the PCI bus as a primary interface into the programmable registers. In this case the software can be run on the host and read and write operations from the software can be translated into PCI bus transactions in the hardware execution engine. This is a good example of when it is useful to abstract the software execution to the host and link it to hardware execution at the PCI interface.

Host-code mode requires the embedded software to be modified to perform function calls when it accesses the hardware design through the bus functional model. This process of putting in specific function calls can either be a pain if a lot of embedded software already exists or be little or no problem if the code is being written from scratch and all memory accesses are coded to go through a common function call. Examples of C library calls that are used for host-code execution are shown in Figure 6.12.

```
ret_val = CoverRead(address, &data, size, options);

ret_val = CoverWrite(address, data, size, options);
```

Figure 6.12: Host-Code Mode Example Function Calls

Inserting these C calls into the software is called explicit access because the user must explicitly put in the references to the hardware design. The other way to use host-code mode is to use implicit access. Implicit access does not require the user to put in special calls, but automatically figures out when the software is accessing the hardware based on the load and store instructions being run. This technique will be covered in more detail in another chapter, but with implicit access, the user can use ordinary C code to access hardware via pointers as shown in Figure 6.13.

```
unsigned long *ptr;
unsigned long data;
ptr = 0xff0000000 /* address of ASIC control registers */
data = *ptr; /* read the control register */
data |= 1;   /* set bit 0 to 1 */
*ptr = data; /* write new value back to control register */
```

Figure 6.13: Example of Implicit Access

Host-code mode can also be used to integrate an RTOS simulator such as VxSim as discussed above. A diagram of host-code execution in the context of an RTOS simulator is shown in Figure 6.14.

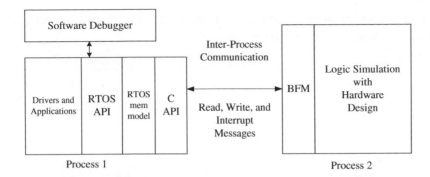

Figure 6.14: RTOS Simulation and Host-Code Execution

6.4.4.2   *Instruction Set Simulation with Logic Simulation*

Another way to perform co-verification is to compile the embedded system software for the target processor and run it on an instruction set simulator. An ISS allows not only C code but also assembly language of the target processor to be run. This allows more realistic simulation of things normally coded in assembly language such as initialization sequences, cache and MMU configuration and simulation, and exception handlers. This mode of operation is referred to as target-code mode.

As with host-code mode, some type of inter-process communication (IPC) is required to exchange information between the instruction set simulator and the logic simulator.

Target-code mode is not limited to using a logic simulator as the hardware execution engine. Any hardware execution engine can be used, but since the instruction set simulator will likely run slower than a host code program it is important to make sure the speed of the instruction set simulator is not too slow to see benefits from a hardware execution engine such as an accelerator.

The bus functional models used with an ISS are the same or similar to those used in host code mode. The main difference is that with an ISS it may be possible to understand the context of the bus transactions better. In host code mode, only a single bus transaction is considered at a time. On a bus that supports address pipelining, such as AHB, there is no way to determine the next bus cycle that will be done by the host code program, so only a single transaction would be simulated and there is no pipelining. The ISS can utilize knowledge of what will be the next bus transaction to occur and can supply the bus functional model with the next address so that it can model the address pipelining correctly.

This is a major benefit of using a good ISS for co-verification. Target-code mode also enables instruction fetches to be verified.

Like host-code mode, software engineers can debug code in a familiar environment. In target-code mode, the debugger is not a host debugger, but rather a debugger that can work with the ISS and debug programs cross-compiled for the embedded processor. Figure 6.15 shows the architecture.

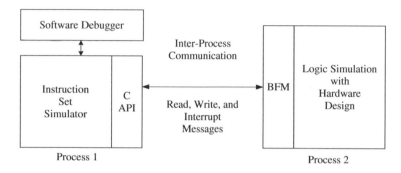

Figure 6.15: Instruction Set Simulator Connected to Logic Simulation

To integrate an ISS with a bus functional model, the memory interface to the ISS must be modified to run logic simulation to satisfy the memory accesses. Instruction set simulators as used by software engineers normally have a flat memory model that is a simple C model allowing the program to be loaded and run. Some instruction set simulators have the ability to customize this memory model so the users can add their C models (stubs) to provide some rudimentary model of the hardware. Without at least the ability to put in stub models, most embedded system code will not run on a flat memory model since it deals with memory-mapped hardware registers that should have nonzero values after reset. Doing co-verification with an ISS is really just a simple extension of the use of stubs to instead turn memory transactions into calls to the logic simulator for execution on the bus functional model. The other thing that must be reported to the ISS is interrupts. When an interrupt occurs on the bus, the ISS must know about it so it can model the exception processing and start the service routine. Most commercial co-verification tools provide many more features that just gluing the memory model of the ISS to a bus functional model and reporting interrupts, but this description is easy to understand and has been used by many users to construct their own co-verification environment using an ISS.

Some instruction set simulators keep statistics and account for the simulation cycles that have been used to satisfy memory requests. This allows useful features such as performance estimation and profiling to be used to find out details of software execution. In the simple ISS integration description above, the read and write activity would have to report a number of bus clocks that were consumed to satisfy the transaction. The ISS may be able to use this clock cycle count and update its internal notion of time. Unfortunately, this is not always easy to do since the time domain of the ISS is now out-of-step with that of the logic simulator. Synchronization between the software execution environment and the hardware execution environment are types of issues that have led to the shift from a transaction-based interface to one that is cycle based.

One way to think of a cycle-based ISS is to say that it exchanges pin values between the ISS and logic simulator on every bus clock cycle. This is equivalent to moving the bus functional model state machine into the ISS and just applying the signal values in logic simulation. Another way to view it is as a transaction-based interface where the logic simulator has the ability to report wait states to the ISS and the ISS will return with the same memory transaction until it completes. This approach is better suited for cases where better accuracy is needed. It is also better suited for multiprocessor designs since it can keep all processors synchronized with the logic simulator on a cycle-by-cycle basis. Figure 6.16 shows the architecture of a cycle-based instruction set simulator.

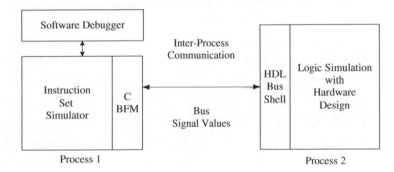

Figure 6.16: Cycle-based Instruction Set Simulator Connected to Logic Simulation

6.4.4.3   *C Simulation*

The logic simulation and acceleration techniques discussed so far evolved from the hardware simulation domain. One complaint about co-verification developed by extending the hardware simulation platform to include software engineers includes limited availability of the platform. For example, to perform co-verification using logic simulation requires a logic simulation license for each software engineer that is running and debugging software. Most companies purchase logic simulation licenses based on the demand for hardware verification and don't have extras available for the purposes of co-verification. Similarly, higher performance hardware execution engines such as simulation acceleration and emulation are even more difficult to acquire for software development. Most companies have only one or two such machines that must be shared by verification engineers and software engineers. This limited scalability often leaves engineers wondering if there is a way to do co-verification that doesn't require traditional logic simulation.

The natural conclusion is to think about using a C or C++ simulation environment to eliminate the need for logic simulation. At the same time, there is a perception that C simulation is faster than Verilog and VHDL simulation. SystemC is one such environment that is gaining momentum as a modeling language that can provide C++ simulation of the design without requiring logic simulation, and at the same time can also co-simulate with an HDL simulator when needed. SystemC by itself is not a co-verification method, but rather an alternative hardware execution environment or even an alternative modeling language to be used instead of Verilog and VHDL. Model-based methods require a library of models to be created, and missing models are a common source of difficulty.

The question with any C simulation environment, SystemC or homegrown, has always been the development of the design model. Like the primitive hardware stub methods used by software engineers, somebody must create the simulation model of the hardware design. Since this model creation is not yet a mainstream path to design implementation, any work to create an alternative model that is not in the critical path of design implementation is usually a lower priority that may never become reality. Contrast this to logic simulation where RTL code for the design must be developed for implementation so using this RTL code in a logic simulator is always a model that is readily available.

Tools are now available to take the Verilog and VHDL code for the design and turn it into a C model by translating it into C or SystemC or even directly to an executable program that is not a traditional logic simulator. Of course, such tools must do more than just eliminate the need for the logic simulator license; they also must offer some performance gain to satisfy the perception that somehow C should be faster than Verilog or VHDL; a tough job considering the optimization already being done by today's logic simulators. By doing nothing more than eliminating the logic simulator license the price would have to be dramatically lower than that of a simulator to be compelling, which is very difficult since the simulation market is mature and prices will only come down as time progresses. The approach of these Verilog to C translators is to turn the Verilog into a cycle-based simulation by eliminating timing. Cycle-based simulation has never been a mainstream methodology, so it is not clear that converting Verilog code into a cycle-based executable will succeed; only time will tell. A common post on newsgroups related to Verilog simulation is from the engineer looking for the Verilog to C translator. There are many of them, and a couple of them are shown in Figure 6.17. The answer usually comes back that the best Verilog to C translator is the VCS logic simulator. Most engineers asking for the translator are not clear on how it would benefit them. In fact, many of the products mentioned are no longer available as commercial products.

```
> Was wondering if anyone could point me in the direction of a
> Verilog to C translator ... if such a thing exists.
>

> Hi all,
> I am looking for a Verilog to C converter.
```

Figure 6.17: Verilog-to-C Translator Requests

The only real way to gain higher performance from C or SystemC simulation is to raise the abstraction level of the model. Instead of modeling the design at RTL, more abstract models must be developed that eliminate the detail of the model and as a result enable it to run faster. The theory on high-level modeling is that an engineer can make an abstract model in about 1/10 the time it takes to develop an RTL model and the model should run 100 to 1000 times faster in a C or SystemC environment. Engineers are looking for a minimum of 100 kHz performance, and 1 MHz is more desirable. Some tools translating HDL into C are starting to show about 10x performance speedup over logic simulation by eliminating some of the detailed timing of logic simulation without requiring the user to make any changes to the RTL code. Raising the level of abstraction holds promise for running software before the RTL for the hardware design is available.

Co-verification utilizing C simulation environments is very much the same as with traditional logic simulators. Instruction set simulators and host code execution methods can be used to run the embedded system software and perform software debug. The compelling reason to look into co-verification based on C simulation is the ability to scale co-verification to many software engineers. Once a C model of the design is in place and co-verification is available, then every software engineer can use it by simply making copies of the software model. This also makes it possible to give the model and environment to software engineers that are outside the company to start developing software and doing such tasks as porting an RTOS without waiting for hardware and without the need to use logic simulation. I have never confirmed it, but I can guess that software companies such as Wind River have a need to port vxWorks to new processors and custom hardware designs before chips and boards are available. I can also guess they don't have a Verilog simulator and even if they could get a simulator they probably don't want to learn how to use it.

Companies that started out developing co-verification tools that allow users to create their own C models and combine them with microprocessor models and debugging tools to form a representation of the design face a difficult modeling dilemma about who will create the models. To enable wider use of the technology and go beyond focusing on the creation of models for custom designs, some products shifted toward the use of a C model as a replacement for the common tool that all software engineers know and love, the evaluation board. The all-software virtual evaluation board is an alternative to buying hardware, cables, power supplies, and JTAG (joint test action group) tools. When many engineers need access to the board, it becomes much more cost effective to deploy a software version of it. In addition to basic microprocessor evaluation boards, C models can be created for reference designs and platforms that are often used as starting points for adding custom hardware. This type of virtual board enables debugging that is not possible on a real piece of hardware. Value is derived from being able to monitor hardware states and have easy access to performance information. By constraining support to off-the-shelf boards it is easier to serve the market, but does not address custom designs. Model based methods always seem to face model availability questions.

Co-verification revolving around C simulation is an interesting area that will continue to evolve as engineers start to look at top down design methodology that could leverage such a model for high-speed simulation and also use it for the design implementation.

6.4.4.4   RTL Model of CPU with Software Debugging

As we have seen, there are benefits and drawbacks of using software models of microprocessors and other hardware. This section and the next discuss techniques that avoid model creation issues by using a representation of the microprocessor that doesn't depend on an engineer coding a model of its behavior.

As the world of SoC design has evolved, the design flows used for microprocessor and DSP IP have changed. In the beginning, most IP for critical blocks such as the embedded microprocessor were in the form of hard IP. The company creating the IP wanted to make sure the user realized the maximum benefit in terms of optimized performance and area. The hard macro also allows the IP to be used without revealing all of the source code of the design. As an example, most of the ARM7TDMI designs use a hard macro licensed from

ARM. Today, most SoC designs don't use hard macros but instead use a soft macro in the form of synthesizable Verilog or VHDL. Soft macros offer better flexibility and eliminate portability issues in the physical design and fabrication process.

Now that the RTL code for the CPU is available and can easily be run in a logic simulator or emulation system, everybody wants to know the best way to perform co-verification. Is a separate model like the instruction set simulator really needed? It does not seem natural to most engineers (especially hardware engineers) to replace the golden RTL of the CPU, the representation of the design that will be implemented in the silicon, with something else. The reality is that the RTL code can be used for co-verification and has successfully been used by project teams.

The drawback of using the RTL code is that it can only execute as fast as the hardware execution engine it is running on. Since it is totally inside the hardware execution engine, there is no chance to take any simulation short cuts that are possible (or automatic) with host-code execution or instruction set simulation. Historically, logic simulation has always been too slow to make the investigation of this technique interesting. After all, a simulation environment for a large SoC typically runs less than 100 cycles/sec and running at this speed it is not possible to use a software debugger to perform interactive debugging.

The primary area where this technique has seen success is with simulation acceleration and emulation systems that are capable of running at much higher speeds. With a hardware execution engine that runs a few hundred kHz up to 1 MHz it is possible to interactively debug software running on the RTL model of the CPU.

To perform co-verification with an RTL model of the microprocessor, a software debugger must be able to communicate with the CPU RTL. To debug software programs, a software debugger requires only a few primitive operations to control execution of a microprocessor. This can best be seen in a summary of the GNU debugger (gdb) remote protocol requirements. To communicate with a target CPU gdb requires the target to perform the following functions:

- Read and write registers

- Read and write memory

- Continue execution

- Single step

- Retrieve the current status of the program (stopped, exited, and so forth)

In fact, gdb provides an interface and specification called the remote protocol interface that implements a communication channel between the debugger and the target CPU to implement the necessary functionality to enable gdb to debug a program.

On a silicon target where a chip is placed on a board, the only way to communicate with gdb to send and receive the protocol information is by adding some special software to the user's software running on the embedded processor that will communicate with gdb to send information such as the register contents and memory contents. The piece of code added to the software is called a gdb stub. The stub (running on the target) communicates with gdb running on a different machine (the host) using a serial port or an Ethernet connection. While this may seem complicated, it is the easiest way to debug without requiring the CPU to provide provisions in silicon for debugging.

The good news is that for simulation acceleration and emulation applications there is much greater flexibility since it is really a simulation of the CPU RTL code and not a piece of silicon. The difference is visibility. In silicon there is no visibility. There is no way to see the values of the registers inside without the aid of software to export the values or special purpose hardware to scan out the values. Simulation, on the other hand, has very good visibility. In a simulation acceleration or emulation platform, all of the values of the registers and wires are visible at all times. This visibility makes the use of the gdb remote protocol even better than its original intent since a special stub is no longer needed by the user in the embedded system code. Now the solution is totally transparent to the user. Now gdb can use the remote protocol specification to talk to the simulation, both of which are programs running on a PC or workstation. This technique requires no changes to gdb, and the work to implement it is contained in the simulation environment to bridge the gap between gdb and the data it is requesting from the simulation. The architecture of using the gdb remote protocol with simulation acceleration and emulation is shown in Figure 6.18.

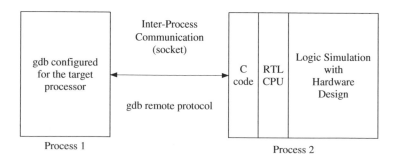

Figure 6.18: gdb Connected to the RTL Code of the Microprocessor

6.4.4.5 *Hardware Model with Logic Simulation*

Another way to eliminate the issues associated with microprocessor models is to use the concept of a "hardware model." A hardware model uses the silicon of the microprocessor as a model for Verilog and VHDL simulation. A custom socket holds the silicon and captures the outputs from the silicon, sends them to a logic simulator and applies the inputs from the simulator to the input pins of the silicon. The communication mechanism between the hardware modeler and the simulator must involve software to talk to the simulator so a network connection is most natural. The concept is much like that of a tester where the stimulus and response is provided by a logic simulator. The architecture of using the hardware model for co-verification is shown in Figure 6.19.

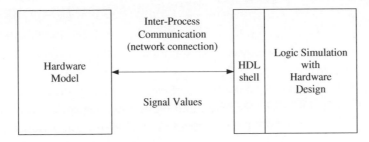

Figure 6.19: Hardware Model of the Microprocessor

Software debugging with the hardware model can be accomplished in multiple ways. In the previous section, the gdb stub was presented. This is a technique that can be used on the hardware model to debug software. Unlike the RTL model in a simulation environment, the hardware model cannot provide visibility of the internal registers so the user must integrate the stub with the other software running on the microprocessor. The other technique for debugging software is a JTAG connection for those microprocessors that support this type of debugging by providing dedicated silicon to connect to the JTAG probe and debugger. In both cases, performance of the environment can limit the utility of the hardware model for software debugging.

The hardware model can also provide local memory in the hardware to service some memory requests that are not required to be simulated. For pure software development, software engineers are interested in high performance and less interested in simulation detail. By servicing some of the memory requests locally on the hardware modeler and avoiding simulation, the software can run at a much higher speed. Hardware modelers can run at speeds of up to 100 kHz when running independently of the logic simulator.

Of course, in the lock step mode they will only run as fast as the logic simulator and exchange pin information every cycle.

With the hardware model, co-verification is no longer completely virtual since a real sample of the microprocessor is used, but for those engineers that have negative experiences with poor simulation models in the past, the concept is very easy to understand and very appealing. What could be a better model than the chip itself?

Clocking limitations are one of the main drawbacks of the hardware model. To do interactive software debugging, the CPU must be capable of running slowly and maintaining its state. Early hardware modeling products were developed at a time when many microprocessor chips started using phase-locked loops and could not be slowed down because the PLLs don't work at slow speeds. To get around this problem, the hardware modeler would reset the device and replay the previous n vectors to get to vector $n + 1$. This allowed the device to be clocked at speeds high enough to support PLL operation, but made software debugging impossible, except by using waveforms from the logic simulator. As we have seen, today's microprocessors come in two flavors, the high-performance variety with PLLs and those more focused on low power. The high-performance variety usually have mechanisms to bypass the PLL to enable static operation and the low-power variety are meant for static design and are very flexible in terms of slow clocking and even stopping the clock. Unfortunately, experiments with such processors have revealed that when bypassing the PLL, device behavior is no longer 100% identical to behavior with the PLL. For low-power cores like ARM, irregular clocking can also be trouble since it requires the clock input to be treated more like a data input since it must be sampled in simulation and is not required to be regular.

With the RTL core becoming more common, there are now products that provide an FPGA for the synthesizable CPU and link to the logic simulator in the same way as the more traditional hardware modeler. Using the CPU in an FPGA gives some benefit by allowing JTAG debugging products to be used, but performance is still likely to be a concern. If the JTAG clock can run independently of the logic simulator, high performance can be obtained for good JTAG debugging.

6.4.4.6 *Evaluation Board with Logic Simulation*

The microprocessor evaluation board is a popular way for software engineers to test code before hardware is available. These boards are readily available for a reasonable cost. To extend the use of the evaluation board for co-verification, the board can serve a similar purpose as the instruction set simulator. Since most boards have networking support, a socket connection between the board and the logic simulator can be developed. A bus functional model residing in the logic simulator can interface the board to the rest of the hardware design. The architecture of using the evaluation board for co-verification is shown in Figure 6.20.

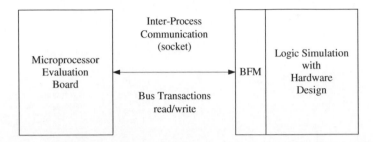

Figure 6.20: Microprocessor Evaluation Board with Logic Simulation

This combination of a CPU board connected to logic simulation via a socket connection and BFM is most appealing to software engineers since the performance of the board is very good. Since each is running independently, there is no synchronization or correlation between the two time domains of the board and the logic simulator.

The drawback to this type of environment is the need to add custom software to the code running on the CPU board to handle the socket connection to the logic simulator. Some commercial co-verification vendors provide such a library that may be suitable, but must always be modified since each board is different and the software operating environment is different for different real-time operating systems. Although the solution requires a lot of customization, it has been used successfully on projects.

6.4.4.7   In-Circuit Emulation

In-circuit emulation involves using external hardware connected to an emulation system that runs at much higher speeds than a logic simulator. Emulation is an attractive platform to do co-verification since the higher speed enables software to run faster. This section discusses three different ways to perform co-verification with an emulation system.

The first method is useful for microprocessor cores that are available in RTL form. As we have seen, there is a trend for the IP vendors to provide RTL code to the user for the purposes of simulation and synthesis. If this is available, the microprocessor can be mapped directly into the emulation system. Most cores used in SoC design today support some kind of JTAG interface for software debugging. To perform co-verification a software engineer can connect a JTAG probe to the I/O pins of the emulator and communicate with the CPU that is mapped inside the emulator. The architecture of using a JTAG connection to an emulator for co-verification is shown in Figure 6.21.

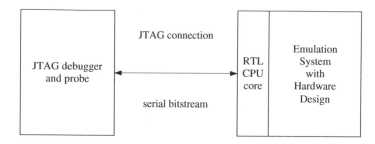

Figure 6.21: JTAG Connection to an Emulation System

In this mode of operation, the CPU runs at the speed of the emulation system, in lock-step with the rest of the design. The main issues in performing co-verification are the overall speed of the emulator and its ability to maintain the JTAG connection reliably at speeds that are lower than most hardware boards.

A second way to perform co-verification with an emulation system is to use a board with the microprocessor test chip and connect the pins of the chip to the I/O pins of the emulator. This technique is useful for hard macro microprocessor IP such as the ARM7TDMI that cannot be mapped into the emulation system. JTAG debugging can also be done by connecting to the JTAG port on the chip. The architecture of using a JTAG connection to an emulator for co-verification is shown in Figure 6.22.

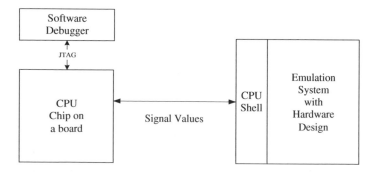

Figure 6.22: JTAG Connection with Test Chip and Emulation System

Like the previous method, the CPU core will run at the speed of the emulation system. Signal values will be updated on each clock cycle. The result is a cycle-accurate simulation of the connection between the test chip and the rest of the design. The cycle-accurate lock-step simulation is desired for hardware engineers that want to model the system exactly and want to run faster using emulation technology for long software tests and regression tests.

In both of the previous techniques, the user must make sure to confirm that the JTAG software and hardware being used for debugging can tolerate slow clock speeds. Most emulation systems run in the 250 kHz to 1 MHz range depending on the emulation technology and the design being run on the emulator. While this is much faster than a logic simulator, it is much slower than what the developers of the JTAG tools probably expected. Most JTAG tools have built in time-outs, either in the hardware or in the software debugger (or both) for situations when the design is not responding. It is crucial to verify that these time-outs can be turned off. Emulation, like simulation, allows the user to stop the test by pressing **Ctrl+c**, waiting for some unspecified amount of time, and then restarting operation. If time-outs exist in the JTAG solution, this will certainly cause a disconnect and result in the loss of software debugging. The best way to provide a stable JTAG connection is to use a feedback clock to the JTAG hardware to help it adapt its speed based on the speed of the emulation system.

The third co-verification method commonly used with emulation is to use a speed bridge between hardware containing a microprocessor device and the emulation system. The classic case for this application is for verification of a chip that connects to the PCI bus. A common setup is for software engineers that are developing device drivers for operating systems such as Windows or Linux and the board they are writing the driver for sits on the PCI bus. Since the PCI board is not yet available, they can use a PC to test the software and the emulation system provides a PCI board that plugs into the PC and bridges the speed differences between the real speed of the PCI bus in the PC (33 or 66 MHz) and the slower speed of the emulator. The PC will run at full speed until the device driver makes a memory or I/O access to the slot with the hardware being developed. When this occurs, the bridge to the emulator will detect the PCI transaction and send it over to the emulator. While the emulator is executing the PCI transaction, the bridge card will continuously respond with a retry response to stall the PC until the emulator is ready. Eventually, the emulator will complete the PCI transaction and the bridge card will complete the transaction on the PC. This method is shown in Figure 6.23.

Similar environments are common for embedded systems where a board containing a microprocessor can run an RTOS such as VxWorks and communicate with the emulator through a speed bridge for a bus such as PCI or AHB.

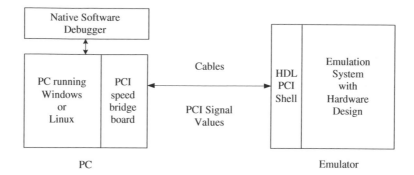

Figure 6.23: JTAG Connection Speed Bridge and Emulation System

6.4.4.8   FPGA Prototype

I always get a laugh when the FPGA prototype is discussed as a co-verification technique. Prototyping is really just building the system out of programmable logic and using the debugger just as if the final hardware was constructed. The only difference may be ASICs are substituted for FPGA, and as a result the performance is lower than the final implementation. Since hardware debugging is very difficult, prototyping barely qualifies as co-verification, but since the representation of the hardware is not the final product it is a useful way for software engineers to get early access to the hardware to debug software.

Recent advances in FPGA technology have caused many projects to reexamine hardware prototyping. With FPGAs from Altera and Xilinx now exceeding 250 k to 500 k ASIC gates, custom prototyping has become a possibility for hardware and software integration. Until now design flow issues, tool issues, and the great density differences between ASIC and FPGA have limited the use of prototyping. With the latest FPGA devices, most ASICs can now be mapped into a set of one to six FPGAs. New partitioning tools have also been introduced that work at the RT level and do not require changes to the RTL code or difficult gate-level, post-synthesis partitioning.

Although prototyping is easier than it has ever been it is still not a trivial task. Prototyping issues fall into two categories: FPGA resource issues and ASIC/FPGA technology differences. Common resource issues can be the limited number of I/O pins available on the FPGA or the number of clock domains available in an FPGA. Technology differences can be related to differences in synthesis tools forcing the user to modify the design to map to the FPGA technology. Another common technology issue is gated clocks that are difficult to handle in FPGA technology. If resource and technology issues can be overcome, prototyping can provide the highest performance co-verification solution that is scalable to large numbers of software engineers. Before committing to prototyping is it important to clearly understand the issues as well as the cost. On the surface, prototyping appears cheap compared to alternatives, but like all engineering projects cost should be measured not only in hardware but also in engineering time to create a working solution.

6.4.5 Co-Verification Metrics

Many metrics can be used to determine which co-verification methods are best for a particular project. Following is a list of some of them:

- Performance (speed)

- Accuracy

- Synchronization

- Type of software to be verified

- Ability to do hardware debugging (visibility)

- Ability to do performance analysis

- Specific versus general-purpose solutions

- Software only (simulated hardware) versus hardware methods

- Time to create and integrate models: bus interface, cache, peripherals, RTOS

- Time to integrate software debug tools

- Pre-silicon compared to post-silicon

6.4.5.1 Performance

It is common to see numbers thrown out about cycles/sec and instructions/sec related to co-verification. While some projects may indeed achieve very high performance using co-verification, it is difficult to predict performance of a co-verification solution. Of course, every vendor will say that performance is "design dependent," but with a good understanding of co-verification methods it is possible to get a good feel for what kind of performance can be achieved. The general unpredictability is a result of two factors; first, many co-verification methods use a dual-process architecture to execute hardware and software. Second, the size of the design, the level of detail of the simulation, and the performance of the hardware verification platform results in very different performance levels.

6.4.5.2   Verification Accuracy

While performance issues are the number one objection to co-verification from software engineers, accuracy is the number one concern of hardware engineers. Some common questions to think about when evaluating co-verification accuracy are listed here. The key to successful hardware/software co-verification is the microprocessor model.

- *How is the model verified to guarantee it behaves identically to the device silicon?* Software models can be verified by using manufacturing test vectors from the microprocessor vendor or running a side-by-side comparison with the microprocessor RTL design database. Metrics such as code coverage can also provide information about software model testing. Alternatively, not all co-verification techniques rely on separately developed models. Techniques based on RTL code for the CPU can eliminate this question altogether. Make sure the model comes with a documented verification plan. Anybody can make a model, but the effort required to make a good model should not be underestimated.

- *Does the model contain complete functionality, including all peripherals?* Using bus functional models was a feasible modeling method before so many peripherals were integrated with the microprocessor. For chips with high integration, it becomes very difficult to model all of the peripherals. Even if a device appears to have no integrated peripherals, look for things like cache controllers and write buffers.

- *Is the model cycle accurate?* Do all parts of the model take into account the internal clock of the microprocessor? This includes things such as the microprocessor pipeline timing and the correlation of bus transaction times with instruction execution. This may or may not be necessary depending on the goals of co-verification. A noncycle accurate model can run at a higher speed and more may be more suitable for software development.

- *Are all features of the bus protocol modeled?* Many microprocessors use more complex bus protocols to improve performance. Techniques such as bus pipelining, bursting, out-of-order transaction completion, write posting, and write reordering are usually a source of design errors. Simple read and write transactions by themselves rarely bring out hardware design errors. It is the sequence of many transactions of different types that bring out most design errors.

There is nothing more frustrating than trying to use a model for a CPU that has multiple modes of operation only to find out that the mode that is used by the design suffers from the most dreaded word in modeling, "unsupported."

I once worked on a project involving a design with the ARM920T CPU. This core uses separate clocks for the bus clock (BCLK) and the internal core clock (FCLK). The clocking has three modes of operation:

- *FastBus Mode:* The internal CPU is clocked directly from the bus clock (BCLK) and FCLK is not used.

- *Synchronous Mode:* The internal CPU is clocked from FCLK, which must be faster and a synchronous integer multiple of the bus clock (BCLK).

- *Asynchronous Mode:* The internal CPU is clocked from FCLK, which can be totally asynchronous to BCLK as long as it is faster than BCLK.

As you can probably guess, the asynchronous mode caused this particular problem. The ARM920T starts off using FastBus mode after reset until which time software can change a bit in coprocessor 15 to switch to one of the other clocking modes to get higher performance. When the appropriate bit in cp15 was changed to enable asynchronous mode, a mysterious message comes out: "Set to Asynch mode, WARNING this is not supported"

It is quite disheartening to learn this information only after a long campaign to convince the project team that co-verification is useful.

Following are some other things to pay attention to when evaluating models used for co-verification:

- *Can performance data be gathered to ensure system design meets requirements?*
  If the model is not cycle accurate, the answer is NO. Both hardware and software engineers are interested in using co-verification to obtain measurements about bus throughput, cache hit rates, and software performance. A model that is not cycle accurate cannot provide this information.

- *Will the model accurately model hardware and software timing issues?*
  Like the bus protocol, the more difficult to find software errors are brought out by the timing interactions between software and hardware. Examples include interrupt latency, timers, and polling.

When it comes to modeling and accuracy issues there are really two different groups. One set of engineers is mostly interested in the value of co-verification for software development purposes. If co-verification provides a way to gain early access to the hardware and the code runs at a high enough speed the software engineer is quite happy and the details of accuracy are not that important. The other group is the hardware engineers and verification engineers that insist that if the simulation is not exact then there is no reason to even bother to run it. Simulating something that is not reality provides no benefit to these engineers. The following examples demonstrate the difficulty in satisfying both groups.

6.4.5.3   AHB Arbitration and Cycle Accuracy Issues

I once worked with a project that used a single-layer AHB implementation for the ARM926EJ-S. One way to perform co-verification is to replace a full-functional logic simulation model of the ARM CPU by a bus functional model and an instruction set simulator. Since the ARM926 bus functional model will actually implement two bus interfaces, a question was raised by the project team about arbitration and the ordering of the transfers on the bus between the IAHB and the DAHB. The order in which the arbiter will grant the bus is greatly dependent on the timing of the bus request signals from each AHB master. From the discussion of AHB the **HREADY** signal plays a key role in arbitration since the bus is only granted when both **HGRANT** and **HREADY** are high. The particular bus functional model being used decided that since **HREADY** is required to be high for arbitration and data transfer, it can be used more like an enable for the bus interface model state machine since nothing can happen without it being high. To optimize performance the bus functional model decided to do nothing during the time when **HREADY** was low. This assumption about the function of **HREADY** is nearly correct, but not exactly. The CPU indication to the bus interface unit of the ARM926 that it needs to request the bus has nothing to do with **HREADY** or the bus interface clock, it uses the CPU clock. This produced a situation where the **IHBUSREQ** and **DHBUSREQ** were artificially linked to the **HREADY** and the timing of these was incorrect. The result to the user was the arbiter granting the IAHB to use the bus instead of the DAHB. Since the two busses are independent, except for the few exceptions we discussed, there is no harm in running the transactions in a different order on the single-layer AHB. Functionally, this makes no difference and all verification tests and software will execute just fine, but to hardware engineers seeking accuracy the situation is no good. This case does bring up some interesting questions related to performance:

- Does arbitration priority affect system performance?

- What are the performance differences between single-layer AHB versus multilayer AHB versus separate AHB interfaces?

Figure 6.24 shows the correct bus request timing. The sequence shows the IAHB reading addresses 0x44 and 0x48 followed by the DAHB reading from address 0x90. It is difficult to see, but the next transaction is the IAHB reading from address 0x4c. Notice the timing of **DHBUSREQ** at the start of the waveform. It transitions high before the first **IHREADY** on the waveform. This demonstrates that the timing of **DHBUSREQ** is not related to **HREADY**.

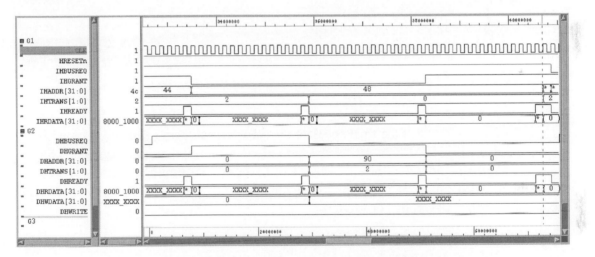

Figure 6.24: Correct Timing of Bus Request

Figure 6.25 shows the incorrect ordering of bus transfers caused by the difference in the timing of **DHBUSREQ**. The sequence start the same way with IAHB reads from 0x44 and 0x48, but the read from 0x4c comes before the DAHB read from address 0x90. The reason is the timing of **DHBUSREQ**. Notice **DHBUSREQ** transitions high AFTER the first **IHREADY** on the waveform. This difference results in out-of-order transactions.

Contrast this pursuit of accuracy with a software engineer I met once that didn't care anything about the detail of the hardware simulation. Skipping the majority of the simulation activity just to run fast was the best way to go. He had no desire to run a detailed, cycle-accurate simulation. Actually, he was interested in making sure the software ran on the cycle-accurate simulation, but once it had been debugged using a noncycle accurate

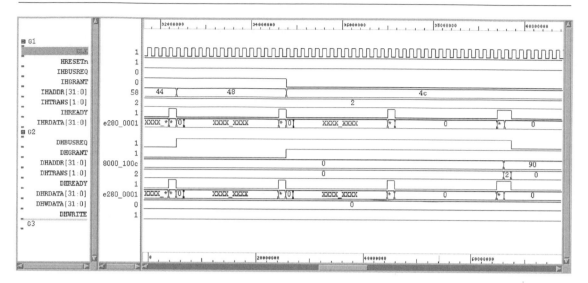

Figure 6.25: Incorrect Timing of Bus Request

co-verification environment the final check of the software was better suited for a long batch simulation using the ARM RTL model and farm of workstations that was maintained by the hardware engineers, not him. Since the chance of finding a software bug was low there was no reason to worry about the problem of debugging the software in a pure logic simulation environment using waveforms or logfiles.

6.4.5.4   Modeling Summary

Modeling is always painful. There is no way around it. No matter what kind of checks and balances are available to compare the model to the actual implementation, there are always differences. One of the common debates is about what represents the golden view of the IP. In the case of ARM microprocessors, there are three possible representations that are considered "golden":

- RTL code (for synthesizable ARM designs)

- Design sign-off model (DSM) derived from the implementation

- Silicon in the form of a test chip or FPGA

Engineers view these three as golden, even more so than the specification. Co-verification techniques that use models that do not come from one of these golden sources are at a disadvantage since any problems are always blamed on the "nongolden" model. I have seen cases where the user's design does not match the bus specification, but does work when simulated with a golden model. Since a specification is not executable, engineers feel strongly that the design working with the golden model is most important, not the specification. When alternative models are used for co-verification a model that conforms to the specification is still viewed as a buggy model in any places it differs from the golden model. It is not always easy to convince engineers that a design that runs with many different models and adheres to the specification is better than a design that runs only with the golden model.

6.4.5.5 *Synchronization*

Most co-verification tools operate by hiding cycles from the slower logic simulation environment. Because of this, issues related to synchronization of the microprocessor with the rest of the simulated design often arise. This situation is also true using in-circuit emulation with a processor linked to the emulator via a speed bridge. In co-verification there are two distinct time domains, the microprocessor model running outside of the logic simulator and the logic simulator itself. Understanding the correlation of these two time domains is important to achieving success with co-verification. Co-verification uses mainly the spatial memory references to decide when software meets the hardware simulation. Synchronization is defined by what happens to the logic simulator when there is no bus transaction occurring in the logic simulator. It could be stopped until a new bus transaction is received. It could just "drift" forward in time executing other parts of the logic simulation hardware (even with an idle microprocessor bus). In either case the amount of time simulated in the logic simulator and the microprocessor model is different. Another alternative is to advance the logic simulation time the proper number of clock cycles to account for the hidden bus transaction but don't run the transaction on the bus. Now the correction of the time domains is maintained at the expense of performance. Synchronization is also important for those temporal activities where the hardware design communicates with software such as interrupts and DMA transfers. Without proper synchronization, things like system timers and DMA transfers may not work correctly because of differences in the two time domains.

6.4.5.6 *Types of Software*

The type of software to be verified also has a major impact on which co-verification methods to deploy. There are different types of software that engineers see as candidates for co-verification: system diagnostics, device drivers, RTOS, and application code. Different co-verification methods are better suited to different types of software. Usually the lower level software requires a more accurate co-verification environment and higher-level software is less interested in accuracy and more focused on performance because of the code size. Running an RTOS such as VxWorks has been shown to be viable by multiple co-verification methods including in-circuit emulation, an ISS, and the RTOS simulator, VxSIM. Even with the marketing claims that software does not have to be modified, expect some modification to optimize such things like long memory tests and UART accesses. The major confusion today exists because of the many types of software and the many methods of hardware execution. Often, different levels of performance will enable different levels of software to be verified using co-verification. A quick sanity check to calculate the number of cycles required to run a given type of software and the speed of the environment will ensure engineers can remain productive. If it takes 1 hour to run a software program to get to the new software this means the software engineer will have only a handful of chances per day to run the code and debug any problems found.

6.4.5.7   Other Metrics

Besides performance and accuracy, there are some other metrics worth thinking about. Project teams should also determine if a general-purpose solution is important versus a project specific solution. General-purpose solutions can be reused on future projects and only one set of tools needs to be learned. Unfortunately, general-purpose solutions are not general if the model used on the next project is not available. Methods using the evaluation board or prototyping are more specific and may not be applicable on the next project. For many engineers, especially software engineers, a solution that consists of simulation only is preferred over one that contains hardware. Another important distinction is whether the solution is available pre- or post-silicon. Many leading edge projects use microprocessors that are not yet available and a pre-silicon method is required. All of these variables should be considered when deciding on a co-verification strategy.

Understanding all of these metrics will avoid committing to a co-verification solution that will not meet the project needs. Remember, the goal of co-verification is to save time in the project schedule.

CHAPTER 7

Techniques for Embedded Media Processing

David J. Katz
Rick Gentile

With the multimedia revolution in full swing, we're becoming accustomed to toting around cell phones, PDAs, cameras, and MP3 players, concentrating our daily interactions into the palms of our hands. But given the usefulness of each gadget, it's surprising how often we upgrade to "the latest and greatest" device. This is, in part, due to the fact that the cell phone we bought last year can't support the new video clip playback feature touted in this year's TV ads.

After all, who isn't frustrated after discovering that his portable audio player gets tangled up over the latest music format? In addition, which overworked couple has the time, much less the inclination, to figure out how to get the family vacation travelogue off their mini-DV camcorder and onto a DVD or hard disk?

As Figure 7.1 implies, we've now reached the point where a single gadget can serve as a phone, a personal organizer, a camera, an audio player, and a web-enabled portal to the rest of the world.

But still, we're not happy.

Let's add a little perspective: we used to be satisfied just to snap a digital picture and see it on our computer screen. Just 10 years ago, there were few built-in digital camera features, the photo resolution was comparatively low, and only still pictures were an option. Not that we were complaining, since previously our only digital choice involved scanning 35-mm prints into the computer.

In contrast, today we expect multimegapixel photos, snapped several times per second, which are automatically white-balanced and color-corrected. What's more, we demand seamless transfer between our camera and other media nodes, a feature made practical only because the camera can compress the images before moving them.

Clearly, consumer appetites demand steady improvement in the "media experience." That is, people want high-quality video and audio streams in small form factors, with low power requirements (for improved battery life) and at low cost. This desire leads to constant development of better compression algorithms that reduce storage requirements while increasing audio/video resolution and frame rates.

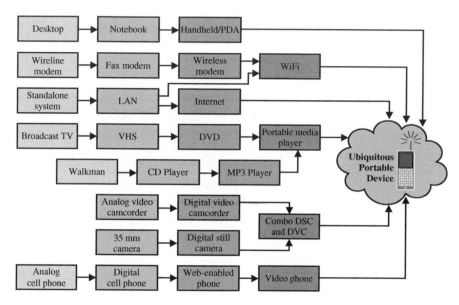

Figure 7.1: The "Ultimate" Portable Device Is Almost within Our Grasp

To a large extent, the Internet drives this evolution. After all, it made audio, images, and streaming video pervasive, forcing transport algorithms to become increasingly clever at handling ever-richer media across the limited bandwidth available on a network. As a result, people today want their portable devices to be net-connected, high-speed conduits for a never-ending information stream and media show. Unfortunately, networking infrastructure is upgraded at a much slower rate than bandwidth demands grow, and this underscores the importance of excellent compression ratios for media-rich streams.

It may not be readily apparent, but behind the scenes, processors have had to evolve dramatically to meet these new and demanding requirements. They now need to run at very high clock rates (to process video in real time), be very power efficient (to prolong battery life), and comprise very small, inexpensive single-chip solutions (to save board real estate and keep end products price-competitive). What's more, they need to be software-reprogrammable, in order to adapt to the rapidly changing multimedia standards environment.

7.1 A Simplified Look at a Media Processing System

Consider the components of a typical media processing system, shown in Figure 7.2. Here, an input source presents a data stream to a processor's input interface, where it is manipulated appropriately and sent to a memory subsystem. The processor core(s) then interact with the memory subsystem in order to process the data, generating intermediate data buffers in the process. Ultimately, the final data buffer is sent to its destination via an output subsystem. Let's examine each of these components in turn.

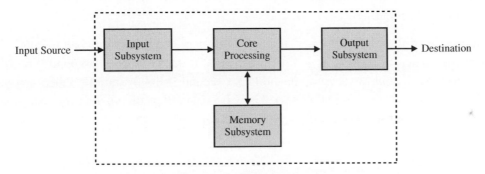

Figure 7.2: Components of a Typical Media Processing System

7.1.1 Core Processing

Multimedia processing—that is, the actual work done by the media processor core—boils down into three main categories: format coding, decision operating, and overlaying.

Software *format coders* separate into three classifications. *Encoders* convert raw video, image, audio and/or voice data into a compressed format. A digital still camera (DSC) provides a good example of an encoding framework, converting raw image sensor data into compressed JPEG format. *Decoders*, on the other hand, convert a compressed stream into an approximation (or exact duplicate) of the original uncompressed content. In playback mode, a DSC decodes the compressed pictures stored in its file system and displays them on the camera's LCD screen. *Transcoders* convert one media format into another one, for instance MP3 into Windows Media Audio 9 (WMA9).

Unlike the coders mentioned above, *decision operators* process multimedia content and arrive at some result, but do not require the original content to be stored for later retrieval. For instance, a pick-and-place machine vision system might snap pictures of electronic components and, depending on their orientation, size and location, rotate the parts for proper placement on a circuit board. However, the pictures themselves are not saved for later viewing or processing. Decision operators represent the fastest growing segment of image and video processing, encompassing applications as diverse as facial recognition, traffic light control, and security systems.

Finally, *overlays* blend multiple media streams together into a single output stream. For example, a time/date stamp might be instantiated with numerous views of surveillance footage to generate a composited output onto a video monitor. In another instance, graphical menus and icons might be blended over a background video stream for purposes of annotation or user input.

Considering all of these system types, the input data varies widely in its bandwidth requirements. Whereas raw audio might be measured in tens of kilobits/second (kb/s), compressed video could run several megabits per second (Mbps), and raw video could entail tens of megabytes per second (Mbytes/s). Thus, it is clear that the media processor needs to handle different input formats in different ways. That's where the processor's peripheral set comes into play.

7.1.2   Input/Output Subsystems—Peripheral Interfaces

Peripherals are classified in many ways, but a particularly useful generalization is to stratify them into functional groups like those in Table 7.1. Basically, these interfaces act to help control a subsystem, assist in moving and storing data, or enable connectivity with other systems or modules in an application.

Table 7.1: Classes of Peripherals and Representative Examples

Programming Models

	Asymmetric	Homogenous	Master-Slave	Pipelined
Processor				
Asymmetric	✓			
Symmetric	✓	✓	✓	✓

Let's look now at some examples of each interface category.

7.1.2.1   Subsystem Control—Low-Speed Serial Interfaces

UART (Universal Asynchronous Receiver/Transmitter)—As its name suggests, this full-duplex interface needs no separate clock or frame synchronization lines. Instead, these are decoded from the bit stream in the form of start bit, data bits, stop bits, and optional parity bits.

UARTs are fairly low-speed (kbps to Mbps) and have high overhead, since every data word has control and error checking bits associated with it. UARTs can typically support RS-232 modem implementations, as well as IrDA functionality for close-range infrared transfer.

SPI (Serial Peripheral Interface)—This is a synchronous, moderate-speed (tens of Mbps), full-duplex master/slave interface developed by Motorola. The basic interface consists of a clock line, an enable line, a data input ("Master In, Slave Out") and a data output ("Master Out, Slave In"). SPI supports both multimaster and multislave environments. Many video and audio codecs have SPI control interfaces, as do many EEPROMs.

I²C (Inter-IC Bus)—Developed by Philips, this synchronous interface requires only two wires (clock and data) for communication. The phase relationships between the two lines determines the start and completion of data transfer. There are primarily three speed levels: 100 kbps, 400 kbps and 3.4 Mbps. Like SPI, I²C is very commonly used for the control channel in video and audio converters, as well as in some ROM-based memories.

Programmable Timers—These multifunction blocks can generate programmable pulse-width modulated (PWM) outputs that are useful for one-shot or periodic timing waveform generation, digital-to-analog conversion (with an external resistor/capacitor network, for instance), and synchronizing timing events (by starting several PWM outputs simultaneously). As inputs, they'll typically have a width-capture capability that allows precise measurement of an external pulse, referenced to the processor's system clock or another time base. Finally, they can act as event counters, counting external events or internal processor clock cycles (useful for operating system ticks, for instance).

Real-Time Clock (RTC)—This circuit is basically a timer that uses a 32.768 kHz crystal or oscillator as a time base, where every 2^{15} ticks equals one second. In order to use more stable crystals, sometimes higher frequencies are employed instead; the most common are 1.048 MHz and 4.194 MHz. The RTC can track seconds, minutes, hours, days, and even years—with the functionality to generate a processor alarm interrupt at a particular day, hour, minute, second combination, or at regular intervals (say, every minute). For instance, a real-time clock might wake up a temperature sensor to sample the ambient environment and relay information back to the MCU via I/O pins. Then, a timer's pulse-width modulated (PWM) output could increase or decrease the speed of a fan motor accordingly.

Programmable Flags/GPIO (General Purpose Inputs/Outputs)—These all-purpose pins are the essence of flexibility. Configured as inputs, they convey status information from the outside world, and they can be set to interrupt upon receiving an edge-based or level-based signal of a given polarity. As outputs, they can drive high or low to control external circuitry. GPIO can be used in a "bit-banging" approach to simulate interfaces like I^2C, detect a key press through a key matrix arrangement, or send out parallel chunks of data via block writes to the flag pins.

Watchdog Timer (WDT)—This peripheral provides a way to detect if there's a system software malfunction. It's essentially a counter that is reset by software periodically with a count value such that, in normal system operation, it never actually expires. If, for some reason, the counter reaches 0, it will generate a processor reset, a nonmaskable interrupt, or some other system event.

Host Interface—Often in multimedia applications an external processor will need to communicate with the media processor, even to the point of accessing its entire internal/external memory and register space. Usually, this external host will be the conduit to a network, storage interface, or other data stream, but it won't have the performance characteristics that allow it to operate on the data in real time. Therefore, the need arises or a relatively high-bandwidth "host port interface" on the media processor. This port can be anywhere from 8 bits to 32 bits wide and is used to control the media processor and transfer data to/from an external processor.

7.1.2.2 Storage

External Memory Interface (Asynchronous and SDRAM)—An external memory interface can provide both asynchronous memory and SDRAM memory controllers. The asynchronous memory interface facilitates connection to FLASH, SRAM, EEPROM, and peripheral bridge chips, whereas SDRAM provides the necessary storage for computationally intensive calculations on large data frames. It should be noted that, while some designs may employ the external memory bus as a means to read in raw multimedia data, this is often a suboptimal solution. Because the external bus is intimately involved in processing intermediate frame buffers, it will be hard pressed to manage the real-time requirements of reading in a raw data stream while writing and reading intermediate data blocks to and from L1 memory. This is why the video port needs to be decoupled from the external memory interface, with a separate data bus.

ATAPI/Serial ATA—These are interfaces used to access mass storage devices like hard disks, tape drives, and optical drives (CD/DVD). Serial ATA is a newer standard that encapsulates the venerable ATAPI protocol, yet in a high-speed serialized form, for increased throughput, better noise performance, and easier cabling.

Flash Storage Card Interfaces—These peripherals originally started as memory cards for consumer multimedia devices, like cameras and PDAs. They allow very small footprint, high density storage, and connectivity, from mass storage to I/O functions like wireless networking, Bluetooth, and Global Positioning System (GPS) receivers. They include CompactFlash, Secure Digital (SD), MemoryStick, and many others. Given their rugged profile, small form factor, and low power requirements, they're perfect for embedded media applications.

7.1.2.3   Connectivity

Interfacing to PCs and PC peripherals remains essential for most portable multimedia devices, because the PC constitutes a source of constant Internet connectivity and near-infinite storage. Thus, a PC's 200-Gbyte hard drive might serve as a "staging ground" and repository for a portable device's current song list or video clips. To facilitate interaction with a PC, a high-speed port is mandatory, given the substantial file sizes of multimedia data. Conveniently, the same transport channel that allows portable devices to converse in a peer-to-peer fashion often lets them dock with the "mother ship" as a slave device.

Universal Serial Bus (USB) 2.0—Universal Serial Bus is intended to simplify communication between a PC and external peripherals via high-speed serial communication. USB 1.1 operated only up to 12 Mbps, and USB 2.0 was introduced in 2000 to compete with IEEE 1394, another high-speed serial bus standard. USB 2.0 supports Low Speed (1.5 Mbps), Full Speed (12 Mbps), and High Speed (480 Mbps) modes, as well as Host and On-the-Go (OTG) functionality. Whereas, a USB 2.0 Host can master up to 127 peripheral connections simultaneously, OTG is meant for a peer-to-peer host/device capability, where the interface can act as an ad hoc host to a single peripheral connected to it. Thus, OTG is well-suited to embedded applications where a PC isn't needed. Importantly, USB supports *Plug-and-Play* (automatic configuration of a plugged-in device), as well as *hot pluggability* (the ability to plug in a device without first powering down). Moreover, it allows for bus-powering of a plugged-in device from the USB interface itself.

PCI (Peripheral Component Interconnect)—This is a local bus standard developed by Intel Corporation and used initially in personal computers. Many media processors use PCI as a general-purpose "system bus" interface to bridge to several different types of devices via external chips (e.g., PCI to hard drive, PCI to 802.11, and so on). PCI can offer the extra benefit of providing a separate internal bus that allows the PCI bus master to send or retrieve data from an embedded processor's memory without loading down the processor core or peripheral interfaces.

Network Interface—In wired applications, Ethernet (IEEE 802.3) is the most popular physical layer for networking over a LAN (via TCP/IP, UDP, and the like), whereas IEEE 802.11a/b/g is emerging as the prime choice for wireless LANs. Many Ethernet solutions are available either on-chip or bridged through another peripheral (like asynchronous memory or USB).

IEEE 1394 ("Firewire")—IEEE 1394, better known by its Apple Computer trademark "Firewire," is a high-speed serial bus standard that can connect with up to 63 devices at once. 1394a supports speeds up to 400 Mbps, and 1394b extends to 800 Mbps. Like USB, IEEE 1394 features hot pluggability and Plug-and-Play capabilities, as well as bus-powering of plugged-in devices.

7.1.2.4   Data Movement

Synchronous Serial Audio/Data Port—Sometimes called a "SPORT," this interface can attain full-duplex data transfer rates above 65 Mbps. The interface itself includes a data line (receive or transmit), clock, and frame sync. A SPORT usually supports many configurations of frame synchronization and clocking (for instance, "receive mode with internally generated frame sync and externally supplied clock"). Because of its high operating speeds, the SPORT is quite suitable for DSP applications like connecting to high-resolution audio codecs. It also features a multichannel mode that allows data transfer over several time-division-multiplexed channels, providing a very useful mode for high-performance telecom interfaces. Moreover, the SPORT easily supports transfer of compressed video streams, and it can serve as a convenient high bandwidth control channel between processors.

Parallel Video/Data Port—This is a parallel port available on some high-performance processors. Although implementations differ, this port can, for example, gluelessly transmit and receive video streams, as well as act as a general-purpose 8- to 16-bit I/O port for high-speed analog-to-digital (A/D) and digital-to-analog (D/A) converters. Moreover, it can act as a video display interface, connecting to video encoder chips or LCD displays. On the Blackfin processor, this port is known as the "Parallel Peripheral Interface," or "PPI."

7.1.3   Memory Subsystem

As important as it is to get data into (or send it out from) the processor, even more important is the structure of the memory subsystem that handles the data during processing. It's essential that the processor core can access data in memory at rates fast enough to meet the demands of the application. Unfortunately, there's a trade-off between memory access speed and physical size of the memory array.

Because of this, memory systems are often structured with multiple tiers that balance size and performance. Level 1 (L1) memory is closest to the core processor and executes instructions at the full core-clock rate. L1 memory is often split between Instruction and Data segments for efficient utilization of memory bus bandwidth. This memory is usually configurable as either SRAM or cache. Additional on-chip L2 memory and off-chip L3 memory provide additional storage (code and data)—with increasing latency as the memory gets further from the processor core.

In multimedia applications, on-chip memory is normally insufficient for storing entire video frames, although this would be the ideal choice for efficient processing. Therefore, the system must rely on L3 memory to support relatively fast access to large buffers. The processor interface to off-chip memory constitutes a major factor in designing efficient media frameworks, because L3 access patterns must be planned to optimize data throughput.

7.2   System Resource Partitioning and Code Optimization

In an ideal situation, we can select an embedded processor for our application that provides maximum performance for minimum extra development effort. In this utopian environment, we could code everything in a high-level language like C, we wouldn't need an intimate knowledge of our chosen device, it wouldn't matter where we placed our data and code, we wouldn't need to devise any data movement subsystem, the performance of external devices wouldn't matter. In short, everything would just work.

Alas, this is only the stuff of dreams and marketing presentations. The reality is, as embedded processors evolve in performance and flexibility, their complexity also increases. Depending on the time-to-market for your application, you will have to walk a fine line to reach your performance targets. The key is to find the right balance between getting the application to work and achieving optimum performance. Knowing when the performance is "good enough" rather than optimal can mean getting your product out on time versus missing a market window.

In this chapter, we want to explain some important aspects of processor architectures that can make a real difference in designing a successful multimedia system. Once you understand the basic mechanics of how the various architectural sections behave, you will be able to gauge where to focus your efforts, rather than embark on the noble yet unwieldy goal of becoming an expert on all aspects of your chosen processor. For our example processor, we will use Analog Devices' Blackfin. Here, we'll explore in detail some Blackfin processor architectural constructs. Again, keep in mind that much of our discussion generalizes to other processor families from different vendors as well.

We will begin with what should be key focal points in any complex application: interrupt and exception handling and response times.

7.3   Event Generation and Handling

Nothing in an application should make you think "performance" more than event management. If you have used a microprocessor, you know that "events" encompass two categories: interrupts and exceptions. An interrupt is an event that happens asynchronous to processor execution. For example, when a peripheral completes a transfer, it can generate an interrupt to alert the processor that data is ready for processing.

Exceptions, on the other hand, occur synchronously to program execution. An exception occurs based on the instruction about to be executed. The change of flow due to an exception occurs prior to the offending instruction actually being executed. Later in this chapter, we'll describe the most widely used exception handler in an embedded processor—the handler that manages pages describing memory attributes. Now, however, we will focus on interrupts rather than exceptions, because managing interrupts plays such a critical role in achieving peak performance.

7.3.1   System Interrupts

System level interrupts (those that are generated by peripherals) are handled in two stages—first in the system domain, and then in the core domain. Once the system interrupt controller (SIC) acknowledges an interrupt request from a peripheral, it compares the peripheral's assigned priority to all current activity from other peripherals to decide when to service this particular interrupt request. The most important peripherals in an application should be mapped to the highest priority levels. In general, the highest bandwidth peripherals need the highest priority. One "exception" to this rule (pardon the pun!) is where an external processor or supervisory circuit uses a nonmaskable interrupt (NMI) to indicate the occurrence of an important event, such as powering down.

When the SIC is ready, it passes the interrupt request information to the core event controller (CEC), which handles all types of events, not just interrupts. Every interrupt from the SIC maps into a priority level at the CEC that regulates how to service interrupts with respect to one another, as Figure 7.3 shows. The CEC checks the "vector" assignment for the current interrupt request, to find the address of the appropriate interrupt service routine (ISR). Finally, it loads this address into the processor's execution pipeline to start executing the ISR.

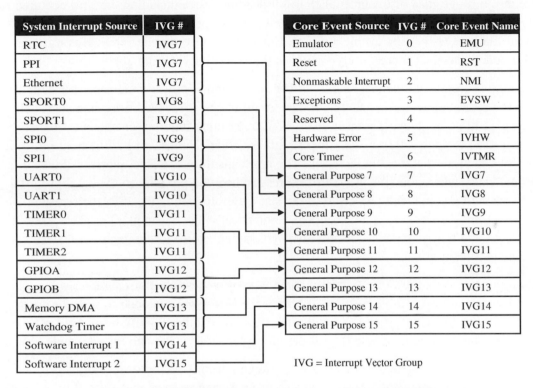

System Interrupt Source	IVG #
RTC	IVG7
PPI	IVG7
Ethernet	IVG7
SPORT0	IVG8
SPORT1	IVG8
SPI0	IVG9
SPI1	IVG9
UART0	IVG10
UART1	IVG10
TIMER0	IVG11
TIMER1	IVG11
TIMER2	IVG11
GPIOA	IVG12
GPIOB	IVG12
Memory DMA	IVG13
Watchdog Timer	IVG13
Software Interrupt 1	IVG14
Software Interrupt 2	IVG15

Core Event Source	IVG #	Core Event Name
Emulator	0	EMU
Reset	1	RST
Nonmaskable Interrupt	2	NMI
Exceptions	3	EVSW
Reserved	4	-
Hardware Error	5	IVHW
Core Timer	6	IVTMR
General Purpose 7	7	IVG7
General Purpose 8	8	IVG8
General Purpose 9	9	IVG9
General Purpose 10	10	IVG10
General Purpose 11	11	IVG11
General Purpose 12	12	IVG12
General Purpose 13	13	IVG13
General Purpose 14	14	IVG14
General Purpose 15	15	IVG15

IVG = Interrupt Vector Group

Figure 7.3: Sample System-to-Core Interrupt Mapping

There are two key interrupt-related questions you need to ask when building your system. The first is, "How long does the processor take to respond to an interrupt?" The second is, "How long can any given task afford to wait when an interrupt comes in?"

The answers to these questions will determine what your processor can actually perform within an interrupt or exception handler.

For the purposes of this discussion, we define interrupt response time as the number of cycles it takes from when the interrupt is generated at the source (including the time it takes for the current instruction to finish executing) to the time that the first instruction is executed in the interrupt service routine. In our experience, the most common method software engineers use to evaluate this interval for themselves is to set up a programmable flag to generate an interrupt when its pin is triggered by an externally generated pulse.

The first instruction in the interrupt service routine then performs a write to a different flag pin. The resulting time difference is then measured on an oscilloscope. This method only provides a rough idea of the time taken to service interrupts, including the time required to latch an interrupt at the peripheral, propagate the interrupt through to the core, and then vector the core to the first instruction in the interrupt service routine. Thus, it is important to run a benchmark that more closely simulates the profile of your end application.

Once the processor is running code in an ISR, other higher priority interrupts are held off until the return address associated with the current interrupt is saved off to the stack. This is an important point, because even if you designate all other interrupt channels as higher priority than the currently serviced interrupt, these other channels will all be held off until you save the return address to the stack. The mechanism to re-enable interrupts kicks in automatically when you save the return address. When you program in C, any register the ISR uses will automatically be saved to the stack. Before exiting the ISR, the registers are restored from the stack. This also happens automatically, but depending on where your stack is located and how many registers are involved, saving and restoring data to the stack can take a significant amount of cycles.

Interrupt service routines often perform some type of processing. For example, when a line of video data arrives into its destination buffer, the ISR might run code to filter or down sample it. For this case, when the handler does the work, other interrupts are held off (provided that nesting is disabled) until the processor services the current interrupt.

When an operating system or kernel is used, however, the most common technique is to service the interrupt as soon as possible, release a semaphore, and perhaps make a call to a callback function, which then does the actual processing. The semaphore in this context provides a way to signal other tasks that it is okay to continue or to assume control over some resource.

For example, we can allocate a semaphore to a routine in shared memory. To prevent more than one task from accessing the routine, one task takes the semaphore while it is using the routine, and the other task has to wait until the semaphore has been relinquished before it can use the routine. A Callback Manager can optionally assist with this activity by allocating a callback function to each interrupt. This adds a protocol layer on top of the lowest layer of application code, but in turn it allows the processor to exit the ISR as soon as possible and return to a lower-priority task. Once the ISR is exited, the intended processing can occur without holding off new interrupts.

We already mentioned that a higher-priority interrupt could break into an existing ISR once you save the return address to the stack. However, some processors (like Blackfin) also support self-nesting of core interrupts, where an interrupt of one priority level can interrupt an ISR of the same level, once the return address is saved. This feature can be useful for building a simple scheduler or kernel that uses low-priority software-generated interrupts to preempt an ISR and allow the processing of ongoing tasks.

There are two additional performance-related issues to consider when you plan out your interrupt usage. The first is the placement of your ISR code. For interrupts that run most frequently, every attempt should be made to locate these in L1 instruction memory. On Blackfin processors, this strategy allows single-cycle access time. Moreover, if the processor were in the midst of a multicycle fetch from external memory, the fetch would be interrupted, and the processor would vector to the ISR code.

Keep in mind that before you re-enable higher priority interrupts, you have to save more than just the return address to the stack. Any register used inside the current ISR must also be saved. This is one reason why the stack should be located in the fastest available memory in your system. An L1 "scratchpad" memory bank, usually smaller in size than the other L1 data banks, can be used to hold the stack. This allows the fastest context switching when taking an interrupt.

7.4   Programming Methodology

It's nice not to have to be an expert in your chosen processor, but even if you program in a high-level language, it's important to understand certain things about the architecture for which you're writing code.

One mandatory task when undertaking a signal-processing-intensive project is deciding what kind of programming methodology to use. The choice is usually between assembly language and a high-level language (HLL) like C or C++. This decision revolves around many factors, so it's important to understand the benefits and drawbacks each approach entails.

The obvious benefits of C/C++ include modularity, portability, and reusability. Not only do the majority of embedded programmers have experience with one of these high-level languages, but also a huge code base exists that can be ported from an existing processor domain to a new processor in a relatively straightforward manner. Because assembly language is architecture-specific, reuse is typically restricted to devices in the same processor family. Also, within a development team it is often desirable to have various teams coding different system modules, and an HLL allows these cross-functional teams to be processor-agnostic.

One reason assembly has been difficult to program is its focus on actual data flow between the processor register sets, computational units and memories. In C/C++, this manipulation occurs at a much more abstract level through the use of variables and function/procedure calls, making the code easier to follow and maintain.

The C/C++ compilers available today are quite resourceful, and they do a great job of compiling the HLL code into tight assembly code. One common mistake happens when programmers try to "outsmart" the compiler. In trying to make it easier for the compiler, they in fact make things more difficult! It's often best to just let the optimizing compiler do its job. However, the fact remains that compiler performance is tuned to a specific set of features that the tool developer considered most important. Therefore, it cannot exceed handcrafted assembly code performance in all situations.

The bottom line is that developers use assembly language only when it is necessary to optimize important processing-intensive code blocks for efficient execution. Compiler features can do a very good job, but nothing beats thoughtful, direct control of your application data flow and computation.

7.5   Architectural Features for Efficient Programming

In order to achieve high performance media processing capability, you must understand the types of core processor structures that can help optimize performance. These include the following capabilities:

- Multiple operations per cycle

- Hardware loop constructs

- Specialized addressing modes

- Interlocked instruction pipelines

These features can make an enormous difference in computational efficiency. Let's discuss each one in turn.

7.5.1   Multiple Operations per Cycle

Processors are often benchmarked by how many millions of instructions they can execute per second (MIPS). However, for today's processors, this can be misleading because of the confusion surrounding what actually constitutes an instruction. For example, multi-issue instructions, which were once reserved for use in higher-cost parallel processors, are now also available in low-cost, fixed-point processors. In addition to performing multiple ALU/MAC operations each core processor cycle, additional data loads, and stores can be completed in the same cycle. This type of construct has obvious advantages in code density and execution time.

An example of a Blackfin multi-operation instruction is shown in Figure 7.4. In addition to two separate MAC operations, a data fetch and data store (or two data fetches) can also be accomplished in the same processor clock cycle. Correspondingly, each address can be updated in the same cycle that all of the other activities are occurring.

Instruction:
R1.H=(A1+=R0.H*R2.H), R1.L=(A0+=R0.L*R2.L) ‖ R2 = [I0--] ‖ [I1++] = R1;

R1.H=(A1+=R0.H*R2.H), R1.L=(A0+=R0.L*R2.L)
- multiply R0.H*R2.H, accumulate to A1, store to R1.H
- multiply R0.L*R2.L, accumulate to A0, store to R1.L

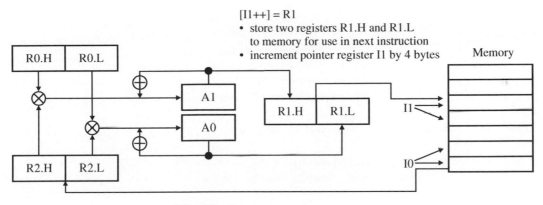

[I1++] = R1
- store two registers R1.H and R1.L to memory for use in next instruction
- increment pointer register I1 by 4 bytes

R2 = [I0 - -]
- load two 16-bit registers R2.H and R2.L from memory for use in next instruction
- decrement pointer register I0 by 4 bytes

Figure 7.4: Example of Singe-cycle, Multi-issue Instruction

7.5.2   Hardware Loop Constructs

Looping is a critical feature in real-time processing algorithms. There are two key looping-related features that can improve performance on a wide variety of algorithms: *zero-overhead hardware loops* and *hardware loop buffers*.

Zero-overhead loops allow programmers to initialize loops simply by setting up a count value and defining the loop bounds. The processor will continue to execute this loop until the count has been reached. In contrast, a software implementation would add overhead that would cut into the real-time processing budget.

Many processors offer zero-overhead loops, but hardware loop buffers, which are less common, can really add increased performance in looping constructs. They act as a kind of cache for instructions being executed in the loop. For example, after the first time through a loop, the instructions can be kept in the loop buffer, eliminating the need to re-fetch the same code each time through the loop. This can produce a significant savings in cycles by keeping several loop instructions in a buffer where they can be accessed in a single cycle. The use of the hardware loop construct comes at no cost to the HLL programmer, since the compiler should automatically use hardware looping instead of conditional jumps.

Let's look at some examples to illustrate the concepts we've just discussed.

Example 7.1: Dot Product

The dot product, or scalar product, is an operation useful in measuring orthogonality of two vectors. It's also a fundamental operator in digital filter computations. Most C programmers should be familiar with the following implementation of a dot product:

```
short dot(const short  a[ ], const short  b[ ], int size) {

/* Note: It is important to declare the input buffer arrays as const, because this
gives the compiler a guarantee that neither "a" nor "b" will be
modified by the function. */

    int i;
    int output = 0;

    for(i=0; i<size; i++) {
      output += (a[i] * b[i]);
    }
    return output;
}
```

Below is the main portion of the equivalent assembly code:

```
/* P0 = Loop Count, P1 & I0 hold starting addresses of a & b
   arrays */

A1 = A0 = 0;              /* A0 & A1 are accumulators */
LSETUP (loop1,loop1) LC0 = P0 ;     /* Set up hardware loop
   starting and ending at label loop1 */
loop1: A1 += R1.H * R0.H , A0 += R1.L * R0.L || R1 = [ P1 ++ ]
   || R0 = [ I0 ++ ] ;
```

The following points illustrate how a processor's architectural features can facilitate this tight coding.

Hardware loop buffers and loop counters eliminate the need for a jump instruction at the end of each iteration. Since a dot product is a summation of products, it is implemented in a loop. Some processors use a JUMP instruction at the end of each iteration in order to process the next iteration of the loop. This contrasts with the assembly program above, which shows the LSETUP instruction as the only instruction needed to implement a loop.

Multi-issue instructions allow computation and two data accesses with pointer updates in the same cycle. In each iteration, the values a[i] and b[i] must be read, then multiplied, and finally written back to the running summation in the variable output. On many microcontroller platforms, this effectively amounts to four instructions. The last line of the assembly code shows that all of these operations can be executed in one cycle.

Parallel ALU operations allow two 16-bit instructions to be executed simultaneously. The assembly code shows two accumulator units (A0 and A1) used in each iteration. This reduces the number of iterations by 50%, effectively halving the original execution time.

7.5.3   Specialized Addressing Modes

7.5.3.1   Byte Addressability

Allowing the processor to access multiple data words in a single cycle requires substantial flexibility in address generation. In addition to the more signal-processing-centric access sizes along 16- and 32-bit boundaries, byte addressing is required for the most efficient processing. This is important for multimedia processing because many video-based systems operate on 8-bit data. When memory accesses are restricted to a single boundary, the processor may spend extra cycles to mask off relevant bits.

7.5.3.2   Circular Buffering

Another beneficial addressing capability is *circular buffering*. For maximum efficiency, this feature must be supported directly by the processor, with no special management overhead. Circular buffering allows a programmer to define buffers in memory and stride through them automatically. Once the buffer is set up, no special software interaction is required to navigate through the data. The address generator handles nonunity strides and, more importantly, handles the "wraparound" feature illustrated in Figure 7.5. Without this automated address generation, the programmer would have to manually keep track of buffer pointer positions, thus wasting valuable processing cycles.

Many optimizing compilers will automatically use hardware circular buffering when they encounter array addressing with a modulus operator.

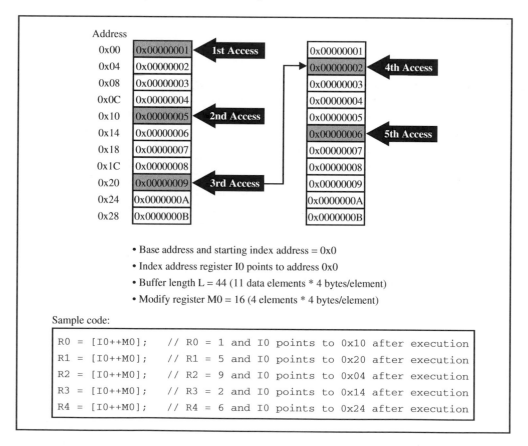

• Base address and starting index address = 0x0
• Index address register I0 points to address 0x0
• Buffer length L = 44 (11 data elements * 4 bytes/element)
• Modify register M0 = 16 (4 elements * 4 bytes/element)

Sample code:

```
R0 = [I0++M0];    // R0 = 1 and I0 points to 0x10 after execution
R1 = [I0++M0];    // R1 = 5 and I0 points to 0x20 after execution
R2 = [I0++M0];    // R2 = 9 and I0 points to 0x04 after execution
R3 = [I0++M0];    // R3 = 2 and I0 points to 0x14 after execution
R4 = [I0++M0];    // R4 = 6 and I0 points to 0x24 after execution
```

Figure 7.5: Circular Buffer in Hardware

Example 7.2: Single-Sample FIR

The finite impulse response filter is a very common filter structure equivalent to the convolution operator. A straightforward C implementation follows:

```
// sample the signal into a circular buffer
x[cur] = sampling_function();
cur = (cur+1)%TAPS; // advance cur pointer in circular fashion

// perform the multiply-addition
y = 0;
for (k=0; k<TAPS; k++) {
  y += h[k] * x[(cur+k)%TAPS];
}
```

The essential part of an FIR kernel written in assembly is shown below.

```
/* the samples are stored in the R0 register, while the
   coefficients are stored in the R1 register */
LSETUP (loop_begin, loop_end) LC0 = P0; /* loop counter set to
   traverse the filter */
loop_begin: A1+=R0.H*R1.L, A0+=R0.L*R1.L || R0.L = [I0++] ;
   /* perform MAC and fetch next data */
loop_end: A1+=R0.L*R1.H, A0+=R0.H*R1.H || R0.H = [I0++] || R1 =
   [I1++];/* perform MAC and fetch next data */
```

In the C code snippet, the % (modulus) operator provides a mechanism for circular buffering. As shown in the assembly kernel, this modulus operator does not get translated into an additional instruction inside the loop. Instead, the Data Address Generator registers I0 and I1 are configured outside the loop to automatically wraparound to the beginning upon hitting the buffer boundary.

7.5.3.3 Bit Reversal

An essential addressing mode for efficient signal-processing operations, such as the FFT and DCT, is bit reversal. Just as the name implies, bit reversal involves reversing the bits in a binary address. That is, the least significant bits are swapped in position with the most significant bits. The data ordering required by a radix-2 butterfly is in "bit-reversed" order, so bit-reversed indices are used to combine FFT stages. It is possible to calculate these bit-reversed indices in software, but this is very inefficient. An example of bit reversal address flow is shown in Figure 7.6.

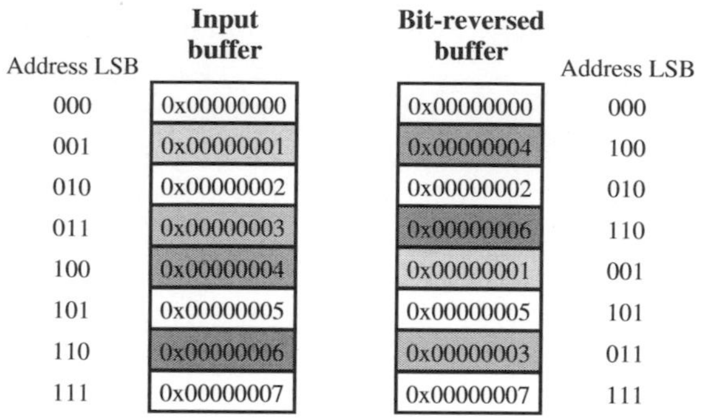

	Input buffer	Bit-reversed buffer	
Address LSB			Address LSB
000	0x00000000	0x00000000	000
001	0x00000001	0x00000004	100
010	0x00000002	0x00000002	010
011	0x00000003	0x00000006	110
100	0x00000004	0x00000001	001
101	0x00000005	0x00000005	101
110	0x00000006	0x00000003	011
111	0x00000007	0x00000007	111

Sample code:

```
LSETUP(start,end) LC0 = P0;        //Loop count = 8
start: R0 = [I0] || I0 += M0 (BREV);  // I0 points to input buffer, automatically incremented in
                                   //bit-reversed progression
end:[I2++] = R0;                   // I2 points to bit-reversed buffer
```

Figure 7.6: Bit Reversal in Hardware

Since bit reversal is very specific to algorithms like fast Fourier transforms and discrete Fourier transforms, it is difficult for any HLL compiler to employ hardware bit reversal. For this reason, comprehensive knowledge of the underlying architecture and assembly language are key to fully utilizing this addressing mode.

Example 7.3: FFT

A fast Fourier transform is an integral part of many signal-processing algorithms. One of its peculiarities is that if the input vector is in sequential time order, the output comes out in bit-reversed order. Most traditional general-purpose processors require the programmer to implement a separate routine to unscramble the bit-reversed output. On a media processor, bit reversal is often designed into the addressing engine.

Allowing the hardware to automatically bit-reverse the output of an FFT algorithm relieves the programmer from writing additional utilities, and thus improves performance.

7.5.4 *Interlocked Instruction Pipelines*

As processors increase in speed, it is necessary to add stages to the processing pipeline. For instances where a high-level language is the sole programming language, the compiler is responsible for dealing with instruction scheduling to maximize performance through the pipeline. That said, the following information is important to understand even if you're programming in C.

On older processor architectures, pipelines are usually not interlocked. On these architectures, executing certain combinations of neighboring instructions can yield incorrect results. Interlocked pipelines like the one in Figure 7.7, on the other hand, make assembly programming (as well as the life of compiler engineers) easier by automatically inserting stalls when necessary. This prevents the assembly programmer from scheduling instructions in a way that will produce inaccurate results. It should be noted that, even if the pipeline is interlocked, instruction rearrangement could still yield optimization improvements by eliminating unnecessary stalls.

Let's take a look at stalls in more detail. Stalls will show up for one of four reasons:

1. The instruction in question may itself take more than one cycle to execute. When this is the case, there isn't anything you can do to eliminate the stall. For example, a 32-bit integer multiply might take three core-clock cycles to execute on a 16-bit processor. This will cause a "bubble" in two pipeline stages for a three-cycle instruction.

IF1-3: Instruction Fetch
DC: Decode
AC: Address Calculation
EX1-4: Execution
WB: Writeback

	Pipeline Stage									
	IF1	**IF2**	**IF3**	**DC**	**AC**	**EX1**	**EX2**	**EX3**	**EX4**	**WB**
1	Inst1									
2	Inst2	Inst1								
3	Inst3	Inst2	Inst1							
4	Inst4	Inst3	Inst2	Inst1						
5	Inst5	Inst4	Inst3	Inst2	Inst1					
6	Branch	Inst5	Inst4	Inst3	Inst2	Inst1				
7	Stall	Branch	Inst5	Inst4	Inst3	Inst2	Inst1			
8	Stall	Stall	Branch	Inst5	Inst4	Inst3	Inst2	Inst1		
9	Stall	Stall	Stall	Branch	Inst5	Inst4	Inst3	Inst2	Inst1	
10	Stall	Stall	Stall	Stall	Branch	Inst5	Inst4	Inst3	Inst2	Inst1

Time →

Figure 7.7: Example of Interlocked Pipeline Architecture with Stalls Inserted

2. The second case involves the location of one instruction in the pipeline with respect to an instruction that follows it. For example, in some instructions, a stall may exist because the result of the first instruction is used as an operand of the following instruction. When this happens and you are programming in assembly, it is often possible to move the instruction so that the stall is not in the critical path of execution.

    Here are some simple examples on Blackfin processors that demonstrate these concepts.

Register Transfer/Multiply latencies (One stall, due to R0 being used in the multiply):

```
R0 = R4; /* load R0 with contents of R4 */
<STALL>
R2.H = R1.L * R0.H; /* R0 is used as an operand */
```

In this example, any instruction that does not change the value of the operands can be placed in-between the two instructions to hide the stall.

When we load a pointer register and try to use the content in the next instruction, there is a latency of three stalls:

```
P3 = [SP++]; /* Pointer register loaded from stack */
<STALL>
<STALL>
<STALL>
R0 = P3; /* Use contents of P3 after it gets its value
    from earlier
```

---

3. The third case involves a change of flow. While a deeper pipeline allows increased clock speeds, any time a change of flow occurs, a portion of the pipeline is flushed, and this consumes core-clock cycles. The branching latency associated with a change of flow varies based on the pipeline depth. Blackfin's 10-stage pipeline yields the following latencies:

    Instruction flow dependencies (Static Prediction):

    Correctly predicted branch (4 stalls)

    Incorrectly predicted branch (8 stalls)

Unconditional branch (8 stalls)

"Drop-through" conditional branch (0 stalls)

The term "predicted" is used to describe what the sequencer does as instructions that will complete ten core-clock cycles later enter the pipeline. You can see that when the sequencer does not take a branch, and in effect "drops through" to the next instruction after the conditional one, there are no added cycles. When an unconditional branch occurs, the maximum number of stalls occurs (eight cycles). When the processor predicts that a branch occurs and it actually is taken, the number of stalls is four. In the case where it predicted no branch, but one is actually taken, it mirrors the case of an unconditional branch.

One more note here. The maximum number of stalls is eight, while the depth of the pipeline is ten. This shows that the branching logic in an architecture does not implicitly have to match the full size of the pipeline.

4. The last case involves a conflict when the processor is accessing the same memory space as another resource (or simply fetching data from memory other than L1). For instance, a core fetch from SDRAM will take multiple core-clock cycles. As another example, if the processor and a DMA channel are trying to access the same memory bank, stalls will occur until the resource is available to the lower-priority process.

7.6   *Compiler Considerations for Efficient Programming*

Since the compiler's foremost task is to create correct code, there are cases where the optimizer is too conservative. In these cases, providing the compiler with extra information (through pragmas, built-in keywords, or command-line switches) will help it create more optimized code.

In general, compilers can't make assumptions about what an application is doing. This is why pragmas exist—to let the compiler know it is okay to make certain assumptions. For example, a pragma can instruct the compiler that variables in a loop are aligned and that they are not referenced to the same memory location. This extra information allows the compiler to optimize more aggressively, because the programmer has made a guarantee dictated by the pragma.

In general, a four-step process can be used to optimize an application consisting primarily of HLL code:

1. Compile with an HLL-optimizing compiler.

2. Profile the resulting code to determine the "hot spots" that consume the most processing bandwidth.

3. Update HLL code with pragmas, built-in keywords, and compiler switches to speed up the "hot spots."

4. Replace HLL procedures/functions with assembly routines in places where the optimizer did not meet the timing budget.

For maximum efficiency, it is always a good idea to inspect the most frequently executed compiler-generated assembly code to make a judgment on whether the code could be more vectorized. Sometimes, the HLL program can be changed to help the compiler produce faster code through more use of multi-issue instructions. If this still fails to produce code that is fast enough, then it is up to the assembly programmer to fine-tune the code line-by-line to keep all available hardware resources from idling.

7.6.1   Choosing Data Types

It is important to remember how the standard data types available in C actually map to the architecture you are using. For Blackfin processors, each type is shown in Table 7.2.

Table 7.2: C Data Types and Their Mapping to Blackfin Registers

C type	Blackfin equivalent
char	8-bit signed
unsigned char	8-bit unsigned
short	16-bit signed integer
unsigned short	16-bit unsigned integer
int	32-bit signed integer
unsigned int	32-bit unsigned integer
long	32-bit signed integer
unsigned long	32-bit unsigned integer

The `float`(32-bit), `double`(32-bit), `long long`(64-bit) and `unsigned long long` (64-bit) formats are not supported natively by the processor, but these can be emulated.

7.6.1.1   Arrays versus Pointers

We are often asked whether it is better to use arrays to represent data buffers in C, or whether pointers are better. Compiler performance engineers always point out that arrays are easier to analyze. Consider the example:

```
void array_example(int a[], ` b[], int sum[], int n)
{

     int i;
     for (i = 0; i < n; ++i)
     sum[i] = a[i] + b[i];
}
```

Even though we chose a simple example, the point is that these constructs are very easy to follow.

Now let's look at the same function using pointers. With pointers, the code is "closer" to the processor's native language.

```
void pointer_example(int a[], int b[], int sum[], int n) {
     int i;
     for (i = 0; i < n; ++i)
          *out++ = *a++ + *b++ ;
}
```

Which produces the most efficient code? Actually, there is usually very little difference. It is best to start by using the array notation because it is easier to read. An array format can be better for "alias" analysis in helping to ensure there is no overlap between elements in a buffer. If performance is not adequate with arrays (for instance, in the case of tight inner loops), pointers may be more useful.

7.6.1.2   Division

Fixed-point processors often do not support division natively. Instead, they offer division primitives in the instruction set, and these help accelerate division.

The "cost" of division depends on the range of the inputs. There are two possibilities: You can use division primitives where the result and divisor each fit into 16 bits. On Blackfin processors, this results in an operation of ~40 cycles. For more precise, bitwise 32-bit division, the result is ~10x more cycles.

If possible, it is best to avoid division, because of the additional overhead it entails. Consider the example:

if (X/Y > A/B)

This can easily be rewritten as:

if (X * B > A * Y)

to eliminate the division.

Keep in mind that the compiler does not know anything about the data precision in your application. For example, in the context of the above equation rewrite, two 12-bit inputs are "safe," because the result of the multiplication will be 24 bits maximum. This quick check will indicate when you can take a shortcut, and when you have to use actual division.

7.6.1.3 Loops

We already discussed hardware looping constructs. Here we'll talk about software looping in C. We will attempt to summarize what you can do to ensure best performance for your application.

1. Try to keep loops short. Large loop bodies are usually more complex and difficult to optimize. Additionally, they may require register data to be stored in memory, decreasing code density and execution performance.

2. Avoid loop-carried dependencies. These occur when computations in the present iteration depend on values from previous iterations. Dependencies prevent the compiler from taking advantage of loop overlapping (i.e., nested loops).

3. Avoid manually unrolling loops. This confuses the compiler and cheats it out of a job at which it typically excels.

4. Don't execute loads and stores from a noncurrent iteration while doing computations in the current loop iteration. This introduces loop-carried dependencies. This means avoiding loop array writes of the form:

```
for (i = 0; i < n; ++i)
    a[i] = b[i] * a[c[i]]; /* has array dependency*/
```

5. Make sure that inner loops iterate more than outer loops, since most optimizers focus on inner loop performance.

6. Avoid conditional code in loops. Large control-flow latencies may occur if the compiler needs to generate conditional jumps.

   As an example,

```
for {
      if { ... } else {...}
    }
```

   should be replaced, if possible, by:

```
if {
      for {...}
    } else {
      for   {...}
          }
```

7. Don't place function calls in loops. This prevents the compiler from using hardware loop constructs, as we described earlier in this chapter.

8. Try to avoid using variables to specify stride values. The compiler may need to use division to figure out the number of loop iterations required, and you now know why this is not desirable!

7.6.1.4   Data Buffers

It is important to think about how data is represented in your system. It's better to pre-arrange the data in anticipation of "wider" data fetches—that is, data fetches that optimize the amount of data accessed with each fetch. Let's look at an example that represents complex data.

One approach that may seem intuitive is:

```
short Real_Part[ N ];
short Imaginary_Part [ N ];
```

While this is perfectly adequate, data will be fetched in two separate 16-bit accesses. It is often better to arrange the array in one of the following ways:

```
short Complex [ N*2 ];
   or
long Complex [ N ];
```

Here, the data can be fetched via one 32-bit load and used whenever it's needed. This single fetch is faster than the previous approach.

On a related note, a common performance-degrading buffer layout involves constructing a 2D array with a column of pointers to `malloc`'d rows of data. While this allows complete flexibility in row and column size and storage, it may inhibit a compiler's ability to optimize, because the compiler no longer knows if one row follows another, and therefore it can see no constant offset between the rows.

7.6.1.5   *Intrinsics and In-lining*

It is difficult for compilers to solve all of your problems automatically and consistently. This is why you should, if possible, avail yourself of "in-line" assembly instructions and intrinsics.

In-lining allows you to insert an assembly instruction into your C code directly. Sometimes this is unavoidable, so you should probably learn how to in-line for the compiler you're using.

In addition to in-lining, most compilers support intrinsics, and their optimizers fully understand intrinsics and their effects. The Blackfin compiler supports a comprehensive array of 16-bit intrinsic functions, which must be programmed explicitly. Below is a simple example of an intrinsic that multiplies two 16-bit values.

```
#include <fract.h>
fract32 fdot(fract16 *x, fract16 *y, int n)
{
    fract32 sum = 0;
    int i;
    for (i = 0; i < n; i++)
        sum = add_fr1x32(sum, mult_fr1x32(x[i], y[i]));
    return sum;
}
```

Here are some other operations that can be accomplished through intrinsics:

- Align operations

- Packing operations

- Disaligned loads

- Unpacking

- Quad 8-bit add/subtract

- Dual 16-bit add/clip

- Quad 8-bit average

- Accumulator extract with addition

- Subtract/absolute value/accumulate

The intrinsics that perform the above functions allow the compiler to take advantage of video-specific instructions that improve performance but that are difficult for a compiler to use natively.

When should you use in-lining, and when should you use intrinsics? Well, you really don't have to choose between the two. Rather, it is important to understand the results of using both, so that they become tools in your programming arsenal. With regard to in-lining of assembly instructions, look for an option where you can include in the in-lining construct the registers you will be "touching" in the assembly instruction. Without this information, the compiler will invariably spend more cycles, because it's limited in the assumptions it can make and therefore has to take steps that can result in lower performance. With intrinsics, the compiler can use its knowledge to improve the code it generates on both sides of the intrinsic code. In addition, the fact that the intrinsic exists means someone who knows the compiler and architecture very well has already translated a common function to an optimized code section.

7.6.1.6 Volatile Data

The `volatile` data type is essential for peripheral-related registers and interrupt-related data.

Some variables may be accessed by resources not visible to the compiler. For example, they may be accessed by interrupt routines, or they may be set or read by peripherals.

The `volatile` attribute forces all operations with that variable to occur exactly as written in the code. This means that a variable is read from memory each time it is needed, and it's written back to memory each time it's modified. The exact order of events is preserved. Missing a `volatile` qualifier is the largest single cause of trouble when engineers port from one C-based processor to another. Architectures that don't require `volatile` for hardware-related accesses probably treat all accesses as volatile by default and thus may perform at a lower performance level than those that require you to state this explicitly. When a C program works with optimization turned off but doesn't work with optimization on, a missing `volatile` qualifier is usually the culprit.

7.7   System and Core Synchronization

Earlier we discussed the importance of an interlocked pipeline, but we also need to discuss the implications of the pipeline on the different operating domains of a processor. On Blackfin devices, there are two synchronization instructions that help manage the relationship between when the core and the peripherals complete specific instructions or sequences. While these instructions are very straightforward, they are sometimes used more than necessary. The `CSYNC` instruction prevents any other instructions from entering the pipeline until all pending core activities have completed. The `SSYNC` behaves in a similar manner, except that it holds off new instructions until all pending system actions have completed. The performance impact from a `CSYNC` is measured in multiple `CCLK` cycles, while the impact of an `SSYNC` is measured in multiple `SCLK`s. When either of these instructions is used too often, performance will suffer needlessly.

So when do you need these instructions? We'll find out in a minute. But first we need to talk about memory transaction ordering.

7.7.1   Load/Store Synchronization

Many embedded processors support the concept of a Load/Store data access mechanism. What does this mean, and how does it impact your application? "Load/Store" refers to the characteristic in an architecture where memory operations (loads and stores) are intentionally separated from the arithmetic functions that use the results of fetches from memory operations. The separation is made because memory operations, especially instructions that access off-chip memory or I/O devices, take multiple cycles to complete and would normally halt the processor, preventing an instruction execution rate of one instruction per core-clock cycle. To avoid this situation, data is brought into a data register from a source memory location, and once it is in the register, it can be fed into a computation unit.

In write operations, the "store" instruction is considered complete as soon as it executes, even though many clock cycles may occur before the data is actually written to an external memory or I/O location. This arrangement allows the processor to execute one instruction per clock cycle, and it implies that the synchronization between when writes complete and when subsequent instructions execute is not guaranteed. This synchronization is considered unimportant in the context of most memory operations. With the presence of a write buffer that sits between the processor and external memory, multiple writes can, in fact, be made without stalling the processor.

For example, consider the case where we write a simple code sequence consisting of a single write to L3 memory surrounded by five NOP ("no operation") instructions. Measuring the cycle count of this sequence running from L1 memory shows that it takes six cycles to execute. Now let's add another write to L3 memory and measure the cycle count again. We will see the cycle count increase by one cycle each time, until we reach the limits of the write buffer, at which point it will increase substantially until the write buffer is drained.

7.7.2 Ordering

The relaxation of synchronization between memory accesses and their surrounding instructions is referred to as "weak ordering" of loads and stores. Weak ordering implies that the timing of the actual completion of the memory operations—even the order in which these events occur—may not align with how they appear in the sequence of a program's source code.

In a system with weak ordering, only the following items are guaranteed:

- Load operations will complete before a subsequent instruction uses the returned data.

- Load operations using previously written data will use the updated values, even if they haven't yet propagated out to memory.

- Store operations will eventually propagate to their ultimate destination.

Because of weak ordering, the memory system is allowed to prioritize reads over writes. In this case, a write that is queued anywhere in the pipeline, but not completed, may be deferred by a subsequent read operation, and the read is allowed to be completed before the write. Reads are prioritized over writes because the read operation has a dependent operation waiting on its completion, whereas the processor considers the write operation complete, and the write does not stall the pipeline if it takes more cycles to propagate the value out to memory.

For most applications, this behavior will greatly improve performance. Consider the case where we are writing to some variable in external memory. If the processor performs a write to one location followed by a read from a different location, we would prefer to have the read complete before the write.

This ordering provides significant performance advantages in the operation of most memory instructions. However, it can cause side effects—when writing to or reading from nonmemory locations such as I/O device registers, the order of how read and write operations complete is often significant. For example, a read of a status register may depend on a write to a control register. If the address in either case is the same, the read would return a value from the write buffer rather than from the actual I/O device register, and the order of the read and write at the register may be reversed. Both of these outcomes could cause undesirable side effects. To prevent these occurrences in code that requires precise (strong) ordering of load and store operations, synchronization instructions like CSYNC or SSYNC should be used.

The CSYNC instruction ensures all pending core operations have completed and the core buffer (between the processor core and the L1 memories) has been flushed before proceeding to the next instruction. Pending core operations may include any pending interrupts, speculative states (such as branch predictions) and exceptions. A CSYNC is typically required after writing to a control register that is in the core domain. It ensures that whatever action you wanted to happen by writing to the register takes place before you execute the next instruction.

The SSYNC instruction does everything the CSYNC does, and more. As with CSYNC, it ensures all pending operations have to be completed between the processor core and the L1 memories. SSYNC further ensures completion of all operations between the processor core, external memory, and the system peripherals. There are many cases where this is important, but the best example is when an interrupt condition needs to be cleared at a peripheral before an interrupt service routine (ISR) is exited. Somewhere in the ISR, a write is made to a peripheral register to "clear" and, in effect, acknowledge the interrupt. Because of differing clock domains between the core and system portions of the processor, the SSYNC ensures the peripheral clears the interrupt before exiting the ISR. If the ISR were exited before the interrupt was cleared, the processor might jump right back into the ISR.

Load operations from memory do not change the state of the memory value itself. Consequently, issuing a speculative memory-read operation for a subsequent load instruction usually has no undesirable side effect. In some code sequences, such as a conditional branch instruction followed by a load, performance may be improved by speculatively issuing the read request to the memory system before the conditional branch is resolved. For example,

```
IF CC JUMP away_from_here
RO = [P2];
...
away_from_here:
```

If the branch is taken, then the load is flushed from the pipeline, and any results that are in the process of being returned can be ignored. Conversely, if the branch is not taken, the memory will have returned the correct value earlier than if the operation were stalled until the branch condition was resolved.

However, this could cause an undesirable side effect for a peripheral that returns sequential data from a FIFO or from a register that changes value based on the number of reads that are requested. To avoid this effect, use an SSYNC instruction to guarantee the correct behavior between read operations.

Store operations never access memory speculatively, because this could cause modification of a memory value before it is determined whether the instruction should have executed.

7.7.3   Atomic Operations

We have already introduced several ways to use semaphores in a system. While there are many ways to implement a semaphore, using atomic operations is preferable, because they provide noninterruptible memory operations in support of semaphores between tasks.

The Blackfin processor provides a single atomic operation: TESTSET. The TESTSET instruction loads an indirectly addressed memory word, tests whether the low byte is zero, and then sets the most significant bit of the low memory byte without affecting any other bits. If the byte is originally zero, the instruction sets a status bit. If the byte is originally nonzero, the instruction clears the status bit. The sequence of this memory transaction is atomic—hardware bus locking insures that no other memory operation can occur between the test and set portions of this instruction. The TESTSET instruction can be interrupted by the core. If this happens, the TESTSET instruction is executed again upon return from the interrupt. Without something like this TESTSET facility, it is difficult to ensure true protection when more than one entity (for example, two cores in a dual-core device) vies for a shared resource.

7.8 Memory Architecture—the Need for Management

7.8.1 Memory Access Trade-offs

Embedded media processors usually have a small amount of fast, on-chip memory, whereas microcontrollers usually have access to large external memories. A hierarchical memory architecture combines the best of both approaches, providing several tiers of memory with different performance levels. For applications that require the most determinism, on-chip SRAM can be accessed in a single core-clock cycle. Systems with larger code sizes can utilize bigger, higher-latency on-chip and off-chip memories.

Most complex programs today are large enough to require external memory, and this would dictate an unacceptably slow execution speed. As a result, programmers would be forced to manually move key code in and out of internal SRAM. However, by adding data and instruction caches into the architecture, external memory becomes much more manageable. The cache reduces the manual movement of instructions and data into the processor core, thus greatly simplifying the programming model.

Figure 7.8 demonstrates a typical memory configuration where instructions are brought in from external memory as they are needed. Instruction cache usually operates with some type of least recently used (LRU) algorithm, insuring that instructions that run more often get replaced less often. The figure also illustrates that having the ability to configure some on-chip data memory as cache and some as SRAM can optimize performance. DMA controllers can feed the core directly, while data from tables can be brought into the data cache as they are needed.

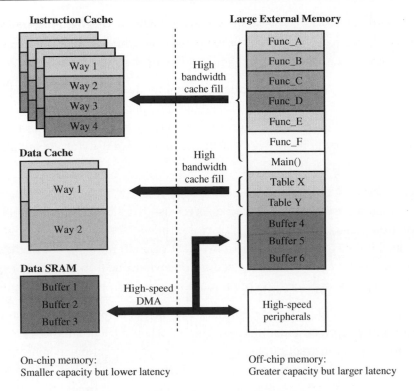

Figure 7.8: Typical Memory Configuration

When porting existing applications to a new processor, "out-of-the-box" performance is important. As we saw earlier, there are many features compilers exploit that require minimal developer involvement. Yet, there are many other techniques that, with a little extra effort by the programmer, can have a big impact on system performance.

Proper memory configuration and data placement always pays big dividends in improving system performance. On high-performance media processors, there are typically three paths into a memory bank. This allows the core to make multiple accesses in a single clock cycle (e.g., a load and store, or two loads). By laying out an intelligent data flow, a developer can avoid conflicts created when the core processor and DMA vie for access to the same memory bank.

7.8.2   Instruction Memory Management—to Cache or to DMA?

Maximum performance is only realized when code runs from internal L1 memory. Of course, the ideal embedded processor would have an unlimited amount of L1 memory, but this is not practical. Therefore, programmers must consider several alternatives to take advantage of the L1 memory that exists in the processor, while optimizing memory and data flows for their particular system. Let's examine some of these scenarios.

The first, and most straightforward, situation is when the target application code fits entirely into L1 instruction memory. For this case, there are no special actions required, other than for the programmer to map the application code directly to this memory space. It thus becomes intuitive that media processors must excel in code density at the architectural level.

In the second scenario, a caching mechanism is used to allow programmers access to larger, less expensive external memories. The cache serves as a way to automatically bring code into L1 instruction memory as needed. The key advantage of this process is that the programmer does not have to manage the movement of code into and out of the cache. This method is best when the code being executed is somewhat linear in nature. For nonlinear code, cache lines may be replaced too often to allow any real performance improvement.

The instruction cache really performs two roles. For one, it helps pre-fetch instructions from external memory in a more efficient manner. That is, when a cache miss occurs, a cache-line fill will fetch the desired instruction, along with the other instructions contained within the cache line. This ensures that, by the time the first instruction in the line has been executed, the instructions that immediately follow have also been fetched. In addition, since caches usually operate with an LRU algorithm, instructions that run most often tend to be retained in cache.

Some strict real-time programmers tend not to trust cache to obtain the best system performance. Their argument is that if a set of instructions is not in cache when needed for execution, performance will degrade. Taking advantage of cache-locking mechanisms can offset this issue. Once the critical instructions are loaded into cache, the cache lines can be locked, and thus not replaced. This gives programmers the ability to keep what they need in cache and to let the caching mechanism manage less-critical instructions.

In a final scenario, code can be moved into and out of L1 memory using a DMA channel that is independent of the processor core. While the core is operating on one section of memory, the DMA is bringing in the section to be executed next. This scheme is commonly referred to as an overlay technique.

While overlaying code into L1 instruction memory via DMA provides more determinism than caching it, the trade-off comes in the form of increased programmer involvement. In other words, the programmer needs to map out an overlay strategy and configure the DMA channels appropriately. Still, the performance payoff for a well-planned approach can be well worth the extra effort.

7.8.3 Data Memory Management

The data memory architecture of an embedded media processor is just as important to the overall system performance as the instruction clock speed. Because multiple data transfers take place simultaneously in a multimedia application, the bus structure must support both core and DMA accesses to all areas of internal and external memory. It is critical that arbitration between the DMA controller and the processor core be handled automatically, or performance will be greatly reduced. Core-to-DMA interaction should only be required to set up the DMA controller, and then again to respond to interrupts when data is ready to be processed.

A processor performs data fetches as part of its basic functionality. While this is typically the least efficient mechanism for transferring data to or from off-chip memory, it provides the simplest programming model. A small, fast scratch pad memory is sometimes available as part of L1 data memory, but for larger, off-chip buffers, access time will suffer if the core must fetch everything from external memory. Not only will it take multiple cycles to fetch the data, but the core will also be busy doing the fetches.

It is important to consider how the core processor handles reads and writes. As we detailed above, Blackfin processors possess a multislot write buffer that can allow the core to proceed with subsequent instructions before all posted writes have completed. For example, in the following code sample, if the pointer register P0 points to an address in external memory and P1 points to an address in internal memory, line 50 will be executed before R0 (from line 46) is written to external memory:

```
...
Line 45: R0 =R1+R2;
Line 46: [P0] = R0; /* Write the value contained in R0 to slower
   external memory */
Line 47: R3 = 0x0 (z);
Line 48: R4 = 0x0 (z);
Line 49: R5 = 0x0 (z);
Line 50: [P1] = R0; /* Write the value contained in R0 to faster
   internal memory */
```

In applications where large data stores constantly move into and out of external DRAM, relying on core accesses creates a difficult situation. While core fetches are inevitably needed at times, DMA should be used for large data transfers, in order to preserve performance.

7.8.3.1   What about Data Cache?

The flexibility of the DMA controller is a double-edged sword. When a large C/C++ application is ported between processors, a programmer is sometimes hesitant to integrate DMA functionality into already-working code. This is where data cache can be very useful, bringing data into L1 memory for the fastest processing. The data cache is attractive because it acts like a mini-DMA, but with minimal interaction on the programmer's part.

Because of the nature of cache-line fills, data cache is most useful when the processor operates on consecutive data locations in external memory. This is because the cache doesn't just store the immediate data currently being processed; instead, it prefetches data in a region contiguous to the current data. In other words, the cache mechanism assumes there's a good chance that the current data word is part of a block of neighboring data about to be processed. For multimedia streams, this is a reasonable conjecture.

Since data buffers usually originate from external peripherals, operating with data cache is not always as easy as with instruction cache. This is due to the fact that coherency must be managed manually in "nonsnooping" caches. Nonsnooping means that the cache is not aware of when data changes in source memory unless it makes the change directly. For these caches, the data buffer must be invalidated before making any attempt to access the new data. In the context of a C-based application, this type of data is "volatile." This situation is shown in Figure 7.9.

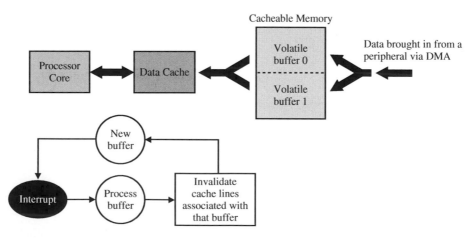

Figure 7.9: Data Cache and DMA Coherency

In the general case, when the value of a variable stored in cache is different from its value in the source memory, this can mean that the cache line is "dirty" and still needs to be written back to memory. This concept does not apply for volatile data. Rather, in this case the cache line may be "clean," but the source memory may have changed without the knowledge of the core processor. In this scenario, before the core can safely access a volatile variable in data cache, it must invalidate (but not flush!) the affected cache line.

This can be performed in one of two ways. The cache tag associated with the cache line can be directly written, *or* a "Cache Invalidate" instruction can be executed to invalidate the target memory address. Both techniques can be used interchangeably, but the direct method is usually a better option when a large data buffer is present (e.g., one greater in size than the data cache size). The Invalidate instruction is always preferable when the buffer size is smaller than the size of the cache. This is true even when a loop is required, since the Invalidate instruction usually increments by the size of each cache line instead of by the more typical 1-, 2- or 4-byte increment of normal addressing modes.

From a performance perspective, this use of data cache cuts down on improvement gains, in that data has to be brought into cache each time a new buffer arrives. In this case, the benefit of caching is derived solely from the pre-fetch nature of a cache-line fill. Recall that the prime benefit of cache is that the data is present the second time through the loop.

One more important point about volatile variables, regardless of whether or not they are cached, if they are shared by both the core processor and the DMA controller, the programmer must implement some type of semaphore for safe operation. In sum, it is best to keep volatiles out of data cache altogether.

7.8.4   System Guidelines for Choosing between DMA and Cache

Let's consider three widely used system configurations to shed some light on which approach works best for different system classifications.

7.8.4.1   Instruction Cache, Data DMA

This is perhaps the most popular system model, because media processors are often architected with this usage profile in mind. Caching the code alleviates complex instruction flow management, assuming the application can afford this luxury. This works well when the system has no hard real-time constraints, so that a cache miss would not wreak havoc on the timing of tightly coupled events (for example, video refresh or audio/video synchronization).

Also, in cases where processor performance far outstrips processing demand, caching instructions is often a safe path to follow, since cache misses are then less likely to cause bottlenecks. Although it might seem unusual to consider that an "oversized" processor would ever be used in practice, consider the case of a portable media player that can decode and play both compressed video and audio. In its audio-only mode, its performance requirements will be only a fraction of its needs during video playback. Therefore, the instruction/data management mechanism could be different in each mode.

Managing data through DMA is the natural choice for most multimedia applications, because these usually involve manipulating large buffers of compressed and uncompressed video, graphics, and audio. Except in cases where the data is quasi-static (for instance, a graphics icon constantly displayed on a screen), caching these buffers makes little sense, since the data changes rapidly and constantly. Furthermore, as discussed above, there are usually multiple data buffers moving around the chip at one time—unprocessed blocks headed for conditioning, partly conditioned sections headed for temporary storage, and completely processed segments destined for external display or storage. DMA is the logical management tool for these buffers, since it allows the core to operate on them without having to worry about how to move them around.

7.8.4.2 Instruction Cache, Data DMA/Cache

This approach is similar to the one we just described, except in this case part of L1 data memory is partitioned as cache, and the rest is left as SRAM for DMA access. This structure is very useful for handling algorithms that involve a lot of static coefficients or lookup tables. For example, storing a sine/cosine table in data cache facilitates quick computation of FFTs. Or, quantization tables could be cached to expedite JPEG encoding or decoding.

Keep in mind that this approach involves an inherent trade-off. While the application gains single-cycle access to commonly used constants and tables, it relinquishes the equivalent amount of L1 data SRAM, thus limiting the buffer size available for single-cycle access to data. A useful way to evaluate this trade-off is to try alternate scenarios (Data DMA/Cache versus only DMA) in a Statistical Profiler (offered in many development tools suites) to determine the percentage of time spent in code blocks under each circumstance.

7.8.4.3   Instruction DMA, Data DMA

In this scenario, data and code dependencies are so tightly intertwined that the developer must manually schedule when instruction and data segments move through the chip. In such hard real-time systems, determinism is mandatory, and thus cache isn't ideal.

Although this approach requires more planning, the reward is a deterministic system where code is always present before the data needed to execute it, and no data blocks are lost via buffer overruns. Because DMA processes can link together without core involvement, the start of a new process guarantees that the last one has finished, so that the data or code movement is verified to have happened. This is the most efficient way to synchronize data and instruction blocks.

The Instruction/Data DMA combination is also noteworthy for another reason. It provides a convenient way to test code and data flows in a system during emulation and debug. The programmer can then make adjustments or highlight "trouble spots" in the system configuration.

An example of a system that might require DMA for both instructions and data is a video encoder/decoder. Certainly, video and its associated audio need to be deterministic for a satisfactory user experience. If the DMA signaled an interrupt to the core after each complete buffer transfer, this could introduce significant latency into the system, since the interrupt would need to compete in priority with other events. What's more, the context switch at the beginning and end of an interrupt service routine would consume several core processor cycles. All of these factors interfere with the primary objective of keeping the system deterministic.

Figures 7.10 and 7.11 provide guidance in choosing between cache and DMA for instructions and data, as well as how to navigate the trade-off between using cache and using SRAM, based on the guidelines we discussed previously.

As a real-world illustration of these flowchart choices, Tables 7.3 and 7.4 provide actual benchmarks for G.729 and GSM AMR algorithms running on a Blackfin processor under various cache and DMA scenarios. You can see that the best performance can be obtained when a balance is achieved between cache and SRAM.

In short, there is no single answer as to whether cache or DMA should be the mechanism of choice for code and data movement in a given multimedia system. However, once developers are aware of the trade-offs involved, they should settle into the "middle ground," the perfect optimization point for their system.

Instruction Cache versus Code Overlay decision flow

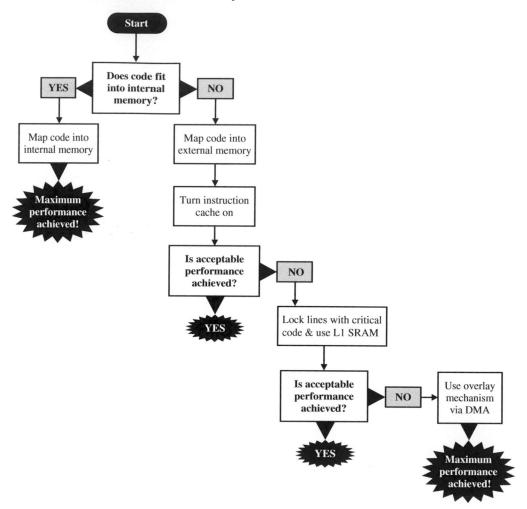

Figure 7.10: Checklist for Choosing between Instruction Cache and DMA

Data Cache versus DMA decision flow

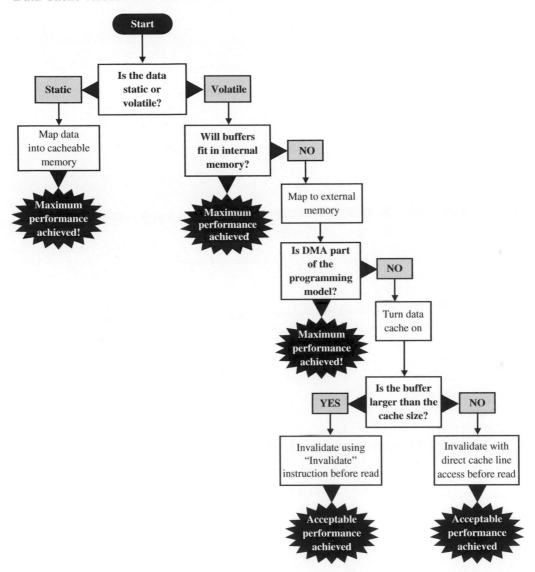

Figure 7.11: Checklist for Choosing between Data Cache and DMA

Table 7.3: Benchmarks (Relative Cycles per Frame) for G.729a Algorithm with Cache Enabled

	L1 banks configured as SRAM		L1 banks configured as cache			Cache + SRAM
	All L2	L1	Code only	Code + DataA	Code + DataB	DataA cache, DataB SRAM
Coder	1.00	0.24	0.70	0.21	0.21	0.21
Decoder	1.00	0.19	0.80	0.20	0.19	0.19

Table 7.4: Benchmarks (Relative Cycles per Frame) for GSM aMr Algorithm with Cache Enabled

	L1 banks configured as SRAM		L1 banks configured as cache			Cache + SRAM
	AllL2	L1	Code	Code + DataA	Code + DataB	DataA cache, DataB SRAM
Coder	1.00	0.34	0.74	0.20	0.20	0.20
Decoder	1.00	0.42	0.75	0.23	0.23	0.23

7.8.5 *Memory Management Unit (MMU)*

An MMU in a processor controls the way memory is set up and accessed in a system. The most basic capabilities of an MMU provides for memory protection, and when cache is used, it also determines whether or not a memory page is cacheable. Explicitly using the MMU is usually optional, because you can default to the standard memory properties on your processor.

On Blackfin processors, the MMU contains a set of registers that can define the properties of a given memory space. Using something called cacheability protection look-aside buffers (CPLBs), you can define parameters such as whether or not a memory page is cacheable, and whether or not a memory space can be accessed. Because the 32-bit-addressable external memory space is so large, it is likely that CPLBs will have to be swapped in and out of the MMU registers.

7.8.5.1   CPLB Management

Because the amount of memory in an application can greatly exceed the number of available CPLBs, it may be necessary to use a CPLB manager. If so, it's important to tackle some issues that could otherwise lead to performance degradation. First, whenever CPLBs are enabled, any access to a location without a valid CPLB will result in an exception being executed prior to the instruction completing. In the exception handler, the code must free up a CPLB and reallocate it to the location about to be accessed. When the processor returns from the exception handler, the instruction that generated the exception then executes.

If you take this exception too often, it will impact performance, because every time you take an exception, you have to save off the resources used in your exception handler. The processor then has to execute code to reprogram the CPLB. One way to alleviate this problem is to profile the code and data access patterns. Since the CPLBs can be "locked," you can protect the most frequently used CPLBs from repeated page swaps.

Another performance consideration involves the search method for finding new page information. For example, a "nonexistent CPLB" exception handler only knows the address where an access was attempted. This information must be used to find the corresponding address "range" that needs to be swapped into a valid page. By locking the most frequently used pages and setting up a sensible search based on your memory access usage (for instructions and/or data), exception-handling cycles can be amortized across thousands of accesses.

7.8.5.2   Memory Translation

A given MMU may also provide memory translation capabilities, enabling what's known as *virtual memory*. This feature is controlled in a manner that is analogous to memory protection. Instead of CPLBs, *translation look-aside buffers* (TLBs) are used to describe physical memory space. There are two main ways in which memory translation is used in an application. As a holdover from older systems that had limited memory resources, operating systems would have to swap code in and out of a memory space from which execution could take place.

A more common use on today's embedded systems still relates to operating system support. In this case, all software applications run thinking they are at the same physical memory space, when, of course, they are not. On processors that support memory translation, operating systems can use this feature to have the MMU translate the actual physical memory address to the same virtual address based on which specific task is running. This translation is done transparently, without the software application getting involved.

7.9 Physics of Data Movement

So far, we've seen that the compiler and assembler provide a bunch of ways to maximize performance on code segments in your system. Using of cache and DMA provide the next level for potential optimization. We will now review the third tier of optimization in your system—it's a matter of physics.

Understanding the "physics" of data movement in a system is a required step at the start of any project. Determining if the desired throughput is even possible for an application can yield big performance savings without much initial investment.

For multimedia applications, on-chip memory is almost always insufficient for storing entire video frames. Therefore, the system must usually rely on L3 DRAM to support relatively fast access to large buffers. The processor interface to off-chip memory constitutes a major factor in designing efficient media frameworks, because access patterns to external memory must be well planned in order to guarantee optimal data throughput. There are several high-level steps that can ensure that data flows smoothly through memory in any system. Some of these are discussed below and play a key role in the design of system frameworks.

7.9.1 Grouping Like Transfers to Minimize Memory Bus Turnarounds

Accesses to external memory are most efficient when they are made in the same direction (e.g., consecutive reads or consecutive writes). For example, when accessing off-chip synchronous memory, 16 reads followed by 16 writes is always completed sooner than 16 individual read/write sequences. This is because a write followed by a read incurs latency. Random accesses to external memory generate a high probability of bus turnarounds. This added latency can easily halve available bandwidth. Therefore, it is important to take advantage of the ability to control the number of transfers in a given direction. This can be done either automatically (as we'll see here) or by manually scheduling your data movements, which we'll review.

A DMA channel garners access according to its priority, signified on Blackfin processors by its channel number. Higher priority channels are granted access to the DMA bus(es) first. Because of this, you should always assign higher priority DMA channels to peripherals with the highest data rates or with requirements for lowest latency.

To this end, MemDMA streams are always lower in priority than peripheral DMA activity. This is due to the fact that, with Memory DMA no external devices will be held off or starved of data. Since a Memory DMA channel requests access to the DMA bus as long as the channel is active, efficient use of any time slots unused by a peripheral DMA are applied to MemDMA transfers. By default, when more than one MemDMA stream is enabled and ready, only the highest priority MemDMA stream is granted.

When it is desirable for the MemDMA streams to share the available DMA bus bandwidth, however, the DMA controller can be programmed to select each stream in turn for a fixed number of transfers.

This "Direction Control" facility is an important consideration in optimizing use of DMA resources on each DMA bus. By grouping same-direction transfers together, it provides a way to manage how frequently the transfer direction changes on the DMA buses. This is a handy way to perform a first level of optimization without real-time processor intervention. More importantly, there's no need to manually schedule bursts into the DMA streams.

When direction control features are used, the DMA controller preferentially grants data transfers on the DMA or memory buses that are going in the same read/write direction as in the previous transfer, until either the direction control counter times out, or until traffic stops or changes direction on its own. When the direction counter reaches zero, the DMA controller changes its preference to the opposite flow direction.

In this case, reversing direction wastes no bus cycles other than any physical bus turnaround delay time. This type of traffic control represents a trade-off of increased latency for improved utilization (efficiency). Higher block transfer values might increase the length of time each request waits for its grant, but they can dramatically improve the maximum attainable bandwidth in congested systems, often to above 90%.

Here's an example that puts these concepts into some perspective:

Example 7.4:

First, we set up a memory DMA from L1 to L3 memory, using 16-bit transfers that takes about 1100 system clock (SCLK) cycles to move 1024 16-bit words. We then begin a transfer from a different bank of external memory to the video port (PPI). Using 16-bit unpacking in the PPI, we continuously feed an NTSC video encoder with 8-bit data. Since the PPI sends out an 8-bit quantity at a 27 MHz rate, the DMA bus bandwidth required for the PPI transfer is roughly 13.5M transfers/second.

When we measure the time it takes to complete the same 1024-word MemDMA transfer with the PPI transferring simultaneously, it now takes three times as long.

Why is this? It's because the PPI DMA activity takes priority over the MemDMA channel transactions. Every time the PPI is ready for its next sample, the bus effectively reverses direction. This translates into cycles that are lost both at the external memory interface and on the various internal DMA buses.

When we enable Direction Control, the performance increases because there are fewer bus turnarounds.

As a rule of thumb, it is best to maximize same direction contiguous transfers during moderate system activity. For the most taxing system flows, however, it is best to select a block transfer value in the middle of the range to ensure no one peripheral gets locked out of accesses to external memory. This is especially crucial when at least two high-bandwidth peripherals (like PPIs) are used in the system.

In addition to using direction control, transfers among MDMA streams can be alternated in a "round-robin" fashion on the bus as the application requires. With this type of arbitration, the first DMA process is granted access to the DMA bus for some number of cycles, followed by the second DMA process, and then back to the first. The channels alternate in this pattern until all of the data is transferred. This capability is most useful on dual-core processors (for example, when both core processors have tasks that are awaiting a data stream transfer). Without this "round-robin" feature, the first set of DMA transfers will occur, and the second DMA process will be held off until the first one completes. Round-robin prioritization can help insure that both transfer streams will complete back-to-back.

Another thing to note: using DMA and/or cache will always help performance because these types of transactions transfer large data blocks in the same direction. For example, a DMA transfer typically moves a large data buffer from one location to another. Similarly, a cache-line fill moves a set of consecutive memory locations into the device, by utilizing block transfers in the same direction.

Buffering data bound for L3 in on-chip memory serves many important roles. For one, the processor core can access on-chip buffers for preprocessing functions with much lower latency than it can by going off-chip for the same accesses. This leads to a direct increase in system performance. Moreover, buffering this data in on-chip memory allows more efficient peripheral DMA access to this data. For instance, transferring a video frame on-the-fly through a video port and into L3 memory creates a situation where other peripherals might be locked out from accessing the data they need, because the video transfer is a high-priority process. However, by transferring lines incrementally from the video port into L1 or L2 memory, a Memory DMA stream can be initiated that will quietly transfer this data into L3 as a low-priority process, allowing system peripherals access to the needed data.

This concept will be further demonstrated in the "Performance-based Framework" later in this chapter.

7.9.2   Understanding Core and DMA SDRAM Accesses

Consider that on a Blackfin processor, core reads from L1 memory take one *core*-clock cycle, whereas core reads from SDRAM consume eight *system* clock cycles. Based on typical CCLK/SCLK ratios, this could mean that eight SCLK cycles equate to 40 CCLKs. Incidentally, these eight SCLKs reduce to only one SCLK by using a DMA controller in a burst mode instead of direct core accesses.

There is another point to make on this topic. For processors that have multiple data fetch units, it is better to use a dual-fetch instruction instead of back-to-back fetches. On Blackfin processors with a 32-bit external bus, a dual-fetch instruction with two 32-bit fetches takes nine SCLKs (eight for the first fetch and one for the second). Back-to-back fetches in separate instructions take 16 SCLKs (eight for each). The difference is that, in the first case, the request for the second fetch in the single instruction is pipelined, so it has a head start.

Similarly, when the external bus is 16 bits in width, it is better to use a 32-bit access rather than two 16-bit fetches. For example, when the data is in consecutive locations, the 32-bit fetch takes nine SCLKs (eight for the first 16 bits and one for the second). Two 16-bit fetches take 16 SCLKs (eight for each).

7.9.3 Keeping SDRAM Rows Open and Performing Multiple Passes on Data

Each access to SDRAM can take several SCLK cycles, especially if the required SDRAM row has not yet been activated. Once a row is active, it is possible to read data from an entire row without reopening that row on every access. In other words, it is possible to access any location in memory on every SCLK cycle, as long as those locations are within the same row in SDRAM. Multiple SDRAM clock cycles are needed to close a row, and therefore constant row closures can severely restrict SDRAM throughput. Just to put this into perspective, an SDRAM page miss can take 20–50 CCLK cycles, depending on the SDRAM type.

Applications should take advantage of open SDRAM banks by placing data buffers appropriately and managing accesses whenever possible. Blackfin processors, as an example, keep track of up to four open SDRAM rows at a time, so as to reduce the setup time—and thus increase throughput—for subsequent accesses to the same row within an open bank. For example, in a system with one row open, row activation latency would greatly reduce the overall performance. With four rows open at one time, on the other hand, row activation latency can be amortized over hundreds of accesses.

Let's look at an example that illustrates the impact this SDRAM row management can have on memory access bandwidth. Figure 7.12 shows two different scenarios of data and code mapped to a single *external* SDRAM bank. In the first case, all of the code and data buffers in external memory fit in a single bank, but because the access patterns of each code and data line are random, almost every access involves the activation of a new row. In the second case, even though the access patterns are randomly interspersed between code and data accesses, each set of accesses has a high probability of being within the same row. For example, even when an instruction fetch occurs immediately before and after a data access, two rows are kept open and no additional row activation cycles are incurred.

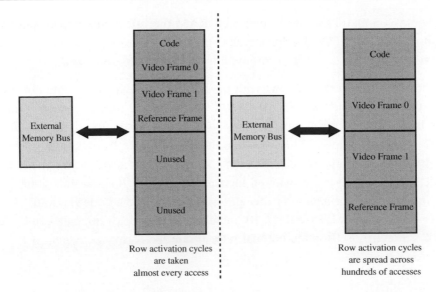

Figure 7.12: Taking Advantage of Code and Data Partitioning in External Memory

When we ran an MPEG-4 encoder from external memory (with both code and data in SDRAM), we gained a 6.5% performance improvement by properly spreading out the code and data in external memory.

7.9.4 Optimizing the System Clock Settings and Ensuring Refresh Rates Are Tuned for the Speed at Which SDRAM Runs

External DRAM requires periodic refreshes to ensure that the data stored in memory retains its proper value. Accesses by the core processor or DMA engine are held off until an in-process refresh cycle has completed. If the refresh occurs too frequently, the processor can't access SDRAM as often, and throughput to SDRAM decreases as a result.

On the Blackfin processor, the SDRAM Refresh Rate Control register provides a flexible mechanism for specifying the Auto-Refresh timing. Since the clock frequency supplied to the SDRAM can vary, this register implements a programmable refresh counter. This counter coordinates the supplied clock rate with the SDRAM device's required refresh rate.

Once the desired delay (in number of SDRAM clock cycles) between consecutive refresh counter time-outs is specified, a subsequent refresh counter time-out triggers an Auto-Refresh command to all external SDRAM devices.

Not only should you take care not to refresh SDRAM too often, but also be sure you're refreshing it often enough. Otherwise, stored data will start to decay because the SDRAM controller will not be able to keep corresponding memory cells refreshed.

Table 7.5 shows the impact of running with the best clock values and optimal refresh rates. Just in case you were wondering, RGB, CYMK, and YIQ are imaging/video formats. Conversion between the formats involves basic linear transformation that is common in video-based systems. Table 7.5 illustrates that the performance degradation can be significant with a nonoptimal refresh rate, depending on your actual access patterns. In this example, CCLK is reduced to run with an increased SCLK to illustrate this point. Doing this improves performance for this algorithm because the code fits into L1 memory and the data is partially in L3 memory. By increasing the SCLK rate, data can be fetched faster. What's more, by setting the optimal refresh rate, performance increases a bit more.

Table 7.5: Using the Optimal Refresh Rate

	Suboptimal SDRAM refresh rate		Optimal SDRAM refresh rate	
CCLK (MHz)	594 MHz	526 MHz	526 MHz	
SCLK (MHz)	119 MHz	132 MHz	132 MHz	
RGB to CMYK Conversion (iterations per second)	226	244	250	
RGB to YIQ Conversion (iterations per second)	266	276	282	Total
Cumulative Improvement		5%	2%	7%

7.9.5   Exploiting Priority and Arbitration Schemes between System Resources

Another important consideration is the priority and arbitration schemes that regulate how processor subsystems behave with respect to one another. For instance, on Blackfin processors, the core has priority over DMA accesses, by default, for transactions involving L3 memory that arrive at the same time. This means that if a core read from L3 occurs at the same time a DMA controller requests a read from L3, the core will win, and its read will be completed first.

Let's look at a scenario that can cause trouble in a real-time system. When the processor has priority over the DMA controller on accesses to a shared resource like L3 memory, it can lock out a DMA channel that also may be trying to access the memory. Consider the case where the processor executes a tight loop that involves fetching data from external memory. DMA activity will be held off until the processor loop has completed. It's not only a loop with a read embedded inside that can cause trouble. Activities like cache line fills or nonlinear code execution from L3 memory can also cause problems because they can result in a series of uninterruptible accesses.

There is always a temptation to rely on core accesses (instead of DMA) at early stages in a project, for a number of reasons. The first is that this mimics the way data is accessed on a typical prototype system. The second is that you don't always want to dig into the internal workings of DMA functionality and performance. However, with the core and DMA arbitration flexibility, using the memory DMA controller to bring data into and out of internal memory gives you more control of your destiny early on in a project. We will explore this concept in more detail in the following section.

7.10 Media Processing Frameworks

As more applications transition from PC-centric designs to embedded solutions, software engineers need to port media-based algorithms from prototype systems where memory is an "unlimited" resource (such as a PC or a workstation) to embedded systems where resource management is essential to meet performance requirements. Ideally, they want to achieve the highest performance for a given application without increasing the complexity of their "comfortable" programming model. Figure 7.13 shows a summary of the challenges they face in terms of power consumption, memory allocation, and performance.

A small set of programming frameworks are indispensable in navigating through key challenges of multimedia processing, like organizing input and output data buffer flows, partitioning memory intelligently, and using semaphores to control data movement. While reading this chapter, you should see how the concepts discussed in the previous chapters fit together into a cohesive structure. Knowing how audio and video work within the memory and DMA architecture of the processor you select will help you build your own framework for your specific application.

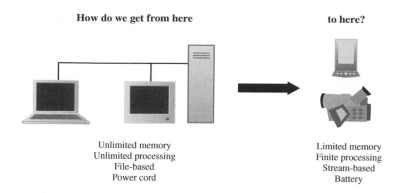

How do we get from here

to here?

Unlimited memory
Unlimited processing
File-based
Power cord

Limited memory
Finite processing
Stream-based
Battery

Figure 7.13: Moving an Application to an Embedded Processor

7.10.1   What Is a Framework?

Typically, a project starts out with a prototype developed in a high-level language such as C, or in a simulation and modeling tool such as Matlab or LabView. This is a particularly useful route for a few reasons. First, it's easy to get started, both psychologically and materially. Test data files, such as video clips or images, are readily available to help validate an algorithm's integrity. In addition, no custom hardware is required, so you can start development immediately, without waiting for, and debugging, a test setup. Optimal performance is not a focus at this stage because processing and memory resources are essentially unlimited on a desktop system. Finally, you can reuse code from other projects as well as from specialized toolbox packages or libraries.

The term "framework" has a wide variety of meanings, so let's define exactly what we mean by the word. It's important to harness the memory system architecture of an embedded processor to address performance and capacity trade-offs in ways that enable system development. Unfortunately, if we were to somehow find embedded processors with enough single-cycle memory to fit complicated systems on-chip, the cost and power dissipation of the device would be prohibitive. As a result, the embedded processor needs to use internal and external memory in concert to construct an application.

To this end, "framework" is the term we use to describe the complete code and data movement infrastructure within the embedded system. If you're working on a prototype development system, you can access data as if it were in L1 memory all of the time. In an embedded system, however, you need to choreograph data movement to meet the required real-time budget. A framework specifies how data moves throughout the system, configuring and managing all DMA controllers and related descriptors. In addition, it controls interrupt management and the execution of the corresponding interrupt service routines. Code movement is an integral part of the framework. We'll soon review some examples that illustrate how to carefully place code so that it peacefully coexists with the data movement tasks.

So, the first deliverable to tackle on an embedded development project is defining the framework. At this stage of the project, it is not necessary to integrate the actual algorithm yet. Your project can start off on the wrong foot if you add the embedded algorithm before architecting the basic data and coding framework!

7.11 Defining Your Framework

There are many important questions to consider when defining your framework. Hopefully, by answering these questions early in the design process, you'll be able to avoid common tendencies that could otherwise lead you down the wrong design path.

Q: At what rate does data come into the system, and how often does data leave the system?

Comment: This will help bound your basic system. For example, is there more data to move around than your processor can handle? How closely will you approach the limits of the processor you select, and will you be able to handle future requirements as they evolve?

Q: What is the smallest collection of data operated on at any time? Is it a line of video? Is it a macroblock? Is it a frame or field of video? How many audio samples are processed at any one time?

Comment: This will help you focus on the worst-case timing scenario. Later, we will look at some examples to help you derive these numbers. All of the data buffering parameters (size of each buffer and number of buffers) will be determined from this scenario.

Q: How much code will your application comprise? What is the execution profile of this code? Which code runs the most often?

Comment: This will help determine if your code fits into internal memory, or whether you have to decide between cache and overlays. When you have identified the code that runs most frequently, answering these questions will help you decide which code is allocated to the fastest memory.

Q: How will data need to move into and out of the system? How do the receive and transmit data paths relate to each other?

Comment: Draw out the data flow, and understand the sizes of the data blocks you hope to process simultaneously. Sketch the flows showing how input and output data streams are related.

Q: What are the relative priorities for peripheral and memory DMA channels? Do the default priorities work, or do these need to be reprogrammed? What are your options for data packing in the peripherals and the DMA?

Comment: This will help you lay out your DMA and interrupt priority levels between channels. It will also ensure that data transfers use internal buses optimally.

Q: Which data buffers will be accessed at any one time? Will multiple resources try to access the same memory space? Is the processor polling memory locations or manually moving data within memory?

Comment: This will help you organize where data and code are placed in your system to minimize conflicts.

Q: How many cycles can you budget for real-time processing? If you take the number of pixels (or audio samples, or both) being processed each collection interval, how many processor core-clock and system-clock cycles can you allocate to each pixel?

Comment: This will set your processing budget and may force you to, for example, reduce either your frame rate or image size.

We have already covered most of these topics in previous chapters, and it is important to reexamine these items before you lay out your own framework. We will now attack a fundamental issue related to the above questions: understanding your worst-case situation in the application timeline.

7.11.1   *The Application Timeline*

Before starting to code your algorithm, you need to identify the timeline requirements for the smallest processing interval in your application. This is best characterized as the minimum time between data collection intervals. In a video-based system, this interval typically relates to a macroblock within an image, a line of video data, or perhaps an entire video frame. The processor must complete its task on the current buffer before the next data set overwrites the same buffer. In some applications, the processing task at hand will simply involve making a decision on the incoming data buffer. This case is easier to handle because the processed data does not have to be transferred out. When the buffer is processed and the results still need to be stored or displayed, the processing interval calculation must include the data transfer time out of the system as well.

Figure 7.14 shows a summary of the minimum timelines associated with a variety of applications. The timeline is critical to understand because in the end, it is the foremost benchmark that the processor must meet.

An NTSC-based application that processes data on a frame-by-frame basis takes 33 ms to collect a frame of video. Let's assume that at the instant the first frame is received, the video port generates an interrupt. By the time the processor services the interrupt, the beginning of the next frame is already entering the FIFO of the video port. Because the processor needs to access one buffer while the next is on its way in, a second buffer needs to be maintained. Therefore, the time available to process the frame is 33 ms. Adding additional buffers can help to some extent, but if your data rates overwhelm your processor, this only provides short-term relief.

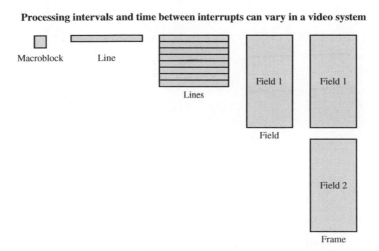

Figure 7.14: Minimum Timeline Examples

7.11.1.1   *Evaluating Bandwidth*

When selecting a processor, it's easy to oversimplify the estimates for overall bandwidth required. Unfortunately, this mistake is often realized after a processor has been chosen, or after the development schedule has been approved by management!

Consider the viewfinder subsystem of a digital video camera. Here, the raw video source feeds the input of the processor's video port. The processor then down samples the data, converts the color space from YCbCr to RGB, packs each RGB word into a 16-bit output, and sends it to the viewfinder's LCD. The process is shown in Figure 7.15.

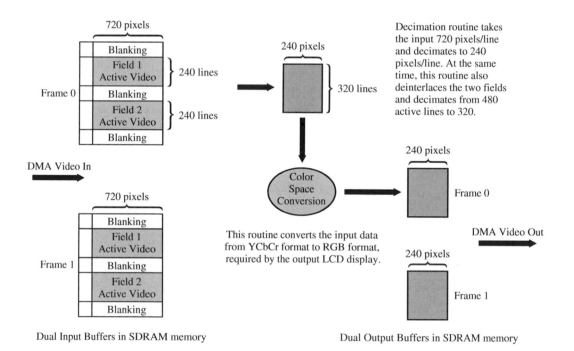

Figure 7.15: Block Diagram of Video Display System

The system described above provides a good platform to discuss design budgets within a framework. Given a certain set of data flows and computational routines, how can we determine if the processor is close to being "maxed out"?

Let's assume here we're using a single processor core running at 600 MHz, and the video camera generates NTSC video at the rate of 27 Mbytes per second.

So the basic algorithm flow is:

A. Read in an NTSC frame of input data (1716 bytes/row $\times$ 525 rows).

B. Down sample it to a QVGA image containing (320 $\times$ 240 pixels) $\times$ (2 bytes/ pixel).

C. Convert the data to RGB format.

D. Add a graphics overlay, such as the time of day or an "image tracker" box.

E. Send the final output buffer to the QVGA LCD at its appropriate refresh rate.

We'd like to get a handle on the overall performance of this algorithm. Is it taxing the processor too much, or barely testing it? Do we have room left for additional video processing, or for higher input frame rates or resolutions?

In order to measure the performance of each step, we need to gather timing data. It's convenient to do this with a processor's built-in cycle counters, which use the core-clock (CCLK) as a time base. Since in our example CCLK = 600 MHz, each tick of the cycle counter measures 1/(600 MHz), or 1.67 ns.

OK, so we've done our testing, and we find the following:

Step A: (27 million cycles per second/30 frames per second), or 900,000 cycles to collect a complete frame of video.

Steps B/C: 5 million CCLK cycles to down sample and color-convert that frame.

Steps D/E: 11 million CCLK cycles to add a graphics overlay and send that one processed frame to the LCD panel.

Keep in mind that these processes don't necessarily happen sequentially. Instead, they are pipelined for efficient data flow. But measuring them individually gives us insight into the ultimate limits of the system.

Given these timing results, we might be misled into thinking, "Wow, it only takes 5 million CCLK cycles to process a frame (because all other steps are allocated to the inputting and outputting of data), so 30 frames per second would only use up about 30 $\times$ 5 = 150 MHz of the core's 600 MHz performance. We could even do 60 frames/sec and still have 50% of the processor bandwidth left."

This type of surface analysis belies the fact that there are actually three important bandwidth studies to perform in order to truly get your hands around a system:

- Processor bandwidth

- DMA bandwidth

- Memory bandwidth

Bottlenecks in any one of these can prevent your application from working properly. More importantly, the combined bandwidth of the overall system can be very different than the sum of each individual bandwidth number, due to interactions between resources.

Processor Bandwidth

In our example, in Steps B and C the processor core needs to spend 5 M cycles operating on the input buffer from Step A. However, this analysis does not account for the total time available for the core to work on each input buffer. In processor cycles, a 600 MHz core can afford to spend around 20 M cycles (600 MHz/30 fps) on each input frame of data, before the next frame starts to arrive.

Viewed from this angle, then, Steps B and C tell us that the processor core is 5 M/20 M, or 25%, loaded. That's a far cry from the "intuitive" ratio of 5 M/600 M, or 0.8%, but it's still low enough to allow for a considerable amount of additional processing on each input buffer.

What would happen if we doubled the frame rate to 60 frames/second, keeping the identical resolution? Even though there are twice as many frames, it would still take only 5 M cycles to do the processing of Steps B and C, since the frame size has not changed. But now our 600 MHz core can only afford to spend 10 M cycles (600 MHz/60 frames/sec) on each input frame. Therefore, the processor is 50% loaded (5 M processing cycles/10 M available cycles) in this case.

Taking a different slant, let's dial back the frame rate to 30 frames/sec, but double the resolution of each frame. Effectively, this means there are twice as many pixels per frame. Now, it should take twice as long to read in a single frame and twice as long (10 M cycles) to process each frame as in the last case. However, since there are only 30 frames/second, If CCLK remains at 600 MHz, then the core can afford to spend 20 M cycles on each frame. As in the last case, the processor is 50% loaded (10 M processing cycles/20 M available cycles). It's good to see that these last two analyses matched up, since the total input data rate is identical.

DMA Bandwidth

Let's forget about the processor core for a minute and concentrate on the DMA controller. On a dual-core Blackfin processor, each 32-bit peripheral DMA channel (such as one used for video in/out functionality) can transfer data at clock speeds up to half the system clock (SCLK) rate, where SCLK maxes out at 133 MHz. This means that a given DMA channel can transfer data on every other SCLK cycle. Other DMA channels can use the free slots on a given DMA bus. In fact, for transfers in the same direction (e.g., into or out of the same memory space), every bus cycle can be utilized. For example, if the video port (PPI) is transferring data from external memory, the audio port (SPORT) can interleave its transfers from external memory to an audio codec without spending a cycle of latency for turning around the bus.

This implies that the maximum bandwidth on a given DMA bus is 133 MHz × 4 bytes, or 532 Mbytes/sec. As an aside, keep in mind that a processor might have multiple DMA buses available, thus allowing multiple transfers to occur at the same time.

In an actual system, however, it is not realistic to assume every transfer will occur in the same direction. Practically speaking, it is best to plan on a maximum transfer rate of one half of the theoretical bus bandwidth. This bus "derating" is important in an analogous manner to that of hardware component selection. In any system design, the more you exceed a 50% utilization factor, the more care you must take during software integration and future software maintenance efforts. If you plan on using 50% from the beginning, you'll leave yourself plenty of breathing room for coping with interactions of the various DMA channels and the behavior of the memory to which you're connecting. Of course, this value is not a hard limit, as many systems exist where every cycle is put to good use. The 50% derating factor is simply a useful guideline to allow for cycles that may be lost from bus turnarounds or DMA channel conflicts.

Memory Bandwidth

Planning the memory access patterns in your application can mean the difference between crafting a successful project and building a random number generator! Determining up front if the desired throughput is even possible for an application can save lots of headaches later.

As a system designer, you'll need to balance memory of differing sizes and performance levels at the onset of your project.

For multimedia applications involving image sizes above QCIF (176 × 144 pixels), on-chip memory is almost always insufficient for storing entire video frames. Therefore, the system must rely on L3 DRAM to support relatively fast access to large buffers. The processor interface to off-chip memory constitutes a major factor in designing efficient media frameworks, because access patterns to external memory must be well thought out in order to guarantee optimal data throughput. There are several high-level steps to ensure that data flows smoothly through memory in any system.

Once you understand the actual bandwidth needs for the processor, DMA and memory components, you can return to the issue at hand: what is the minimum processing interval that needs to be satisfied in your application?

Let's consider a new example where the smallest collection interval is defined to be a line of video. Determining the processing load under ideal conditions (when all code and data are in L1 memory) is easy. In the case where we are managing two buffers at a time, we must look at the time it takes to fill each buffer. The DMA controller "ping-pongs" between buffers to prevent a buffer from being overwritten while processing is underway on it. While the computation is done "in place" in Buffer 0, the peripheral fills Buffer 1. When Buffer 1 fills, Buffer 0 again becomes the destination. Depending on the processing timeline, an interrupt can optionally signal when each buffer has been filled.

So far, everything seems relatively straightforward. Now, consider what happens when the code is not in internal memory, but instead is executing from external memory. If instruction cache is enabled to improve performance, a fetch to external memory will result in a cache-line fill whenever there is not a match in L1 instruction memory (i.e., a cache-line miss occurs). The resulting fill will typically return at least 32 bytes. Because a cache-line fill is not interruptible—once it starts, it continues to completion—all other accesses to external memory are held off while it completes. From external memory, a cache-line fill can result in a fetch that takes 8 SCLKs (on Blackfin processors) for the first 32-bit word, followed by 7 additional SCLKs for the next seven 32-bit fetches (1 SCLK for each 32-bit fetch). This may be okay when the code being brought in is going to be executed. But now, what if one of the instructions being executed is a branch instruction, and this instruction, in turn, also generates a cache-line miss because it is more than a cache-line fill width away in memory address space? Code that is fetched from the second cache-line fill might also contain dual accesses that again are both data cache misses. What if these misses result in accesses to a page in external memory that is not active? Additional cycles can continue to hold off the competing resources. In a multimedia system, this situation can cause clicking sounds or video artifacts.

By this time, you should see the snowballing effect of the many factors that can reduce the performance of your application if you don't consider the interdependence of every framework component. Figure 7.16 illustrates one such situation.

The scenario described in Figure 7.16 demonstrates the need to, from the start, plan the utilization on the external bus. Incidentally, it is this type of scenario that drives the need for FIFOs in media processor peripherals, to insure that each interface has a cushion against the occurrence of these hard-to-manage system events. When you hear a click or see a glitch, what may be happening is that one of the peripherals has encountered an overrun (when it is receiving) or underrun (when it is transmitting) condition. It is important to set up error interrupt service routines to trap these conditions. This sounds obvious, but it's an often overlooked step that can save loads of debugging time.

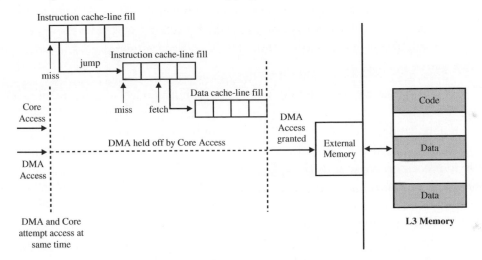

Figure 7.16: A Line of Video with Cache-line Misses Overlayed onto It

The question is, what kinds of tasks will happen at the worst possible point in your application? In the scenario we just described with multiple cache-line fills happening at the wrong time, eliminating cache may solve the problem on paper, but if your code will not fit into L1 memory, you will have to decide between shutting off cache and using the available DMA channels to manage code and data movement into and out of L1 memory. Even when system bandwidth seems to meet your initial estimates, the processor has to be able to handle the ebbs and flows of data transfers for finite intervals in any set of circumstances.

7.12 Asymmetric and Symmetric Dual-Core Processors

So far we've defaulted to talking about single-core processors for embedded media applications. However, there's a lot to be said about dual-core approaches. A processor with two cores (or more, if you're really adventurous) can be very powerful, yet along with the extra performance can come an added measure of complexity. As it turns out, there are a few common and quite useful programming models that suit a dual-core processor, and we'll examine them here.

There are two types of dual-core architectures available today. The first we'll call an "asymmetric" dual-core processor, meaning that the two cores are architecturally different. This means that, in addition to possessing different instruction sets, they also run at different operating frequencies and have different memory and programming models.

The main advantage of having two different architectures in the same physical package is that each core can optimize a specific portion of the processing task. For example, one core might excel at controller functionality, while the second one might target higher-bandwidth processing.

As you may figure, there are several disadvantages with asymmetric arrangements. For one, they require two sets of development tools and two sets of programming skill sets in order to build an application. Secondly, unused processing resources on one core are of little use to a fully loaded second core, since their competencies are so divergent. What's more, asymmetric processors make it difficult to scale from light to heavy processing profiles. This is important, for instance, in battery-operated devices, where frequency and voltage may be adjusted to meet real-time processing requirements; asymmetric cores don't scale well because the processing load is divided unevenly, so that one core might still need to run at maximum frequency while the other could run at a much lower clock rate. Finally, as we will see, asymmetric processors don't support many different programming models, which limits design options (and makes them much less exciting to talk about!).

In contrast to the asymmetric processor, a symmetric dual-core processor (extended to "symmetric multiprocessor," or SMP) consists of two identical cores integrated into a single package. The dual-core Blackfin ADSP-BF561 is a good example of this device class. An SMP requires only a single set of development tools and a design team with a single architectural knowledge base. Also, since both cores are equivalent, unused processing resources on one core can often be leveraged by the other core. Another very important benefit of the SMP architecture is the fact that frequency and voltage can more easily be modified together, improving the overall energy usage in a given application. Lastly, while the symmetric processor supports an asymmetric programming model, it also supports many other models that are very useful for multimedia applications.

The main challenge with the symmetric multiprocessor is splitting an algorithm across two processor cores without complicating the programming model.

7.13   Programming Models

There are several basic programming models that designers employ across a broad range of applications. We described an asymmetric processor in the previous discussion; we will now look at its associated programming model.

7.13.1   Asymmetric Programming Model

The traditional use of an asymmetric dual-core processor involves discrete and often different tasks running on each of the cores, as shown in Figure 7.17. For example, one of the cores may be assigned all of the control-related tasks. These typically include graphics and overlay functionality, as well as networking stacks and overall flow control. This core is also most often where the operating system or kernel will reside. Meanwhile, the second core can be dedicated to the high-intensity processing functions of the application. For example, compressed data may come over the network into the first core. Received packets can feed the second core, which in turn might perform some audio and video decode function.

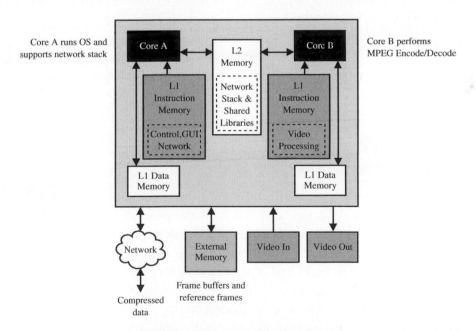

Figure 7.17: Asymmetric Model

In this model, the two processor cores act independently from each other. Logically, they are more like two stand-alone processors that communicate through the interconnect structures between them. They don't share any code and share very little data. We refer to this as the Asymmetric Programming Model. This model is preferred by developers who employ separate teams in their software development efforts. The ability to allocate different tasks to different processors allows development to be accomplished in parallel, eliminating potential critical path dependencies in the project. This programming model also aids the testing and validation phases of the project. For example, if code changes on one core, it does not necessarily invalidate testing efforts already completed on the other core.

Also, by having a dedicated processor core available for a given task, code developed on a single-core processor can be more easily ported to "half" of the dual-core processor. Both asymmetric and symmetric multiprocessors support this programming model. However, having identical cores available allows for the possibility of re-allocating any unused resources across functions and tasks. As we described earlier, the symmetric processor also has the advantage of providing a common, integrated environment.

Another important consideration of this model relates to the fact that the size of the code running the operating system and control tasks is usually measured in megabytes. As such, the code must reside in external memory, with instruction cache enabled. While this scheme is usually sufficient, care must be taken to prevent cache line fills from interfering with the overall timeline of the application. A relatively small subset of code runs the most often, due to the nature of algorithm coding. Therefore, enabling instruction cache is usually adequate in this model.

7.13.2 Homogeneous Programming Model

Because there are two identical cores in a symmetric multiprocessor, traditional processing-intensive applications can be split equally across each core. We call this a Homogeneous Model. In this scheme, code running on each core is identical. Only the data being processed is different. In a streaming multichannel audio application, for example, this would mean that one core processes half of the audio channels, and the other core processes the remaining half. Extending this concept to video and imaging applications, each core might process alternate frames. This usually translates to a scenario where all code fits into internal memory, in which case instruction cache is probably not used.

The communication flow between cores in this model is usually pretty basic. A mailbox interrupt (or on the Blackfin processor, a supplemental interrupt between cores) can signal the other core to check for a semaphore, to process new data or to send out processed data.

Usually, an operating system or kernel is not required for this model; instead, a "super loop" is implemented. We use the term "super loop" to indicate a code segment that just runs over and over again, of the form:

```
While (1)
      {
      Process_data();
      Send_results();
      Idle();
      }
```

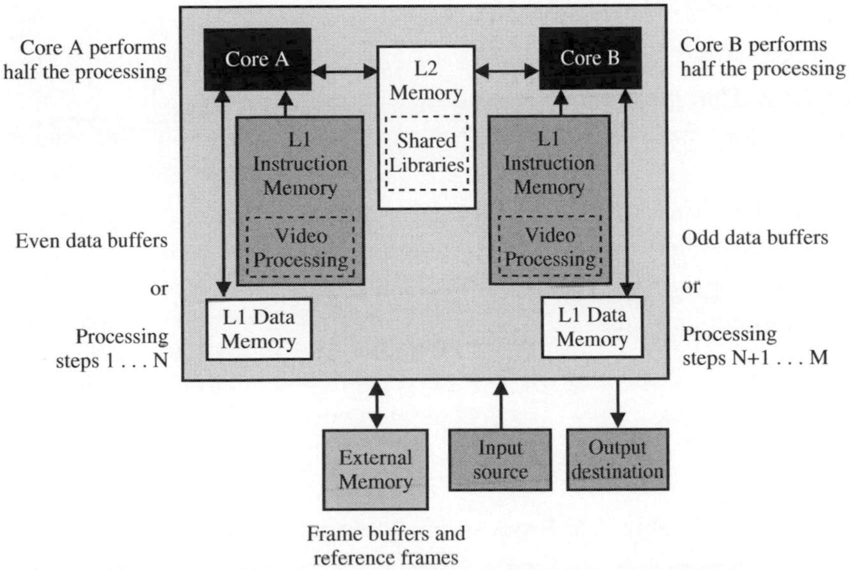

Figure 7.18: Master-Slave and Pipelined Model Representations

7.13.2.1   Master-Slave Programming Model

In the Master-Slave usage model, both cores perform intensive computation in order to achieve better utilization of the symmetric processor architecture. In this arrangement, one core (the master) controls the flow of the processing and actually performs at least half the processing load. Portions of specific algorithms are split and handled by the slave, assuming these portions can be parallelized. This situation is represented in Figure 7.18.

A variety of techniques, among them interrupts and semaphores, can be used to synchronize the cores in this model. The slave processor usually takes less processing time than the master does. Thus, the slave can poll a semaphore in shared memory when it is ready for more work. This is not always a good idea, though, because if the master core is still accessing the bus to the shared memory space, a conflict will arise. A more robust solution is for the slave to place itself in idle mode and wait for the master to interrupt it with a request to perform the next block of work.

A scheduler or simple kernel is most useful in this model, as we'll discuss later in the chapter.

7.13.2.2 Pipelined Programming Model

Also depicted in Figure 7.18, a variation on the Master-Slave model allocates processing steps to each core. That is, one core is assigned one or more serial steps, and the other core handles the remaining ones. This is analogous to a manufacturing pipeline where one core's output is the next core's input. Ideally, if the processing task separation is optimized, this will achieve a performance advantage greater than that of the other models. The task separation, however, is heavily dependent on the processor architecture and its memory hierarchy. For this reason, the Pipelined Model isn't as portable across processors as the other programming models are.

As Table 7.6 illustrates, the symmetric processor supports many more programming models than the asymmetric processor does, so you should consider all of your options before starting a project!

Table 7.6: Programming Model Summary

Processor	Asymmetric	Homogenous	Master-Slave	Pipelined
Asymmetric	✓			
Symmetric	✓	✓	✓	✓

7.14 Strategies for Architecting a Framework

We have discussed how tasks can be allocated across multiple cores when necessary. We have also described the basic ways a programming model can take shape. We are now ready to discuss several types of multimedia frameworks that can ride on top of either a single or dual-core processor. Regardless of the programming model, a framework is necessary in all but the simplest applications.

While they represent only a subset of all possible strategies, the categories shown below provide a good sampling of the most popular resource management situations. For illustration, we'll continue to use video-centric systems as a basis for these scenarios, because they incorporate the transfer of large amounts of data between internal and external memory, as well as the movement of raw data into the system and processed data out of the system. Here are the categories we will explore:

1. A system where data is processed as it is collected

2. A system where programming ease takes precedence over performance

3. A processor-intensive application where performance supersedes programming ease

7.14.1 Processing Data On-the-Fly

We'll first discuss systems where data is processed on-the-fly, as it is collected. Two basic categories of this class exist: low "system latency" applications and systems with either no external memory or a reduced external memory space.

This scenario strives for the absolute lowest system latency between input data and output result. For instance, imagine the camera-based automotive object avoidance system of Figure 7.19 tries to minimize the chance of a collision by rapidly evaluating successive video frames in the area of view. Because video frames require a tremendous amount of storage capacity (recall that one NTSC active video frame alone requires almost 700 Kbytes of memory), they invariably need external memory for storage. But if the avoidance system were to wait until an entire road image were buffered into memory before starting to process the input data, 33 ms of valuable time would be lost (assuming a 30-Hz frame rate). This is in contrast to the time it takes to collect a single line of data, which is only 63 μs.

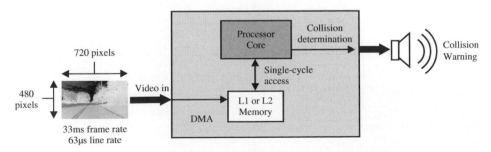

Figure 7.19: Processing Data as It Enters the System

To ensure lowest latency, video can enter L1 or L2 memory for processing on a line-by-line basis, rendering quick computations that can lead to quicker decisions. If the algorithm operates on only a few lines of video at a time, the frame storage requirements are much less difficult to meet. A few lines of video can easily fit into L2 memory, and since L2 memory is closer to the core processor than off-chip DRAM, this also improves performance considerably when compared to accessing off-chip memory.

Under this framework, the processor core can directly access the video data in L1 or L2 memory. In this fashion, the programming model matches the typical PC-based paradigm. In order to guarantee data integrity, the software needs to insure that the active video frame buffer is not overwritten with new data until processing on the current frame completes. As shown in Figure 7.20, this can be easily managed through a "ping-pong" buffer, as well through the use of a semaphore mechanism. The DMA controller in this framework is best configured in a descriptor mode, where Descriptor 0 points to Descriptor 1 when its corresponding data transfer completes. In turn, Descriptor 1 points back to Descriptor 0. This looks functionally like an Autobuffer scheme, which is also a realistic option to employ. What happens when the processor is accessing a buffer while it is being output to a peripheral? In a video application, you will most likely see some type of smearing between frames. This will show up as a blurred image, or one that appears to jump around.

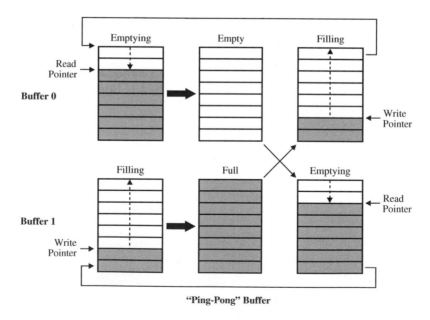

Figure 7.20: "Ping-Pong" Buffer

In our collision-avoidance system, the result of processing each frame is a decision—is a crash imminent or not? Therefore, in this case there is no output display buffer that needs protection against being overwritten. The size of code required for this type of application most likely will support execution from on-chip memory. This is helpful—again, it's one less thing to manage.

In this example, the smallest processing interval is the time it takes to collect a line of video from the camera. There are similar applications where multiple lines are required—for example, a 3×3 convolution kernel for image filtering.

Not all of the applications that fit this model have low system-latency requirements. Processing lines on-the-fly is useful for other situations as well. JPEG compression can lend itself to this type of framework, where image buffering is not required because there is no motion component to the compression. Here, macroblocks of 16 pixels × 16 pixels form a compression work unit. If we double-buffer two sets of 16 active-video lines, we can have the processor work its way through an image as it is generated. Again, a double-buffer scheme can be set up where two sets of 16 lines are managed. That is, one set of 16 lines is compressed while the next set is transferred into memory.

7.14.2 Programming Ease Trumps Performance

The second framework we'll discuss focuses entirely on using the simplest programming model at the possible expense of some performance. In this scenario, time to market is usually the most important factor. This may result in overspecifying a device, just to be sure there's plenty of room for inefficiencies caused by nonoptimal coding or some small amount of redundant data movements. In reality, this strategy also provides an upgrade platform, because processor bandwidth can ultimately be freed up once it's possible to focus on optimizing the application code. A simple flow diagram is shown in Figure 7.21.

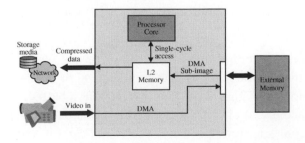

Figure 7.21: Framework that Focuses on Ease of Use

We used JPEG as an example in the previous framework because no buffering was required. For this framework any algorithm that operates on more than one line of data at a time, and is not an encoder or decoder, is a good candidate. Let's say we would like to perform a 3×3 two-dimensional convolution as part of an edge detection routine. For optimal operation, we need to have as many lines of data in internal memory as possible. The typical kernel navigates from left to right across an image, and then starts at the left side again (the same process used when reading words on a page). This convolution path continues until the kernel reaches the end of the entire image frame.

It is very important for the DMA controller to always fetch the next frame of data while the core is crunching on the current frame. That said, care should be taken to insure that DMA can't get too far ahead of the core, because then unprocessed data would be overwritten. A semaphore mechanism is the only way to guarantee that this process happens correctly. It can be provided as part of an operating system or in some other straightforward implementation.

Consider that, by the time the core finishes processing its first subframe of data, the DMA controller either has to wrap back around to the top of the buffer, or it has to start filling a second buffer. Due to the nature of the edge detection algorithm, it will most certainly require at least two buffers. The question is, is it better to make the algorithm library aware of the wrap-around, or to manage the data buffer to hide this effect from the library code?

The answer is, it is better not to require changes to an algorithm that has already been tested on another platform. Remember, on a C-based application on the PC, you might simply pass a pointer to the beginning of an image frame in memory when it is available for processing. The function may return a pointer to the processed buffer.

On an embedded processor, that same technique would mean operating on a buffer in external memory, which would hurt performance. That is, rather than operations at 30 frames per second, it could mean a maximum rate of just a few frames per second. This is exactly the reason to use a framework that preserves the programming model and achieves enough performance to satisfy an application's needs, even if requirements must be scaled back somewhat.

Let's return to our edge detection example. Depending on the size of the internal buffer, it makes sense to copy the last few lines of data from the end of one buffer to the beginning of the next one. Take a look at Figure 7.22. Here we see that a buffer of 120×120 pixels is brought in from L3 memory. As the processor builds an output buffer 120×120 pixels at a time, the next block comes in from L3. But if you're not careful, you'll have trouble in the output buffer at the boundaries of the processed blocks. That is, the convolution kernel needs to have continuity across consecutive lines, or visual artifacts will appear in the processed image.

One way to remedy this situation is to repeat some data lines (i.e., bring them into the processor multiple times). This allows you to present the algorithm with "clean" frames to work on, avoiding wraparound artifacts. You should be able to see that the added overhead associated with checking for a wraparound condition is circumvented by instead moving some small amount of data twice. By taking these steps, you can then maintain the programming model you started with by passing a pointer to the smaller subimage, which now resides in internal memory.

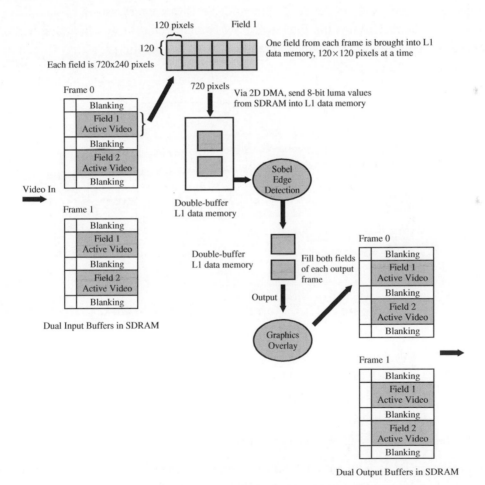

Figure 7.22: Edge Detection

7.14.3 Performance-based Framework

The third framework we'll discuss is often important for algorithms that push the limits of the target processor. Typically, developers will try to right-size their processor to their intended application, so they won't have to pay a cost premium for an overcapable device. This is usually the case for extremely high-volume, cost-sensitive applications. As such, the "performance-based" framework focuses on attaining best performance at the expense of a possible increase in programming complexity. In this framework, implementation may take longer and integration may be a bit more challenging, but the long-term savings in designing with a less expensive device may justify the extra development time. The reason there's more time investment early in the development cycle is that every aspect of data flow needs to be carefully planned. When the final data flow is architected, it will be much harder to reuse, because the framework was hand-crafted to solve a specific problem. An example is shown in Figure 7.23.

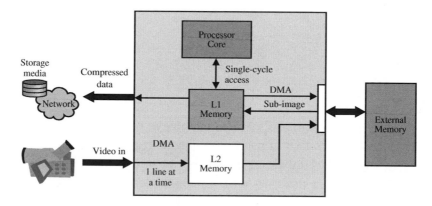

Figure 7.23: Performance-based Framework

The examples in this category are many and varied. Let's look at two particular cases: a design where an image pipe and compression engine must coexist, and a high-performance video decoder.

7.14.3.1 Image Pipe and Compression Example

Our first example deals with a digital camera where the processor connects to a CMOS sensor or CCD module that outputs a Bayer RGB pattern. This application often involves a software image pipe that preprocesses the incoming image frame. In this design, we'll also want to perform JPEG compression or a limited-duration MPEG-4 encode at 30 fps. It's almost as if we have two separate applications running on the same platform.

This design is well-suited to a dual-core development effort. One core can be dedicated to implementing the image pipe, while the other performs compression. Because the two processors may share some internal and external memory resources, it is important to plan out accesses so that all data movement is choreographed. While each core works on separate tasks, we need to manage memory accesses to ensure that one of the tasks doesn't hold off any others. The bottom line is that both sides of the application have to complete before the next frame of data enters the system.

Just as in the "Processing on-the-Fly" framework example, lines of video data are brought into L2 memory, where the core directly accesses them for preprocessing as needed, with lower latency than accessing off-chip memory. While the lower core data access time is important, the main purpose of using L2 memory is to buffer up a set of lines in order to make group transfers in the same direction, thus maximizing bus performance to external memory.

A common (but incorrect) assumption made early in many projects is to consider only individual benchmarks when comparing transfers to/from L2 and with transfers to/from L3 memory. The difference in transfer times does not appear to be dramatic when the measurements are taken individually, but the interaction of multiple accesses can make a big difference.

Why is this the case? Because if the video port feeds L3 memory directly, the data bus turns around more times than necessary. Let's assume we have 8-bit video data packed into 32-bit DMA transfers. As soon as the port collects 4 bytes of sensor input, it will perform a DMA transfer to L3. For most algorithms, a processor makes more reads than writes to data in L3 memory. This, of course, is application-dependent, but in media applications there are usually at least three reads for every write. Since the video port is continuously writing to external memory, turnarounds on the external memory bus happen frequently, and performance suffers as a result.

By the time each line of a video frame passes into L2 memory and back out to external memory, the processor has everything it needs to process the entire frame of data. Very little bandwidth has been wasted by turning the external bus around more than necessary. This scheme is especially important when the image pipe runs in parallel with the video encoder. It ensures the least conflict when the two sides of the application compete for the same resources.

To complete this framework requires a variety of DMA flows. One DMA stream reads data in from external memory, perhaps in the form of video macroblocks. The other flow sends compressed data out—over a network or to a storage device, for instance. In addition, audio streams are part of the overall framework. But, of course, video is the main flow of concern, from both memory traffic and DMA standpoints.

7.14.3.2   High Performance Decoder Example

Another sample flow in the "performance-based" framework involves encoding or decoding audio and video at the highest frame rate and image size possible. For example, this may correspond to implementing a decoder (MPEG-4, H.264 or WMV9) that operates on a D-1 video stream on a single-core processor.

Designing for this type of situation conveys an appreciation of the intricacies of a system that is more complex than the ones we have discussed so far. Once the processor receives the encoded bit stream, it parses and separates the header and data payloads from the stream. The overall processing limit for the decoder can be determined by:

(# of cycles/pixel)$\times$(# of pixels/frame) $\times$ (# of frames/second) < (Budgeted # of cycles/second)

At least 10% of the available processing bandwidth must be reserved for steps like audio decode and transport layer processing. For a D-1 video running on a 600 MHz device, we have to process around 10 Mpixels per second. Considering only video processing, this allows ~58 cycles per pixel. However, reserving 10% for audio and transport stream processing, we are left with just over 50 cycles per pixel as our processing budget.

When you consider the number of macroblocks in a D-1 frame, you may ask, "Do I need an interrupt after each of these transfers?" The answer, thankfully, is "No." As long as you time the transfers and understand when they start and stop, there is no need for an interrupt.

Now let's look at the data flow of the video decoder shown in Figure 7.24.

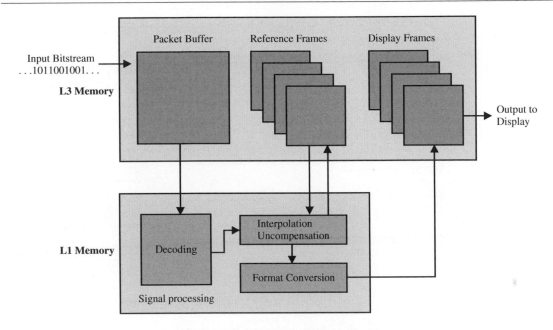

Figure 7.24: Typical Video Decoder

Figure 7.25 shows the data movement involved in this example. We use a 2D-to-1D DMA to bring the buffers into L1 memory for processing. Figure 7.26 shows the data flow required to send buffers back out to L3 memory.

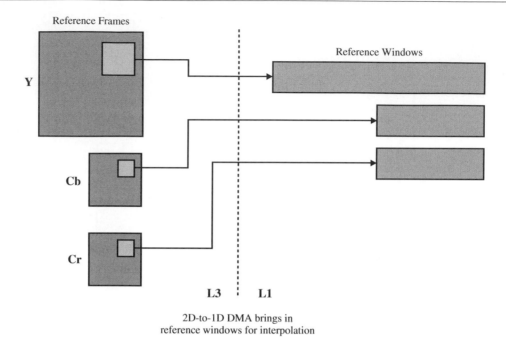

2D-to-1D DMA brings in
reference windows for interpolation

Figure 7.25: Data Movement (L3 to L1 Memory)

On the DMA side of the framework, we need DMA streams for the incoming encoded bit stream. We also need to account for the reference frame being DMAed into L1 memory, a reconstructed frame sent back out to L3 memory, and the process of converting the frame into 4:2:2 YCbCr format for ITU-R BT.656 output. Finally, another DMA is required to output the decoded video frame through the video port.

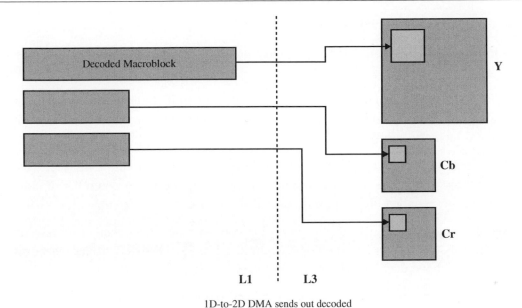

1D-to-2D DMA sends out decoded
macroblock to L3 to build reference frame

Figure 7.26: Data Movement (L1 to L3)

For this scheme, larger buffers are staged in L3 memory, while smaller buffers, including lookup tables and building blocks of the larger buffers, reside in L1 memory. When we add up the total bandwidth to move data into and out of the processor, it looks something like the following:

Input data stream: 1 Mbyte/sec

Reference frame in: 15 Mbyte/sec

Loop filter (input and output): 30 Mbyte/sec

Reference data out: 15 Mbyte/sec

Video out: 27 Mbyte/sec

The percentage of bandwidth consumed will depend on the software implementation. One thing, however, is certain, you cannot simply add up each individual transfer rate to arrive at the system bandwidth requirements. This will only give you a rough indication, not tell you whether the system will work.

7.14.4 Framework Tips

Aside from what we've mentioned above, there are some additional items that you may find useful.

1. Consider using L2 memory as a video line buffer. Even if it means an extra pass in your system, this approach conserves bandwidth where it is the most valuable, at the external memory interface.

2. Avoid polling locations in L2 or L3 memory. Polling translates into a tight loop by the core processor that can then lock out other accesses by the DMA controller, if core accesses are assigned higher priority than DMA accesses.

3. Avoid moving large blocks of memory using core accesses. Consecutive accesses by the core processor can lock out the DMA controller. Use the DMA controller whenever possible to move data between memory spaces.

4. Profile your code with the processor's software tools suite, shooting for at least 97% of your code execution to occur in L1 memory. This is best accomplished through a combination of cache and strategic placement of the most critical code in L1 SRAM. It should go without saying, but place your event service routines in L1 memory.

5. Interrupts are not mandatory for every transfer. If your system is highly deterministic, you may choose to have most transfers finish without an interrupt. This scheme reduces system latency, and it's the best guarantee that high-bandwidth systems will work. Sometimes, adding a control word to the stream can be useful to indicate the transfer has occurred. For example, the last word of the transfer could be defined to indicate a macroblock number that the processor could then use to set up new DMA transfers.

6. Taking shortcuts is sometimes okay, especially when these shortcuts are not visually or audibly discernable. For example, as long as the encoded output stream is compliant to a standard, shortcuts that impact the quality only matter if you can detect them. This is especially helpful to consider when the display resolution is the limiting factor or the weak link in a system.

7.15   Other Topics in Media Frameworks

7.15.1   Audio-Video Synchronization

We haven't talked too much about audio processing in this chapter because it makes up a small subset of the bandwidth in a video-based system. Data rates are measured in kilobytes/sec, versus megabytes/sec for even the lowest-resolution video systems.

Where audio does become important in the context of video processing is when we try to synchronize the audio and video streams in a decoder/encoder system. While we can take shortcuts in image quality in some applications when the display is small, it is hard to take shortcuts on the synchronization task, because an improperly synchronized audio/video output is quite annoying to end users.

For now let's assume we have already decoded an audio and video stream. Figure 7.27 shows the format of an MPEG-2 transport stream. There are multiple bit stream options for MPEG-2, but we will consider the MPEG-2 transport stream (TS).

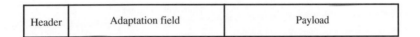

Figure 7.27: MPEG-2 Encoded Transport Stream Format

The header shown in Figure 7.27 includes a Packet ID code and a sequence number to ensure decode is performed in the proper order. The adaptation field is used for additional control information. One of these control words is the program clock reference, or PCR. This is used as the timing reference for the communication channel.

Video and audio encoders put out packet elementary streams (PES) that are split into the transport packets shown. When a PES packet is split to fit into a set of transport packets, the PES header follows the 4-byte Transport Header. A presentation time stamp is added to the packet. A second time stamp is also added when frames are sent out of order, which is done intentionally for things like anchor video frames. This second time stamp is used to control the order in which data is fed to the decoder.

Let's take a slight tangent to discuss some data buffer basics, to set the stage for the rest of our discussion. Figure 7.28 shows a generic buffer structure, with high and low watermarks, as well as read and write pointers. The locations of the high and low watermarks are application-dependent, but they should be set appropriately to manage data flow in and out of the buffer. The watermarks determine the hysteresis of the buffer data flow. For example, the high watermark indicates a point in the buffer that triggers some processor action when the buffer is filling up (like draining it down to the low watermark). The low watermark also provides a trigger point that signals a task that some processor action needs to be taken (like transferring enough samples into the buffer to reach the high watermark). The read and write pointers in any specific implementation must be managed with respect to the high and low watermarks to ensure data is not lost or corrupted.

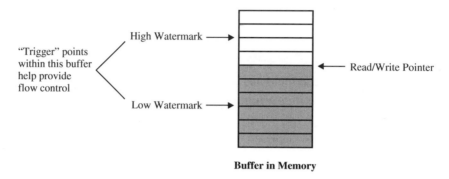

Buffer in Memory

Figure 7.28: Buffer Basics

In a video decoder, audio buffers and video frames are created in external memory. As these output buffers are written to L3, a time stamp from the encoded stream is assigned to each buffer and frame. In addition, the processor needs to track its own time base. Then, before each decoded video frame and audio buffer is sent out for display, the processor performs a time check and finds the appropriate data match from each buffer. There are multiple ways to accomplish this task via DMA, but the best way is to have the descriptors already assembled and then, depending on which packet time matches the current processor time, adjust the write pointer to the appropriate descriptor.

Figure 7.29 shows a conceptual illustration of what needs to occur in the processor. As you can probably guess, skipping a video frame or two is usually not catastrophic to the user experience. Depending on the application, even skipping multiple frames may go undetected. On the other hand, not synchronizing audio properly, or skipping audio samples entirely, is much more objectionable to viewers and listeners. The synchronization process of comparing times of encoded packets and matching them with the appropriate buffer is not computationally intensive. The task of parsing the encoded stream takes up the majority of MIPS in this framework, and this number will not vary based on image size.

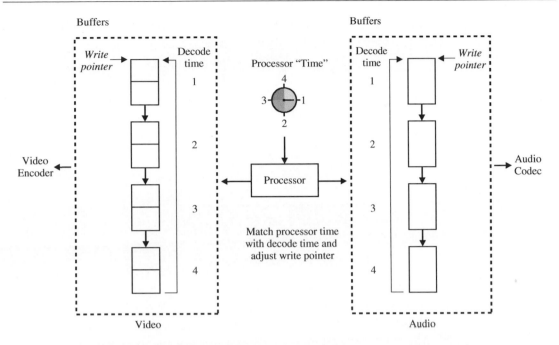

Figure 7.29: Conceptual Diagram of Audio-Video Synchronization

7.15.2 Managing System Flow

We have already discussed applications where no operating system is used. We referred to this type of application as a ***super loop*** because there is a set order of processing that repeats every iteration. This is most common in the highest-performing systems, as it allows the programmer to retain most of the control over the order of processing. As a result, the block diagram of the data flow is usually pretty simple, but the intensity of the processing (image size, frame rate, or both) is usually greater.

Having said this, even the most demanding application normally requires some type of ***system services***. These allow a system to take advantage of some kernel-like features without actually using an OS or a kernel. In addition to system services, a set of device drivers also works to control the peripherals. Figure 7.30 shows the basic services that are available with the Blackfin VisualDSP++ tool chain.

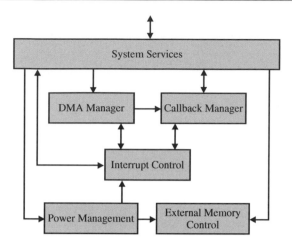

Figure 7.30: System Services

Of those shown, the external memory and power management services are typically initialization services that configure the device or change operating parameters. On the other hand, the Interrupt, DMA, and Callback Managers all provide ways to manage system flow.

As part of the DMA services, you can move data via a standard API, without having to configure every control register manually. A manager is also provided that accepts DMA work requests. These requests are handled in the order they are received by application software. The DMA Manager simplifies a programming model by abstracting data transfers.

The Interrupt Manager allows the processor to service interrupts quickly. The idea is that the processor leaves the higher-priority interrupt and spawns an interrupt at the lowest priority level. The higher-priority interrupt is serviced, and the lower-priority interrupt is generated via a software instruction. Once this happens, new interrupts are no longer held off.

When the processor returns from the higher-priority interrupt, it can execute a callback function to perform the actual processing. The Callback Manager allows you to respond to an event any way you choose. It passes a pointer to the routine you want to execute when the event occurs. The key is that the basic event is serviced, and the processor runs in a lower-priority task. This is important, because otherwise you run the risk of lingering in a higher-level interrupt, which can then delay response time for other events.

As we mentioned at the beginning of this section, device drivers provide the software layer to various peripherals, such as the video and audio ports. Figure 7.31 shows how the device drivers relate to a typical application. The device driver manages communications between memory and a given peripheral. The device drivers provided with VisualDSP++, for example, can work with or without an operating system.

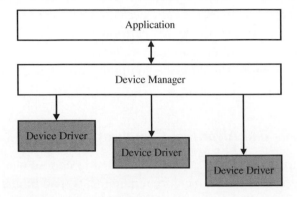

Figure 7.31: Application with Device Drivers

Finally, an OS can be an integral part of your application. If it is, Figure 7.32 shows how all of these components can be connected together. There are many OS and kernel options available for a processor. Typically, the products span a range of strengths and focus areas, for example, security, performance or code footprint. There is no "silver bullet" when it comes to these parameters. That is, if an OS has more security features, for instance, it may sacrifice on performance and/or kernel size.

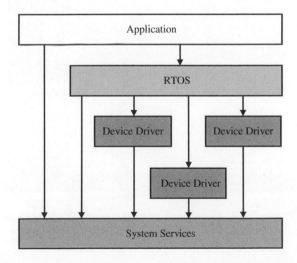

Figure 7.32: Application with Device Drivers and OS

In the end, you'll have to make a trade-off between performance and flexibility. One of the best examples of this trade-off is with uCLinux, an embedded instantiation of Linux that has only partial memory management unit capabilities. Here, the kernel size is measured in megabytes and must run from larger, slower external memory. As such, the instruction cache plays a key role in ensuring best kernel performance. While uCLinux's performance will never rival that of smaller, more optimized systems, its wealth of open-source projects available, with large user bases, should make you think twice before dismissing it as an option.

7.15.3   Frameworks and Algorithm Complexity

In this section we've tried to provide guidance on when to choose one framework over another, largely based on data flows, memory requirements, and timing needs. Figure 7.33 shows another slant on these factors. It conveys a general idea of how complexity increases exponentially as data size grows. Moreover, as processing moves from being purely spatial in nature to having a temporal element as well, complexity (and the resulting need for a well-managed media framework) increases even further.

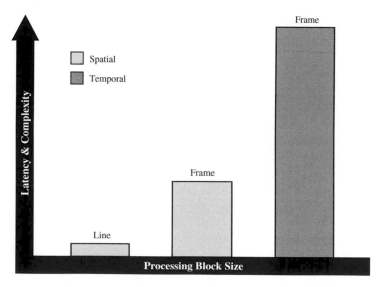

Figure 7.33: Relative Complexity of Applications

CHAPTER 8

DSP in Embedded Systems

Robert Oshana

In order to understand the usefulness of programmable Digital Signal Processing, I will first draw an analogy and then explain the special environments where DSPs are used.

A DSP is really just a special form of microprocessor. It has all of the same basic characteristics and components; a CPU, memory, instruction set, buses, and so on. The primary difference is that each of these components is customized slightly to perform certain operations more efficiently. We'll talk about specifics in a moment, but in general, a DSP has hardware and instruction sets that are optimized for high-speed numeric processing applications and rapid, real-time processing of analog signals from the environment. The CPU is slightly customized, as is the memory, instruction sets, buses, and so forth.

I like to draw an analogy to society. We, as humans, are all processors (cognitive processors) but each of us is specialized to do certain things well; engineering, nursing, finance, and so forth. We are trained and educated in certain fields (specialized) so that we can perform certain jobs efficiently. When we are specialized to do a certain set of tasks, we expend less energy doing those tasks. It is not much different for microprocessors. There are hundreds to choose from and each class of microprocessor is specialized to perform well in certain areas. A DSP is a specialized processor that does signal processing very efficiently. And, like our specialty in life, because a DSP specializes in signal processing, it expends less energy getting the job done. DSPs, therefore, consume less time, energy, and power than a general-purpose microprocessor when carrying out signal processing tasks.

When you specialize a processor, it is important to specialize those areas that are commonly and frequently put to use. It doesn't make sense to make something efficient at doing things that are hardly ever needed! Specialize those areas that result in the biggest bang for the buck!

However, before I go much further, I need to give a quick summary of what a processor must do to be considered a digital signal processor. It must do two things very well. First, it must be good at math and be able to do millions (actually billions) of multiplies and adds per second. This is necessary to implement the algorithms used in digital signal processing.

The second thing it must do well is to guarantee real time. Let's go back to our real-life example. I took my kids to a movie recently, and when we arrived, we had to wait in line to purchase our tickets. In effect, we were put into a queue for processing, standing in line behind other moviegoers. If the line stays the same length and doesn't continue to get longer and longer, then the queue is real-time in the sense that the same number of customers are being processed as there are joining the queue. This queue of people may get shorter or grow a bit longer but does not grow in an unbounded way. If you recall the evacuation from Houston as Hurricane Rita approached, that was a queue that was growing in an unbounded way! This queue was definitely not real time and it grew in an unbounded way, and the system (the evacuation system) was considered a failure. Real-time systems that cannot perform in real time are failures.

If the queue is really big (meaning, if the line I am standing in at the movies is really long) but not growing, the system may still not work. If it takes me 50 minutes to move to the front of the line to buy my ticket, I will probably be really frustrated, or leave altogether before buying my ticket to the movies (my kids will definitely consider this a failure). Real-time systems also need to be careful of large queues that can cause the system to fail. Real-time systems can process information (queues) in one of two ways: either one data element at a time, or by buffering information and then processing the "queue." The queue length cannot be too long or the system will have significant latency and not be considered real time.

If real time is violated, the system breaks and must be restarted. To further the discussion, there are two aspects to real time. The first is the concept that for every sample period, one input piece of data must be captured, and one output piece of data must be sent out. The second concept is latency. Latency means that the delay from the signal being input into the system and then being output from the system must be preserved as immediate.

Keep in mind the following when thinking of real-time systems: producing the correct answer too late is wrong! If I am given the right movie ticket and charged the correct amount of money after waiting in line, but the movie has already started, then the system is still broke (unless I arrived late to the movie to begin with). Now go back to our discussion.

So what are the "special" things that a DSP can perform? Well, like the name says, DSPs do signal processing very well. What does "signal processing" mean? Really, it's a set of algorithms for processing signals in the digital domain. There are analog equivalents to these algorithms, but processing them digitally has been proven to be more efficient. This has been a trend for many many years. Signal processing algorithms are the basic building blocks for many applications in the world; from cell phones to MP3 players, digital still cameras, and so on. A summary of these algorithms is shown in Table 8.1.

Table 8.1

Algorithm	Equation
Finite Impulse Response Filter	$y(n) = \sum\limits_{k=0}^{M} a_k x(n-k)$
Infinite Impulse Response Filter	$y(n) = \sum\limits_{k=0}^{M} a_k x(n-k) + \sum\limits_{k=1}^{N} b_k y(n-k)$
Convolution	$y(n) = \sum\limits_{k=0}^{N} x(k) h(n-k)$
Discrete Fourier Transform	$X(k) = \sum\limits_{n=0}^{N-1} x(n) \exp[-j(2\pi/N)nk]$
Discrete Cosine Transform	$F(u) = \sum\limits_{x=0}^{N-1} c(u) \cdot f(x) \cdot \cos\left[\dfrac{\pi}{2N} u(2x+1)\right]$

One or more of these algorithms are used in almost every signal processing application. Finite Impulse Response Filters and Infinite Impulse Response Filters are used to remove unwanted noise from signals being processed, convolution algorithms are used for looking for similarities in signals, discrete Fourier transforms are used for representing signals in formats that are easier to process, and discrete cosine transforms are used in image processing applications. We'll discuss the details of some of these algorithms later, but there are some things to notice about this entire list of algorithms. First, they all have a summing operation, the function. In the computer world, this is equivalent to an accumulation of a large number of elements that is implemented using a "for" loop. DSPs are designed to have large accumulators because of this characteristic. They are specialized in this way. DSPs also have special hardware to perform the "for" loop operation so that the programmer does not have to implement this in software, which would be much slower.

The algorithms above also have multiplications of two different operands. Logically, if we were to speed up this operation, we would design a processor to accommodate the multiplication and accumulation of two operands like this very quickly. In fact, this is what has been done with DSPs. They are designed to support the multiplication and accumulation of data sets like this very quickly; for most processors, in just one cycle. Since these algorithms are very common in most DSP applications, tremendous execution savings can be obtained by exploiting these processor optimizations.

There are also inherent structures in DSP algorithms that allow them to be separated and operated on in parallel. Just as in real life, if I can do more things in parallel, I can get more done in the same amount of time. As it turns out, signal processing algorithms have this characteristic as well. Therefore, we can take advantage of this by putting multiple orthogonal (nondependent) execution units in our DSPs and exploit this parallelism when implementing these algorithms.

DSPs must also add some reality to the mix of these algorithms shown above. Take the IIR filter described above. You may be able to tell just by looking at this algorithm that there is a feedback component that essentially feeds back previous outputs into the calculation of the current output. Whenever you deal with feedback, there is always an inherent stability issue. IIR filters can become unstable just like other feedback systems. Careless implementation of feedback systems like the IIR filter can cause the output to oscillate instead of asymptotically decaying to zero (the preferred approach). This problem is compounded in the digital world where we must deal with finite word lengths, a key limitation in all digital systems. We can alleviate this using saturation checks in software or use a specialized instruction to do this for us. DSPs, because of the nature of signal processing algorithms, use specialized saturation underflow/overflow instructions to deal with these conditions efficiently.

There is more I can say about this, but you get the point. Specialization is really all it's about with DSPs; these devices are specifically designed to do signal processing really well. DSPs may not be as good as other processors when dealing with nonsignal processing centric algorithms (that's fine; I'm not any good at medicine either). Therefore, it's important to understand your application and pick the right processor.

With all of the special instructions, parallel execution units, and so on designed to optimize signal-processing algorithms, there is not much room left to perform other types of general-purpose optimizations. General-purpose processors contain optimization logic such as branch prediction and speculative execution, which provide performance improvements in other types of applications. But some of these optimizations don't work as well for signal processing applications. For example, branch prediction works really well when there are a lot of branches in the application. But DSP algorithms do not have a lot of branches. Much signal processing code consists of well-defined functions that execute off a single stimulus, not complicated state machines requiring a lot of branch logic.

Digital signal processing also requires optimization of the software. Even with the fancy hardware optimizations in a DSP, there is still some heavy-duty tools support required—specifically, the compiler—that makes it all happen. The compiler is a nice tool for taking a language like C and mapping the resultant object code onto this specialized microprocessor. Optimizing compilers perform a very complex and difficult task of producing code that fully "entitles" the DSP hardware platform.

There is no black magic in DSPs. As a matter of fact, over the last couple of years, the tools used to produce code for these processors have advanced to the point where you can write much of the code for a DSP in a high level language like C or C++ and let the compiler map and optimize the code for you. Certainly, there will always be special things you can do, and certain hints you need to give the compiler to produce the optimal code, but it's really no different from other processors.

The environment in which a DSP operates is important as well, not just the types of algorithms running on the DSP. Many (but not all) DSP applications are required to interact with the real world. This is a world that has a lot of stuff going on; voices, light, temperature, motion, and more. DSPs, like other embedded processors, have to *react* in certain ways within this real world. Systems like this are actually referred to as *reactive* systems. When a system is reactive, it needs to respond and control the *real* world, not too surprisingly, in *real*-time. Data and signals coming in from the real world must be processed in a timely way. The definition of timely varies from application to application, but it requires us to keep up with what is going on in the environment.

Because of this timeliness requirement, DSPs, as well as other processors, must be designed to respond to real-world events quickly, get data in and out quickly, and process the data quickly. We have already addressed the processing part of this. But believe it or not, the bottleneck in many real-time applications is not getting the data processed, but getting the data in and out of the processor quickly enough. DSPs are designed to support this real-world requirement. High speed I/O ports, buffered serial ports, and other peripherals are designed into DSPs to accommodate this. DSPs are, in fact, often referred to as data pumps, because of the speed in which they can process streams of data. This is another characteristic that makes DSPs unique.

DSPs are also found in many embedded applications. I'll discuss the details of embedded systems later in this chapter. However, one of the constraints of an embedded application is scarce resources. Embedded systems, by their very nature, have scarce resources. The main resources I am referring to here are processor cycles, memory, power, and I/O. It has always been this way, and always will. Regardless of how fast embedded processors run, how much memory can be fit on chip, and so on, there will always be applications that consume all available resources and then look for more! In addition, embedded applications are very application-specific, not like a desktop application that is much more general-purpose.

At this point, we should now understand that a DSP is like any other programmable processor, except that it is specialized to perform signal processing really efficiently. So now the only question should be; why program anything at all? Can't I do all this signal processing stuff in hardware? Well, actually you can. There is a fairly broad spectrum of DSP implementation techniques, with corresponding trade-offs in flexibility, as well as cost, power, and a few other parameters. Figure 8.1 summarizes two of the main trade-offs in the programmable versus fixed-function decision: flexibility and power.

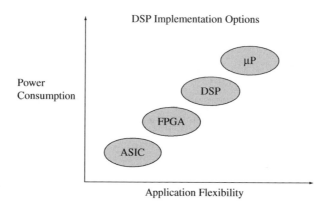

Figure 8.1

An application-specific integrated circuit (ASIC) is a hardware only implementation option. These devices are programmed to perform a fixed-function or set of functions. Being a hardware only solution, an ASIC does not suffer from some of the programmable von Neumann-like limitations, such as loading and storing of instructions and data. These devices run exceedingly fast in comparison to a programmable solution, but they are not as flexible. Building an ASIC is like building any other microprocessor, to some extent. It's a rather complicated design process, so you have to make sure the algorithms you are designing into the ASIC work and won't need to be changed for a while! You cannot simply recompile your application to fix a bug or change to a new wireless standard. (Actually, you could, but it will cost a lot of money and take a lot of time.) If you have a stable, well-defined function that needs to run really fast, an ASIC may be the way to go.

Field-programmable gate arrays (FPGAs) are one of those in-between choices. You can program them and reprogram them in the field, to a certain extent. These devices are not as flexible as true programmable solutions, but they are more flexible than an ASIC. Since FPGAs are hardware they offer similar performance advantages to other hardware-based solutions. An FPGA can be "tuned" to the precise algorithm, which is great for performance. FPGAs are not truly application specific, unlike an ASIC. Think of an FPGA as a large sea of gates where you can turn on and off different gates to implement your function. In the end, you get your application implemented, but there are a lot of spare gates laying around, kind of going along for the ride. These take up extra space as well as cost, so you need to do the trade-offs; are the cost, physical area, development cost, and performance all in line with what you are looking for?

DSP and μP (microprocessor): We have already discussed the difference here, so there is no need to rehash it. Personally, I like to take the flexible route: programmability. I make a lot of mistakes when I develop signal processing systems; it's very complicated technology! Therefore, I like to know that I have the flexibility to make changes when I need to in order to fix a bug, perform an additional optimization to increase performance or reduce power, or change to the next standard. The entire signal-processing field is growing and changing so quickly—witness the standards that are evolving and changing all the time—that I prefer to make the rapid and inexpensive upgrades and changes only a programmable solution can afford.

The general answer, as always, lies somewhere in between. In fact, many signal processing solutions are partitioned across a number of different processing elements. Certain parts of the algorithm stream—those that have a pretty good probability of changing in the near future—are mapped to a programmable DSP. Signal processing functions that will remain fairly stable for the foreseeable future are mapped into hardware gates (either an ASIC, an FPGA, or other hardware acceleration). Those parts of the signal processing system that control the input, output, user interface, and overall management of the system heartbeat may be mapped to a more general-purpose processor. Complicated signal processing systems need the right combination of processing elements to achieve true system performance/cost/power trade-offs.

Signal processing is here to stay. It's everywhere. Any time you have a signal that you want to know more about, communicate in some way, make better or worse, you need to process it. The digital part is just the process of making it all work on a computer of some sort. If it's an embedded application you must do this with the minimal amount of resources possible. Everything costs money; cycles, memory, power—so everything must be conserved. This is the nature of embedded computing; be application specific, tailor to the job at hand, reduce cost as much as possible, and make things as efficient as possible. This was the way things were done in 1982 when I started in this industry, and the same techniques and processes apply today. The scale has certainly changed; computing problems that required supercomputers in those days are on embedded devices today!

This chapter will touch on these areas and more as it relates to digital signal processing. There is a lot to discuss and I'll take a practical rather than theoretical approach to describe the challenges and processes required to do DSP well.

8.1   Overview of Embedded Systems and Real-Time Systems

Nearly all real-world DSP applications are part of an embedded real-time system. While this chapter will focus primarily on the DSP-specific portion of such a system, it would be naive to pretend that the DSP portions can be implemented without concern for the real-time nature of DSP or the embedded nature of the entire system.

The next several sections will highlight some of special design considerations that apply to embedded real-time systems. I will look first at real-time issues, then some specific embedded issues, and finally, at trends and issues that commonly apply to both real-time and embedded systems.

8.2   Real-Time Systems

A real-time system is a system that is required to react to stimuli from the environment (including the passage of physical time) within time intervals dictated by the environment. The *Oxford Dictionary* defines a real-time system as "any system in which the time at which output is produced is significant." This is usually because the input corresponds to some movement in the physical world, and the output has to relate to that same movement. The lag from input time to output time must be sufficiently small for acceptable timeliness. Another way of thinking of real-time systems is any information processing activity or system that has to respond to externally generated input stimuli within a finite and specified period. Generally, real-time systems are systems that maintain a *continuous timely* interaction with their environment (Figure 8.2).

8.2.1   Types of Real-Time Systems—Soft and Hard

Correctness of a computation depends not only on its results but also on the time at which its outputs are generated. A real-time system must satisfy response time constraints or suffer significant system consequences. If the consequences consist of a degradation of performance, but not failure, the system is referred to as a soft real-time system. If the consequences are system failure, the system is referred to as a hard real-time system (for instance, antilock braking systems in an automobile).

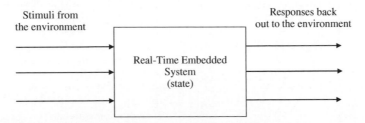

Figure 8.2: A Real-Time System Reacts to Inputs from the Environment and Produces Outputs that Affect the Environment

8.3   Hard Real-Time and Soft Real-Time Systems

8.3.1   Introduction

A system function (hardware, software, or a combination of both) is considered hard real-time if, and only if, it has a hard deadline for the completion of an action or task. This deadline must always be met, otherwise the task has failed. The system may have one or more hard real-time tasks as well as other nonreal-time tasks. This is acceptable, as long as the system can properly schedule these tasks in such a way that the hard real-time tasks always meet their deadlines. Hard real-time systems are commonly also embedded systems.

8.3.2   Differences between Real-Time and Time-Shared Systems

Real-time systems are different from time-shared systems in the three fundamental areas (Table 8.2). These include predictably fast response to urgent events:

High degree of schedulability—Timing requirements of the system must be satisfied at high degrees of resource usage.

Worst-case latency—Ensuring the system still operates under worst-case response time to events.

Stability under transient overload—When the system is overloaded by events and it is impossible to meet all deadlines, the deadlines of selected critical tasks must still be guaranteed.

Table 8.2: Real-Time Systems Are Fundamentally Different from Time-Shared Systems

Characteristic	Time-Shared Systems	Real-Time Systems
System capacity	High throughput	Schedulability and the ability of system tasks to meet all deadlines
Responsiveness	Fast average response time	Ensured worst-case latency, which is the worst-case response time to events
Overload	Fairness to all	Stability—When the system is overloaded, important tasks must meet deadlines while others may be starved

8.3.3   DSP Systems Are Hard Real-Time

Usually, DSP systems qualify as hard real-time systems. As an example, assume that an analog signal is to be processed digitally. The first question to consider is how often to *sample* or measure an analog signal in order to represent that signal accurately in the digital domain. The sample rate is the number of samples of an analog event (like sound) that are taken per second to represent the event in the digital domain. Based on a signal processing rule called the Nyquist rule, the signal must be sampled at a rate at least equal to twice the highest frequency that we wish to preserve. For example, if the signal contains important components at 4 kilohertz (kHZ), then the sampling frequency would need to be at least 8 KHz. The sampling period would then be:

$$T = 1/8000 = 125 \text{ microseconds} = 0.000125 \text{ seconds}$$

8.3.3.1   Based on Signal Sample, Time to Perform Actions Before Next Sample Arrives

This tells us that, for this signal being sampled at this rate, we would have 0.000125 seconds to perform all the processing necessary before the next sample arrives. Samples are arriving on a continuous basis, and the system cannot fall behind in processing these samples and still produce correct results—it is hard real-time.

8.3.3.2   Hard Real-Time Systems

The collective timeliness of the hard real-time tasks is binary—that is, either they will all always meet their deadlines (in a correctly functioning system), or they will not (the system is infeasible). In all hard real-time systems, collective timeliness is deterministic.

This determinism does not imply that the actual individual task completion times, or the task execution ordering, are necessarily known in advance.

A computing system being hard real-time says nothing about the magnitudes of the deadlines. They may be microseconds or weeks. There is a bit of confusion with regards to the usage of the term "hard real-time." Some relate hard real-time to response time magnitudes below some arbitrary threshold, such as 1 msec. This is not the case. Many of these systems actually happen to be soft real-time. These systems would be more accurately termed "real fast" or perhaps "real predictable." However, certainly not hard real-time.

The feasibility and costs (for example, in terms of system resources) of hard real-time computing depend on how well known a priori are the relevant future behavioral characteristics of the tasks and execution environment. These task characteristics include:

- timeliness parameters, such as arrival periods or upper bounds

- deadlines

- resource utilization profiles

- worst-case execution times

- precedence and exclusion constraints

- ready and suspension times

- relative importance, and so on

There are also pertinent characteristics relating to the execution environment:

- system loading

- service latencies

- resource interactions

- interrupt priorities and timing

- queuing disciplines

- caching

- arbitration mechanisms, and so on

Deterministic collective task timeliness in hard (and soft) real-time computing requires that the future characteristics of the relevant tasks and execution environment be deterministic—that is, known absolutely in advance. The knowledge of these characteristics must then be used to preallocate resources so all deadlines will always be met.

Usually, the task's and execution environment's future characteristics must be adjusted to enable a schedule and resource allocation that meets all deadlines. Different algorithms or schedules that meet all deadlines are evaluated with respect to other factors. In many real-time computing applications, it is common that the primary factor is maximizing processor utilization.

Allocation for hard real-time computing has been performed using various techniques. Some of these techniques involve conducting an offline enumerative search for a static schedule that will deterministically always meet all deadlines. Scheduling algorithms include the use of priorities that are assigned to the various system tasks. These priorities can be assigned either offline by application programmers, or online by the application or operating system software. The task priority assignments may either be static (fixed), as with rate monotonic algorithms[1] or dynamic (changeable), as with the earliest deadline first algorithm.[2]

8.3.4   Real-Time Event Characteristics—Real-Time Event Categories

Real-time events fall into one of three categories: asynchronous, synchronous, or isochronous.

Asynchronous events are entirely unpredictable. An example of this is a cell phone call arriving at a cellular base station. As far as the base station is concerned, the action of making a phone call cannot be predicted.

Synchronous events are predictable and occur with precise regularity. For example, the audio and video in a camcorder take place in synchronous fashion.

Isochronous events occur with regularity within a given window of time. For example, audio data in a networked multimedia application must appear within a window of time when the corresponding video stream arrives. Isochronous is a subclass of asynchronous.

---

[1] Rate monotonic analysis (RMA) is a collection of quantitative methods and algorithms that allow engineers to specify, understand, analyze, and predict the timing behavior of real-time software systems, thus improving their dependability and evolvability.

[2] A strategy for CPU or disk access scheduling. With EDF, the task with the earliest deadline is always executed first.

In many real-time systems, task and future execution environment characteristics are hard to predict. This makes true hard real-time scheduling infeasible. In hard real-time computing, deterministic satisfaction of the collective timeliness criterion is the driving requirement. The necessary approach to meeting that requirement is static (that is, a priori)[3] scheduling of deterministic task and execution environment characteristic cases. The requirement for advance knowledge about each of the system tasks and their future execution environment to enable offline scheduling and resource allocation significantly restricts the applicability of hard real-time computing.

8.4 Efficient Execution and the Execution Environment

8.4.1 Efficiency Overview

Real-time systems are time critical, and the efficiency of their implementation is more important than in other systems. Efficiency can be categorized in terms of processor cycles, memory or power. This constraint may drive everything from the choice of processor to the choice of the programming language. One of the main benefits of using a higher level language is to allow the programmer to abstract away implementation details and concentrate on solving the problem. This is not always true in the embedded system world. Some higher-level languages have instructions that are an order of magnitude slower than assembly language. However, higher-level languages can be used in real-time systems effectively, using the right techniques.

8.4.2 Resource Management

A system operates in real time as long as it completes its time-critical processes with acceptable timeliness. *Acceptable timeliness* is defined as part of the behavioral or "nonfunctional" requirements for the system. These requirements must be objectively quantifiable and measurable (stating that the system must be "fast," for example, is not quantifiable). A system is said to be real-time if it contains some model of real-time resource management (these resources must be explicitly managed for the purpose of operating in real time). As mentioned earlier, resource management may be performed statically, offline, or dynamically, online.

---

[3] Relating to or derived by reasoning from self-evident propositions (formed or conceived beforehand), as compared to a posteriori that is presupposed by experience (www.wikipedia.org).

Real-time resource management comes at a cost. The degree to which a system is required to operate in real time cannot necessarily be attained solely by hardware over-capacity (such as, high processor performance using a faster CPU). To be cost effective, there must exist some form of real-time resource management. Systems that must operate in real time consist of both real-time resource management and hardware resource capacity. Systems that have interactions with physical devices require higher degrees of real-time resource management. These computers are referred to as embedded systems, which we spoke about earlier. Many of these embedded computers use very little real-time resource management. The resource management that is used is usually static and requires analysis of the system prior to it executing in its environment. In a real-time system, physical time (as opposed to logical time) is necessary for real-time resource management in order to relate events to the precise moments of occurrence. Physical time is also important for action time constraints as well as measuring costs incurred as processes progress to completion. Physical time can also be used for logging history data.

All real-time systems make trade-offs of scheduling costs versus performance in order to reach an appropriate balance for attaining acceptable timeliness between the real-time portion of the scheduling optimization rules and the offline scheduling performance evaluation and analysis.

Types of Real-Time Systems—Reactive and Embedded

There are two types of real-time systems: reactive and embedded. A reactive real-time system has constant interaction with its environment (such as a pilot controlling an aircraft). An embedded real-time system is used to control specialized hardware that is installed within a larger system (such as a microprocessor that controls anti-lock brakes in an automobile).

8.5   Challenges in Real-Time System Design

Designing real-time systems poses significant challenges to the designer. One of these challenges comes from the fact that real-time systems must interact with the environment. The environment is complex and changing and these interactions can become very complex. Many real-time systems don't just interact with one, but many different entities in the environment, with different characteristics and rates of interaction. A cell phone base station, for example, must be able to handle calls from literally thousands of cell phone subscribers at the same time. Each call may have different requirements for processing and be in different sequences of processing. All of this complexity must be managed and coordinated.

8.5.1   Response Time

Real-time systems must respond to external interactions in the environment within a predetermined amount of time. Real-time systems must produce the correct result and produce it in a timely way. This implies that response time is as important as producing correct results. Real-time systems must be engineered to meet these response times. Hardware and software must be designed to support response time requirements for these systems. Optimal partitioning of the system requirements into hardware and software is also important.

Real-time systems must be architected to meet system response time requirements. Using combinations of hardware and software components, engineering makes architecture decisions such as interconnectivity of the system processors, system link speeds, processor speeds, memory size, I/O bandwidth, and so on. Key questions to be answered include:

Is the architecture suitable?—To meet the system response time requirements, the system can be architected using one powerful processor or several smaller processors. Can the application be partitioned among the several smaller processors without imposing large communication bottlenecks throughout the system? If the designer decides to use one powerful processor, will the system meet its power requirements? Sometimes a simpler architecture may be the better approach—more complexity can lead to unnecessary bottlenecks that cause response time issues.

Are the processing elements powerful enough?—A processing element with high utilization (greater than 90%) will lead to unpredictable run time behavior. At this utilization level, lower priority tasks in the system may get starved. As a general rule, real-time systems that are loaded at 90% take approximately twice as long to develop, due to the cycles of optimization and integration issues with the system at these utilization rates. At 95% utilization, systems can take three times longer to develop, due to these same issues. Using multiple processors will help, but the interprocessor communication must be managed.

Are the communication speeds adequate?—Communication and I/O are a common bottleneck in real-time embedded systems. Many response time problems come not from the processor being overloaded but in latencies in getting data into and out of the system. On other cases, overloading a communication port (greater than 75%) can cause unnecessary queuing in different system nodes and this causes delays in messages passing throughout the rest of the system.

Is the right scheduling system available?—In real-time systems, tasks that are processing real-time events must take higher priority. But, how do you schedule multiple tasks that are all processing real-time events? There are several scheduling approaches available, and the engineer must design the scheduling algorithm to accommodate the system priorities in order to meet all real-time deadlines. Because external events may occur at any time, the scheduling system must be able to preempt currently running tasks to allow higher priority tasks to run. The scheduling system (or real-time operating system) must not introduce a significant amount of overhead into the real-time system.

8.5.2 Recovering from Failures

Real-time systems interact with the environment, which is inherently unreliable. Therefore, real-time systems must be able to detect and overcome failures in the environment. Also, since real-time systems are often embedded into other systems and may be hard to get at (such as a spacecraft or satellite) these systems must also be able to detect and overcome internal failures (there is no "reset" button in easy reach of the user!). In addition, since events in the environment are unpredictable, it's almost impossible to test for every possible combination and sequence of events in the environment. This is a characteristic of real-time software that makes it somewhat nondeterministic in the sense that it is almost impossible in some real-time systems to predict the multiple paths of execution based on the nondeterministic behavior of the environment. Examples of internal and external failures that must be detected and managed by real-time systems include:

- Processor failures

- Board failures

- Link failures

- Invalid behavior of external environment

- Interconnectivity failure

8.5.3 Distributed and Multiprocessor Architectures

Real-time systems are becoming so complex that applications are often executed on multiprocessor systems distributed across some communication system. This poses challenges to the designer that relate to the partitioning of the application in a multiprocessor system. These systems will involve processing on several different nodes. One node may be a DSP, another node a more general-purpose processor, some specialized hardware processing elements, and so forth. This leads to several design challenges for the engineering team:

Initialization of the system—Initializing a multiprocessor system can be very complicated. In most multiprocessor systems, the software load file resides on the general-purpose processing node. Nodes that are directly connected to the general-purpose processor, for example, a DSP, will initialize first. After these nodes complete loading and initialization, other nodes connected to them may then go through this same process until the system completes initialization.

Processor interfaces—When multiple processors must communicate with each other, care must be taken to ensure that messages sent along interfaces between the processors are well defined and consistent with the processing elements. Differences in message protocol, including endianness, byte ordering, and other padding rules, can complicate system integration, especially if there is a system requirement for backwards compatibility.

Load distribution—As mentioned earlier, multiple processors lead to the challenge of distributing the application, and possibly developing the application to support efficient partitioning of the application among the processing elements. Mistakes in partitioning the application can lead to bottlenecks in the system and this degrades the full capability of the system by overloading certain processing elements and leaving others under utilized. Application developers must design the application to be partitioned efficiently across the processing elements.

Centralized Resource Allocation and Management—In systems of multiple processing elements, there is still a common set of resources including peripherals, cross bar switches, memory, and so on that must be managed. In some cases the operating system can provide mechanisms like semaphores to manage these shared resources. In other cases there may be dedicated hardware to manage the resources. Either way, important shared resources in the system must be managed in order to prevent more system bottlenecks.

8.5.4   *Embedded Systems*

An embedded system is a specialized computer system that is usually integrated as part of a larger system. An embedded system consists of a combination of hardware and software components to form a computational engine that will perform a specific function. Unlike desktop systems that are designed to perform a general function, embedded systems are constrained in their application. Embedded systems often perform in reactive and time-constrained environments as described earlier. A rough partitioning of an embedded system consists of the hardware that provides the performance necessary for the application (and other system properties, like security) and the software, which provides a majority of the features and flexibility in the system. A typical embedded system is shown in Figure 8.3.

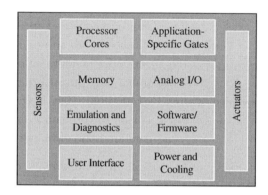

Figure 8.3: Typical Embedded System Components

- **Processor core**—At the heart of the embedded system is the processor core(s). This can be a simple inexpensive 8 bit microcontroller to a more complex 32 or 64 bit microprocessor. The embedded designer must select the most cost sensitive device for the application that can meet all of the functional and nonfunctional (timing) requirements.

- **Analog I/O**—D/A and A/D converters are used to get data from the environment and back out to the environment. The embedded designer must understand the type of data required from the environment, the accuracy requirements for that data, and the input/output data rates in order to select the right converters for the application. The external environment drives the reactive nature of the embedded system. Embedded systems have to be at least fast enough to keep up with the environment. This is where the analog information such as light or sound pressure or acceleration are sensed and input into the embedded system (see Figure 8.4).

- **Sensors and Actuators**—Sensors are used to sense analog information from the environment. Actuators are used to control the environment in some way.

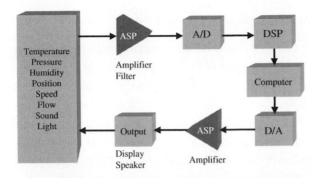

Figure 8.4: Analog Information of Various Types Is Processed by Embedded System

- **Embedded systems** also have user interfaces. These interfaces may be as simple as a flashing LED to a sophisticated cell phone or digital still camera interface.

- **Application-specific gates**—Hardware acceleration like ASICs or FPGA are used for accelerating specific functions in the application that have high performance requirements. The embedded designer must be able to map or partition the application appropriately using available accelerators to gain maximum application performance.

- **Software** is a significant part of embedded system development. Over the last several years, the amount of embedded software has grown faster than Moore's law, with the amount doubling approximately every 10 months. Embedded software is usually optimized in some way (performance, memory, or power). More and more embedded software is written in a high level language like C/C++ with some of the more performance critical pieces of code still written in assembly language.

- **Memory** is an important part of an embedded system and embedded applications can either run out of RAM or ROM depending on the application. There are many types of volatile and nonvolatile memory used for embedded systems and we will talk more about this later.

- **Emulation and diagnostics**—Many embedded systems are hard to see or get to. There needs to be a way to interface to embedded systems to debug them. Diagnostic ports such as a JTAG (joint test action group) port are used to debug embedded systems. On-chip emulation is used to provide visibility into the behavior of the application. These emulation modules provide sophisticated visibility into the runtime behavior and performance, in effect replacing external logic analyzer functions with onboard diagnostic capabilities.

8.5.4.1   *Embedded Systems Are Reactive Systems*

A typical embedded system responds to the environment via sensors and controls the environment using actuators (Figure 8.5). This imposes a requirement on embedded systems to achieve performance consistent with that of the environment. This is why embedded systems are referred to as reactive systems. A reactive system must use a combination of hardware and software to respond to events in the environment, within defined constraints. Complicating the matter is the fact that these external events can be periodic and predictable or aperiodic and hard to predict. When scheduling events for processing in an embedded system, both periodic and aperiodic events must be considered and performance must be guaranteed for worst-case rates of execution. This can be a significant challenge. Consider the example in Figure 8.6. This is a model of an automobile airbag deployment system showing sensors including crash severity and occupant detection. These sensors monitor the environment and could signal the embedded system at any time. The embedded control unit (ECU) contains accelerometers to detect crash impacts. In addition, rollover sensors, buckle sensors and weight sensors (Figure 8.8) are used to determine how and when to deploy airbags. Figure 8.7 shows the actuators in this same system. These include Thorax bags actuators, pyrotechnic buckle pretensioner with load limiters, and the central airbag control unit. When an impact occurs, the sensors must detect and send a signal to the ECU, which must deploy the appropriate airbags within a hard real-time deadline for this system to work properly.

The previous example demonstrates several key characteristics of embedded systems:

- **Monitoring and reacting to the environment**—Embedded systems typically get input by reading data from input sensors. There are many different types of sensors that monitor various analog signals in the environment, including temperature, sound pressure, and vibration. This data is processed using embedded system algorithms. The results may be displayed in some format to a user or simply used to control actuators (like deploying the airbags and calling the police).

- **Control the environment**—Embedded systems may generate and transmit commands that control actuators, such as airbags, motors, and so on.

- **Processing of information**—Embedded systems process the data collected from the sensors in some meaningful way, such as data compression/decompression, side impact detection, and so on.

- **Application-specific**—Embedded systems are often designed for applications, such as airbag deployment, digital still cameras, or cell phones. Embedded systems may also be designed for processing control laws, finite state machines, and signal processing algorithms. Embedded systems must also be able to detect and react appropriately to faults in both the internal computing environment as well as the surrounding systems.

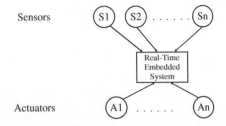

Figure 8.5: A Model of Sensors and Actuators in Embedded Systems

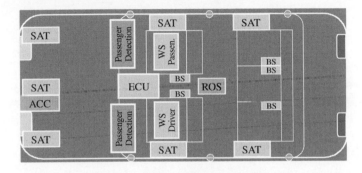

Figure 8.6: Airbag System: Possible Sensors (Including Crash Severity and Occupant Detection)
Source: Courtesy of Texas Instruments

SAT = satellite with serial communication interface

ECU = central airbag control unit (including accelerometers)

ROS = roll over sensing unit

WS = weight sensor

BS = buckle switch

TB = thorax bag

PBP = pyrotechnic buckle pretensioner with load limiter

ECU = central airbag control unit

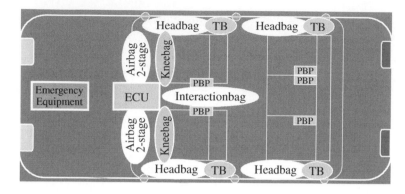

Figure 8.7: Airbag System: Possible Sensors (Including Crash Severity and Occupant Detection)
Source: Courtesy of Texas Instruments

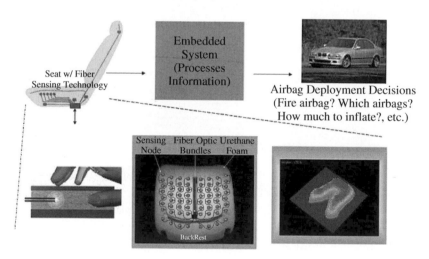

Figure 8.8: Automotive Seat Occupancy Detection
Source: Courtesy of Texas Instruments

Figure 8.9 shows a block diagram of a digital still camera (DSC). A DSC is an example of an embedded system. Referring back to the major components of an embedded system shown in Figure 8.3, we can see the following components in the DSC:

- The charge-coupled device analog front-end (CCD AFE) acts as the primary sensor in this system.

- The digital signal processor is the primary processor in this system.

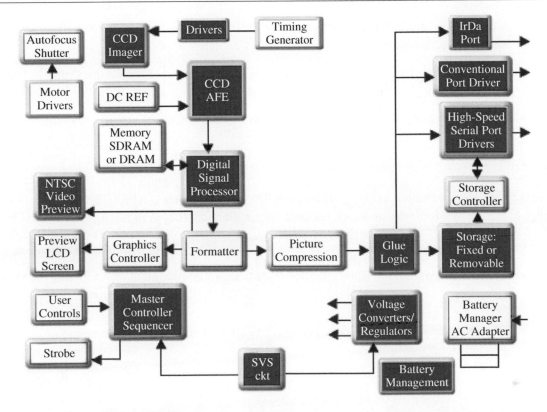

Figure 8.9: Analog Information of Various Types Is Processed by Embedded System

- The battery management module controls the power for this system.

- The preview LCD screen is the user interface for this system.

- The Infrared port and serial ports are actuators in this system that interface to a computer.

- The graphics controller and picture compression modules are dedicated application-specific gates for processing acceleration.

- The signal processing software runs on the DSP.

- The antenna is one of the sensors in this system. The microphone is another sensor. The keyboard also provides aperiodic events into the system.

- The voice codec is an application-specific acceleration in hardware gates.

- The DSP is one of the primary processor cores that runs most of the signal processing algorithms.

- The ARM processor is the other primary system processor running the state machines, controlling the user interface, and other components in this system.

- The battery/temp monitor controls the power in the system along with the supply voltage supervisor.

- The display is the primary user interface in the system.

Figure 8.10 shows another example of an embedded system. This is a block diagram of a cell phone. In this diagram, the major components of an embedded system are again obvious:

- The antenna is one of the sensors in this system. The microphone is another sensor.

- The keyboard also provides aperiodic events into the system.

- The voice codec is an application-specific acceleration in hardware gates.

- The DSP is one of the primary processor cores that runs most of the signal processing algorithms.

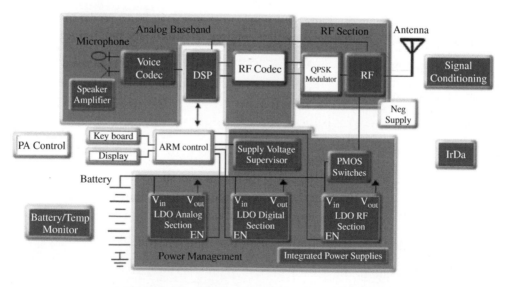

Figure 8.10: Block Diagram of a Cell Phone
Source: Courtesy of Texas Instrument

- The ARM processor is the other primary system processor running the state machines, controlling the user interface, and other components in this system.

- The battery/temp monitor controls the power in the system along with the supply voltage supervisor.

- The display is the primary user interface in the system.

8.6   Summary

Many of the items that we interface with or use on a daily basis contain an embedded system. An embedded system is a system that is "hidden" inside the item we interface with. Systems such as cell phones, answering machines, microwave ovens, VCRs, DVD players, video game consoles, digital cameras, music synthesizers, and cars all contain embedded processors. A late model car contains more than 60 embedded microprocessors. These embedded processors keep us safe and comfortable by controlling such tasks as antilock braking, climate control, engine control, audio system control, and airbag deployment.

Embedded systems have the added burden of reacting quickly and efficiently to the external "analog" environment. That may include responding to the push of a button, a sensor to trigger an air bag during a collision, or the arrival of a phone call on a cell phone. Simply put, embedded systems have deadlines that can be hard or soft. Given the "hidden" nature of embedded systems, they must also react to and handle unusual conditions without the intervention of a human.

DSPs are useful in embedded systems principally for one reason: signal processing. The ability to perform complex signal processing functions in real time gives DSP the advantage over other forms of embedded processing. DSPs must respond in real time to analog signals from the environment, convert them to digital form, perform value added processing to those digital signals, and, if required, convert the processed signals back to analog form to send back out to the environment.

Programming embedded systems requires an entirely different approach from that used in desktop or mainframe programming. Embedded systems must be able to respond to external events in a very predictable and reliable way. Real-time programs must not only execute correctly, they must execute on time. A late answer is a wrong answer. Because of this requirement, we will be looking at issues such as concurrency, mutual exclusion, interrupts, hardware control, and processing. Multitasking, for example, has proven to be a powerful paradigm for building reliable and understandable real-time programs.

8.7 Overview of Embedded Systems Development Life Cycle Using DSP

As mentioned earlier, an embedded system is a specialized computer system that is integrated as part of a larger system. Many embedded systems are implemented using digital signal processors. The DSP will interface with the other embedded components to perform a specific function. The specific embedded application will determine the specific DSP to be used. For example, if the embedded application is one that performs video processing, the system designer may choose a DSP that is customized to perform media processing, including video and audio processing. An example of an application specific DSP for this function is shown in Figure 8.11. This device contains dual channel video ports that are software configurable for input or output, as well as video filtering and automatic horizontal scaling and support of various digital TV formats such as HDTV, multichannel audio serial ports, multiple stereo lines, and an Ethernet peripheral to connect to IP packet networks. It is obvious that the choice of a DSP "system" depends on the embedded application.

In this chapter, we will discuss the basic steps to develop an embedded application using DSP.

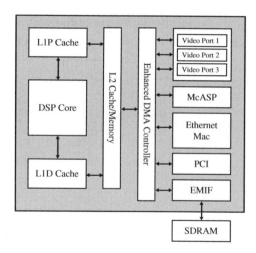

Figure 8.11: Example of a DSP-based "System" for Embedded Video Applications

8.8 The Embedded System Life Cycle Using DSP

In this section we will overview the general embedded system life cycle using DSP. There are many steps involved in developing an embedded system—some are similar to other system development activities and some are unique. We will step through the basic process of embedded system development, focusing on DSP applications.

8.8.1   Step 1—Examine the Overall Needs of the System

Choosing a design solution is a difficult process. Often the choice comes down to emotion or attachment to a particular vendor or processor, inertia based on prior projects and comfort level. The embedded designer must take a positive logical approach to comparing solutions based on well defined selection criteria. For DSP, specific selection criteria must be discussed. Many signal processing applications will require a mix of several system components as shown in Figure 8.12.

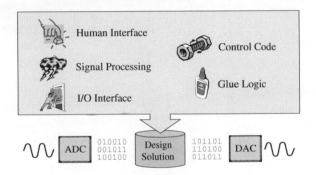

Figure 8.12: Most Signal Processing Applications Will Require a Mix of Various System Components
Source: Courtesy of Texas Instruments

8.8.1.1   What Is a DSP Solution?

A typical DSP product design uses the digital signal processor itself, analog/mixed signal functions, memory, and software, all designed with a deep understanding of overall system function. In the product, the analog signals of the real world, signals representing anything from temperature to sound and images, are translated into digital bits—zeros and ones—by an analog/mixed signal device. Then the digital bits or signals are processed by the DSP. Digital signal processing is much faster and more precise than traditional analog processing. This type of processing speed is needed for today's advanced communications devices where information requires instantaneous processing, and in many portable applications that are connected to the Internet.

There are many selection criteria for embedded DSP systems. Some of these are shown in Figure 8.13. These are the major selection criteria defined by Berkeley Design Technology Incorporated (bdti.com). Other selection criteria may be "ease of use," which is closely linked to "time-to-market" and also "features." Some of the basic rules to consider in this phase are:

- For a fixed cost, maximize performance.

- For a fixed performance, minimize cost.

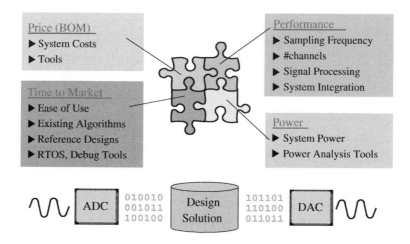

Figure 8.13: The Design Solution Will Be Influenced by These Major Criteria and Others
Source: Courtesy of Texas Instruments

8.8.2   Step 2—Select the Hardware Components Required for the System

In many systems, a general-purpose processor (GPP), field-programmable gate array (FPGA), microcontroller (mC) or DSP is not used as a single-point solution. This is because designers often combine solutions, maximizing the strengths of each device (Figure 8.14).

One of the first decisions that designers often make when choosing a processor is whether they would like a software-programmable processor in which functional blocks are developed in software using C or assembly, or a hardware processor in which functional blocks are laid out logically in gates. Both FPGAs and application specific integrated circuits (ASICs) may integrate a processor core (very common in ASICs).

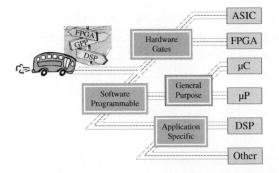

Figure 8.14: Many Applications, Multiple Solutions
Source: *Courtesy of Texas Instruments*

8.8.3   Hardware Gates

Hardware gates are logical blocks laid out in a flow, therefore any degree of parallelization of instructions is theoretically possible. Logical blocks have very low latency, therefore FPGAs are more efficient for building peripherals than "bit-banging" using a software device.

If a designer chooses to design in hardware, he or she may design using either an FPGA or ASIC. FPGAs are termed "field programmable" because their logical architecture is stored in a nonvolatile memory and booted into the device. Thus, FPGAs may be reprogrammed in the field simply by modifying the nonvolatile memory (usually FLASH or EEPROM). ASICs are not field-programmable. They are programmed at the factory using a mask that cannot be changed. ASICs are often less expensive and/or lower power. They often have sizable nonrecurring engineering (NRE) costs.

8.8.4   Software-Programmable

In this model, instructions are executed from memory in a serial fashion (that is, one per cycle). Software-programmable solutions have limited parallelization of instructions; however, some devices can execute multiple instructions in parallel in a single cycle. Because instructions are executed from memory in the CPU, device functions can be changed without having to reset the device. Also, because instructions are executed from memory, many different functions or routines may be integrated into a program without the need to lay out each individual routine in gates. This may make a software-programmable device more cost efficient for implementing very complex programs with a large number of subroutines.

If a designer chooses to design in software, there are many types of processors available to choose from. There are a number of general-purpose processors, but in addition, there are processors that have been optimized for specific applications. Examples of such application specific processors are graphics processors, network processors, and digital signal processors (DSPs). Application specific processors usually offer higher performance for a target application, but are less flexible than general-purpose processors.

8.8.5 General-Purpose Processors

Within the category of general-purpose processors are microcontrollers (mC) and microprocessors (mP) (Figure 8.15).

Microcontrollers usually have control-oriented peripherals. They are usually lower cost and lower performance than microprocessors. Microprocessors usually have communications-oriented peripherals. They are usually higher cost and higher performance than microcontrollers.

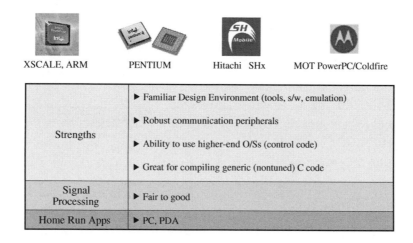

XSCALE, ARM	PENTIUM
Hitachi SHx	MOT PowerPC/Coldfire

Strengths	▶ Familiar Design Environment (tools, s/w, emulation)
	▶ Robust communication peripherals
	▶ Ability to use higher-end O/Ss (control code)
	▶ Great for compiling generic (nontuned) C code
Signal Processing	▶ Fair to good
Home Run Apps	▶ PC, PDA

Figure 8.15: General-Purpose Processor Solutions
Source: Courtesy of Texas Instruments

Note that some GPPs have integrated MAC units. It is not a "strength" of GPPs to have this capability because all DSPs have MACs—but it is worth noting because a student might mention it. Regarding performance of the GPP's MAC, it is different for each one.

8.8.6 Microcontrollers

A microcontroller is a highly integrated chip that contains many or all of the components comprising a controller. This includes a CPU, RAM and ROM, I/O ports, and timers. Many general-purpose computers are designed the same way. But a microcontroller is usually designed for very specific tasks in embedded systems. As the name implies, the specific task is to control a particular system, hence the name microcontroller. Because of this customized task, the device's parts can be simplified, which makes these devices very cost effective solutions for these types of applications.

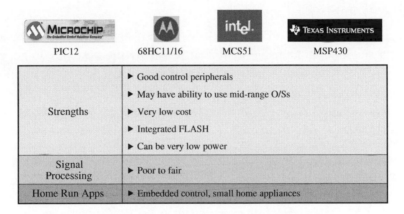

Figure 8.16: Microcontroller Solutions
Source: Courtesy of Texas Instruments

Some microcontrollers can actually do a multiply and accumulate (MAC) in a single cycle. But that does not necessarily make it a DSP. True DSPs can allow two 16x16 MACS in a single cycle including bringing the data in over the buses, and so on. It is this that truly makes the part a DSP. So, devices with hardware MACs might get a "fair" rating. Others get a "poor" rating. In general, microcontrollers can do DSP but they will generally do it slower.

8.8.7 FPGA Solutions

An FPGA is an array of logic gates that are hardware-programmed to perform a user-specified task. FPGAs are arrays of programmable logic cells interconnected by a matrix of wires and programmable switches. Each cell in an FPGA performs a simple logic function. These logic functions are defined by an engineer's program. FPGA contain large numbers of these cells (1000–100,000) available to use as building blocks in DSP applications. The advantage of using FPGAs is that the engineer can create special purpose functional units that can perform limited tasks very efficiently. FPGAs can be reconfigured dynamically as well (usually 100–1000 times per second depending on the device). This makes it possible to optimize FPGAs for complex tasks at speeds higher than what can be achieved using a general-purpose processor. The ability to manipulate logic at the gate level means it is possible to construct custom DSP-centric processors that efficiently implement the desired DSP function. This is possible by simultaneously performing all of the algorithm's subfunctions. This is where the FPGA can achieve performance gains over a programmable DSP processor.

The DSP designer must understand the trade-offs when using an FPGA. If the application can be done in a single programmable DSP, that is usually the best way to go since talent for programming DSPs is usually easier to find than FPGA designers. In addition, software design tools are common, cheap, and sophisticated, which improves development time and cost. Most of the common DSP algorithms are also available in well packaged software components. It's harder to find these same algorithms implemented and available for FPGA designs.

An FPGA is worth considering, however, if the desired performance cannot be achieved using one or two DSPs, or when there may be significant power concerns (although a DSP is also a power efficient device—benchmarking needs to be performed) or when there may be significant programmatic issues when developing and integrating a complex software system.

Typical applications for FPGAs include radar/sensor arrays, physical system and noise modeling, and any really high I/O and high-bandwidth application.

8.8.8   Digital Signal Processors

A DSP is a specialized microprocessor used to perform calculations efficiently on digitized signals that are converted from the analog domain. One of the big advantages of DSP is the programmability of the processor, which allows important system parameters to be changed easily to accommodate the application. DSPs are optimized for digital signal manipulations.

DSPs provide ultra-fast instruction sequences, such as shift and add, and multiply and add. These instruction sequences are common in many math-intensive signal processing applications. DSPs are used in devices where this type of signal processing is important, such as sound cards, modems, cell phones, high-capacity hard disks, and digital TVs (Figure 8.17).

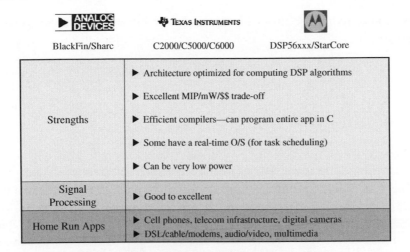

Figure 8.17: DSP Processor Solutions
Source: Courtesy of Texas Instruments

8.8.9 A General Signal Processing Solution

The solution shown in Figure 8.18 allows each device to perform the tasks it's best at, achieving a more efficient system in terms of cost/power/performance. For example, in Figure 8.18, the system designer may put the system control software (state machines and other communication software) on the general-purpose processor or microcontroller, the high performance, single dedicated fixed functions on the FPGA, and the high I/O signal processing functions on the DSP.

When planning the embedded product development cycle, there are multiple opportunities to reduce cost and/or increase functionality using combinations of GPP/uC, FPGA, and DSP. This becomes more of an issue in higher-end DSP applications. These are applications that are computationally intensive and performance critical. These applications require more processing power and channel density than can be provided by GPPs alone. For these high-end applications, there are software/hardware alternatives that the system designer must consider. Each alternative provides different degrees of performance benefits and must also be weighed against other important system parameters including cost, power consumption, and time-to-market.

The system designer may decide to use an FPGA in a DSP system for the following reasons:

- A decision to extend the life of a generic, lower-cost microprocessor or DSP by offloading computationally intensive work to a FPGA.

- A decision to reduce or eliminate the need for a higher-cost, higher performance DSP processor.

- To increase computational throughput. If the throughput of an existing system must increase to handle higher resolutions or larger signal bandwidths, an FPGA may be an option. If the required performance increases are computational in nature, an FPGA may be an option.

- For prototyping new signal processing algorithms; since the computational core of many DSP algorithms can be defined using a small amount of C code, the system designer can quickly prototype new algorithmic approaches on FPGAs before committing to hardware or other production solutions, like an ASIC.

- For implementing "glue" logic; various processor peripherals and other random or "glue" logic are often consolidated into a single FPGA. This can lead to reduced system size, complexity, and cost.

By combining the capabilities of FPGAs and DSP processors, the system designer can increase the scope of the system design solution. Combinations of fixed hardware and programmable processors are a good model for enabling flexibility, programmability, and computational acceleration of hardware for the system.

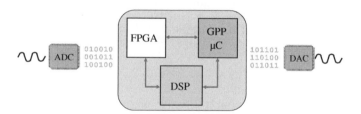

Figure 8.18: General Signal Processing Solution
Source: Courtesy of Texas Instruments

8.8.10   DSP Acceleration Decisions

In DSP system design, there are several things to consider when determining whether a functional component should be implemented in hardware or software:

Signal processing algorithm parallelism—Modern processor architectures have various forms of instruction level parallelism (ILP). One example is the 64x DSP that has a very long instruction word (VLIW) architecture. The 64x DSP exploits ILP by grouping multiple instructions (adds, multiplies, loads, and stores) for execution in a single processor cycle. For DSP algorithms that map well to this type of instruction parallelism, significant performance gains can be realized. But not all signal processing algorithms exploit such forms of parallelism. Filtering algorithms such as finite impulse response (FIR) algorithms are recursive and are suboptimal when mapped to programmable DSPs. Data recursion prevents effective parallelism and ILP. As an alternative, the system designer can build dedicated hardware engines in an FPGA.

Computational complexity—Depending on the computational complexity of the algorithms, these may run more efficiently on a FPGA instead of a DSP. It may make sense, for certain algorithmic functions, to implement in a FPGA and free up programmable DSP cycles for other algorithms. Some FPGAs have multiple clock domains built into the fabric, which can be used to separate different signal processing hardware blocks into separate clock speeds based on their computational requirements. FPGAs can also provide flexibility by exploiting data and algorithm parallelism using multiple instantiations of hardware engines in the device.

Data locality—The ability to access memory in a particular order and granularity is important. Data access takes time (clock cycles) due to architectural latency, bus contention, data alignment, direct memory access (DMA) transfer rates, and even the type of memory being used in the system. For example, static RAM (SRAM), which is very fast but much more expensive than dynamic RAM (DRAM), is often used as cache memory due to its speed. Synchronous DRAM (SDRAM), on the other hand, is directly dependent on the clock speed of the entire system (that's why they call it synchronous). It basically works at the same speed as the system bus. The overall performance of the system is driven in part by which type of memory is being used. The physical interfaces between the data unit and the arithmetic unit are the primary drivers of the data locality issue.

Data parallelism—Many signal processing algorithms operate on data that is highly capable of parallelism, such as many common filtering algorithms. Some of the more advanced high-performance DSPs have single instruction multiple data (SIMD) capability in the architectures and/or compilers that implement various forms of vector processing operations. FPGA devices are also good at this type of parallelism. Large amounts of RAM are used to support high bandwidth requirements. Depending on the DSP processor being used, an FPGA can be used to provide this SIMD processing capability for certain algorithms that have these characteristics.

A DSP-based embedded system could incorporate one, two, or all three of these devices depending on various factors:

▶ # signal processing tasks/channels	▶ Amount of control code
▶ Sampling rate	▶ Development environment
▶ Memory/peripherals needed	▶ Operating system (O/S or RTOS)
▶ Power requirements	▶ Debug capabilities
▶ Availability of desired algorithms	▶ Form factor, system cost

The trend in embedded DSP development is moving more towards programmable solutions as shown in Figure 8.19. There will always be a trade-off depending on the application but the trend is moving towards software and programmable solutions.

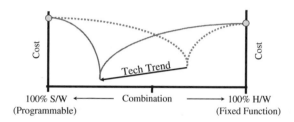

Figure 8.19: Hardware/Software Mix in an Embedded System; the Trend Is Towards More Software
Source: Courtesy of Texas Instruments

"Cost" can mean different things to different people. Sometimes, the solution is to go with the lowest "device cost." However, if the development team then spends large amounts of time redoing work, the project may be delayed; the "time-to-market" window may extend, which, in the long run, costs more than the savings of the low-cost device.

The first point to make is that a 100% software or hardware solution is usually the most expensive option. A combination of the two is the best. In the past, more functions were done in hardware and less in software. Hardware was faster, cheaper (ASICs), and good C compilers for embedded processors just weren't available. However, today, with better compilers, faster and lower-cost processors available, the trend is toward more of a software-programmable solution. A software-only solution is not (and most likely never will be) the best overall cost. Some hardware will still be required. For example, let's say you have ten functions to perform and two of them require extreme speed. Do you purchase a very fast processor (which costs 3–4 times the speed you need for the other eight functions) or do you spend 1x on a lower-speed processor and purchase an ASIC or FPGA to do only those two critical functions? It's probably best to choose the combination.

Cost can be defined by a combination of the following: A combination of software and hardware always gives the lowest cost system design.

8.8.11   Step 3—Understand DSP Basics and Architecture

One compelling reason to choose a DSP processor for an embedded system application is performance. Three important questions to understand when deciding on a DSP are:

- What makes a DSP a DSP?

- How fast can it go?

- How can I achieve maximum performance without writing in assembly?

In this section, we will begin to answer these questions. We know that a DSP is really just an application specific microprocessor. They are designed to do a certain thing, signal processing, very efficiently. We mentioned the types of signal processing algorithms that are used in DSP. They are shown again in Figure 8.20 for reference.

Algorithm	Equation
Finite Impulse Response Filter	$y(n) = \sum\limits_{k=0}^{M} a_k x(n-k)$
Infinite Impulse Response Filter	$y(n) = \sum\limits_{k=0}^{M} a_k x(n-k) + \sum\limits_{k=1}^{N} b_k y(n-k)$
Convolution	$y(n) = \sum\limits_{k=0}^{N} x(k) h(n-k)$
Discrete Fourier Transform	$X(k) = \sum\limits_{n=0}^{N-1} x(n) \exp[-j(2\pi/N)nk]$
Discrete Cosine Transform	$F(u) = \sum\limits_{x=0}^{N-1} c(u) \cdot f(x) \cdot \cos\left[\dfrac{\pi}{2N} u(2x+1)\right]$

Figure 8.20: Typical DSP Algorithms
Source: Courtesy of Texas Instruments

Notice the common structure of each of the algorithms in Figure 8.20:

- They all accumulate a number of computations.

- They all sum over a number of elements.

- They all perform a series of multiplies and adds.

These algorithms all share some common characteristics; they perform multiplies and adds over and over again. This is generally referred to as the sum of products (SOP).

DSP designers have developed hardware architectures that allow the efficient execution of algorithms to take advantage of this algorithmic specialty in signal processing. For example, some of the specific architectural features of DSPs accommodate the algorithmic structure described in Figure 8.20.

As an example, consider the FIR diagram in Figure 8.21 as an example DSP algorithm that clearly shows the multiply/accumulate and shows the need for doing MACs very fast, along with reading at least two data values. As shown in Figure 8.21, the filter algorithm can be implemented using a few lines of C source code. The signal flow diagram shows this

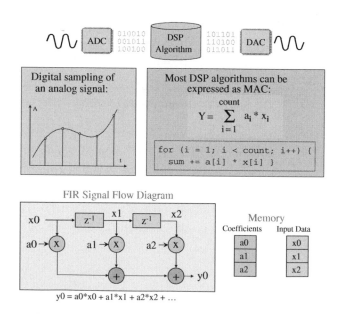

Figure 8.21: DSP Filtering Using a FIR Filter
Source: Courtesy of Texas Instruments

algorithm in a more visual context. Signal flow diagrams are used to show overall logic flow, signal dependencies, and code structure. They make a nice addition to code documentation.

To execute at top speed, a DSP needs to:

- read *at least* two values from memory (minimum),

- multiply coeff * data,

- accumulate (+) answer (an * xn) to running total ...,

- ... and do all of the above in a single cycle (or less).

DSP architectures support the requirements above (Figure 8.22):

- High-speed memory architectures support multiple accesses/cycle.

- Multiple read buses allow two (or more) data reads/cycle from memory.

- The processor pipeline overlays CPU operations allowing one-cycle execution.

All of these things work together to result in the highest possible performance when executing DSP algorithms.

Other DSP architectural features are summarized in Figure 8.23.

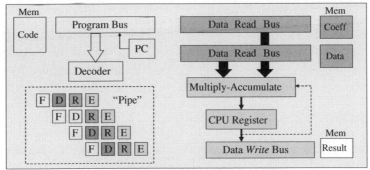

Multiply-accumulate: MAC

◆ Hi-speed memory architecture supports multiple accesses/cycle
◆ Multiple read buses allow two (or more) data reads/cycle from memory
◆ Pipeline overlays CPU operations allowing one-cycle execution

Figure 8.22: Architectural Block Diagram of a DSP
Source: *Courtesy of Texas Instruments*

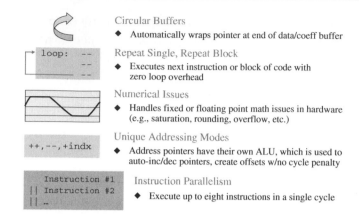

Circular Buffers
◆ Automatically wraps pointer at end of data/coeff buffer

Repeat Single, Repeat Block
◆ Executes next instruction or block of code with zero loop overhead

Numerical Issues
◆ Handles fixed or floating point math issues in hardware (e.g., saturation, rounding, overflow, etc.)

Unique Addressing Modes
◆ Address pointers have their own ALU, which is used to auto-inc/dec pointers, create offsets w/no cycle penalty

Instruction Parallelism
◆ Execute up to eight instructions in a single cycle

Figure 8.23: DSP CPU Architectural Highlights
Source: Courtesy of Texas Instruments

8.8.12   Models of DSP Processing

There are two types of DSP processing models—single sample model and block processing model. In a single sample model of signal processing (Figure 8.24a), the output must result before next input sample. The goal is minimum latency (in-to-out time). These systems tend to be interrupt intensive; interrupts drive the processing for the next sample. Example DSP applications include motor control and noise cancellation.

In the block processing model (Figure 8.24b), the system will output a buffer of results before the next input buffer fills. DSP systems like this use the DMA to transfer samples to the buffer. There is increased latency in this approach as the buffers are filled before

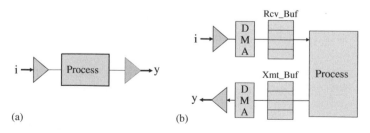

Figure 8.24: Single Sample (a) and Block Processing (b) Models of DSP

processing. However, these systems tend to be computationally efficient. The main types of DSP applications that use block processing include cellular telephony, video, and telecom infrastructure.

An example of stream processing is averaging data sample. A DSP system that must average the last three digital samples of a signal together and output a signal at the same rate as what is being sampled must do the following:

- Input a new sample and store it.

- Average the new sample with the last two samples.

- Output the result.

These three steps must complete before the next sample is taken. This is an example of stream processing. The signal must be processed in real time. A system that is sampling at 1000 samples per second has one thousandth of a second to complete the operation in order to maintain real-time performance. Block processing, on the other hand, accumulates a large number of samples at a time and processes those samples while the next buffer of samples is being collected. Algorithms such as the fast Fourier transform (FFT) operate in this mode.

Block processing (processing a block of data in a tight inner loop) can have a number of advantages in DSP systems:

- If the DSP has an instruction cache, this cache will optimize instructions to run faster the second (or subsequent) time through the loop.

- If the data accesses adhere to a locality of reference (which is quite common in DSP systems) the performance will improve. Processing the data in stages means the data in any given stage will be accessed from fewer areas, and therefore less likely to thrash the data caches in the device.

- Block processing can often be done in simple loops. These loops have stages where only one kind of processing is taking place. In this manner there will be less thrashing from registers to memory and back. In many cases, most if not all of the intermediate results can be kept in registers or in level one cache.

- By arranging data access to be sequential, even data from the slowest level of memory (DRAM) will be much faster because the various types of DRAM assume sequential access.

DSP designers will use one of these two methods in their system. Typically, control algorithms will use single-sample processing because they cannot delay the output very long, such as in the case of block processing. In audio/video systems, block processing is typically used—because there can be some delay tolerated from input to output.

8.8.13 Input/Output Options

DSPs are used in many different systems including motor control applications, performance-oriented applications, and power sensitive applications. The choice of a DSP processor is dependent on not just the CPU speed or architecture but also the mix of peripherals or I/O devices used to get data in and out of the system. After all, much of the bottleneck in DSP applications is not in the compute engine but in getting data in and out of the system. Therefore, the correct choice of peripherals is important in selecting the device for the application. Example I/O devices for DSP include:

GPIO—A flexible parallel interface that allows a variety of custom connections.

UART—Universal asynchronous receiver-transmitter. This is a component that converts parallel data to serial data for transmission and also converts received serial data to parallel data for digital processing.

CAN—Controller area network. The CAN protocol is an international standard used in many automotive applications.

SPI—Serial peripheral interface. A three-wire serial interface developed by Motorola.

USB—Universal serial bus. This is a standard port that enables the designer to connect external devices (digital cameras, scanners, music players, and so on) to computers. The USB standard supports data transfer rates of 12 Mbps (million bits per second).

McBSP—Multichannel buffered serial port. These provide direct full-duplex serial interfaces between the DSP and other devices in a system.

HPI—Host port interface. This is used to download data from a host processor into the DSP.

A summary of I/O mechanisms for DSP application class is shown in Figure 8.25.

Motor	• 12-bit ADC	• CAN 2.0B	• SPI
	• PWM DAC	• GPIO	• SCI
	• McBSP	• EMIF	• I²C
	• UART		

Power	• USB	• EMIF	• MMC/SD serial ports
	• McBSP	• GPIO	• UART
	• HPI	• 10-bit ADC	• I²C

Perf	• PCI	• EMIF	• Video ports
	• McBSP	• GPIO	• Audio ports
	• HPI	• I2C	• McASP
	• Utopia SP		• Ethernet 10/100 MAC

Figure 8.25: Input/Output Options
Source: Courtesy of Texas Instruments

8.8.14   Calculating DSP Performance

Before choosing a DSP processor for a specific application, the system designer must evaluate three key system parameters as shown below:

> ▶ Maximum CPU Performance
> "What is the maximum number of times the CPU can execute your algorithm? (max # channels)
>
> ▶ Maximum I/O Performance
> "Can the I/O keep up with this maximum # channels?"
>
> ▶ Available Hi-Speed Memory
> "Is there enough hi-speed internal memory?"

With this knowledge, the system designer can scale the numbers to meet the application's needs and then determine:

• CPU load (% of maximum CPU).

• At this CPU load, what other functions can be performed?

The DSP system designer can use this process for any CPU they are evaluating. The goal is the find the "weakest link" in terms of performance so that you know what the system constraints are. The CPU might be able to process numbers at sufficient rates, but if the CPU cannot be fed with data fast enough, then having a fast CPU doesn't really matter.

The goal is to determine the maximum number of channels that can be processed given a specific algorithm and then work that number down based on other constraints (maximum input/output speed and available memory).

As an example, consider the process shown in Figure 8.26. The goal is to determine the maximum number of channels that this specific DSP processor can handle given a specific algorithm. To do this, we must first determine the benchmark of the chosen algorithm (in this case, a 200-tap FIR filter). The relevant documentation for an algorithm like this (from a library of DSP functions) gives us the benchmark with two variables: nx (size of buffer) and nh (# coeffs)—these are used for the first part of the computation. This FIR routine takes about 106 K cycles per frame. Now, consider the sampling frequency. A key question to answer at this point is "How many times is a frame FULL per second?" To answer this, divide the sampling frequency (which specifies how often a new data item is sampled) by the size of the buffer. Performing this calculation determines that we fill about 47 frames per second. Next, is the most important calculation—how many MIPS does this algorithm require of a processor? We need to find out how many cycles this algorithm will require per second. Now we multiply frames/second times cycles/frame and perform the calculation using these data to get a throughput rate of about 5 MIPs. Assuming this is the only

Example – Performance Calculation

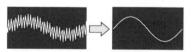

Algorithm: 200-tap (nh) low-pass FIR filter
Frame size: 256 (nx) 16-bit elements
Sampling frequency: 48KHz

How many channels can the DSP handle given this algorithm?

C	FIR benchmark:	(nx/2) (nh+7) = 128 * 207 =	26,496 cyc/frm	
P	#times frm full/s:	(samp freq / frm size) = 48,000/256 =	187.5 frm/s	
U	MIP calc:	(frm/s) (cyc/frm) = 187.5 * 26,496 =	4.97M cyc/s	
	Conclusion:	FIR takes ~5MIPs on a C5502		
	Max #channels:	60 @300MHz		

Max # channels: does not include overhead for interrupts, control code, RTOS, etc.

Are the I/O and memory capable of handling this many channels?

I	Required I/O rate:	48Ksamp/s * #Ch = 48,000 * 16 * 60 =	46.08 Mbps	
/	DSP SP rate:	serial port is full duplex	50.00 Mbps	✔
O	DMA Rate:	(2x16-bit xfrs/cycle) * 300MHz =	9600 Mbps	✔
	Req'd Data Mem	(60 * 200) + (60 * 4 * 256) + (60 * 2 * 199) = 97K x 16-bit		
	Avail int'l mem:		32K x 16-bit	X

Required memory assumes: 60 different filters, 199 element delay buffer, double buffering rcv/xmt

Figure 8.26: Example—Performance Calculation
Source: Courtesy of Texas Instruments

computation being performed on the processor, the channel density (how many channels of simultaneous processing can be performed by a processor) is a maximum of 300/5 = 60 channels. This completes the CPU calculation. This result can not be used in the I/O calculation.

The next question to answer is "Can the I/O interface feed the CPU fast enough to handle 60 channels?" Step one is to calculate the "bit rate" required of the serial port. To do this, the required sampling rate (48 KHz) is multiplied by the maximum channel density (60). This is then multiplied by 16 (assuming the word size is 16—which it is given the chosen algorithm). This calculation yields a requirement of 46 Mbps for 60 channels operating at 48 KHz. In this example what can the 5502 DSP serial port support? The specification says that the maximum bit rate is 50 Mbps (half the CPU clock rate up to 50 Mbps). This tells us that the processor can handle the rates we need for this chosen application. Can the DMA move these samples from the McBSP to memory fast enough? Again, the specification tells us that this should not be a problem.

The next step considers the issue of required data memory. This calculation is somewhat confusing and needs some additional explanation.

Assume that all 60 channels of this application are using different filters—that is, 60 different sets of coefficients and 60 double-buffers (this can be implemented using a ping-pong buffer on both the receive and transmit sides. This is a total of 4 buffers per channel hence the *4 + the delay buffers for each channel (only the receive side has delay buffers ...) so the algorithm becomes:

$$\text{Number of channels } *2* \text{ delay buffer size}$$
$$= 60*2*199$$

This is extremely conservative and the system designer could save some memory if this is not the case. But this is a worst-case scenario. Hence, we'll have 60 sets of 200 coefficients, 60 double-buffers (ping and pong on receive and transmit, hence the *4) and we'll also need a delay buffer of number of coefficients—1 that is 199 for each channel. So, the calculation is:

$$(\text{\#Channels} * \text{\#coefficients}) + (\text{\#Channels} * 4 * \text{frame size})$$
$$+ (\text{\#Channels} * \text{\#delay\_buffers} * \text{delay\_buffer\_size})$$
$$= (60 * 200) + (60 * 4 * 256) + (60 * 26 * 199) = 97{,}320 \text{ bytes of memory}$$

This results in a requirement of 97 K of memory. The 5502 DSP only has 32 K of on-chip memory, so this is a limitation. Again, you can redo the calculation assuming only one type of filter is used, or look for another processor.

Performance Calculation Analysis

	DSP	FIR Benchmark	cyc/frm	frm/s	cyc/s	%CPU	Max Ch
C P U	C2812	(nx/2)(nh+12)+ â	27712	187.5	5.20M	3.5	28
	C5502	(nx/2)(nh+7)	26496	187.5	4.97M	1.7	60
	C6416	(nx/4+15)(nh+11)	16669	187.5	3.13M	0.4	230

â = 36nx/16 = additional time to transfer 16 samples to memory

	DSP	#Ch	Req'd IO rate	Avail SP rate	Avail DMA Rate	Req'd Memory	Avail Int Mem
I / O	C2812	28	21.5 Mbps	50Mbps ✓	None	46K	18K X
	C5502	60	46.1 Mbps	50Mbps ✓	9.6 Gbps	97K	32K X
	C6416	230	176.6 Mbps	100Mbps X	46.1 Gbps	373K	512K ✓

◆ Bandwidth calculations help determine processor's capability
◆ Limiting factors: I/O rate, available memory, CPU performance
◆ Use your system needs (such as 8 Ch) to calculate CPU loading (for example, 3%).

CPU load can help guide your system design …

Figure 8.27: Performance Calculation Analysis
Source: Courtesy of Texas Instruments

Now we extend the calculations to the 2812 and the 6416 processors (Figure 8.27). The following are a couple of things to note.

The 2812 is best used in a single-sample processing mode, so using a block FIR application on a 2812 is not the best fit. However, for example purposes it is done this way to benchmark one processor versus another. Where block processing hurts the 2812 is in relation to getting the samples into on-chip memory. There is no DMA on the 2812 because in single-sample processing, it is not required. The term "beta" in the calculation is the time it takes to move (using CPU cycles) the incoming sampled signals from the A/D to memory. This would be performed by an interrupt service routine and it must be accounted for. Notice that the benchmarks for the 2812 and 5502 are very close.

The 6416 is a high performance machine when doing 16-bit operations—it can do 269 channels given the specific FIR used in this example. Of course, the I/O (on one serial port) can't keep up with this, but it could with 2 serial ports in operation.

Once you've done these calculations, you can "back off" the calculation to the exact number of channels your system requires, determine an initial theoretical CPU load that is expected, and then make some decisions about what to do with any additional bandwidth that is left over (Figure 8.28).

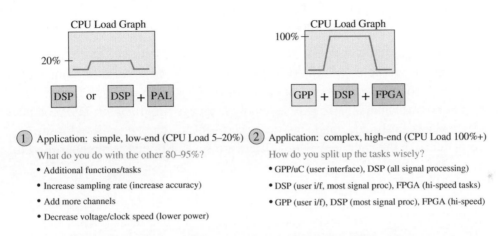

Figure 8.28: Determining What to Do Based on Available CPU Bandwidth
Source: Courtesy of Texas Instruments

Two sample cases that help drive discussion on issues related to CPU load are shown in Figure 8.28. In the first case, the entire application only takes 20% of the CPU's load. What do you do with the extra bandwidth? The designer can add more algorithmic processing, increase the channel density, increase the sampling rate to achieve higher resolution or accuracy, or decrease the clock/voltage so that the CPU load goes up and you save lots of power. It is up to the system designer to determine the best strategy here based on the system requirements.

The second example application is the other side of the fence—where the application takes more processing power than the CPU can handle. This leads the designer to consider a combined solution. The architecture of this again depends on the application's needs.

8.8.15   DSP Software

DSP software development is primarily focused on achieving the performance goals of the system. It's more efficient to develop DSP software using a high-level language like C or C++, but it is not uncommon to see some of the high performance, MIPS intensive algorithms written at least partially in assembly language. When generating DSP algorithm code, the designer should use one or more of the following approaches:

- Find existing algorithms (free code).
- Buy or license algorithms from vendors. These algorithms may come bundled with tools or may be classes of libraries for specific applications (Figure 8.29).
- Write the algorithms in-house. If using this approach, implement as much of the algorithm as possible in C/C++. This usually results in faster time-to-market and requires a common skill found in the industry. It is much easier to find a C programmer than a 5502 DSP assembly language programmer. DSP compiler efficiency is fairly good and significant performance can be achieved using a compiler with the right techniques. There are several tuning techniques used to generate optimal code.

- May be bundled with tools and contains:
  - C-callable highly-optimized assembly routines
  - Documentation on each algorithm
  - Examples: FIR, IIR, FFT, convolution, min/max, log, etc.
- Other libraries available for specific DSPs:
  - Image libraries
  - Other control-specific free libraries
- Use third parties
  - Lists application software by platform, algorithm, and third party
  - Includes specs such as data/code size, performance, licensing fees

Figure 8.29: Reuse Opportunities Using DSP Libraries and Third Parties

To fine-tune code and get the highest efficiency possible, the system designer needs to know three things:

- The architecture.
- The algorithms.
- The compiler.

Figure 8.30 shows some ways to help the compiler generate efficient code. Compilers are pessimistic by nature, so the more information that can be provided about the system algorithms, where data is in memory, the better. The C6000 compiler can achieve 100% efficiency versus hand-coded assembly if the right techniques are used. There are pros and cons to writing DSP algorithms in assembly language as well, so if this must be done, these must be understood from the beginning (Figure 8.31).

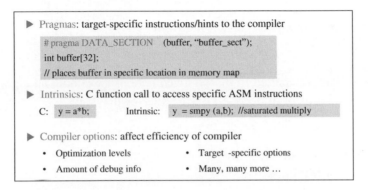

Figure 8.30: Compiler Optimization Techniques for Producing High Performance Code
Source: Courtesy of Texas Instruments

Pros	• Can result in highest possible performance • Access to native instruction set (including application-specific instructions)
Cons	• Usually difficult learning curve (often increases development time) • Usually not portable
Conclusions	• Write in C when possible (most of the time, assembly is not required) • Don't reinvent the wheel – make full use of libraries, third parties, etc.

Figure 8.31: Pros and Cons of Writing DSP Code in Assembly Language
Source: Courtesy of Texas Instruments

8.8.16   DSP Frameworks

All DSP systems have some basic needs—basic requirements for processing high performance algorithms. These include the following.

Input/Output

- Input consists of analog information being converted to digital data.

- Output consists of digital data converted back out to analog format.

- Device drivers to talk to the actual hardware.

Processing

- Algorithms that are applied to the digitized data, for example an algorithm to encrypt secure data streams or to decode an MP3 file for playback.

Control

- Control structures with the ability to make system level decisions, for example to stop or play an MP3 file.

A DSP framework must be developed to connect device drivers and algorithms for correct data flow and processing (Figure 8.32).

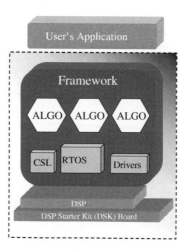

Figure 8.32: A Model of a DSP Framework for Signal Processing

A DSP framework can be custom developed for the application, reused from another application, or even purchased or acquired from a vendor. Since many DSP systems have similar processing frameworks as described above, reuse is a viable option. A framework is system software that uses standardized interfaces to algorithms and software. This includes algorithms as well as hardware drivers. The benefits of using a DSP framework include:

- The development does not have to start from scratch.

- The framework can be used as a starting point for many applications.

- The software components within a framework have well defined interfaces and work well together.

- The DSP designer can focus on the application layer that is usually the main differentiator in the product being developed. The framework can be reused.

An example DSP reference framework is shown in Figure 8.33. This DSP framework consists of:

- I/O drivers for input/output.

- Two processing threads with generic algorithms.

- Split/join threads used to simulate/utilize a stereo codec.

This reference framework has two channels by default. The designer can add and remove channels to suit the applications needs.

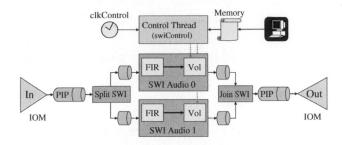

Figure 8.33: An Example DSP Reference Framework
Source: *Courtesy of Texas Instruments*

An example complete DSP Solution is shown in Figure 8.34. There is the DSP as the central processing element. There are mechanisms to get data into and out of the system (the ADC and DAC components). There is a power control module for system power management, a data transmission block with several possible peripherals including USB, FireWire, and so on, some clock generation components and a sensor for the RF component. Of course, this is only one example, but many DSP applications follow a similar structure.

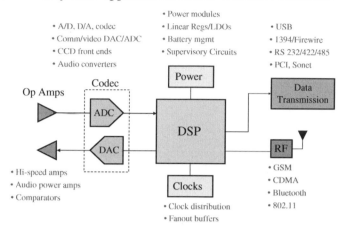

Figure 8.34: An Example DSP Application with Major Building Blocks
Source: Courtesy of Texas Instruments

8.9   Optimizing DSP Software

Many of today's DSP applications are subject to real-time constraints. Many embedded DSP applications will eventually grow to a point where they are stressing the available CPU, memory or power resources. Understanding the workings of the DSP architecture, compiler, and application algorithms can speed up applications, sometimes by an order of magnitude. The following sections will summarize some of the techniques that can improve the performance of your code in terms of cycle count, memory use, and power consumption.

8.10   What Is Optimization?

Optimization is a procedure that seeks to maximize or minimize one or more performance indices. These indices include:

- Throughput (execution speed)

- Memory usage

- I/O bandwidth

- Power dissipation

Since many DSP systems are real-time systems, at least one (and probably more) of these indices must be optimized. It is difficult (and usually impossible) to optimize all these performance indices at the same time. For example, to make the application faster, the developer may require more memory to achieve the goal. The designer must weigh each of these indices and make the best trade-off. The tricky part to optimizing DSP applications is understanding the trade-off between the various performance indices. For example, optimizing an application for speed often means a corresponding decrease in power consumption but an increase in memory usage. Optimizing for memory may also result in a decrease in power consumption due to fewer memory accesses but an offsetting decrease in code performance. The various trade-offs and system goals must be understood and considered before attempting any form of application optimization.

Determining which index or set of indices is important to optimize depends on the goals of the application developer. For example, optimizing for performance means that the developer can use a slow or less expensive DSP to do the same amount of work. In some embedded systems, cost savings like this can have a significant impact on the success of the product. The developer can alternatively choose to optimize the application to allow the addition of more functionality. This may be very important if the additional functionality improves the overall performance of the system, or if the developer can add more capability to the system such as an additional channel of a base station system. Optimizing for memory use can also lead to overall system cost reduction. Reducing the application size leads to a lower demand for memory, which reduces overall system cost. Finally, optimizing for power means that the application can run longer on the same amount of power. This is important for battery powered applications. This type of optimization also reduces the overall system cost with respect to power supply requirements and other cooling functionality required.

8.11   The Process

Generally, DSP optimization follows the 80/20 rule. This rule states that 20% of the software in a typical application uses 80% of the processing time. This is especially true for DSP applications that spend much of their time in tight inner loops of DSP algorithms. Thus, the real issue in optimization isn't how to optimize, but where to optimize. The first rule of optimization is "Don't!." Do not start the optimization process until you have a good understanding of where the execution cycles are being spent.

The best way to determine which parts of the code should be optimized is to profile the application. This will answer the question as to which modules take the longest to execute.

These will become the best candidates for performance-based optimization. Similar questions can be asked about memory usage and power consumption.

DSP application optimization requires a disciplined approach to get the best results. To get the best results out of your DSP optimization effort, the following process should be used:

- **Do your homework**—Make certain you have a thorough understanding of the DSP architecture, the DSP compiler, and the application algorithms. Each target processor and compiler has different strengths and weaknesses and understanding them is critical to successful software optimization. Today's DSP optimizing compilers are advanced. Many allow the developer to use a higher order language such as C and very little, if any, assembly language. This allows for faster code development, easier debugging, and more reusable code. But the developer must understand the "hints" and guidelines to follow to enable the compiler to produce the most efficient code.

- **Know when to stop**—Performance analysis and optimization is a process of diminishing returns. Significant improvements can be found early in the process with relatively little effort. This is the "low hanging fruit." Examples of this include accessing data from fast on-chip memory using the DMA and pipelining inner loops. However, as the optimization process continues, the effort expended will increase dramatically and further improvements and results will fall dramatically.

- **Change one parameter at a time**—Go forward one step at a time. Avoid making several optimization changes at the same time. This will make it difficult to determine what change led to which improvement percentage. Retest after each significant change in the code. Keep optimization changes down to one change per test in order to know exactly how that change affected the whole program. Document these results and keep a history of these changes and the resulting improvements. This will prove useful if you have to go back and understand how you got to where you are.

- **Use the right tools**—Given the complexity of modern DSP CPUs and the increasing sophistication of optimizing compilers, there is often little correlation between what a programmer thinks is optimized code and what actually performs well. One of the most useful tools to the DSP programmer is the profiler. This is a tool that allows the developer to run an application and get a "profile" of where cycles are being used throughput the program. This allows the developer to identify and focus on the core bottlenecks in the program quickly. Without a profiler, gross performance issues as well as minor code modifications can go unnoticed for long periods of time and make the entire code optimization process less disciplined.

- **Have a set of regression tests and use it after each iteration**—Optimization can be difficult. More difficult optimizations can result in subtle changes to the program behavior that lead to wrong answers. More complex code optimizations in the compiler can, at times, produce incorrect code (a compiler, after all, is a software program with its own bugs!). Develop a test plan that compares the expected results to the actual results of the software program. Run the test regression often enough to catch problems early. The programmer must verify that program optimizations have not broken the application. It is extremely difficult to backtrack optimized changes out of a program when a program breaks.

A general code optimization process (see Figure 8.35) consists of a series of iterations. In each iteration, the programmer should examine the compiler generated code and look for optimization opportunities. For example, the programmer may look for an abundance of NOPs or other inefficiencies in the code due to delays in accessing memory and/or another processor resource. These are the areas that become the focus of improvement. The programmer will apply techniques such as software pipelining, loop unrolling, DMA resource utilization, and so on, to reduce the processor cycle count (we will talk more about

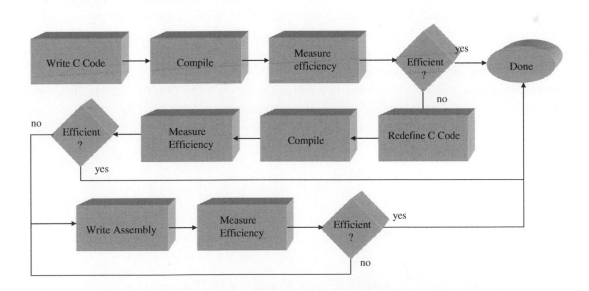

Figure 8.35: A General DSP Code Optimization Process
Source: Courtesy of Texas Instruments

these specific techniques later). As a last resort the programmer can consider hand-tuning the algorithms using assembly language.

Many times, the C code can be modified slightly to achieve the desired efficiency, but finding the right "tweak" for the optimal (or close to optimal) solution can take time and several iterations. Keep in mind that the software engineer/programmer must take responsibility for at least a portion of this optimization. There have been substantial improvements in production DSP compilers with respect to advanced optimization techniques. These optimizing compilers have grown to be quite complex due to the advanced algorithms used to identify optimization opportunities and make these code transformations. With this increased complexity comes the opportunity for errors in the compiler. You still need to understand the algorithms and the tools well enough that you can supply the necessary improvements when the compiler can't. In this chapter we will discuss how to optimize DSP software in the context of this process.

8.12   Make the Common Case Fast

The fundamental rule in computer design as well as programming real-time DSP-based systems is "make the common case fast, and favor the frequent case." This is really just Amdahl's Law that says the performance improvement to be gained using some faster mode of execution is limited by how often you use that faster mode of execution. So don't spend time trying to optimize a piece of code that will hardly ever run. You won't get much out of it, no matter how innovative you are. Instead, if you can eliminate just one cycle from a loop that executes thousands of times, you will see a bigger impact on the bottom line. I will now discuss three different approaches to making the common case fast (by common case, I am referring to the areas in the code that consume the most resources in terms of cycles, memory, or power):

- Understand the DSP architecture.
- Understand the DSP algorithms.
- Understand the DSP compiler.

8.13   Make the Common Case Fast—DSP Architectures

DSP architectures are designed to make the common case fast. Many DSP applications are composed from a standard set of DSP building blocks such as filters, Fourier transforms, and convolutions. Table 8.3 contains a number of these common DSP algorithms. Notice the common structure of each of the algorithms:

- They all accumulate a number of computations.

- They all sum over a number of elements.

- They all perform a series of multiplies and adds.

These algorithms all share some common characteristics; they perform multiplies and adds over and over again. This is generally referred to as the ***sum of products*** (SOP).

As discussed earlier, a DSP is, in many ways, an application specific microprocessor. DSP designers have developed hardware architectures that allow the efficient execution of algorithms to take advantage of the algorithmic specialty in signal processing. For example, some of the specific architectural features of DSPs to accommodate the algorithmic structure of DSP algorithms include:

- Special instructions, such as a single cycle multiple and accumulate (MAC). Many signal processing algorithms perform many of these operations in tight loops. Figure 8.36 shows the savings from computing a multiplication in hardware instead of microde in the DSP processor. A savings of four cycles is significant when multiplications are performed millions of times in signal processing applications.

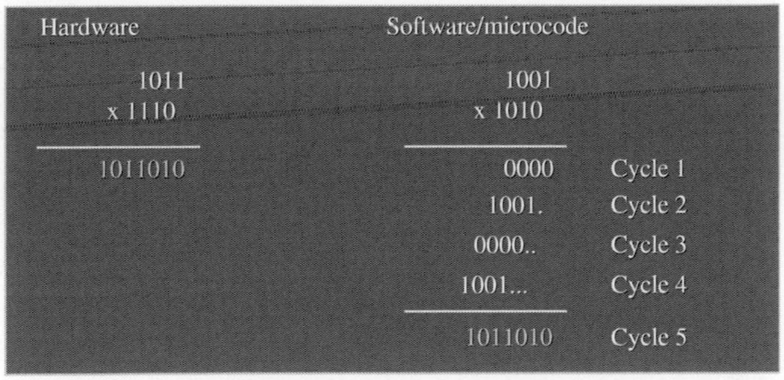

Hardware	Software/microcode	
1011	1001	
x 1110	x 1010	
1011010	0000	Cycle 1
	1001.	Cycle 2
	0000..	Cycle 3
	1001...	Cycle 4
	1011010	Cycle 5

Figure 8.36: Special Multiplication Hardware Speeds Up DSP Processing
Source: Courtesy of Texas Instruments

- Large accumulators to allow for accumulating a large number of elements.

- Special hardware to assist in loop checking so this does not have to be performed in software, which is much slower.

- Access to two or more data elements in the same cycle. Many signal processing algorithms have multiple two arrays of data and coefficients. Being able to access two operands at the same time makes these operations very efficient. The DSP Harvard architecture shown in Figure 8.37 allows for access of two or more data elements in the same cycle.

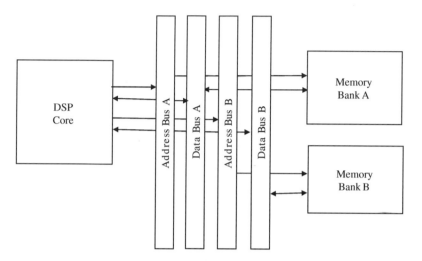

Figure 8.37: DSP Harvard Architecture—Multiple Address and Data Busses Accessing Multiple Banks of Memory Simultaneously

The DSP developer must choose the right DSP architecture to accommodate the signal processing algorithms required for the application as well as the other selection factors such as cost, tools support, and so on.

Table 8.3: DSP Algorithms Share Common Characteristics

Equation
$y(n) = \sum_{k=0}^{M} a_k x(n-k)$
$y(n) = \sum_{k=0}^{M} a_k x(n-k) + \sum_{k=1}^{N} b_k y(n-k)$
$y(n) = \sum_{k=0}^{N} x(k) h(n-k)$
$X(k) = \sum_{n=0}^{N-1} x(n) \exp[-j(2\pi/N)nk]$
$F(u) = \sum_{x=0}^{N-1} c(u) \cdot f(x) \cdot \cos[\frac{\pi}{2N} u(2x+1)]$

8.14 Make the Common Case Fast—DSP Algorithms

DSP algorithms can be made to run faster using techniques of algorithmic transformation. For example, a common algorithm used in DSP applications is the Fourier transform. The Fourier transform is a mathematical method of breaking a signal in the time domain into all of its individual frequency components.[4] The process of examining a time signal broken down into its individual frequency components is also called *spectral analysis* or *harmonic analysis*.

There are different ways to characterize a Fourier transforms:

- The Fourier transform (FT) is a mathematical formula using integrals:

$$F(u) = \int_{-\infty}^{\infty} f(x)e^{-x\pi ixu}d\omega$$

- The discrete Fourier transform (DFT) is a discrete numerical equivalent using sums instead of integrals which maps well to a digital processor like a DSP:

$$F(u) = \frac{1}{N}\sum_{x=0}^{N-1} f(x)e_j^{-x\pi i\omega ufN}$$

- The fast Fourier transform (FFT) is just a computationally fast way to calculate the DFT which reduces many of the redundant computations of the DFT.

How these are implemented on a DSP has a significant impact on overall performance of the algorithm. The FFT, for example, is a fast version of the DFT. The FFT makes use of periodicities in the sines that are multiplied to perform the transform. This significantly reduces the amount of calculations required. A DFT implementation requires N 2 operations to calculate a N point transform. For the same N point data set, using a FFT algorithm requires N * log2(N) operations. The FFT is therefore faster than the DFT by a factor of N/log2(n). The speedup for a FFT is more significant as N increases (Figure 8.38).

Recognizing the significant impact that efficiently implemented algorithms have on overall system performance, DSP vendors and other providers have developed libraries of efficient DSP algorithms optimized for specific DSP architectures. Depending on the type of algorithm, these can downloaded from web sites (be careful of obtaining free software like

---

[4] Brigham, E. Oren, 1988, *The Fast Fourier Transform and Its Applications*, Englewood Cliffs, NJ: Prentice-Hall, Inc., p. 448.

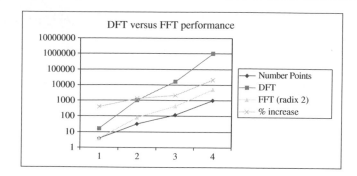

Figure 8.38: FFT versus DFT for Various Sizes of Transforms (Logarithmic Scale)

this—the code may be buggy as there is no guarantee of quality) or bought from DSP solution providers.

8.15 Make the Common Case Fast—DSP Compilers

Just a few years ago, it was an unwritten rule that writing programs in assembly would usually result in better performance than writing in higher level languages like C or C++. The early "optimizing" compilers produced code that was not as good as what one could get by programming in assembly language, where an experienced programmer generally achieves better performance. Compilers have gotten much better and today there are very specific high performance optimizations performed that compete well with even the best assembly language programmers.

Optimizing compilers perform sophisticated program analysis including intraprocedural and interprocedural analysis. These compilers also perform data and control flow analysis as well as dependence analysis and often employ provably correct methods for modifying or transforming code. Much of this analysis is to prove that the transformation is correct in the general sense. Many optimization strategies used in DSP compilers are also strongly heuristic.[5]

One effective code optimization strategy is to write DSP application code that can be **pipelined** efficiently by the compiler. *Software* pipelining is an optimization strategy to

---

[5] Heuristics involves problem solving by experimental and especially trial-and-error methods or relating to exploratory problem-solving techniques that utilize self-educating techniques (as the evaluation of feedback) to improve performance.

schedule loops and functional units efficiently. In modern DSPs there are multiple functional units that are orthogonal and can be used at the same time (Figure 8.39). The compiler is given the burden of figuring out how to schedule instructions so that these functional units can be used in parallel whenever possible. Sometimes this is a matter of a subtle change in the way the C code is structured that makes all the difference. In software pipelining, multiple iterations of a loop are scheduled to execute in parallel. The loop is reorganized in a way that each iteration in the pipelined code is made from instruction sequences selected from different iterations in the original loop. In the example in Figure 8.40, a five-stage loop with three iterations is shown. There is an initial period (cycles n and n+1), called the ***prolog***

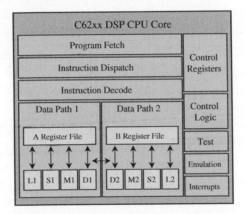

Figure 8.39: DSP Architectures May Have Orthogonal Execution Units and Data Paths Used to Execute DSP Algorithms more efficiently. In This Figure, Units L1, S1, M1, D1, and L2, S2, M2, and D2 Are all Orthogonal Execution Units that Can Have Instructions Scheduled for Execution by the Compiler in the Same Cycle if the Conditions Are Right

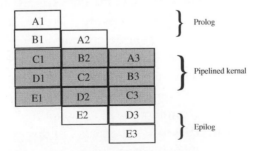

Figure 8.40: A Five-Stage Instruction Pipeline that Is Scheduled to Be Software Pipelined by the Compiler

when the pipes are being "primed" or initially loaded with operations. Cycles n+2 to n+4 is the actual pipelined section of the code. It is in this section that the processor is performing three different operations (C, B, and A) for three different loops (1, 2, and 3). There is a epilog section where the last remaining instructions are performed before exiting the loop. This is an example of a fully utilized set of pipelines that produces the fastest, most efficient code.

Figure 8.41 shows a sample piece of C code and the corresponding assembly language output. In this case the compiler was asked to attempt to pipeline the code. This is evident by the piped loop prolog and piped loop kernel sections in the assembly language output. Keep in mind that the prolog and epilog sections of the code prime the pipe and flush the pipe, respectively, as shown in Figure 8.40. In this case, the pipelined code is not as good as it could be. You can spot inefficient code by looking for how many NOPs are in the piped loop kernel of the code. In this case the piped loop kernel has a total of five NOP cycles, two in line 16, and three in line 20. This loop takes a total of 10 cycles to execute. The NOPs are the first indication that a more efficient loop may be possible. But how short can this loop be? One way to estimate the minimum loop size is to determine what execution unit is being used the most. In this example, the D unit is used more than any other unit, a total of three times (lines 14, 15, and 21). There are two sides to a superscalar device, enabling each unit to be used twice (D1 and D2) per clock for a minimum two clock loop; two D operations in one clock and one D unit in the second clock. The compiler was smart enough to use the D units on both sides of the pipe (lines 14 and 15), enabling it to parallelize the instructions and only use one clock. It should be possible to perform other instructions while waiting for the loads to complete, instead of delaying with NOPs.

In the simple *for* loop, the two input arrays (array1 and array2) may or may not be dependent or overlap in memory. The same with the output array. In a language such as C/C++ this is something that is allowed and, therefore, the compiler must be able to handle this correctly. Compilers are generally pessimistic creatures. They will not attempt an optimization if there is a case where the resultant code will not execute properly. In these situations, the compiler takes a conservative approach and assumes the inputs can be dependent on the previous output each time through the loop. If it is known that the inputs are not dependent on the output, we can hint to the compiler by declaring input1 and input2 as "restrict," indicating that these fields will not change. In this example, "restrict" is a keyword in C that can be used for this purpose. This is also a trigger for enabling software pipelining which can improve throughput. This C code is shown in Figure 8.42 with the corresponding assembly language.

```
1      void example1(float *out, float *input1, float *input2)
2      {
3        int i;
4
5        for(i = 0; i < 100; i++)
6          {
7            out[i] = input1[i] * input2[i];
8          }
9      }


1      _example1:
2      ;** -----------------------------------------------------------*
3              MVK         .S2     0x64,B0
4
5              MVC         .S2     CSR,B6
6      ||      MV          .L1X       B4,A3
7      ||      MV          .L2X       A6,B5
8
9              AND         .L1X       -2,B6,A0
10             MVC          .S2X       A0,CSR
11     ;** -----------------------------------------------------------*
12     L11:         ; PIPED LOOP PROLOG
13     ;** -----------------------------------------------------------*
14     L12:         ; PIPED LOOP KERNEL
15
16             LDW         .D2     *B5++,B4         ;
17     ||      LDW         .D1     *A3++,A0         ;
18
19             NOP                    2
20     [ B0]   SUB         .L2     B0,1,B0          ;
21     [ B0]   B           .S2     L12              ;
22             MPYSP       .M1X    B4,A0,A0         ;
23             NOP                 3
24             STW         .D1     A0,*A4++         ;
25     ;** -----------------------------------------------------------*
26             MVC         .S2     B6,CSR
               B           .S2     B3
               NOP                 5
               ; BRANCH OCCURS
```

Figure 8.41: C Example and the Corresponding Pipelined Assembly Language Output

```
1   void  example2(float  *out, restrict float *input1, restrict float *input2)
2   {
3     int i;
4
5     for(i = 0; i < 100; i++)
6       {
7          out[i] = input1[i] * input2[i];
8       }
9   }
```

```
1   _example2:
2   ;** ----------------------------------------------------------*
3              MVK          .S2          0x64,B0
4
5              MVC          .S2          CSR,B6
6   ||         MV           .L1X         B4,A3
7   ||         MV           .L2X         A6,B5
8
9              AND          .L1X         -2,B6,A0
10
11             MVC          .S2X         A0,CSR
12  ||         SUB          .L2          B0,4,B0
13
14  ;** ----------------------------------------------------------*
15  L8:            ; PIPED LOOP PROLOG
16
17             LDW          .D2          *B5++,B4      ;
18  ||         LDW          .D1          *A3++,A0      ;
19
20             NOP                       1
21
22             LDW          .D2          *B5++,B4      ;@
23  ||         LDW          .D1          *A3++,A0      ;@
24
25      [ B0]  SUB          .L2          B0,1,B0       ;
26
27      [ B0]  B            .S2          L9            ;
28  ||         LDW          .D2          *B5++,B4      ;@@
29  ||         LDW          .D1          *A3++,A0      ;@@
30
31             MPYSP        .M1X         B4,A0,A5      ;
```

Figure 8.42: Corresponding Pipelined Assembly Language Output

```
32      || [ B0]    SUB         .L2         B0,1,B0         ;@
33
34         [ B0]    B           .S2         L9              ;@
35      ||          LDW         .D2         *B5++,B4        ;@@@
36      ||          LDW         .D1         *A3++,A0        ;@@@
37
38               MPYSP          .M1X        B4,A0,A5        ;@
39      || [ B0]    SUB         .L2         B0,1,B0         ;@@
40
41      ;** ----------------------------------------------------------*
42      L9:         ; PIPED LOOP KERNEL
43
44         [ B0]    B           .S2         L9              ;@@
45      ||          LDW         .D2         *B5++,B4        ;@@@@
46      ||          LDW         .D1         *A3++,A0        ;@@@@
47
48               STW            .D1         A5,*A4++        ;
49      ||       MPYSP          .M1X        B4,A0,A5        ;@@
50      || [ B0]    SUB         .L2         B0,1,B0         ;@@@
51
52      ;** ----------------------------------------------------------*
53      L10:        ; PIPED LOOP EPILOG
54               NOP                        1
55
56               STW            .D1         A5,*A4++        ;@
57      ||       MPYSP          .M1X        B4,A0,A5        ;@@@
58
59               NOP                        1
60
61               STW            .D1         A5,*A4++        ;@@
62      ||       MPYSP          .M1X        B4,A0,A5        ;@@@@
63
64               NOP                        1
65               STW            .D1         A5,*A4++        ;@@@
66               NOP                        1
67               STW            .D1         A5,*A4++        ;@@@@
68      ;** ----------------------------------------------------------*
69               MVC            .S2         B6,CSR
70               B              .S2         B3
71               NOP                        5
72               ; BRANCH OCCURS
```

Figure 8.42: Cont'd

There are a few things to notice in looking at this assembly language. First, the piped loop kernel has become smaller. In fact, the loop is now only two cycles long. Lines 44–47 are all executed in one cycle (the parallel instructions are indicated by the ‖ symbol) and lines 48–50 are executed in the second cycle of the loop. The compiler, with the additional dependency information we supplied it with the "restrict" declaration, has been able to take advantage of the parallelism in the execution units to schedule the inner part of the loop very efficiently. The prolog and epilog portions of the code are much larger now. Tighter piped kernels will require more priming operations to coordinate all of the execution based on the various instruction and branching delays. But once primed, the kernel loop executes extremely fast, performing operations on various iterations of the loop. The goal of software pipelining is, like we mentioned earlier, to make the common case fast. The kernel is the common case in this example, and we have made it very fast. Pipelined code may not be worth doing for loops with a small loop count. However, for loops with a large loop count, executing thousands of times, software pipelining produces significant savings in performance while also increasing the size of the code.

In the two cycles the piped kernel takes to execute, there are a lot of things going on. The right hand column in the assembly listing indicates what iteration is being performed by each instruction. (Each "@" symbol is a iteration count. So, in this kernel, line 44 is performing a branch for iteration n+2, lines 45 and 46 are performing loads for iteration n+4, line 48 is storing a result for iteration n, line 49 is performing a multiply for iteration n+2, and line 50 is performing a subtraction for iteration n+3, all in two cycles!) The epilog is completing the operations once the piped kernel stops executing. The compiler was able to make the loop two cycles long, which is what we predicted by looking at the inefficient version of the code.

The code size for a pipelined function becomes larger, as is obvious by looking at the code produced. This is one of the trade-offs for speed that the programmer must make.

Software pipelining does not happen without careful analysis and structuring of the code. Small loops that do not have many iterations may not be pipelined because the benefits are not realized. Loops that are large in the sense that there are many instructions per iteration that must be performed may not be pipelined because there are not enough processors resources (primarily registers) to hold the key data during the pipeline operation. If the compiler has to "spill" data to the stack, precious time will be wasted having to fetch this information from the stack during the execution of the loop.

8.16 An In-Depth Discussion of DSP Optimization

While DSP processors offer tremendous potential throughput, your application won't achieve that potential unless you understand certain important implementation techniques. We will now discuss key techniques and strategies that greatly reduce the overall number of DSP CPU cycles required by your application. For the most part, the main object of these techniques is to fully exploit the potential parallelism in the processor and in the memory subsystem. The specific techniques covered include:

- Direct memory access;

- Loop unrolling; and

- More on software pipelining.

8.17 Direct Memory Access

Modern DSPs are extremely fast; so fast that the processor can often compute results faster than the memory system can supply new operands—a situation known as "data starvation." In other words, the bottleneck for these systems becomes keeping the unit fed with data fast enough to prevent the DSP from sitting idle waiting for data. Direct memory access is one technique for addressing this problem.

Direct memory access (DMA) is a mechanism for accessing memory without the intervention of the CPU. A peripheral device (the DMA controller) is used to write data directly to and from memory, taking the burden off the CPU. The DMA controller is just another type of CPU whose only function is moving data around very quickly. In a DMA capable machine, the CPU can issue a few instructions to the DMA controller, describing what data is to be moved (using a data structure called a *transfer control block* [TCB]), and then go back to what it was doing, creating another opportunity for parallelism. The DMA controller moves the data in parallel with the CPU operation (Figure 8.43), and notifies the CPU when the transfer is complete.

DMA is most useful for copying larger blocks of data. Smaller blocks of data do not have the payoff because the setup and overhead time for the DMA makes it worthwhile just to use the CPU. But when used smartly, the DMA can result in huge time savings. For example, using the DMA to stage data on- and off-chip allows the CPU to access the staged data in a single cycle instead of waiting multiple cycles while data is fetched from slower external memory.

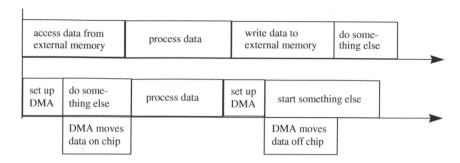

Figure 8.43: Using DMA Instead of the CPU Can Offer Big Performance Improvements Because the DMA Handles the Movement of the Data while the CPU Is Buys Performing Meaningful Operations on the Data

8.18   Using DMA

Because of the large penalty associated with accessing external memory, and the cost of getting the CPU involved, the DMA should be used wherever possible. The code for this is not too overwhelming. The DMA requires a data structure to describe the data it is going to access (where it is, where it's going, how much, and so on). A good portion of this structure can be built ahead of time. Then it is simply a matter of writing to a memory-mapped DMA enable register to start the operation (Figure 8.44). It is best to start the DMA operation well ahead of when the data is actually needed. This gives the CPU something to do in the meantime and does not force the application to wait for the data to be moved. Then, when the data is actually needed, it is already there. The application should check to verify the operation was successful and this requires checking a register. If the operation was done ahead of time, this should be a one time poll of the register, and not a spin on the register, chewing up valuable processing time.

8.18.1   Staging Data

The CPU can access on-chip memory much faster than off-chip or external memory. Having as much data as possible on-chip is the best way to improve performance. Unfortunately, because of cost and space considerations most DSPs do not have a lot of on-chip memory. This requires the programmer to coordinate the algorithms in such a way to efficiently use the available on-chip memory. With limited on-chip memory, data must be staged on- and off-chip using the DMA. All of the data transfers can be happening in the background, while the CPU is actually crunching the data. Once the data is in internal memory, the CPU can access the data in on-chip memory very quickly (Figure 8.46).

```
-----------------------------------------------------------------------
/* Addresses of some of the important DMA registers */
#define DMA_CONTROL_REG   (*(volatile unsigned*)0x40000404)
#define DMA_STATUS_REG    (*(volatile unsigned*)0x40000408)
#define DMA_CHAIN_REG          (*(volatile unsigned*)0x40000414)


/* macro to wait for the DMA to complete and signal the status register */
#define DMA_WAIT              while(DMA_STATUS_REG&1) {}


/* pre-built tcb structure */
typedef struct {

     tcb setup fields

} DMA_TCB;


-----------------------------------------------------------------------
extern DMA_TCB tcb;

/* set up the remaining fields of the tcb structure -
     where you want the data to go, and how much you want to send */
tcb.destination_address = dest_address;
tcb.word_count = word_count;

/* writing to the chain register kicks off the DMA operation */
DMA_CHAIN_REG = (unsigned)&tcb;

Allow the CPU to do other meaningful work....

/* wait for the DMA operation to complete */
DMA_WAIT;
```

Figure 8.44: Code to Set Up and Enable a DMA Operation Is Pretty Simple. The Main Operations Include Setting Up a Data Structure (called a TCB in the Example Above) and Performing a Few Memory Mapped Operations to Initialize and Check the Results of the Operation

Smart layout and utilization of on-chip memory, and judicious use of the DMA can eliminate most of the penalty associated with accessing off-chip memory. In general, the rule is to stage the data in and out of on-chip memory using the DMA and generate the results on chip. Figure 8.45 shows a template describing how to use the DMA to stage blocks of data on and off chip. This technique uses a double-buffering mechanism to stage

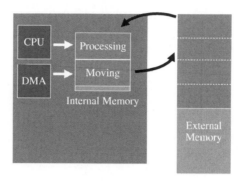

Figure 8.45: Template for Using the DMA to Stage Data On- and Off-chip

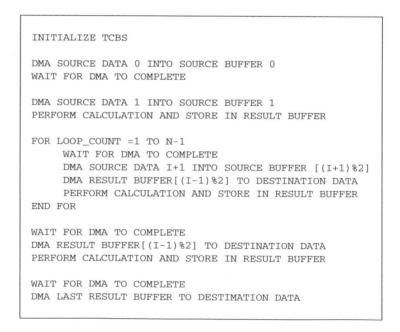

```
INITIALIZE TCBS

DMA SOURCE DATA 0 INTO SOURCE BUFFER 0
WAIT FOR DMA TO COMPLETE

DMA SOURCE DATA 1 INTO SOURCE BUFFER 1
PERFORM CALCULATION AND STORE IN RESULT BUFFER

FOR LOOP_COUNT =1 TO N-1
     WAIT FOR DMA TO COMPLETE
     DMA SOURCE DATA I+1 INTO SOURCE BUFFER [(I+1)%2]
     DMA RESULT BUFFER[(I-1)%2] TO DESTINATION DATA
     PERFORM CALCULATION AND STORE IN RESULT BUFFER
END FOR

WAIT FOR DMA TO COMPLETE
DMA RESULT BUFFER[(I-1)%2] TO DESTINATION DATA
PERFORM CALCULATION AND STORE IN RESULT BUFFER

WAIT FOR DMA TO COMPLETE
DMA LAST RESULT BUFFER TO DESTIMATION DATA
```

Figure 8.46: With Limited On-chip Memory, Data Can Be Staged in and out of On-chip Memory Using the DMA and Leaving the CPU to Perform Other Processing

the data. This way the CPU can be processing one buffer while the DMA is staging the other buffer. Speed improvements over 90% are possible using this technique.

Writing DSP code to use the DMA does have some cost penalties. Code size will increase, depending on how much of the application uses the DMA. Using the DMA also adds increased complexity and synchronization to the application. Code portability is reduced when you add processor specific DMA operations. Using the DMA should only be done in areas requiring high throughput.

8.18.1.1   An Example

As an example of this technique, consider the code in Figure 8.47. This code snippet sums a data field and computes a simple percentage before returning. The code in Figure 8.47 consists of 5 executable lines of code. In this example, the "processed_data" field is assumed to be in external memory of the DSP. Each access of a processed_data element in the loop will cause an external memory access to fetch the data value.

```
int i;
float sum;

/*
** sum data field
*/
sum = 0.0f;
for(i=0; i<num_data_points; i++;)
{
            sum += processed_data[i];
}

/*
** Compute percentage and return
*/
return(MTH_divide(sum,num_data_points));
} /* end */
```

Figure 8.47: A Simple Function Consisting of Five Executable Lines of Code

The code shown in Figure 8.48 is the same function shown in Figure 8.46 but implemented to use the DMA to transfer blocks of data from external memory to internal or on-chip memory. This code consists of 36 executable lines of code, but runs much faster than the code in Figure 8.46. The overhead associated with setting up the DMA, building the transfer packets, initiating the DMA transfers, and checking for completion are relatively small compared to the fast memory accesses of the data from on-chip memory. This code snippet was instrumented following the guidelines in the template of Figure 8.47. This code also performs a loop unrolling operation when summing the data at the end of the computation (we'll talk more about loop unrolling later in Section 8.19). This also adds to the speedup of this code. This code snippet uses semaphores to protect the on-chip memory and DMA resources.

```
/*
** sum data fields to compute percentages
*/
    sum0 = 0.0;
    sum1 = 0.0;
    sum2 = 0.0;
    sum3 = 0.0;
    sum4 = 0.0;
    sum5 = 0.0;
    sum6 = 0.0;
    sum7 = 0.0;
  for (i=0; i<BLOCK_SIZE; i+=8)
  {
    sum0 += internal_processed_data[i    ];
    sum1 += internal_processed_data[i + 1];
    sum2 += internal_processed_data[i + 2];
    sum3 += internal_processed_data[i + 3];
    sum4 += internal_processed_data[i + 4];
    sum5 += internal_processed_data[i + 5];
    sum6 += internal_processed_data[i + 6];
    sum7 += internal_processed_data[i + 7];
  }
    sum += sum0 + sum1 + sum2 + sum3 + sum4 +
           sum5 + sum6 + sum7;

} /* block loop */

/* release on chip memory semaphore */
SEM_post(g_aos_onchip_avail_sem);
```

Figure 8.48: The Same Function Enhanced to Use the DMA. This Function is 36 Executable Lines of Code but Runs Much Faster Than the Code in Figure 8.42

The code in Figure 8.48 runs much faster than the code in Figure 8.47. The penalty is an increased number of lines of code, which takes up memory space. This may or may not be a problem, depending on the memory constraints of the system. Another drawback to the code in Figure 8.48 is that it is a bit less understandable and portable than the code in Figure 8.47. Implementing the code to use the DMA requires the programmer to make the code less readable, which could possibly lead to maintainability problems. The code is now also tuned for a specific DSP. Porting the code to another DSP family may require the programmer to rewrite this code to use the different resources on the new DSP.

8.18.2   Pending versus Polling

The DMA can be considered a resource, just like memory and the CPU. When a DMA operation is in progress, the application can either wait for the DMA transfer to complete or continue processing another part of the application until the data transfer is complete. There are advantages and disadvantages to each approach. If the application waits for the DMA transfer to complete, it must poll the DMA hardware status register until the completion bit is set. This requires the CPU to check the DMA status register in a looping operation that wastes valuable CPU cycles. If the transfer is short enough, this may only require a few cycles to do and may be appropriate. For longer data transfers, the application engineer may want to use a synchronization mechanism like a semaphore to signal when the transfer is complete. In this case, the application will *pend* on a semaphore through the operating system while the transfer is taking place. The application will be swapped with another application that is ready to run. This swapping of tasks incurs overhead as well and should not be performed unless the overhead associated with swapping tasks is less than the overhead associated with simply polling on the DMA completion. The wait time depends on the amount of data being transferred.

```
if (transfer_length < LARGE_TRANSFER)
        IO_DRIVER();
else
        IO_LARGE_DRIVER();
endif
```

Figure 8.49: A Code Snippet that Checks for the Transfer Length and Calls a Driver Function that Will Either Poll the DMA Completion Bit in the DSP Status Register or Pend on an Operating System Semaphore

Figure 8.50 shows some code that checks for the transfer length and performs either a DMA polling operation (if there are only a few words to transfer), or a semaphore pend operation (for larger data transfer sizes). The "break even" length for data size is dependent on the processor and the interface structure and should be prototyped to determine the optimal size.

```
/* wait for port to become available */
while(g_io_channel_status[dir] & ACTIVE_MASK)
{
          /* poll */
}

/* submittcb*/
*(g_io_chain_queue_a[dir]) = (unsigned int)tcb;

/* wait for transfer to complete */
sem_status = SEM_pend(handle, SYS_FOREVER);
```

Figure 8.50: A Code Snippet that Pends on a Semaphore for DMA Completion

The code for a pend operation is shown in Figure 8.50. In this case, the application will perform a SEM_pend operation to wait for the DMA transfer to complete. This allows the application to perform other meaningful work by temporarily suspending the currently executing task and switching to another task to perform other processing. When the operating system suspends one task and begins executing another task, a certain amount of overhead is incurred. The amount of this overhead is dependent on the DSP and operating system.

Figure 8.52 shows the code for the polling operation. In this example, the application will continue polling the DMA completion status register for the operation to complete. This requires the use of the CPU to perform the polling operation. Doing this prevents the CPU from doing other meaningful work. If the transfer is short enough and the CPU only has to poll the status register for a short period of time, this approach may be more efficient. The decision is based on how much data is being transferred and how many cycles the CPU must spend polling. If the poll takes less time than the overhead to go through the operating system to swap out one task and begin executing another, this approach may be more efficient. In that context, the code snippet below checks for the transfer length and, if the length is less than the breakeven transfer length, a function will be called to poll the DMA transfer completion status. If the length is greater than the predetermined cutoff transfer length, a function will be called to set-up the DMA to interrupt on completion of the transfer. This ensures the most efficient processing of the completion of each DMA operation.

```
/* wait for port to become available */
while (g_io_channel_status[dir] & ACTIVE_MASK)
{
            /* poll */
}

/* submittcb*/
*(g_io_chain_queue_a[dir]) = (unsigned int)tcb;

/* wait for transfer to complete by polling the
   DMA status register */

status = *((vol_uint)*)g_io_channel_status[dir];
while  ((status & DMA_ACTIVE_MASK) ==
            DMA_CHANNEL_ACTIVE_MASK)
{
            status = *((vol_uint*)g_io_channel_status[dir];
}
.
.
.
```

Figure 8.51: A Code Snippet that Polls for DMA Completion

8.18.3   *Managing Internal Memory*

One of the most important resources for a DSP is its on-chip or internal memory. This is the area where data for most computations should reside because access to this memory is so much faster than off-chip or external memory. Since many DSPs do not have a data cache because of determinism unpredictability, software designers should think of a DSP internal memory as a sort of programmer managed cache. Instead of the hardware on the processor caching data for performance improvements with no control by the programmer, the DSP internal data memory is under full control of the DSP programmer. Using the DMA, data can be cycled in and out of the internal memory in the background, with little or no intervention by the DSP CPU. If managed correctly and efficiently, this internal memory can be a very valuable resource.

It is important to map out the use of internal memory and manage where data is going in the internal memory at all times. Given the limited amount of internal memory for many applications, not all the program's data can reside in internal memory for the duration of the application timeline. Over time, data will be moved to internal memory, processed, perhaps used again, and moved to external memory when it is no longer needed. Figure 8.53 shows an example of how a memory map of internal DSP memory might look during the timeline of the application. During the execution of the application, different data structures will be moved to on-chip memory, processed to form additional structures on chip, and eventually be moved off-chip to external memory to be saved, or overwritten in internal memory when the data is no longer needed.

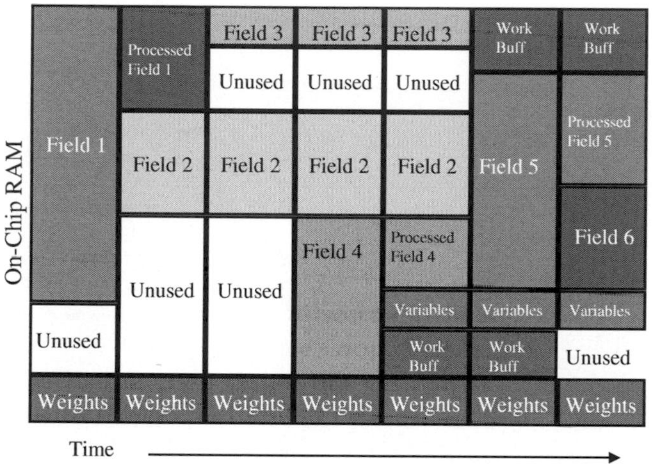

Figure 8.52: Internal Memory of a DSP Must Be Managed by the Programmer

8.19   Loop Unrolling

The standard rule when programming superscalar and VLIW devices is "Keep the pipelines full!" A full pipe means efficient code. In order to determine how full the pipelines are, you need to spend some time inspecting the assembly language code generated by the compiler.

To demonstrate the advantages of parallelism in VLIW-based machines, let's start with a simple looping program shown in Figure 8.53. If we were to write a serial assembly language implementation of this, the code would be similar to that in Figure 8.54. This loop uses one of the two available sides of the superscalar machine. By counting up the instructions and the NOPs, it takes 26 cycles to execute each iteration of the loop. We should be able to do much better.

8.19.1   Filling the Execution Units

There are two things to notice in this example. Many of the execution units are not being used and are sitting idle. This is a waste of processor hardware. Second, there are many delay slots in this piece of assembly (20 to be exact) where the CPU is stalled waiting for data to be loaded or stored. When the CPU is stalled, nothing is happening. This is the worst thing you can do to a processor when trying to crunch large amounts of data.

There are ways to keep the CPU busy while it is waiting for data to arrive. We can be doing other operations that are not dependent on the data we are waiting for. We can also use both sides of the VLIW architecture to help us load and store other data values. The code in Figure 8.55 shows an implementation designed for a CPU with multiple execution units. While the assembly looks very much like conventional serial assembly, the run-time execution is very unconventional. Instead of each line representing a single instruction that is completed before the next instruction is begun, each line in Figure 8.55 represents an individual operation, an operation that might be scheduled to execute in parallel with some other operation. The assembly format has been extended to allow the programmer to specify which execution unit should perform a particular operation and which operations may be scheduled concurrently. The DSP compiler automatically determines which execution unit to use for an operation and indicates this by naming the target unit in the extra column that precedes the operand fields (the column containing D1, D2, and so on) in the assembly listing. To indicate that two or more operations may proceed in parallel, the lines describing the individual operations are "joined" with a parallel bar (as in lines 4 and 5 of Figure 8.55). The parallelism rules are also determined by the compiler. Keep in mind that if the programmer decides to program the application using assembly language, the responsibility for scheduling the instructions on each of the available execution units as well as determining the parallelism rules falls on the programmer. This is a difficult task and should only be done when the compiler generated assembly does not have the required performance.

The code in Figure 8.55 is an improvement over the serial version. We have reduced the number of NOPs from 20 to 5. We are also performing some steps in parallel. Lines 4 and 5 are executing two loads at the same time into each of the two load units (D1 and D2) of the device. This code is also performing the branch operation earlier in the loop and then taking advantage of the delays associated with that operation to complete operations on the current cycle.

The code in Figure 8.55 shows an implementation designed for a CPU with multiple execution units. While the assembly looks very much like conventional serial assembly, the run-time execution is very unconventional. Instead of each line representing a single instruction that is completed before the next instruction is begun, each line in Figure 8.55 represents an individual *operation*, an operation that might be scheduled to execute in parallel with some other operation. The assembly format has been extended to allow the

```
1     void example1(float *out, float *input1, float *input2)
2     {
3      int i;
4
5      for(i = 0; i < 100; i++)
6        {
7          out[i] = input1[i] * input2[i];
8        }
9     }
```

Figure 8.53: Simple for Loop in C

```
1    ;
2    ;     serial implementation of loop (26 cycles per iteration)
3    ;
4    L1:         LDW        *B++,B5       ;load B[i] into B5
5                NOP        4             ; wait for load to complete
6
7                LDW        *A++,A4       ; load A[i] into A4
8                NOP        4             ; wait for load to complete
9
10               MPYSP      B5,A4,A4      ; A4 = A4 * B5
11               NOP        3             ; wait formult to complete
12
13               STW        A4,*C++       ; store A4 in C[i]
14               NOP        4             ; wait got store to complete
15
16               SUB        i,1,i         ; decrement i
17         [i]   B          L1            ; if i != 0, gotoL1
18               NOP        5             ; delay for branch
```

Figure 8.54: Serial Assembly Language Implementation of C Loop

programmer to specify which execution unit should perform a particular operation and which operations may be scheduled concurrently. The programmer specifies which execution unit to use for an operation by naming the target unit in the extra column that precedes the operand fields (the column containing D1, D2, and so on). To indicate that two or more operations may proceed in parallel, the lines describing the individual operations are "joined" with a parallel bar (as in lines 4 and 5 of Figure 8.55).

```
1   ;   using delay slots and duplicate execution units of the device
2   ;   10 cycles per iteration
3
4   L1:         LDW       .D2    *B++,B5     ;load B[i] into B5
5   ||          LDW       .D1    *A++,A4     ;load A[i] into A4
6
7               NOP              2           ;wait load to complete
8               SUB       .L2    i,1,i       ;decrement i
9         [i]   B         .S1    L1          ;if i != 0, goto L1
10
11              MPYSP     .M1X   B5,A4,A4    ;A4 = A4 * B5
12              NOP              3           ;wait mpy to complete
13
14              STW       .D1    A4,*C++     ;store A4 into C[i]
```

Figure 8.55: A More Parallel Implementation of the C Loop

8.19.2 *Reducing Loop Overhead*

Loop unrolling is a technique used to increase the number of instructions executed between executions of the loop branch logic. This reduces the number of times the loop branch logic is executed. Since the loop branch logic is overhead, reducing the number of times this has to execute reduces the overhead and makes the loop body, the important part of the structure, run faster. A loop can be unrolled by replicating the loop body a number of times and then changing the termination logic to comprehend the multiple iterations of the loop body (Figure 8.56). The loops in Figures 8.56a and b each take four cycles to execute, but the loop in Figure 8.56b is doing four times as much work! This is illustrated in Figure 8.57. The assembly language kernel of this loop is shown in Figure 8.57a. The mapping of variables from the loop to the processor is shown in Figure 8.57b. The compiler is able to structure this loop such that all required resources are stored in the register file, and the work is spread across several of the execution units. The work done by cycle for each of these units is shown in Figure 6.57c.

```
a.    for (i = 0;  i < 128;  i ++)
      {
            sum1 + = const[i] * input[128 - i];
      }

b.    for (i = 0;  i < 32;  i ++)
      {
            sum1 += const[i] * input[128 - i];
            sum2 += const[2*i] * input[128 - (2*i)];
            sum3 += const[3*i] * input[128 - (3*i)];
            sum4 += const[4*i] * input[128 - (4*i)];
      }
```

Figure 8.56: Loop Unrolling (a) A Simple Loop (b) The Same Loop Unrolled Four Times

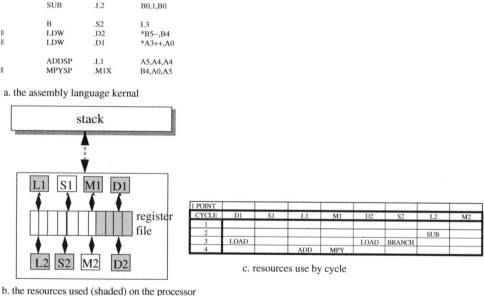

```
NOP                        1
SUB          .L2           B0,1,B0

      B      .S2           L3
||    LDW    .D2           *B5--,B4
||    LDW    .D1           *A3++,A0

      ADDSP  .L1           A5,A4,A4
||    MPYSP  .M1X          B4,A0,A5
```

a. the assembly language kernal

stack

register file

1 POINT								
CYCLE	D1	S1	L1	M1	D2	S2	L2	M2
1								
2						SUB		
3	LOAD				LOAD	BRANCH		
4			ADD	MPY				

b. the resources used (shaded) on the processor

c. resources use by cycle

Figure 8.57: Implementation of a Simple Loop

Now look at the implementation of the loop unrolled four times in Figure 8.58. Again, only the assembly language for the loop kernel is shown. Notice that more of the register file is being used to store the variables needed in the larger loop kernel. An additional execution unit is also being used, as well as a several bytes from the stack in external memory

(Figure 8.58b). In addition, the execution unit utilization shown in Figure 8.58c indicates the execution units are being used more efficiently while still maintaining a four cycle latency to complete the loop. This is an example of using all the available resources of the device to gain significant speed improvements. Although the code size looks bigger, it actually runs faster than the loop in Figure 8.56a.

8.19.3   *Fitting the Loop to Register Space*

Unrolling too much can cause performance problems. In Figure 8.59, the loop is unrolled eight times. At this point, the compiler cannot find enough registers to map all the required variables. When this happens, variables start getting stored on the stack, which is usually in external memory somewhere. This is expensive because instead of a single cycle read, it can now take many cycles to read each of the variables each time it is needed. This causes things to break down, as shown in Figure 8.59. The obvious problems are the number of bytes that are now being stored in external memory (88 versus 8 before) and the lack of parallelism in the assembly language loop kernel. The actual kernel assembly language was

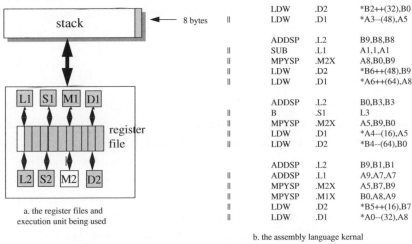

```
          LDW     .D2    *B2++(32),B0
      ||  LDW     .D1    *A3--(48),A5

          ADDSP   .L2    B9,B8,B8
      ||  SUB     .L1    A1,1,A1
      ||  MPYSP   .M2X   A8,B0,B9
      ||  LDW     .D2    *B6++(48),B9
      ||  LDW     .D1    *A6++(64),A8

          ADDSP   .L2    B0,B3,B3
      ||  B       .S1    L3
      ||  MPYSP   .M2X   A5,B9,B0
      ||  LDW     .D1    *A4--(16),A5
      ||  LDW     .D2    *B4--(64),B0

          ADDSP   .L2    B9,B1,B1
      ||  ADDSP   .L1    A9,A7,A7
      ||  MPYSP   .M2X   A5,B7,B9
      ||  MPYSP   .M1X   B0,A8,A9
      ||  LDW     .D2    *B5++(16),B7
      ||  LDW     .D1    *A0--(32),A8
```

a. the register files and
execution unit being used

b. the assembly language kernal

4 POINT								
CYCLE	D1	S1	L1	M1	D2	S2	L2	M2
1	LOAD				LOAD			
2	LOAD		SUB		LOAD		ADD	MPY
3	LOAD	BRANCH			LOAD		ADD	MPY
4	LOAD		ADD	MPY	LOAD		ADD	MPY

c. utilization of the execution units

Figure 8.58: Implementation of a Loop Unrolled Four Times

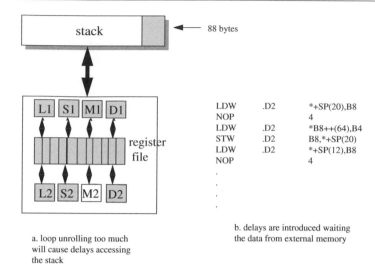

stack ← 88 bytes

L1	S1	M1	D1

register file

L2	S2	M2	D2

```
LDW    .D2    *+SP(20),B8
NOP           4
LDW    .D2    *B8++(64),B4
STW    .D2    B8,*+SP(20)
LDW    .D2    *+SP(12),B8
NOP           4
  .
  .
  .
```

a. loop unrolling too much will cause delays accessing the stack

b. delays are introduced waiting the data from external memory

Figure 8.59: Loop Unrolled Eight Times. Too Much of a Good Thing!

very long and inefficient. A small part of it is shown in Figure 8.59b. Notice the lack of "||" instructions and the new "NOP" instructions. These effectively stalls to the CPU when nothing else can happen. The CPU is waiting for data to arrive from external memory.

8.19.4   Trade-offs

The drawback to loop unrolling is that it uses more registers in the register file as well as execution units. Different registers need to be used for each iteration. Once the available registers are used, the processor starts going to the stack to store required data. Going to the off-chip stack is expensive and may wipe out the gains achieved by unrolling the loop in the first place. Loop unrolling should only be used when the operations in a single iteration of the loop do not use all of the available resources of the processor architecture. Check the assembly language output if you are not sure of this. Another drawback is the code size increase. As you can see in Figure 8.58, the unrolled loop, albeit faster, requires more instructions and, therefore, more memory.

8.20   Software Pipelining

One of the best performance strategies for the DSP programmer is writing code that can be pipelined efficiently by the compiler. Software pipelining is an optimization strategy to schedule loops and functional units efficiently. In software pipelining, operations from different iterations of a software loop are performed in parallel. In each iteration, intermediate results generated by the previous iteration are used. Each iteration will also perform operations whose intermediate results will be used in the next iteration. This technique produces highly optimized code through maximum use of the processor functional units. The advantage to this approach is that most of the scheduling associated with software pipelining is performed by the compiler and not by the programmer (unless the programmer is writing code at the assembly language level). There are certain conditions that must be satisfied for this to work properly and we will talk about that shortly.

DSPs may have multiple functional units available for use while a piece of code is executing. In the case of the TMS320C6X family of VLIW DSPs, there are eight functional units that can be used at the same time, if the compiler can determine how to utilize all of them efficiently. Sometimes, subtle changes in the way the C code is structured can make all the difference. In software pipelining, multiple iterations of a loop are scheduled to execute in parallel. The loop is reorganized so that each iteration in the pipelined code is made from instruction sequences selected from different iterations in the original loop. In the example in Figure 8.60, a five-stage loop with three iterations is shown. As we discussed earlier, the initial period (cycles n and n+1), called the prolog, is when the pipes are being "primed" or initially loaded with operations. Cycles n+2 to n+4 are the actual pipelined section of the code. It is in this section that the processor is performing three different operations (C, B, and A) for three different loops (1, 2, and 3). The epilog section is where the last remaining instructions are performed before exiting the loop. This is an example of a fully utilized set of pipelines that produces the fastest, most efficient code.

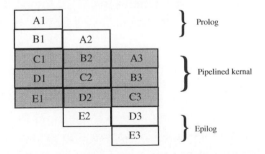

Figure 8.60: A Five Stage Pipe that Is Software Pipelined

We saw earlier how loop unrolling offers speed improvements over simple loops. Software pipelining can be faster than loop unrolling for certain sections of code because, with loop unrolling, the prolog and epilog are only performed once (Figure 8.61).

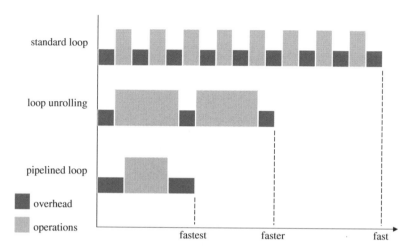

Figure 8.61: Standard Loop Overhead versus Loop Unrolling and Software Pipelining. The Standard Loop Uses a Loop Check Each Iteration through the Loop. This Is Considered Overhead. Loop Unrolling Still Checks the Loop Count but Less often. Software Pipelining Substitutes the Loop Check with Prolog and Epilog Operations that Prime and Empty a Pipelined Loop Operation that Performs Many Iterations of the Loop in Parallel

8.20.1 An Example

To demonstrate this technique, let's look at an example. In Figure 8.62, a simple loop is implemented in C. This loop simply multiplies elements from two arrays and stores the result in another array.

8.20.1.1 A Serial Implementation

The assembly language output for this simple loop is shown in Figure 8.63. This is a serial assembly language implementation of the C loop, serial in the sense that there is no real parallelism (use of the other processor resources) taking place in the code. This is easy to see by the abundance of NOP operations in the code. These NOPs are instructions that do nothing but wait and burn CPU cycles. These are required and inserted in the code when the CPU is waiting for memory fetches or writes to complete. Since a load operation takes five

```
1     void example1(float *out, float *input1, float *input2)
2     {
3      int i;
4
5      for(i = 0; i < 100; i++)
6      {
7        out[i] = input1[i] * input2[i];
8      }
9     }
```

Figure 8.62: A Simple Loop in C

```
1    ;
2    ;      serial implementation of loop (26 cycles per iteration)
3    ;
4    L1:         LDW        *B++,B5    ;load B[i] into B5
5                NOP        4          ; wait for load to complete
6
7                LDW        *A++,A4    ; load A[i] into A4
8                NOP        4          ; wait for load to complete
9
10               MPYSP      B5,A4,A4   ; A4 = A4 * B5
11               NOP        3          ; wait formult to complete
12
13               STW        A4,*C++    ; store A4 in C[i]
14               NOP        4          ; wait got store to complete
15
16               SUB        i,1,i      ; decrement i
17        [i]    B          L1         ; if i != 0, goto L1
18               NOP        5          ; delay for branch
```

Figure 8.63: Assembly Language Output of the Simple Loop

CPU cycles to complete, the load word operation in line 4 (LDW) is followed by a NOP with a length of 4 indicating that the CPU must now wait four additional cycles for the data to arrive in the appropriate register before it can be used. These NOPs are extremely inefficient and the programmer should endeavor to remove as many of these delay slots as possible to improve performance of the system. One way of doing this is to take advantage of the other DSP functional resources.

8.20.1.2 A Minimally Parallel Implementation

A more parallel implementation of the same C loop is shown in Figure 8.64. In this implementation, the compiler is able to use more of the functional units on the DSP. Line 4 shows a load operation, similar to the previous example. However, this load is explicitly loaded into the D2 functional unit on the DSP. This instruction is followed by another load into the D1 unit of the DSP. What is going on here? While the first load is taking place, moving data into the D2 unit, another load is initiated that loads data into the D1 register. These operations can be performed during the same clock cycle because the destination of the data is different. These values can be preloaded for use later and we do not have to waste as many clock cycles waiting for data to move. As you can see in this code listing, there are now only two NOP cycles required instead of four. This is a step in the right direction. The "||" symbol means that the two loads are performed during the same clock cycle.

```
1   ;   using delay slots and duplicate execution units of the device
2   ;   10 cycles per iteration
3
4   L1:          LDW        .D2     *B++,B5      ;load B[i] into B5
5   ||           LDW        .D1     *A++,A4      ;load A[i] into A4
6
7                NOP                2            ;wait load to complete
8                SUB        .L2     i,1,i        ;decrement i
9        [i]     B          .S1     L1           ;if i != 0, goto L1
10
11               MPYSP    .M1X      B5,A4,A4     ;A4 = A4 * B5
12               NOP        3                    ;wait mpy to complete
13
14               STW        .D1     A4,*C++      ;store A4 into C[i]
```

Figure 8.64: Assembly Language Output of the Simple Loop Exploiting the Parallel Orthogonal Execution Units of the DSP

8.20.1.3 Compiler-Generated Pipeline

Figure 8.65 shows the same sample piece of C code and the corresponding assembly language output. In this case, the compiler was asked (via a compile switch) to attempt to pipeline the code. This is evident by the piped loop prolog and piped loop kernel sections in the assembly language output. In this case, the pipelined code is not as good as it could be.

```
1      void example1(float *out, float *input1, float *input2)
2      {
3        int i;
4
5        for(i = 0; i < 100; i++)
6          {
7            out[i] = input1[i] * input2[i];
8          }
9      }


1      _example1:
2      ;** ----------------------------------------------------------*
3                    MVK          .S2          0x64,B0
4
5                    MVC          .S2          CSR,B6
6      ||            MV           .L1X         B4,A3
7      ||            MV           .L2X         A6,B5

8                    AND          .L1X         -2,B6,A0
9                    MVC          .S2X         A0,CSR
10     ;** ---------------------------------------------------------*
11     L11:          ; PIPED LOOP PROLOG
12     ;** ---------------------------------------------------------*
13     L12:          ; PIPED LOOP KERNEL

14                   LDW          .D2          *B5++,B4      ;
15     ||            LDW          .D1          *A3++,A0      ;

16                   NOP                         2
17     [ B0]         SUB          .L2          B0,1,B0       ;
18     [ B0]         B            .S2          L12           ;
19                   MPYSP        .M1X         B4,A0,A0      ;
20                   NOP                       3
21                   STW          .D1          A0,*A4++      ;
22     ;** ----------------------------------------------------------*
23                   MVC          .S2          B6,CSR
24                   B            .S2          B3
25                   NOP                       5
26                   ; BRANCH OCCURS
```

Figure 8.65: C Example and the Corresponding Pipelined Assembly Language Output

Inefficient code can be located by looking for how many NOPs there are in the piped loop kernel of the code. In this case the piped loop kernel has a total of five NOP cycles, two in line 16, and three in line 20. This loop takes a total of ten cycles to execute. The NOPs are the first indication that a more efficient loop may be possible. But how short can this loop be? One way to estimate the minimum loop size is to determine what execution unit is being used the most. In this example, the D unit is used more than any other unit, a total of three times (lines 14, 15, and 21). There are two sides to this VLIW device, enabling each unit to be used twice (D1 and D2) per clock for a minimum two clock loop; two D operations in one clock and one D unit in the second clock. The compiler was smart enough to use the D units on both sides of the pipe (lines 14 and 15), enabling it to parallelize the instructions and only use one clock. It should be possible to perform other instructions while waiting for the loads to complete, instead of delaying with NOPs.

8.20.1.4   *An Implementation with Restrict Keyword*

In the simple "for" loop, it is apparent that the inputs are not dependent on the output. In other words, there are no dependencies. But the compiler does not know that. Compilers are generally pessimistic creatures. They will not optimize something if the situation is not totally understood. The compiler takes the conservative approach and assumes the inputs can be dependent on the previous output each time through the loop. If it is known that the inputs are not dependent on the output, we can hint to the compiler by declaring the input1 and input2 as "restrict," indicating that these fields will not change. This is a trigger for enabling software pipelining and saving throughput. This C code is shown in Figure 8.66 with the corresponding assembly language.

There are a few things to notice in looking at this assembly language. First, the piped loop kernel has become smaller. In fact, the loop is now only two cycles long. Lines 44–47 are all executed in one cycle (the parallel instructions are indicated by the || symbol) and lines 48–50 are executed in the second cycle of the loop. The compiler, with the additional dependency information we supplied with the "restrict" declaration, has been able to take advantage of the parallelism in the execution units to schedule the inner part of the loop very efficiently. But this comes at a price. The prolog and epilog portions of the code are much larger now. Tighter piped kernels will require more priming operations to coordinate all of the execution based on the various instruction and branching delays. But once primed, the kernel loop executes extremely fast, performing operations on various iterations of the loop. The goal of software pipelining is, like we mentioned earlier, to make the common case fast. The kernel is the common case in this example, and we have made it very fast. Pipelined

```
1   void example2(float  *out, restrict float *input1, restrict float *input2)
2   {
3     int i;
4
5     for(i = 0;  i < 100;  i++)
6       {
7           out[i] = input1[i] * input2[i];
8       }
9   }



1   _example2:
2   ;** ----------------------------------------------------------*
3               MVK             .S2             0x64,B0
4
5               MVC             .S2             CSR,B6
6   ||          MV              .L1X            B4,A3
7   ||          MV              .L2X            A6,B5
8
9               AND             .L1X            -2,B6,A0
10
11              MVC             .S2X            A0,CSR
12  ||          SUB             .L2             B0,4,B0
13
14  ;** -----------------------------------------------------------*
15  L8:         ; PIPED LOOP PROLOG
16
17              LDW             .D2             *B5++,B4        ;
18  ||          LDW             .D1             *A3++,A0        ;
19
20              NOP                             1
21
22              LDW             .D2             *B5++,B4        ;@
23  ||          LDW             .D1             *A3++,A0        ;@
24
25      [ B0]   SUB             .L2             B0,1,B0         ;
```

Figure 8.66: Corresponding Pipelined Assembly Language Output

code may not be worth doing for loops with a small loop count. However, for loops with a large loop count, executing thousands of times, software pipelining is the only way to go.

In the two cycles the piped kernel takes to execute, there are a lot of things going on. The right hand column in the assembly listing indicates what iteration is being performed by each instruction. Each "@" symbol is a iteration count. So, in this kernel, line 44 is performing a branch for iteration n+2, lines 45 and 46 are performing loads for iteration n+4, line 48 is storing a result for iteration n, line 49 is performing a multiply for iteration n+2, and line 50 is performing a subtraction for iteration n+3, all in two cycles! The epilog is completing the operations once the piped kernel stops executing. The compiler was able to make the loop two cycles long, which is what we predicted by looking at the inefficient version of the code.

The code size for a pipelined function becomes larger, as is obvious by looking at the code produced. This is one of the trade-offs for speed that the programmer must make.

In summary, when processing arrays of data (which is common in many DSP applications) the programmer must inform the compiler when arrays are not dependent on each other. The compiler must assume that the data arrays can be anywhere in memory, even overlapping each other. Unless informed of array independence, the compiler will assume the next load operation requires the previous store operation to complete (as to not load stale data). Independent data structures allows the compiler to structure the code to load from the input array before storing the last output. Basically, if two arrays are not pointing to the same place in memory, using the "restrict" keyword to indicate this independence will improve performance. Another term for this technique is ***memory disambiguation***.

8.20.2   Enabling Software Pipelining

The compiler must decide what variables to put on the stack (which take longer to access) and which variables to put in the fast on-chip registers. This is part of the register allocator of a compiler. If a loop contains too many operations to make efficient use of the processor registers, the compiler may decide to not pipeline the loop. In cases like that, it may make sense to break up the loop into smaller loops that will enable the compiler to pipeline each of the smaller loops (Figure 8.67).

The compiler will not attempt to software pipeline a loop when there are not enough resources (execution units, registers, and so on) to allow it, or if the compiler determines

```
Instead of:                       Try:

for (expression)                  for (expression) {
{                                           Do A          }
        Do A                      for (expression) {
        Do B                                Do B          }
        Do C                      for (expression) {
        Do D                                Do C          }
}                                 for (expression) {
                                            Do D          }
```

Figure 8.67: Breaking Up Larger Loops into Smaller Loops May Enable Each Loop to Be Pipelined More Efficiently

that it is not worth the effort to pipeline a loop because the benefit does not outweigh the gain (for example, the amount of cycles required to produce the prolog and epilog far outweighs the amount of cycles saved in the loop kernel). But the programmer can intervene in some cases to improve the situation. With careful analysis and structuring of the code, the programmer can make the necessary modification at the high language level to allow certain loops to pipeline. For example, some loops have so many processing requirements inside the loop that the compiler cannot find enough registers and execution units to map all the required data and instructions. When this happens, the compiler will not attempt to pipeline the loop. Also, function calls within a loop will not be pipelined because the compiler has a hard time resolving the function call. Instead, if you want a pipelined loop, replace the function call with an inline expansion of the function.

8.20.3   Interrupts and Pipelined Code

Because an interrupt in the middle of a fully primed pipe destroys the synergy in instruction execution, the compiler may protect a software pipelining operation by disabling interrupts before entering the pipelined section and enabling interrupts on the way out (Figure 8.68). Lines 11 and 69 of Figure 8.66 show interrupts being disabled prior to the prolog and enabled again just after completing the epilog. This means that the price of the efficiency in software pipelining is paid for in a nonpreemptible section of code. The programmer must be able to determine the impact of sections of nonpreemptible code on real time performance. This is not a problem for single task applications. But it may have an impact on systems built using a tasking architecture. Each of the software pipelined sections must be considered a blocking term in the overall tasking equation.

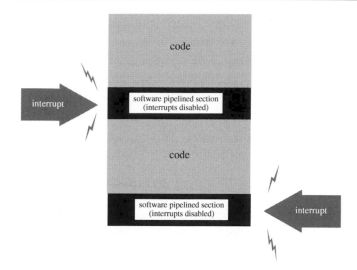

Figure 8.68: Interrupts May Be Disabled during a Software Pipelined Section of Code

8.21   More on DSP Compilers and Optimization

Not too long ago, it was an unwritten rule that writing programs in assembly would usually result in better performance than writing in higher level languages like C or C++. The early "optimizing" compilers solved the problem of "optimization" at too general and simplistic a level. The results were not as good as a good assembly language programmer. Compilers have gotten much better and today there are very specific high performance optimizations that compete well with even the best assembly language programmers.

Optimizing compilers perform sophisticated program analysis including intra-procedural and interprocedural analysis. These compilers also perform data flow and control flow analysis as well as dependence analysis and often require provably correct methods for modifying or transforming code. Much of this analysis is to prove that the transformation is correct in the general sense. Many optimization strategies used in DSP compilers are strongly heuristic.[6]

---

[6] Heuristics involves problem solving by experimental and especially trial-and-error methods or relating to exploratory problem-solving techniques that utilize self-educating techniques (as the evaluation of feedback) to improve performance (*Webster's English Language Dictionary*).

8.21.1 Compiler Architecture and Flow

The general architecture of a modern compiler is shown in Figure 8.69. The front end of the compiler reads in the DSP source code, determines whether the input is legal, detects and reports errors, obtains the meaning of the input, and creates an intermediate representation of the source code. The intermediate stage of the compiler is called the *optimizer*. The optimizer performs a set of optimization techniques on the code including:

- Control flow optimizations.

- Local optimizations.

- Global optimizations.

The back end of the compiler generates the target code from the intermediate code, performs the optimizations on the code for the specific target machine, and performs instruction selection, instruction scheduling and register allocation to minimize memory bandwidth, and finally outputs object code to be run on the target.

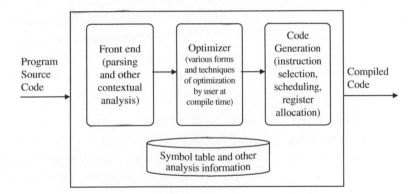

Figure 8.69: General Architecture of a Compiler

8.21.2 Compiler Optimizations

Compilers perform what are called *machine independent* and *machine dependent* optimizations. Machine independent optimizations are those that are not dependent on the architecture of the device. Examples of this are:

- **Branch optimization**—This is a technique that rearranges the program code to minimize branching logic and to combine physically separate blocks of code.

- **Loop invariant code motion**—If variables used in a computation within a loop are not altered within the loop, the calculation can be performed outside of the loop and the results used within the loop (Figure 8.70).

```
do i = 1,100                  j = 100
        j = 10                do i = 1,100
        x(i) = x(i) + j               x(i) = x(i) + j
enddo                         enddo
.......                       .......
........                      ........
```

Figure 8.70: Example of Code Motion Optimization

- **Loop unrolling**—In this technique, the compiler replicates the loop's body and adjusts the logic that controls the number of iterations performed. Now the code effectively performs the same useful work with less comparisons and branches. The compiler may or may not know the loop count. This approach reduces the total number of operations but also increases code size. This may or may not be an issue. If the resultant code size is too large for the device cache (if one is being used), then the resulting cache miss penalty can overcome any benefit in loop overhead. An example of loop unrolling was discussed earlier in the chapter.

- **Common subexpression elimination**—In common expressions, the same value is recalculated in a subsequent expression. The duplicate expression can be eliminated by using the previous value. The goal is to eliminate redundant or multiple computations. The compiler will compute the value once and store it in a temporary variable for subsequent reuse.

- **Constant propagation**—In this technique, constants used in an expression are combined, and new constants are generated. Some implicit conversions between integers and floating-point types may also be done. The goal is to save memory by the removal of these equivalent variables.

- **Dead code elimination**—This approach attempts to eliminate code that cannot be reached or where the results are not subsequently used.

- **Dead store elimination**—This optimization technique will try to eliminate stores to memory when the value stored will never be referenced in the future. An example of this approach is code that performs two stores to the same location without having an intervening load. The compiler will remove the first store because it is unnecessary.

- **Global register allocation**—This optimization technique allocates variables and expressions to available hardware registers using a "graph coloring" algorithm.[7]

- **Inlining**—Inlining replaces function calls with actual program code (Figure 8.71). This can speed up the execution of the software by not having to perform function calls with the associated overhead. The disadvantage to inlining is that the program size will increase.

```
do i = 1,n                      do i = 1,n
    j = k(i)                        j = k(i)
    call subroutine(a(i),j)         temp1 = a(i) * y
    call subroutine(b(i),j)         temp2 = a(i) / y
    call subroutine(c(i),j)         temp3 = temp1 + temp2
.........                            temp1 = b(i) * y
                                    temp2 = b(i) / y
subroutine INL(x,y)                 temp3 = temp1 + temp2
    temp1 = x * y                   temp1 = c(i) * y
    temp2 = x / y                   temp2 = c(i) / y
    temp3 = temp1 + temp2           temp3 = temp1 + temp2
end
```

Figure 8.71: Inlining Replaces Function Calls with Actual Code that Increases Performance but May Increase Program Size

- **Strength reduction**—The basic approach with this form of optimization is to use cheaper operations instead of more expensive ones. A simple example of this is to

[7] The problem of assigning data values to registers is a key challenge for compiler engineers. This problem can be converted into one of graph-coloring. In this approach, attempting to color a graph with N colors is equivalent to attempting to allocate data into N registers. Graph coloring is then the partitioning of the vertices of a graph into a minimum number of independent sets.

use a compound assignment operator instead of an expanded one, since fewer instructions are needed:

```
Instead of:
for ( i=0; i < array_length; i++)
        a[i] = a[i] + constant;

Use:
for ( i=0; i < array_length; i++)
        a[i] += constant;
```

Another example of a strength reduction optimization is using shifts instead of multiplication by powers of two.

- **Alias disambiguation**—Aliasing occurs if two or more symbols, pointer references, or structure references refer to the same memory location. This situation can prevent the compiler from retaining values in registers because it cannot be certain that the register and memory continue to hold the same values over time. Alias disambiguation is a compiler technique that determines when two pointer expressions cannot point to the same location. This allows the compiler to freely optimize these expressions.

- **Inline expansion of runtime-support library functions**—This optimization technique replaces calls to small functions with inline code. This saves the overhead associated with a function call and provides increased opportunities to apply other optimizations.

The programmer has control over the various optimization approaches in a compiler, from aggressive to none at all. Some specific controls are discussed in the next section.

Machine dependent optimizations are those that require some knowledge of the target machine in order to perform the optimization. Examples of this type of optimization include:

- **Implementing special features**—This includes instruction selection techniques that produce efficient code, selecting an efficient combination of machine dependent instructions that implement the intermediate representation in the compiler.

- **Latency**—This involves selecting the right instruction schedules to implement the selected instructions for the target machine. There are a large number of different schedules that can be chosen and the compiler must select one that gives an efficient overall schedule for the code.

- **Resources**—This involves register allocation techniques that includes analysis of program variables and selecting the right combination of registers to hold the variables for the optimum amount of time, such that the optimal memory bandwidth goals can be met. This technique mainly determines which variables should be in which registers at each point in the program.

8.21.2.1 *Instruction Selection*

Instruction selection is important in generating code for a target machine for a number of reasons. As an example, there may be some instructions on the processor that the C compiler cannot implement efficiently. Saturation is a good example. Many DSP applications perform saturation checks on video and image processing applications. To manually write code to saturate requires a lot of code (check sign bits, determine proper limit, and so on). Some DSPs, for example, can do a similar operation in one cycle or as part of another instruction (i.e., replace a multiply instruction, MPY with a saturate multiply, SMUL=1). However, a compiler is often unable to use these algorithm-specific instructions that the DSP provides. Therefore, the programmer often has to force their use. To get the C compiler to use specific assembly language instructions like this, one approach is to use what are called *intrinsics*. Intrinsics are implemented with assembly language instructions on the target processor. Some examples of DSP intrinsics include:

- short _abs(short src); absolute value

- long _labs(long src); long absolute value

- short _norm(short src); normalization

- long _rnd(long src); rounding

- short _sadd(short src1, short src2); saturated add

One benefit to using intrinsics like this is that they are automatically inlined. Since we want to run a processor instruction directly, we would not want to waste the overhead of doing a call. Since intrinsics also require things like the saturation flag to be set, they may be longer than one processor instruction. Intrinsics are better than using assembly language function calls since the compiler is ignorant of the contents of these assembly language functions and may not be able to make some needed optimizations.

```
C Code:                         Compiler Output:
int sadd(int a, int b)          _sadd:
{                                   MOV T1, AR1
  int result;                       XOR T0, T1
  result = a + b;                   BTST @#15, T1, TC1
  if (((a^b) & 0x8000) ==           ADD T0, AR1
0)                                  BCC L2,TC1
  {                                 MOV T0, AR2
    if ((result ^ a) &              XOR AR1, AR2
0x8000)                             BTST @#15, AR2, TC1
        result = ( a < 0)           BCC L2,!TC1
? 0x8000 : 0x7FFF;                  BCC L1,T0 < #0
  }                                 MOV #32767, T0
  return result;                    B L3
}                               L1:    MOV #-32768, AR1
                                L2:    MOV AR1, T0
                                L3:    return
```

Figure 8.72: Code for a Saturated Add Function

Figure 8.72 is an example of using C code to produce a saturated add function. The resulting assembly language is also shown for a C5x DSP. Notice the amount of C code and assembly code required to implement this basic function.

Now look at the same function in Figure 8.73 implemented with intrinsics. Notice the significant code size reduction using algorithm specific special instructions. The DSP designer should carefully analyze the algorithms required for the application and determine

```
C Code                          Compiler Output
int sadd(int a, int            _sadd:
  b)                               BSET ST3_SATA
{                                  ADD T1, T0
  return                           BCLR ST3_SATA
  _sadd(a,b);                      return
}
```

Figure 8.73: Saturated Add Using DSP Intrinsics

whether a specific DSP or family of DSPs supports the class of algorithm with these special instructions. The use of these special instructions in key areas of the application can have a significant impact on the overall performance of the system.

8.21.2.2   *Latency and Instruction Scheduling*

The order in which operations are executed on a DSP has a significant impact on length of time to execute a specific function. Different operations take different lengths of time due to differences in memory access times, and differences in the functional unit of the processor (different functional units in a DSP, for example, may require different lengths of time to complete a specific operation). If the conditions are not right, the processor may delay or stall. The compiler may be able to predict these unfavorable conditions and reorder some of the instructions to get a better schedule. In the worst case, the compiler may have to force the processor to wait for a period of time by inserting delays (sometimes called *NOP* for "no operation") into the instruction stream to force the processor to wait for a cycle or more for something to complete, such as a memory transfer.

Optimizing compilers have instruction schedulers to perform one major function: to reorder operations in the compiled code in an attempt to decrease its running time. DSP compilers have sophisticated schedulers that search for the optimal schedule (within reason; the compiler has to eventually terminate and produce something for the programmer!). The main goals of the scheduler are to preserve the meaning of the code (it can't "break" anything), minimize the overall execution time (by avoiding extra register spills to main memory for example), and operate as efficiently as possible from a usability standpoint.

From a DSP standpoint, loops are critical in many embedded DSP applications. Much of the signal processing performed by DSPs is in loops. Optimizing compilers for DSP often contain specialized loop schedulers to optimize this code. One of the most common examples of this is the function of software pipelining.

An example of software pipelining was given earlier in the chapter. Software pipelining is the execution of operations from different iterations of a software loop in parallel. In each iteration, intermediate results generated by the previous iteration are used and operations are also performed whose intermediate results will be used in the next iteration. This produces highly optimized code and makes maximum use of the processor functional units. Software pipelining is implemented by the compiler if the code structure is suited to making these transformations. In other words, the programmer must produce the right code structure to the compiler such that it can recognize the conditions are right to pipeline the loop. For example, when multiplying two arrays inside a loop, the programmer must inform the compiler when the two arrays do not point to the same space in memory. (Compilers must assume arrays can be anywhere in memory, even overlapping one another. Unless informed of array independence, they will assume the next load requires the previous load to complete. By informing the compiler of this independent structure (something as simple as using a keyword in the C code) allows the compiler to load from the input array before storing last output, as shown in the code snippet below where the "restrict" keyword is used to show this independence.

```
void example (float *out, restrict float *input1, restrict float *input2)
{
int i;
for (i=0; i<100; i++)
{
out[ i ] = input1[ i ] * input2[ i ];
}
}
```

The primary goal of instruction scheduling is to improve running time of generated code. But be careful how this is measured. For example, measuring the quality of the produced code using a simple measure such as "Instructions per second" is misleading. Although this is a common metric in many advertisements, it may not be indicative of the quality of code produced for the specific application running on the DSP. That is why developers should spend time measuring the time to complete a fixed representative task for the application in question. Using industry benchmarks to measure overall system performance is not a good idea because the information is too specific to be used in a broad sense. In reality there is no single metric that can accurately measure quality of code produced by the compiler. We will discuss this more later.

8.21.2.3 Register Allocation

On-chip DSP registers are the fastest locations in the memory hierarchy (see Figure 8.74). The primary responsibility of the register allocator is to make efficient use of the target registers. The register allocator works with the scheduled instructions generated by the instruction scheduler and finds an optimal arrangement for data and variables in the processors registers to avoid "spilling" data into main memory where it is more expensive to access (performance). By minimizing register spills the compiler will generate higher performing code by eliminating expensive reads and writes to main memory.

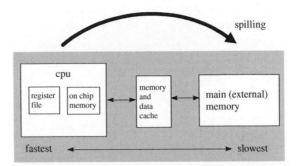

spilling

Figure 8.74: Processor Memory Hierarchy. On-chip Registers and Memory Are the Fastest Way to Access Data, Typically One Cycle per Access. Cache Systems Are Used to Increase the Performance When Requiring Data from Off-chip. Main External Memory Is the Slowest

Sometimes the code structure forces the register allocator to use external memory. For example, the C code in Figure 8.75 (which does not do anything useful) shows what can happen when too many variables are required to perform a specific calculation. This function requires a number of different variables, x0 ... x9. When the compiler attempts to

```
int foo(int a, int b, int c, int d)
{
    int x0, x1, x2, x3, x4, x5, x6, x7, x8, x9;

    x0 = (a&0xa);
    x1 = (b&0xa) + x0;

    x2 = (c&0xb) + x1;
    x3 = (d&0xb) + x2;

    x4 = (a&0xc) + x3;
    x5 = (b&0xc) + x4;

    x6 = (c&0xd) + x5;
    x7 = (d&0xd) + x6;

    x8 = (a&0xe);
    x9 = (b&0xe);

    return (x0&x1&x2&x3)|(x4&x5)|(x6&x7&x8+x9);
}
```

Figure 8.75: C Code Snippet with a Number of Variables

```
        MOV    V4, V9
        AND    V4, V3
        LDR    V3, [SP, #0]    **** this is an example of register spilling. SP indicates
the stack pointer
        AND    V4, V3        ; |21|
        MOV    V3, LR
        AND    V4, V3        ; |21|
        AND    V2, V1
        ORR    V2, V4        ; |21|
        MOV    V1, #14       ; |21|
        AND    A1, V1
        AND    A2, V1
        ADD    A2, A2, A1    ; |21|
        MOV    A1, #13       ; |21|
        AND    A4, A1
        ADD    A1, A3, A4    ; |21|
        AND    A1, A2        ; |21|
        AND    A1, A3        ; |21|
        ORR    A1, V2
        POP    {A4, V1, V2, V3, V4}
        POP    {A3}
        BX     A3
```

Figure 8.76: Register Spilling Caused by Lack of Register Resources

map these into registers, not all variables are accommodated and the compiler must spill some of the variables to the stack. This is shown in the code in Figure 8.76.

One way to reduce or eliminate register spilling is to break larger loops into smaller loops, if the correctness of the code can be maintained. This will enable the register allocator to treat each loop independently, thereby increasing the possibility of finding a suitable register allocation for each of the subfunctions. Figure 8.77 is a simple example of breaking larger loops into smaller loops to enable this improvement.

```
Instead of:                    Try:

for (expression)               for (expression) {
{                                   Do A         }
    Do A                       for (expression) {
    Do B                            Do B         }
    Do C                       for (expression) {
    Do D                            Do C         }
}                              for (expression) {
                                    Do D         }
```

Figure 8.77: Some Loops Are too Large for the Compiler to Pipeline. Reducing the Computational Load within a Loop May Allow the Compiler to Pipeline the Smaller Loops!

Eliminating embedded loops as shown below can also free up registers and allow for more efficient register allocation in DSP code.

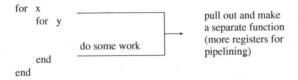

8.21.3   Compile Time Options

Many DSP optimizing compilers offer several options for code size versus performance. Each option allows the programmer to achieve a different level of performance versus code size. These options allow more and more aggressive code size reduction from the compiler. DSP compilers support different levels of code performance. Each option allows the compiler to perform different DSP optimization techniques, for example:

1. **First level of optimization**—*Register level optimizations*. This level of optimization may include techniques such as:

    *   Simplification of control flow.

    *   Allocation of variables to registers.

- Elimination of unused code.

- Simplification of expressions and statements.

- Expand calls to inline functions.

2. **Second level of optimization**—*Local optimization*. This level of optimization may include techniques such as:

   - Local copy/constant propagation.

   - Removal of unused assignments.

   - Elimination of local common expressions.

3. **Third level of optimization**—*Global optimization*. This level of optimization may include techniques such as:

   - Loop optimizations.

   - Elimination of global common subexpressions.

   - Elimination of global unused assignments.

   - Loop unrolling.

4. **Highest level of optimization**—*File optimizations*. This level of optimization may include techniques such as:

   - Removal of functions that are never called.

   - Simplification of functions with return values that are never used.

   - Inlines calls to small functions (regardless of declaration).

   - Reordering of functions so that attributes of called function are known when caller is optimized.

   - Identification of file-level variable characteristics.

The different levels of optimization and the specific techniques used at each level obviously vary by vendor and compiler. The DSP programmer should study the compiler manual to understand what each level does and experiment to see how the code is modified at each level.

8.21.3.1   *Understanding What the Compiler Is Thinking*

There will always be situations where the DSP programmer will need to get pieces of information from the compiler to help understand why an optimization was or was not made. The compiler will normally generate information on each function in the generated assembly code. Information regarding register usage, stack usage, frame size, and how memory is used is usually listed. The programmer can usually get information concerning optimizer decisions (such as to inline or not) by examining information written out by the compiler usually stored in some form of information file (you may have to explicitly ask the compiler to produce this information for you using a compiler command that should be documented in the users manual). This output file contains information as to how optimizations were done or not done (check your compiler—in some situations, asking for this information can sometimes reduce the amount of optimization actually performed).

8.22   *Programmer Helping Out the Compiler*

Part of the job of an optimizing compiler is figuring out what the programmer is trying to do and then helping the programmer achieve that goal as efficiently as possible. That's why well structured code is better for a compiler—it's easier to determine what the programmer is trying to do. This process can be aided by certain "hints" the programmer can provide to the compiler. The proper "hints" allow the compiler to be more aggressive in the optimizations it makes to the source code. Using the standard compiler options can only get you so far towards achieving optimal performance. To get even more optimization, the DSP programmer needs to provide helpful information to the compiler and optimizer, using mechanisms called ***pragmas, intrinsics***, and ***keywords***. We already discussed intrinsics as a method of informing the compiler about special instructions to use in the optimization of certain algorithms. These are special function names that map directly to assembly instructions. Pragmas provide extra information about functions and variables to the preprocessor part of the compiler. Helpful keywords are usually type modifiers that give the compiler information about how a variable is used. ***Inline*** is a special keyword that causes a function to be expanded in place instead of being called.

8.22.1 Pragmas

Pragmas are special instructions to the compiler that tell it how to treat functions and variables. These pragmas must be listed before the function is declared or referenced. Examples of some common pragmas the TI DSP include:

Pragma	Description
CODE_SECTION(symbol, "section name") [;]	This pragma allocates space for a function in a given memory segment
DATA_SECTION(symbol, "section name") [;]	This pragma allocates space for a data variable in a given memory segment
MUST_ITERATE(min, max, multiple) [;]	This pragma gives the optimizer part of the compiler information on the number of times a loop will repeat
UNROLL (n) [;]	This pragma when specified to the optimizer, tells the compiler how many times to unroll a loop.

An example of a pragma to specify the loop count is shown in Figure 8.78. The first code snippet does not have the pragma inserted and is less efficient than the second snippet, which has the pragma for loop count inserted just before the loop in the source code.

```
voidfirFilter(short *x, int f, short *y, int N, int M, QScale)
{ int i, j, sum;
    #pragmaUNROLL(2)          ← Unroll outer loop
    for (j = 0; j < M; j++) {
        sum = 0;
        #pragmaUNROLL(2)      ← Unroll inner loop
        for (i = 0; i < N; i++)
                sum += x[i + j] *filterCoeff[f][i];
        y[j] = sum >>QScale;
        y[j] &= 0xfffe;
    }}
```

Figure 8.78: Example of Using Pragmas to Improve the Efficiency of the Code

8.22.2   *Intrinsics*

Modern optimizing compilers have special functions called intrinsics that map directly to inlined DSP instructions. Intrinsics help to optimize code quickly. They are called in the same way as a function call. Usually intrinsics are specified with some leading indicator such as the underscore (_).

```
Int saturated_add(int a, int b)
{
  int result;

  result = a + b;

// check to see of a and b have the same sign

  if (((a^b) & 0x8000) == 0)
  {
     // if a and b have the same sign, check for underflow or
     // overflow
     if ((result ^ a) & 0x8000)
     {
        // if the result has a different sign than a then
        // underflow or overflow has occurred. If a is negative,
        // result to max positive set result to max negative
        // If a is positive, set result = ( a < 0) ?
           0x8000 : 0x7FFF;

     }
  }
}
return result;
```

Figure 8.79: C Code to Perform Saturated Add

```
Saturated_add:
        SP = SP - #1
                                    ; End Prolog Code
        AR1 = T1                    ; |5|
        AR1 = AR1 + T0              ; |5|
        T1 = T1 ^ T0                ; |7|
        AR2 = T1 & #0x8000          ; |7|
        if (AR2!=#0) goto L2        ; |7|
                                    ; branch occurs ; |7|
        AR2 = T0                    ; |7|
        AR2 = AR2 ^ AR1             ; |7|
        AR2 = AR2 & #0x8000         ; |7|
        if (AR2==#0) goto L2        ; |7|
                                    ; branch occurs ; |7|
        if (T0<#0) goto L1          ; |11|
                                    ; branch occurs ; |11|
        T0 = #32767                 ; |11|
        goto L3                     ; |11|
                        ; branch occurs ; |11|
L1:
        AR1 = #-32768      ; |11|
L2:
        T0 = AR1           ; |14|
L3:
                        ; Begin Epilog Code
        SP = SP + #1       ; |14|
return                     ; |14|
                        ; return occurs ; |14|
```

Figure 8.80: TMS320C55 DSP Assembly Code for the Saturated Add Routine

```
int sadd(int a, int b)
{
return _sadd(a,b);
}
```

**Figure 8.81: TMS320C55 DSP Code for the Saturated Add Routine Using
a Single Call to an Intrinsic**

```
Saturated_add:
      SP = SP - #1
                                            ; End Prolog Code
      bit(ST3, #ST3_SATA) = #1
      T0 = T0 + T1                          ; |3|
                                            ; Begin Epilog Code
      SP = SP + #1                          ; |3|
      bit(ST3, #ST3_SATA) = #0
      return                                ; |3|
                                            ; return occurs ;  |3|
```

Figure 8.82: TMS320C55 DSP Assembly Code for the Saturated Add Routine Using a Single Call to an Intrinsic

Intrinsic	Description
int _sadd(int src1, int src2);	Adds two 16–bit integers, with SATA set, producing a saturated 16–bit result.
int _smpy(int src1, int src2);	Multiplies src1 and src2, and shifts the result left by 1. Produces a saturated 16–bit result. (SATD and FRCT set).
int _abss(int src);	Creates a saturated 16–bit absolute value. _abss(0x8000) => 0x7FFF (SATA set)
int _smpyr(int src1, int src2);	Multiplies src1 and src2, shifts the result left by 1, and rounds by adding 2 15 to the result. (SATD and FRCT set)
int _norm(int src);	Produces the number of left shifts needed to normalize src.
int _sshl(int src1, int src2);	Shifts src1 left by src2 and produces a 16-bit result. The result is saturated if src2 is less than or equal to 8. (SATD set)
long _lshrs(long src1, int src2);	Shifts src1 right by src2 and produces a 32-bit result. Produces a saturated 32–bit result. (SATD set)
long _laddc(long src1, int src2);	Adds src1, src2, and Carry bit and produces a 32-bit result.
long _lsubc(long src1, int src2);	Optimizing C Code Subtracts src2 and logical inverse of sign bit from src1, and produces a 32-bit result.

Figure 8.83: Some Intrinsics for the TMS320C55 DSP
Source: *Courtesy of Texas Instruments*

As an example, if a developer were to write a routine to perform saturated addition[8] in a higher level language such as C, it would look similar to the code in Figure 8.79. The result assembly language for this routine is shown in Figure 8.80. This is quite messy and inefficient. As an alternative, the developer could write a simple routine calling a built in saturated add routine (Figure 8.81), which is much easier and produces cleaner and more efficient assembly code (Figure 8.82). Figure 8.83 shows some of the available intrinsics for the TMS320C55 DSP. Many modern DSPs support intrinsic libraries of this type.

8.22.3   Keywords

Keywords are type modifiers that give the compiler information about how a variable is used. These can be very helpful in helping the optimizer part of the compiler make optimization decisions. Some common keywords in DSP compilers are:

- **Const** —This keyword defines a variable or pointer as having a constant value. The compiler can allocate the variable or pointer into a special data section that can be placed in ROM. This keyword will also provide information to the compiler that allows it to make more aggressive optimization decisions.

- **Interrupt**—This keyword will force the compiler to save and restore context and enable interrupts on exit from a particular pipelined loop or function.

- **Ioport**—This defines a variable as being in I/O space (this keyword is only used with global or static variables).

- **On-chip**—Using this keyword with a variable or structure will guarantee that that memory location is on-chip.

- **Restrict**—This keyword tells the compiler that only this pointer will access the memory location it points to (i.e., no aliasing of this location). This allows the compiler to perform optimization techniques, such as software pipelining.

- **Volatile**—This keyword tells the compiler that this memory location may be changed without compiler's knowledge. Therefore, the memory location should not be stored in a temporary register and, instead, be read from memory before each use.

---

[8] Saturated add is a process by which two operands are added together and, if the result is an overflow, the result is set to the maximum positive value. This is useful in certain multimedia applications where it is more desirable to have a result that is max positive instead of an overflow that effectively becomes a negative number that looks undesirable in an image, for example.

8.22.4   Inlining

For small infrequently called functions it may make sense to paste them directly into code. This eliminates overhead associated with register storage and parameter passing. Inlining uses more program space, but speeds up execution, sometimes significantly. When functions are inlined, the optimizer can optimize the function and the surrounding code in new context. There are two types of inlining: static inlining and normal inlining. With static inlining the function being inlined is only placed in the code where it will be used. Normal inlining also has a function definition that allows the function to be called. The compiler, if specified, will automatically inline functions if the size is small enough. Inlining can also be definition controlled, where the programmer specifies which functions to inline.

8.22.5   Reducing Stack Access Time

When using a real-time operating system (RTOS) for task driven systems, there is overhead to consider that increases with the number of tasks in the system. The overhead in a task switch (or mailbox pend or post, semaphore operation, and so forth) can vary based on where the operating system structures are located. If the structures are in off-chip memory, the access time to perform the operation can be much longer than if the structure was in on-chip memory. The same holds true for the task stack space. If this is in off-chip memory, the performance suffers proportionally to the number of times the stack has to be accessed.

One solution is to allocate the stack in on-chip memory. If the stack is small enough, this may be a viable thing to do. However, if there are many tasks in the system, there will not be enough on-chip memory to store all of the task stacks. However, special code can be written to move the stack on chip when it is needed the most. Before the task (or function) is complete, the stack can be moved back off chip. Figure 8.84 shows the code to do this. Figure 8.85 is a diagrammatic explanation of the steps to perform this operation. The steps are as follows:

1.   Compile the C code to get a .asm file.

2.   Modify the .asm file with the code in the example.

3.   Assemble the new .asm file.

4.   Link the system.

You need to be careful when doing this. This type of modification should not be done in a function that calls other functions. Also, interrupts should be disabled when performing this

```
SSP        .set        0x80001000
SSP2       .set        0x80000FFC


           MVK         SSP, A0
||         MVK         SSP2,B0


           MVKH                SSP,A0
||         MVKH                SSP2,B0


           STW .D1     SP,*A0
||         MV          .L2     B0,SP
```

Figure 8.84: Modifying the Stack Pointer to Point to On-chip Memory

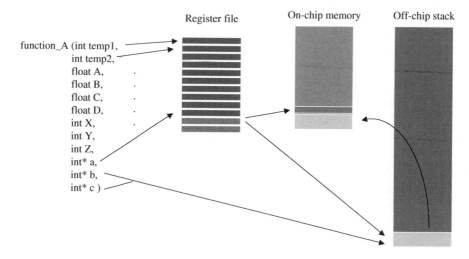

Figure 8.85: Moving the Stack Pointer On-chip to Increase Performance in a DSP Application

type of operation. Finally, the programmer needs to ensure that the secondary stack in on-chip memory does not grow too large, overwriting other data in on-chip memory!

8.22.6   Compilers Helping Out the Programmer

Compilers do their best to optimize applications based on the wishes of their programmers. Compilers also produce output that documents the decisions they were able to make or not make based on the specific source code provided to them by the programmer. (See Figure 8.86.) By analyzing this output, the programmer can understand the specific constraints and decisions and make appropriate adjustments in the source code to improve the performance of the compiler. In other words, if the programmer understands the thought process of the compilation process, they are then able to reorient the application to be more consistent with that thought process. The information below is an example of output generated by the compiler that can be analyzed by the programmer. This output can come in various forms, a simple text output file or a fancier user interface to guide the process of analyzing the compiled output.

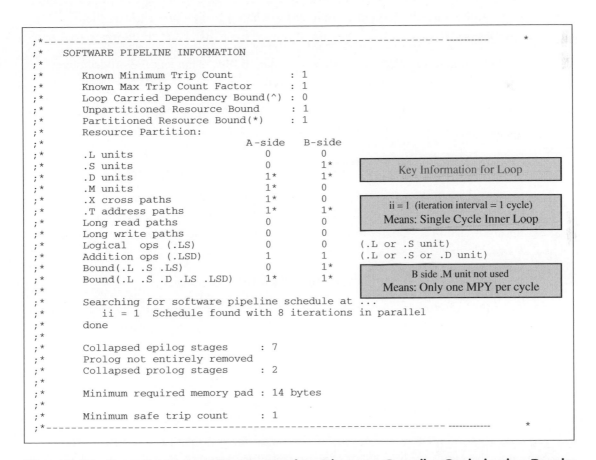

Figure 8.86: Compiler Output Can Be Used to Diagnose Compiler Optimization Results

8.22.7   Summary of Coding Guidelines

Here is a summary of guidelines that the DSP programmer can use to produce the most highly optimized code for the DSP application. Many of these recommendations are general to all DSP compilers. The recommendation is to develop a list like the one below for whatever DSP device and compiler is being used. It will provide a useful reference for the DSP programming team during the software development phase of a DSP project.

General Programming Guidelines

1. Avoid removing registers for C compiler usage. Otherwise valuable compiler resources are being thrown away. There are some cases when it makes sense to use these resources. For example, it is acceptable to preserve a register for an interrupt routine.

2. To optimize functions selectively, place in separate files. This lets the programmer adjust the level of optimization on a file-specific basis.

3. Use the least possible number of *volatile* variables. The compiler cannot allocate registers for these, and also can't inline when variables are declared with the volatile keyword.

Variable Declaration

1. Local variables/pointers are preferred instead of globals. The compiler uses stack-relative addressing for globals, which is not as efficient. If the programmer will frequently be using a global variable in a function, it is better to assign the global to a local variable and then use it.

2. Declare globals in file where they are used the most.

3. Allocate most often used elements of a structure, array or bit-field in the first element, lowest address or LSB; this eliminates the need for extra bytes to specify the address offset.

4. Use unsigned variables instead of int wherever possible; this provides a larger dynamic range and gives extra information to the compiler for optimization.

Variable Declaration (Data Types)

1. Pay attention to data type significance. The better the information provided to the compiler, the better the efficiency of the resulting code.

2. Only use casting if absolutely necessary, casting can use extra cycles, and can invoke wrong RTS functions if done wrong.

3. Avoid common mistakes in data type assumptions. Avoid code that assumes *int* and *long* are the same type. Also, use *int* for fixed-point arithmetic, since *long* requires a call to a library, which is less efficient. Also, avoid code that assumes *char* is 8 bits or *long long* is 64 bits for the same reasons.

4. May be more convenient to define your own data types. *Int16* for 16-bit integer (*int*) and *Int32* for 32-bit integer (*long*). Experiment and see what is best for your DSP device and application.

Initialization of Variables

1. Initialize global variables with constants at load time. This eliminates the need to have code copy values over at run-time.

2. When assigning the same values to global variables, rearrange code if it makes sense to do so. For example use a = b = c = 3; instead of a = 3; b = 3; c = 3. The first uses a register to store the same value to all, the second produces 3 separate long stores;

```
MOV #3, AR1          MOV #3, *(#_a)

MOV AR1, *(#_a)      MOV #3, *(#_b)

MOV AR1, *(#_b)      MOV #3, *(#_c)

MOV AR1, *(#_c)

(17 bytes)           (18 bytes)
```

3. Memory alignment requirements and stack management.

4. Group all like data declarations together. The compiler will usually align a 32-bit data on even boundary, so it will pad an extra 16-bit word in if needed.

5. Use the .align linker directive to guarantee stack alignment on even address. Since the compiler needs 32-bit data aligned on an even boundary, it starts the stack on an even boundary.

Loops

1. Split up loops comprised of two unrelated operations.

2. Avoid function calls and control statements inside loops; the compiler needs to preserve loop context in case of a call to a function. By taking function calls and control statements outside a loop if possible, the compiler can take advantage of the special looping optimizations in the hardware (for example the localrepeat and blockrepeat in the TI DSP) to further optimize the loop (Figure 8.87).

```
for (expression)
{
        Do A
        Call X  ◄── All function calls in inner loops must be
        Do C        inline into the calling function!!
}
```

Figure 8.87: Do Not Call Functions in Inner Loops of Performance Critical Code

3. Keep loop code small to enable compiler use of local repeat optimizations.

4. Avoid deeply nested loops; more deeply nested loops use less efficient types of looping.

5. Use an *int* or *unsigned int* instead of *long* for loop counter. DSP hardware generally uses a 16-bit register for a counter.

6. Use pragmas (if available) to give the compiler better information about loop counts.

Control Code

1. The DSP compiler may generate similar code for nested if-then-else and switch-case statements if the number of cases is less than eight. If greater than eight, the compiler will generate a .switch label section.

2. For highly dense compare code, use switch statements instead of if-then-else.

3. Place the most common case at the start, since the compiler checks in order.

4. For single conditionals, it is always best to test against 0 instead of !0; For example, 'if (a = = 0)' produces more efficient code than 'if (a! = 1)'.

Functions

1. When a function is only called by other functions in same file, make it a *static* function. This will allow the compiler to inline functions better.

2. When a global variable is only used by functions in the same file, make it a *static* variable.

3. Group minor functions in a single file with functions that use them. This makes file-level optimization better.

4. Too many parameters in function calls become inefficient. Once DSP registers are used up, the rest of the parameters go on the stack. Accessing variables from the stack is very inefficient.

5. Parameters that are used frequently in the subroutine should be passed in registers.

Intrinsics

1. There are some instructions on a DSP that the C compiler cannot implement efficiently. For example, the saturation function is hard to implement using standard instructions on many DSPs. To saturate manually requires a lot of code (check sign bits, determine proper limit, and so on). DSPs that support specific instrinsics like saturate will allow the DSP to execute the function much more efficiently.

When developing an application, it is very easy (and sometimes even required) to use generic routines to do various computations. Many time the application developer does not realize how much overhead can be involved in using these generic routines. Often times a more generalized version of an algorithm or function is used because of simple availability instead of creating a more specialized version that better fits the specific need. Creating large numbers of specialized routines is generally a bad programming style as well as a maintenance headache. But strategic use of specialized routines can greatly improve performance in high performance code segments.

Use Libraries

1. Some optimizations are more macro or global level optimizations. These optimizations are performed at the algorithm level. This is somewhat unique to DSP where there are many common routines such as FFT, FIR filters, IIR filters, and so on. Eventually, just about every DSP developer will be required to use one of these functions in the development of a DSP application. For common functions used in DSP, vendors have developed highly efficient implementations of these algorithms that can be easily reused. Many of these algorithms are implemented in C and are tested against standards and well documented. Examples include:

   • FFT

   • Filtering and convolution

   • Adaptive filtering

   • Correlation

- Trigonometric (i.e., sine)

- Math (, max, log, div)

- Matrix computations

Although many of these algorithms are very common routines in DSP, they can be complex to implement. Writing one from scratch would require an in-depth knowledge of how the algorithm works (for example an FFT), in-depth knowledge of the DSP architecture in order to optimize the algorithm, possibly expertise at assembly coding, which is hard to find, and time to get everything working right and optimized.

8.23   Profile-Based Compilation

Because there is a trade-off between code size and higher performance, it is often desirable to compile some functions in a DSP application for performance and others for code size. In fact, the ideal code size and performance for your application needs may be some combination of the different levels of optimization across all of the application functions. The challenge is in determining which functions need which options. In an application with 100 or more functions, each with five possible options, the number of option combinations starts to explode exponentially. Because of this, manual experimentation can take weeks of effort, and the DSP developer may rarely arrive at a solution close to optimal for the particular application needs. Profile-based compilation is one available technique that helps to solve this challenge by automating the entire process.

Profile-based compilation will automatically build and profile multiple compiler option sets. For example, this technique can build the entire application with the highest level of optimization and then profile each function to obtain its resulting code size and cycle count. This process is then repeated using the other compiler options at the remaining code-size reduction levels. The result is a set of different code size and cycle count data points for each function in the application. That data can then be plotted to show the most interesting combinations of functions and compiler options (Figure 8.88). The ideal location for the application is always at the origin of the graph in the lower left hand corner where cycle count and code size are both minimized.

8.23.1   Advantages

An advantage of a profile-based environment is the ability for the application developer to select a profile by selecting the appropriate point on the curve, depending on the overall

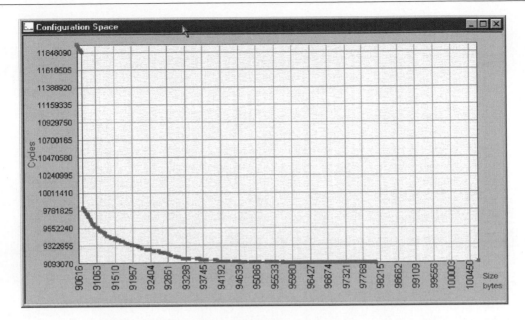

Figure 8.88: Profile-based Compilation Shows the Various Trade-offs between Code Size and Performance

system needs. This automatic approach saves many hours of experimenting manually. The ability to display profiling information (cycle count by module or function, for example) allows the developer to see the specifics of each piece of the application and work on that section independently, if desired (Figure 8.89).

8.23.2   Issues with Debugging Optimized Code

One word of caution: Do not optimize programs that you intend to debug with a symbolic debugger. The compiler optimizer rearranges assembler-language instructions that makes it difficult to map individual instructions to a line of source code. If compiling with optimization options, be aware that this rearrangement may give the appearance that the source-level statements are executed in the wrong order when using a symbolic debugger.

The DSP programmer can ask the compiler to generate symbolic debugging information for use during a debug session. Most DSP compilers have an option for doing this. DSP compiler have directives that will generate symbolic debugging directives used by C source-level debugger. The downside to this is that it forces the compiler to disable many optimizations. The compiler will turn on the maximum optimization compatible with

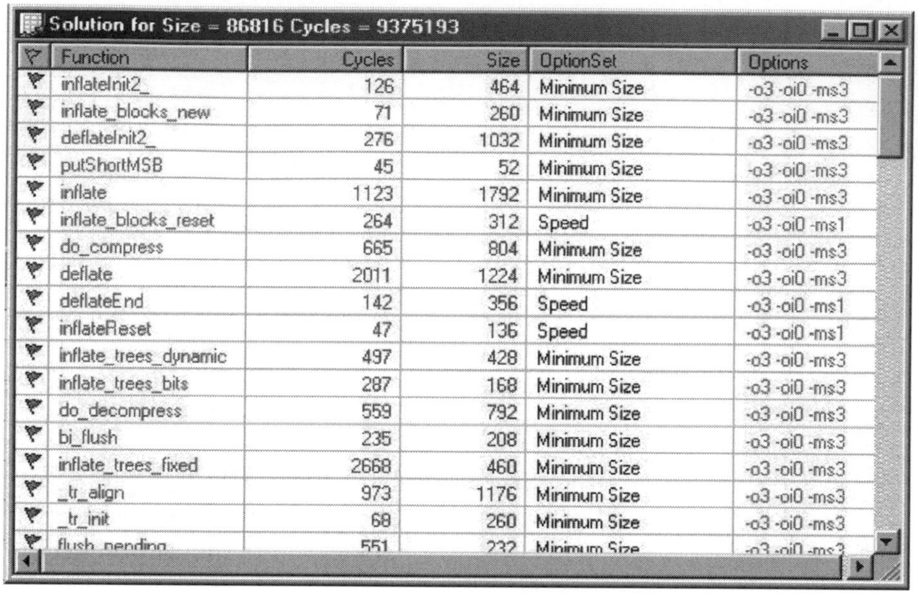
Solution for Size = 86816 Cycles = 9375193

Function	Cycles	Size	OptionSet	Options
inflateInit2_	126	464	Minimum Size	-o3 -oi0 -ms3
inflate_blocks_new	71	260	Minimum Size	-o3 -oi0 -ms3
deflateInit2_	276	1032	Minimum Size	-o3 -oi0 -ms3
putShortMSB	45	52	Minimum Size	-o3 -oi0 -ms3
inflate	1123	1792	Minimum Size	-o3 -oi0 -ms3
inflate_blocks_reset	264	312	Speed	-o3 -oi0 -ms1
do_compress	665	804	Minimum Size	-o3 -oi0 -ms3
deflate	2011	1224	Minimum Size	-o3 -oi0 -ms3
deflateEnd	142	356	Speed	-o3 -oi0 -ms1
inflateReset	47	136	Speed	-o3 -oi0 -ms1
inflate_trees_dynamic	497	428	Minimum Size	-o3 -oi0 -ms3
inflate_trees_bits	287	168	Minimum Size	-o3 -oi0 -ms3
do_decompress	559	792	Minimum Size	-o3 -oi0 -ms3
bi_flush	235	208	Minimum Size	-o3 -oi0 -ms3
inflate_trees_fixed	2668	460	Minimum Size	-o3 -oi0 -ms3
_tr_align	973	1176	Minimum Size	-o3 -oi0 -ms3
_tr_init	68	260	Minimum Size	-o3 -oi0 -ms3
flush_pending	551	232	Minimum Size	-o3 -oi0 -ms3

Figure 8.89: Profiling Information for DSP Functions

debugging. The best solution for debugging DSP code however, is first to verify the program's correctness and then start turning on optimizations to improve performance.

8.23.3 Summary of the Code Optimization Process

Given that this chapter is about managing the DSP software development process, we must now expand our definition of the code optimization process first discussed at the beginning of the chapter. Figure 8.90 shows an expanded software development process for code optimization.

Although this process may vary based on the application, this process is a general flow for all DSP applications. There are 21 steps in this process that will be summarized below.

- **Step 1**—This step involves understanding the key performance scenarios for the application. A performance scenario is a path through the DSP application that will stress the available resources in the DSP. This could be performance, memory, and/or power. Once these key scenarios are understood, the optimization process can focus on these "worst case" performance paths. If the developer can reach performance goals in these conditions, all other scenarios should meet their goals as well.

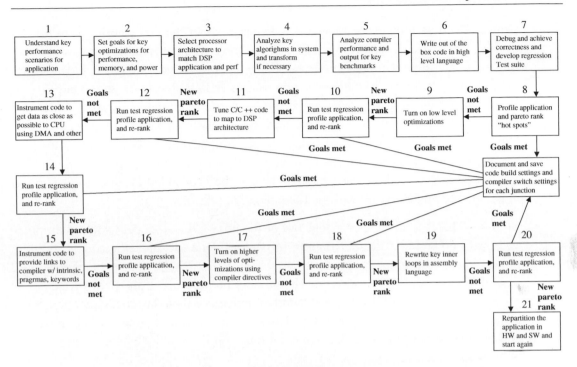

Figure 8.90: The Expanded Code Optimization Process for DSP

- **Step 2**—Once the key performance scenarios are understood, the developer then selects the key goals for each resource being optimized. One goal may be to consume no more than 75% of processor throughput or memory, for example. Once these goals are established, there is something to measure progress towards as well as stopping criteria. Most DSP developers optimize until they reach a certain goal and then stop.

- **Step 3**—Once these goals are selected, the developer, if not done already, selects the DSP to meet the goals of the application. At this point, no code is run, but the processor is analyzed to determine whether it can meet the goals through various modeling approaches.

- **Step 4**—This step involves analyzing key algorithms in the system and making any algorithm transformations necessary to improve the efficiency of the algorithm. This may be in terms of performance, memory, or power. An example of this is selecting a fast Fourier transform instead of a slower discrete Fourier transform.

- **Step 5**—This step involves doing a detailed analysis of the key algorithms in the system. These are the algorithms that will run most frequently in the system or otherwise consume the most resources. These algorithms should be benchmarked in detail, sometimes even to the point of writing these key algorithms in the target language and measuring the efficiency. Given that most of the application cycles may be consumed here, the developer must have detailed data for these key benchmarks. Alternatively, the developer can use industry benchmark data if there are algorithms that are very similar to the ones being used in the application. Examples of these industry benchmarks include the Embedded Processor Consortium at eembc.org and Berkeley Design Technology at bdti.com.

- **Step 6**—This step involves writing "out of the box" C/C++ code, which is simply code with no architecture specific transformations done to "tune" the code to the target. This is the simplest and most portable structure for the code. This is the desired format if possible, and code should only be modified if the performance goals are not met. The developer should not undergo a thought process of "make it as fast as possible." Symptoms of this thought process include excessive optimization and premature optimization. Excessive optimization is when a developer keeps optimizing code even after the performance goals for the application have been met. Premature optimization is when the developer begins optimizing the application before understanding the key areas that should be optimized (following the 80/20 rule). Excessive and premature optimization are dangerous because these consume project resources, delay releases, and compromise good software designs without directly improving performance.

- **Step 7**—This is the "make it work right before you make it work fast" approach. Before starting to optimize the application that could potentially break a working application because of the complexity involved, the developer must make the application work correctly. This is most easily done by turning off optimizations and debugging the application using a standard debugger until the application is working. To get a working application, the developer must also create a test regression that is run to confirm that the application is indeed working. Part of this regression should be used for all future optimizations to ensure that the application is still working correctly after each successive round of optimization.

- **Step 8**—This step involves running the application and collecting profiling information for each function of the application. This profiling information could be cycles consumed for each function, memory used, or power consumed. This data can then be pareto ranked to determine the biggest contributors to the performance bottlenecks. The developer can then focus on these key parts of the application for further optimization. If the goals are met, then no optimizations are needed. If the goals are not met the developer moves on to step 9.

- **Step 9**—In this step, the developer turns on basic compiler optimizations. These include many of the machine independent optimizations that compilers are good at finding. The developer can select options to reduce cycles (increase performance) or to reduce memory size. Power can also be considered during this phase.

- **Steps 10, 12, 14, 16, 18, and 20**—These steps involve re-running the test regression for the application and measuring performance against goals. If the goals are met then the developer is through. If not, the developer reprofiles the application and establishes a new pareto rank of top performance, memory, or power bottlenecks.

- **Step 13**—This step involves restructuring or tuning the C/C++ code to map the code more efficiently to the DSP architecture. Examples of this were discussed earlier in the chapter. Restructuring C or C++ code can lead to significant performance improvements but makes the code potentially less portable and readable. The developer should first attempt this where there are the most significant performance bottlenecks in the system.

- **Step 15**—In this step, the C/C++ code is instrumented with very specific information to help the compiler make more aggressive optimizations based on these "hints" from the developer. Three common forms of this instrumentation include using special instructions with intrinsics, pragmas, and keywords.

- **Step 17**—If the compiler supports multiple levels of optimization, the developer should proceed to turn on higher levels of optimization to allow the compiler to be more aggressive in searching for code transformations to yield more performance gains. These higher levels of optimization are advanced and will cause the compiler to run longer searching for these optimizations. Also, there is the chance that a more aggressive optimization will do something to break the code or otherwise cause it to behave incorrectly. This is where a regression test suite that is run periodically is so important.

- **Step 19**—As final resort, the developer rewrites the key performance bottlenecks in assembly language if result can yield performance improvements above what the compiler can produce. This is a final step, since writing assembly language reduces portability, increases maintainability, decreases readability, and it has other side effects. For advanced architectures, assembly language may mean writing highly complex and parallel code that runs on multiple independent execution units. This can be very difficult to learn to do well and generally there are not many DSP developers that can become expert assembly language programmers without a significant ramp time.

- **Step 21**—Each of the optimization steps described above are iterative. The developer may iterate through these phases multiple times before moving on to the next phase. If the performance goals are not being met by this phase, the developer needs to consider repartitioning the hardware and software for the system. These decisions are painful and costly but they must be considered when goals are not being met by the end of this optimization process.

As shown in Figure 8.91, when moving through the code optimization process, improvements become harder and harder to achieve. When optimizing generic "out of the box" C or C++ code, just a few optimizations can lead to significant performance improvements. Once the code is tuned to the DSP architecture and the right hints are given to the compiler in the form of intrinsics, pragmas, and keywords, additional performance gains are difficult to achieve. The developer must know where to look for these additional improvements. That's why profiling and measuring are so important. Even assembly language programming cannot always yield full entitlement from the DSP device. The developer ultimately has to decide how far down the curve in Figure 8.91 the effort should

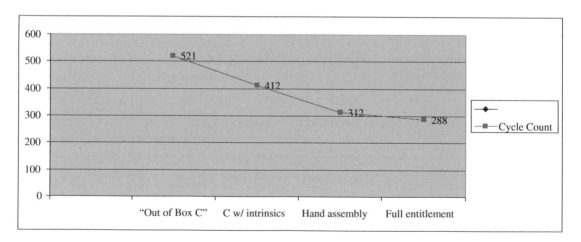

Figure 8.91: Achieving Full Entitlement. Most of the Significant Performance Improvements Are Achieved Early in the Optimization Process

go, as the cost/benefit ratio gets less justifiable given the time required to perform the optimizations at these lower levels of the curve.

8.23.4   Summary

Coding for speed requires the programmer to match the expression of the application's algorithm to the particular resources and capabilities of the processor. Key among these fit issues is how data is staged for the various operations. The iterative nature of DSP algorithms also makes loop efficiency critically important. A full understanding of when to unroll loops and when and how to pipeline loops will be essential if you are to write high-performance DSP code—even if you rely on the compiler to draft most of such code.

Whereas, hand-tuned assembly code was once common for DSP programmers, modern optimizing DSP compilers, with some help, can produce very high performance code.

Embedded real-time applications are an exercise in optimization. There are three main optimization strategies that the embedded DSP developer needs to consider:

- **DSP architecture optimization**—DSPs are optimized microprocessors that perform signal processing functions very efficiently by providing hardware support for common DSP functions.

- **DSP algorithm optimization**—Choosing the right implementation technique for standard and often used DSP algorithms can have a significant impact on system performance.

- **DSP compiler optimization**—DSP compilers are tools that help the embedded programmer exploit the DSP architecture by mapping code onto the resources in such a way as to utilize as much of the processing resources as possible, gaining the highest level of architecture entitlement as possible.

8.24   References

TMS320C62XX Programmers Guide, Texas Instruments, 1997

Computer Architecture, A Quantitative Approach, by John L. Hennesey and David A. Patterson, copyright 1990 by Morgan Kaufmann Publishers, Inc., Palo Alto, CA

The Practice of Programming, by Brian W. Kernighan and Rob Pike, Addison Wesley, 1999

TMS320C55x_DSP_Programmer's_Guide

TMS320C55xx Optimizing C/C++ Compiler User's Guide (SPRU281C)TMS320C55x

DSP Programmer's Guide (SPRU376A)

Generating Efficient Code with TMS320 DSPs: Style Guidelines (SPRA366)

How to Write Multiplies Correctly in C Code (SPRA683)

TMS320C55x DSP Library Programmer's Reference (SPRU422)

CHAPTER 9

Practical Embedded Coding Techniques

Jack Ganssle

9.1 Reentrancy

Virtually every embedded system uses interrupts; many support multitasking or multithreaded operations. These sorts of applications can expect the program's control flow to change contexts at just about any time. When that interrupt comes, the current operation is put on hold and another function or task starts running. What happens if functions and tasks share variables? Disaster surely looms if one routine corrupts the other's data.

By carefully controlling how data is shared, we create **reentrant** functions, those that allow multiple concurrent invocations that do not interfere with each other. The word "pure" is sometimes used interchangeably with "reentrant."

Reentrancy was originally invented for mainframes, in the days when memory was a valuable commodity. System operators noticed that a dozen or hundreds of identical copies of a few big programs would be in the computer's memory array at any time. At the University of Maryland, my old hacking grounds, the monster Univac 1108 had one of the early reentrant FORTRAN compilers. It burned up a breathtaking (for those days) 32 kW of system memory, but being reentrant, it required only 32 k even if 50 users were running it. Everyone executed the same code, from the same set of addresses. Each person had his or her own data area, yet everyone running the compiler quite literally executed identical code. As the operating system changed contexts from user to user it swapped data areas so one person's work didn't affect any other. Share the code, but not the data.

In the embedded world a routine must satisfy the following conditions to be reentrant:

1. It uses all shared variables in an atomic way, unless each is allocated to a specific instance of the function.

2. It does not call nonreentrant functions.

3. It does not use the hardware in a nonatomic way.

9.2   Atomic Variables

Both the first and last rules use the word "atomic," which comes from the Greek word meaning "indivisible." In the computer world "atomic" means an operation that cannot be interrupted. Consider the assembly language instruction:

```
mov ax,bx
```

Since nothing short of a reset can stop or interrupt this instruction it's atomic. It will start and complete without any interference from other tasks or interrupts.

The first part of rule 1 requires the atomic use of shared variables. Suppose two functions each share the global variable "foobar." Function A contains:

```
temp=foobar;
temp+=1;
foobar=temp;
```

This code is not reentrant, because foobar is used nonatomically. That is, it takes three statements to change its value, not one. The foobar handling is not indivisible; an interrupt can come between these statements, switch context to the other function, which then may also try and change foobar. Clearly there's a conflict, foobar will wind up with an incorrect value, the autopilot will crash, and hundreds of screaming people will wonder, "Why didn't they teach those developers about reentrancy?"

Suppose, instead, function A looks like:

```
foobar+=1;
```

Now the operation is atomic, an interrupt will not suspend processing with foobar in a partially changed state, so the routine is reentrant.

Except . . . do you really know what your C compiler generates? On an x86 processor the code might look like:

```
movax,[foobar]
incax
mov[foobar],ax
```

which is clearly not atomic, and so not reentrant. The atomic version is:

```
inc[foobar]
```

The moral is to be wary of the compiler; assume it generates atomic code and you may find *60 Minutes* knocking at your door.

The second part of the first reentrancy rule reads " . . . unless each is allocated to a specific instance of the function." This is an exception to the atomic rule that skirts the issue of shared variables.

An *instance* is a path through the code. There's no reason a single function can't be called from many other places. In a multitasking environment it's quite possible that several copies of the function may indeed be executing concurrently. (Suppose the routine is a driver that retrieves data from a queue; many different parts of the code may want queued data more or less simultaneously.) Each execution path is an "instance" of the code.

Consider:

```
int foo;
void some_function(void){
foo++;
}
```

foo is a global variable whose scope exists beyond that of the function. Even if no other routine uses foo, `some_function` can trash the variable if more than one instance of it runs at any time.

C and C++ can save us from this peril. Use automatic variables. That is, declare foo inside of the function. Then, each instance of the routine will use a new version of foo created from the stack, as follows:

```
void some_function(void){
int foo;
foo++;
}
```

Another option is to dynamically assign memory (using malloc), again so each incarnation uses a unique data area. The fundamental reentrancy problem is thus avoided, as it's impossible for multiple instances to stamp on a common version of the variable.

9.3   Two More Rules

The rest of the rules are very simple.

Rule 2 tells us a calling function inherits the reentrancy problems of the callee. That makes sense, if other code inside the function trashes shared variables; the system is going to crash. Using a compiled language, though, there's an insidious problem. Are you sure—really sure—that the runtime package is reentrant? Obviously string operations and a lot of other complicated things use runtime calls to do the real work. An awful lot of compilers also generate runtime calls to do, for instance, long math, or even integer multiplications and divisions.

If a function must be reentrant, talk to the compiler vendor to ensure that the entire runtime package is pure. If you buy software packages (like a protocol stack) that may be called from several places, take similar precautions to ensure the purchased routines are also reentrant.

Rule 3 is a uniquely embedded caveat. Hardware looks a lot like a variable, if it takes more than a single I/O operation to handle a device, reentrancy problems can develop.

Consider Zilog's SCC serial controller. Accessing any of the device's internal registers requires two steps: first write the register's address to a port, then read or write the register from the same port, the same I/O address. If an interrupt comes between setting the port and accessing the register another function might take over and access the device. When control returns to the first function the register address you set will be incorrect.

9.4 Keeping Code Reentrant

What are our best options for eliminating nonreentrant code? The first rule of thumb is to avoid shared variables. Globals are the source of no end of debugging woes and failed code. Use automatic variables or dynamically allocated memory.

Yet globals are also the fastest way to pass data around. It's not entirely possible to eliminate them from real time systems. Therefore, when using a shared resource (variable or hardware) we must take a different sort of action.

The most common approach is to disable interrupts during nonreentrant code. With interrupts off the system suddenly becomes a single-process environment. There will be no context switches. Disable interrupts, do the nonreentrant work, and then turn interrupts back on.

Most times this sort of code looks like:

```
long i;
void do_something(void){
   disable_interrupts();
   i+=0x1234;
   enable_interrupts();
}
```

This solution *does not work*. If do_something() is a generic routine, perhaps called from many places, and is invoked with interrupts disabled, it returns after turning them back on. The machine's context is changed, probably in a very dangerous manner.

Don't use the old excuse "yeah, but I wrote the code and I'm careful. I'll call the routine only when I know that interrupts will be on." A future programmer probably does not know about this restriction, and may see do_something() as just the ticket needed to solve some other problem . . . perhaps when interrupts are off.

Better code looks like:

```
long i;
void do_something(void){
   push interrupt state;
   disable_interrupts();
   i+=0x1234;
   pop interrupt state;
}
```

Shutting interrupts down does increase system latency, reducing its ability to respond to external events in a timely manner. A kinder, gentler approach is to use a semaphore to indicate when a resource is busy. Semaphores are simple on-off state indicators whose processing is inherently atomic, often used as "in-use" flags to have routines idle when a shared resource is not available.

Nearly every commercial real time operating system includes semaphores; if this is your way of achieving reentrant code, by all means use an RTOS.

Don't have an RTOS? Sometimes I see code that assigns an in-use flag to protect a shared resource, like this:

```
while (in_use);    //wait till resource free

in_use=TRUE;       //set resource busy

Do non-reentrant stuff

in_use=FALSE;      //set resource available
```

If some other routine has access to the resource it sets `in_use` true, causing this routine to idle until `in_use` gets released. Seems elegant and simple, but it does not work. An interrupt that occurs after the `while` statement will preempt execution. This routine feels it now has exclusive access to the resource, yet hasn't had a chance to set `in_use` true. Some other routine can now get access to the resource.

Some processors have a test-and-set instruction, which acts like the in-use flag, but that is interrupt-safe. It'll always work. The instruction looks something like:

```
Tset variable ; if (variable==0){
              ;       variable=1;
              ;       returns TRUE;}
              ; else {returns FALSE;}
```

If you're not so lucky to have a test-and-set, try the following:

```
loop:   mov     al,0            ; 0 means "in use"
        xchg    al,variable
        cmp     al,0
        je      loop            ; loop if in use
```

If al = 0, we swapped 0 with zero; nothing changed, but the code loops since someone else is using the resource. If al = 1, we put a 0 into the "in use" variable, marking the resource as busy. We fall out of the loop, now having control of the resource. It'll work every time.

9.5 Recursion

No discussion of reentrancy is complete without mentioning recursion, if only because there's so much confusion between the two.

A function is recursive if it calls itself. That's a classic way to remove iteration from many sorts of algorithms. Given enough stack space this is a perfectly valid, though tough to debug, way to write code. Since a recursive function calls itself, clearly it must be reentrant to avoid trashing its variables. So all recursive functions must be reentrant, but not all reentrant functions are recursive.

9.6 Asynchronous Hardware/Firmware

But there are subtler issues that result from the interaction of hardware and software. These may not meet the classical definition of reentrancy, but pose similar risks, and require similar solutions.

We work at that fuzzy interface between hardware and software, which creates additional problems due to the interactions of our code and the device. Some cause erratic and quite impossible-to-diagnose crashes that infuriate our customers. The worst bugs of all are those that appear infrequently, that can't be reproduced. Yet a reliable system just cannot tolerate any sort of defect, especially the random one that passes our tests, perhaps dismissed with the "ah, it's just a glitch" behavior.

Potential evil lurks whenever hardware and software interact asynchronously. That is, when some physical device runs at its own rate, sampled by firmware running at some different speed.

I was poking through some open-source code and came across a typical example of asynchronous interactions. The RTEMS real time operating system provided by OAR Corporation (ftp://ftp.oarcorp.com/pub/rtems/releases/4.5.0/ for the code) is a nicely written, well organized product with a lot of neat features. But the timer handling routines, at least for the 68302 distribution, are flawed in a way that will fail infrequently but possibly catastrophically. This is just one very public example of the problem I constantly see buried in proprietary firmware.

The code is simple and straightforward, and looks much like any other timer handler.

```
int timer_hi;
interrupt timer(){
    ++timer_hi;}

long read_timer(void){
    unsigned int low, high;
    low =inword(hardware_register);
    high=timer_hi;
    return (high<<16 + low);}
```

There's an interrupt service routine invoked when the 16 bit hardware timer overflows. The ISR services the hardware, increments a global variable named `timer_hi`, and returns. Therefore, `timer_hi` maintains the number of times the hardware counted to 65536.

Function `read_timer` returns the current "time" (the elapsed time in microseconds as tracked by the ISR and the hardware timer). It, too, is delightfully free of complications. Like most of these sorts of routines it reads the current contents of the hardware's timer register, shifts `Timer_hi` left 16 bits, and adds in the value read from the timer. That is, the current time is the concatenation the timer's current value and the number of overflows.

Suppose the hardware rolled over 5 times, creating five interrupts. `timer_hi` equals 5. Perhaps the internal register is, when we call `read_timer`, 0x1000. The routine returns a value of 0x51000. Simple enough and seemingly devoid of problems.

9.7 Race Conditions

But let's think about this more carefully. There are really two things going on at the same time. *Not* concurrently, which means "apparently at the same time," as in a multitasking environment where the RTOS doles out CPU resources so all tasks *appear* to be running simultaneously. No, in this case the code in `read_timer` executes whenever called, and the clock-counting timer runs at its own rate. The two are asynchronous.

A fundamental rule of hardware design is to panic whenever asynchronous events suddenly synchronize. For instance, when two different processors share a memory array there's quite a bit of convoluted logic required to ensure that only one gets access at any time. If the CPUs use different clocks the problem is much trickier, since the designer may find the two requesting exclusive memory access within fractions of a nanosecond of each other.

This is called a "race" condition and is the source of many gray hairs and dramatic failures.

One of `read_timer`'s race conditions might be:

- It reads the hardware and gets, let's say, a value of 0xffff.

- Before having a chance to retrieve the high part of the time from variable `timer_hi`, the hardware increments again to 0x0000.

- The overflow triggers an interrupt. The ISR runs `timer_hi` is now 0x0001, not 0 as it was just nanoseconds before.

- The ISR returns, our fearless `read_timer` routine, with no idea an interrupt occurred, blithely concatenates the new 0x0001 with the previously read timer value of 0xffff, and returns 0x1ffff—a hugely incorrect value.

Alternatively, suppose `read_timer` is called during a time when interrupts are disabled—say, if some other ISR needs the time. One of the few perils of writing encapsulated code and drivers is that you're never quite sure what state the system is in when the routine gets called. In this case:

- `read_timer` starts. The timer is 0xffff with no overflows.

- Before much else happens it counts to 0x0000. With interrupts off the pending interrupt gets deferred.

- `read_timer` returns a value of 0x0000 instead of the correct 0x10000, or the reasonable 0xffff.

So the algorithm that seemed so simple has quite subtle problems, necessitating a more sophisticated approach. The RTEMS RTOS, at least in its 68 k distribution, will likely create infrequent but serious errors.

Sure, the odds of getting a misread are small. In fact, the chance of getting an error plummets as the frequency we call `read_timer` decreases. How often will the race condition surface? Once a week? Monthly?

Many embedded systems run for years without rebooting. Reliable products must *never* contain fragile code. Our challenge as designers of robust systems is to identify these sorts of issues and create alternative solutions that work correctly, every time.

9.8 Options

Fortunately a number of solutions do exist. The easiest is to stop the timer before attempting to read it. There will be no chance of an overflow putting the upper and lower halves of the data out of sync. This is a simple and guaranteed solution.

We will lose time. Since the hardware generally counts the processor's clock, or clock divided by a small number, it may lose quite a few ticks during the handful of instructions executed to do the reads. The problem will be much worse if an interrupt causes a context switch after disabling the counting. Turning interrupts off during this period will eliminate unwanted tasking, but increases both system latency and complexity.

I just *hate* disabling interrupts, system latency goes up, and sometimes the debugging tools get a bit funky. When reading code a red flag goes up if I see a lot of disable interrupt instructions sprinkled about. Though not necessarily bad, it's often a sign that either the code was beaten into submission (made to work by heroic debugging instead of careful design), or there's something quite difficult and odd about the environment.

Another solution is to read the `timer_hi` variable, then the hardware timer, and then reread `timer_hi`. An interrupt occurred if both variable values aren't identical. Iterate until the two variable reads are equal. The upside: correct data, interrupts stay on, and the system doesn't lose counts.

The downside: in a heavily loaded, multitasking environment, it's possible that the routine could loop for rather a long time before getting two identical reads. The function's execution time is nondeterministic. We've gone from a very simple timer reader to somewhat more complex code that could run for milliseconds instead of microseconds.

Another alternative might be to simply disable interrupts around the reads. This will prevent the ISR from gaining control and changing `timer_hi` after we've already read it, but creates another issue.

We enter `read_timer` and immediately shut down interrupts. Suppose the hardware timer is at our notoriously-problematic 0xffff, and `timer_hi` is zero. Now, before the code has a chance to do anything else, the overflow occurs. With context switching shut down we miss the rollover. The code reads a zero from both the timer register and from `timer_hi`, returning zero instead of the correct 0x10000, or even a reasonable 0x0ffff.

Yet disabling interrupts is probably indeed a good thing to do, despite my rant against this practice. With them on there's always the chance our reading routine will be suspended by higher priority tasks and other ISRs for perhaps a very long time. Maybe long enough for the timer to roll over several times. So let's try to fix the code. Consider the following:

```
long Read_timer(void){
    unsigned int low, high;
    push_interrupt_state;
    disable_interrupts;
    low=inword(Timer_register);
    high=timer_hi;
    if(inword(timer_overflow))
        {++high;
         low=inword(timer_register);}
    pop_interrupt_state;
    return (((ulong)high)<<16 + (ulong)low);
}
```

We've made three changes to the RTEMS code. First, interrupts are off, as described. Second, you'll note that there's no explicit interrupt re-enable. Two new pseudo-C statements have appeared, which push and pop the interrupt state. Trust me for a moment—this is just a more sophisticated way to manage the state of system interrupts. The third change is a new test that looks at something called "`timer_overflow`," an input port that is part of the hardware. Most timers have a testable bit that signals an overflow took place. We check this to see if an overflow occurred between turning interrupts off and reading the low part of the time from the device. With an inactive ISR variable `timer_hi` won't properly reflect such an overflow.

We test the status bit and reread the hardware count if an overflow had happened. Manually incrementing the high part corrects for the suspended ISR. The code then concatenates the two fixed values and returns the correct result—every time.

With interrupts off we have increased latency. However, there are no loops; the code's execution time is entirely deterministic.

9.9 Other RTOSes

Unhappily, race conditions occur anytime we're need more than one read to access data that's changing asynchronously to the software. If you're reading X and Y coordinates, even with just 8 bits of resolution, from a moving machine there's some peril they could be seriously out of sync if two reads are required. A ten-bit encoder managed through byte-wide ports potentially could create a similar risk.

Having dealt with this problem in a number of embedded systems over the years, I wasn't too shocked to see it in the RTEMS RTOS. It's a pretty obscure issue, after all, though terribly real and potentially deadly. For fun I looked through the source of uC/OS, another very popular operating system whose source is on the net (see www.ucos-ii.com). uC/OS never reads the timer's hardware. It only counts overflows as detected by the ISR, as there's no need for higher resolution. There's no chance of an incorrect value.

Some of you, particularly those with hardware backgrounds, may be clucking over an obvious solution I've yet to mention. Add an input capture register between the timer and the system; the code sets a "lock the value into the latch" bit, then reads this safely unchanging data. The register is nothing more than a parallel latch, as wide as the input data. A single clock line drives each flip-flop in the latch, when strobed it locks the data into the register. The output is fed to a pair of processor input ports.

When it's time to read a safe, unchanging value the code issues a "hold the data now" command, which strobes encoder values into the latch. So all bits are stored and can be read by the software at any time, with no fear of things changing between reads.

Some designers tie the register's clock input to one of the port control lines. The I/O read instruction then automatically strobes data into the latch, assuming one is wise enough to ensure the register latches data on the leading edge of the clock.

The input capture register is a very simple way to suspend moving data during the duration of a couple of reads. At first glance it seems perfectly safe. However, a bit of analysis shows that for asynchronous inputs it *is not reliable*. We're using hardware to fix a software problem, so we must be aware of the limitations of physical logic devices.

To simplify things for a minute, let's zoom in on that input capture register and examine just one of its bits. Each gets stored in a flip-flop, a bit of logic that might have only three connections: data in, data out, and clock. When the input is a one, strobing clock puts a one at the output.

However, suppose the input changes at about the same time clock cycles. What happens? The short answer is that no one knows.

9.10   Metastable States

Every flip-flop has two critical specifications we violate at our peril. "Set-up time" is the minimum number of nanoseconds that input data must be stable before clock comes. "Hold time" tells us how long to keep the data present after clock transitions. These specs vary depending on the logic device. Some might require tens of nanoseconds of set-up and/or hold time; others need an order of magnitude less.

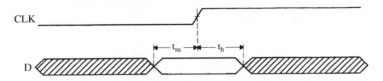

Figure 9.1: Setup and Hold Times

If we tend to our knitting we'll respect these parameters and the flip-flop will always be totally predictable. But when things are asynchronous—say, the wrist rotates at it's own rate and the software does a read whenever it needs data—there's a chance the we'll violate set-up or hold time.

Suppose the flip-flop requires 3 nanoseconds of set-up time. Our data changes within that window, flipping state perhaps a single nanosecond before clock transitions. The device will go into a metastable state where the output gets very strange indeed.

By violating the specification the device really doesn't know if we presented a zero or a one. It's output goes, not to a logic state, but to either a half-level (in between the digital norms) or it will oscillate, toggling wildly between states. The flip-flop is metastable.

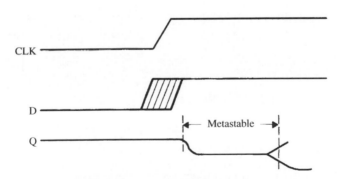

Figure 9.2: A Metastable State

This craziness doesn't last long; typically after a few to 50 nanoseconds the oscillations damp out or the half-state disappears, leaving the output at a valid one or zero. But which one is it? This is a digital system, and we expect ones to be ones, and zeroes zeroes.

The output is *random*. Bummer, that. You cannot predict which level it will assume. That sure makes it hard to design predictable digital systems!

Hardware folks feel that the random output isn't a problem. Since the input changed at almost exactly the same time the clock strobed, either a zero or a one is reasonable. If we had clocked just a hair ahead or behind we'd have gotten a different value, anyway. Philosophically, who knows which state we measured? Is this really a big deal? Maybe not to the EEs, but this impacts our software in a big way, as we'll see shortly.

Metastability occurs only when clock and data arrive almost simultaneously; the odds increase as clock rates soar. An equally important factor is the type of logic component used: slower logic (like 74HCxx) has a much wider metastable window than faster devices (say, 74FCTxx). Clearly at reasonable rates the odds of the two asynchronous signals arriving closely enough in time to cause a metastable situation are low, measurable, yes, important, certainly. With a 10 MHz clock and 10 KHz data rate, using typical but not terribly speedy logic, metastable errors occur about once a minute. Though infrequent, no reliable system can stand that failure rate.

The classic metastable fix uses two flip-flops connected in series. Data goes to the first; its output feeds the data input of the second. Both use the same clock input. The second flop's output will be "correct" after two clocks, since the odds of two metastable events occurring back-to-back are almost nil. With two flip-flops, at reasonable data rates errors occur millions or even billions of years apart, good enough for most systems.

However "correct" means the second stage's output will not be metastable: it's not oscillating, nor is it at an illegal voltage level. There's still an equal chance the value will be in either legal logic state.

9.11   Firmware, Not Hardware

To my knowledge there's no literature about how metastability affects software, yet it poses very real threats to building a reliable system.

Hardware designers smugly cure their metastability problem using the two stage flops described. Their domain is that of a single bit, whose input changed just about the same time the clock transitioned. Thinking in such narrow terms it's indeed reasonable to accept the inherent random output the flops generate.

However, we software folks are reading parallel I/O ports, each perhaps 8 bits wide. That means there are 8 flip-flops in the input capture register, all driven by the same clock pulse.

Let's look at what might happen. The encoder changes from 0xff to 0x100. This small difference might represent just a tiny change in angle. We request a read at just about the same time the data changes, our input operation strobes the capture register's clock creating a violation of set-up or hold time. Every input bit changes, each of the flip-flops inside the register goes metastable. After a short time the oscillations die out, but now every bit in the register is random. Though the hardware folks might shrug and complain that no one knows what the right value was, since everything changed as clock arrived, in fact the data was around 0xff or 0x100. A random result of, say, 0x12 is absurd and totally unacceptable, and may lead to crazy system behavior.

The case where data goes from 0xff to 0x100 is pathological since every bit changes at once. The system faces the same peril whenever lots of bits change. 0x0f to 0x10. 0x1f to 0x20. The upper, unchanging data bits will always latch correctly, but every changing bit is at risk.

Why not use the multiple flip-flop solution? Connect two input capture registers in series, both driven by the same clock. Though this will eliminate the illegal logic states and oscillations, the second stage's output will be random as well.

One option is to ignore metastability and hope for the best. Or use very fast logic with very narrow set-up/hold time windows to reduce the odds of failure. If the code samples in the inputs infrequently it's possible to reduce metastability to one chance in millions or even billions. Building a safety critical system? Feeling lucky?

It is possible to build a synchronizer circuit that takes a request for a read from the processor, combines it with a data available bit from the I/O device, responding with a data-OK signal back to the CPU. This is nontrivial and prone to errors.

An alternative is to use a different coding scheme for the I/O device. Buy an encoder with Gray code output, for example (if you can find one). Gray code is a counting scheme where only a single bit changes between numbers, as follows:

```
0    000
1    001
2    011
3    010
4    110
5    111
6    101
7    100
```

Gray code makes sense if, and only if, your code reads the device faster than it's likely to change, and if the changes happen in a fairly predictable fashion—like counting up. Then there's no real chance of more than a single bit changing between reads, if the inputs go metastable only one bit will be wrong. The result will still be reasonable.

Another solution is to compute a parity or checksum of the input data before the capture register. Latch that, as well, into the register. Have the code compute parity and compare it to that read, if there's an error do another read.

Although I've discussed adding an input capture register, please don't think that this is the root cause of the problem. Without that register—if you just feed the asynchronous inputs directly into the CPU—it's quite possible to violate the processor's innate set-up/hold times. There's no free lunch, all logic has physical constraints we must honor.

Some designs will never have a metastability problem. It always stems from violating set-up or hold times, which in turn comes from either poor design or asynchronous inputs.

All of this discussion has revolved around asynchronous inputs, when the clock and data are unrelated in time. Be wary of anything not slaved to the processor's clock. Interrupts are a notorious source of problems. If caused by, say, someone pressing a button, be sure that the interrupt itself, and the vector-generating logic, don't violate the processor's set-up and hold times.

However, in computer systems most things do happen synchronously. If you're reading a timer that operates from the CPU's clock, it is inherently synchronous to the code. From a metastability standpoint it's totally safe.

Bad design, though, can plague any electronic system. Every logic component takes time to propagate data; when a signal traverses many devices the delays can add up significantly. If the data then goes to a latch it's quite possible that the delays may cause the input to transition at the same time as the clock. Instant metastability.

Designers are pretty careful to avoid these situations, though. Do be wary of FPGAs and other components where the delays vary depending on how the software routes the device. In addition, when latching data or clocking a counter it's not hard to create a metastability problem by using the wrong clock edge. Pick the edge that gives the device time to settle before it's read.

What about analog inputs? Connect a 12 bit A/D converter to two 8 bit ports and we'd seem to have a similar problem: the analog data can wiggle all over, changing during the time we read the two ports. However, there's no need for an input capture register because the converter itself generally includes a "sample and hold" block, which stores the analog signal while the A/D digitizes. Most A/Ds then store the digital value till we start the next conversion.

Other sorts of inputs we use all share this problem. Suppose a robot uses a 10 bit encoder to monitor the angular location of a wrist joint. As the wrist rotates the encoder sends back a binary code, 10 bits wide, representing the joint's current position. An 8 bit processor requires two distinct I/O instructions—two byte-wide reads—to get the data. No matter how fast the computer might be there's a finite time between the reads during which the encoder data may change.

The wrist is rotating. A "`get_position`" routine reads 0xff from the low part of the position data. Then, before the next instruction, the encoder rolls over to 0x100. "`get_position`" reads the high part of the data—now 0x1—and returns a position of 0x1ff, clearly in error and perhaps even impossible.

This is a common problem, handling input from a two-axis controller. If the hardware continues to move during our reads, then the X and Y data will be slightly uncorrelated, perhaps yielding impossible results. One friend tracked a rare autopilot failure to the way the code read a flux-gate compass, whose output is a pair of related quadrature signals. Reading them at disparate times, while the vessel continued to move, yielded impossible heading data.

9.12 Interrupt Latency

My dad was a mechanical engineer, who spent his career designing spacecraft. I remember back even in the early days of the space program how he and his colleagues analyzed seemingly every aspect of their creations' behavior. Center of gravity calculations insured that the vehicles were always balanced. Thermal studies guaranteed nothing got too hot or too cold. Detailed structural mode analysis even identified how the system would vibrate, to avoid destructive resonances induced by the brutal launch phase.

Though they were creating products that worked in a harsh and often unknown environment, their detailed computations profiled how the systems would behave.

Think about civil engineers. Today no one builds a bridge without "doing the math." That delicate web of cables supporting a thin dancing roadway is simply going to work. Period. The calculations proved it long before contractors started pouring concrete.

Airplane designers also use quantitative methods to predict performance. When was the last time you heard of a new plane design that wouldn't fly? Yet wing shapes are complex and notoriously resistant to analytical methods. In the absence of adequate theory, the engineers rely on extensive tables acquired over decades of wind tunnel experiments. The engineers can still understand how their product will work—in general—before bending metal.

Compare this to our field. Despite decades of research, formal methods to prove software correctness are still impractical for real systems. We embedded engineers build, then test, with no real proof that our products will work. When we pick a CPU, clock speed, memory size, we're betting that our off-the-cuff guesses will be adequate when, a year later, we're starting to test 100,000+ lines of code. Experience plays an important role in getting the resource requirements right. All too often luck is even more critical. However, hope is our chief tool, and the knowledge that generally, with enough heroics, we can overcome most challenges.

In my position as embedded gadfly, looking into thousands of projects, I figure some 10–15% are total failures due simply to the use of inadequate resources. The 8051 just can't handle that fire hose of data. The PowerPC part was a good choice but the program grew to twice the size of available Flash, and with the new cost model the product is not viable.

Recently I've been seeing quite a bit written about ways to make our embedded systems more predictable, to insure they react fast enough to external stimuli, to guarantee processes complete on time. To my knowledge there is no realistically useful way to calculate predictability. In most cases we build the system and start changing stuff if it runs too slowly. Compared to aerospace and civil engineers we're working in the dark.

It's especially hard to predict behavior when asynchronous activities alter program flow. Multitasking and interrupts both lead to impossible-to-analyze problems. Recent threads on USENET, as well as some discussions at the Embedded Systems Conference, suggest banning interrupts altogether! I guess this does lead to a system that's easier to analyze, but the solution strikes me as far too radical.

I've built polled systems. Yech. Worse are applications that must deal with several different things, more or less concurrently, without using multitasking. The software in both situations is invariably a convoluted mess. About 20 years ago I naively built a steel thickness gauge without an RTOS, only to later have to shoehorn one in. There were too many async things going on, the in-line code grew to outlandish complexity. I'm still trying to figure out how to explain that particular sin to St. Peter.

A particularly vexing problem is to ensure the system will respond to external inputs in a timely manner. How can we guarantee that an interrupt will be recognized and processed fast enough to keep the system reliable?

Let's look in some detail at the first of the requirements: that an interrupt be recognized in time. Simple enough, it seems. Page through the processor's data book and you'll find a specification called "latency," a number always listed at submicrosecond levels. No doubt a footnote defines latency as the longest time between when the interrupt occurs and when the CPU suspends the current processing context. That would seem to be the interrupt response time—but it ain't.

Latency as defined by CPU vendors varies from zero (the processor is ready to handle an interrupt RIGHT NOW) to the maximum time specified. It's a product of what sort of instruction is going on. Obviously it's a bad idea to change contexts in the middle of executing an instruction, so the processor generally waits until the current instruction is complete before sampling the interrupt input. Now, if it's doing a simple register-to-register move that may be only a single clock cycle, a mere 50 nsec on a zero wait state 20-MHz processor. Not much of a delay at all.

Other instructions are much slower. Multiplies can take dozens of clocks. Read-modify-write instructions (like "increment memory") are also inherently pokey. Maximum latency numbers come from these slowest of instructions.

Many CPUs include looping constructs that can take hundreds, even thousands, of microseconds. A block memory-to-memory transfer, for instance, initiated by a single instruction, might run for an awfully long time, driving latency figures out of sight. All processors I'm aware of will accept an interrupt in the middle of these long loops to keep interrupt response reasonable. The block move will be suspended, but enough context is saved to allow the transfer to resume when the ISR (Interrupt Service Routine) completes.

Therefore, the latency figure in the datasheet tells us the longest time the processor can't service interrupts. The number is totally useless to firmware engineers.

OK, if you're building an extreme cycle-countin', nanosecond-poor, gray-hair-inducing system then perhaps that 300 nsec latency figure is indeed a critical part of your system's performance. For the rest of us, real latency—the 99% component of interrupt response—comes not from what the CPU is doing, but from our own software design. And that, my friend, is hard to predict at design time. Without formal methods we need empirical ways to manage latency.

If latency is time between getting an interrupt and entering the ISR, then surely most occurs because we've disabled interrupts! It's because of the way we wrote the darn code. Turn interrupts off for even a few C statements and latency might run to hundreds of microseconds, far more than those handful of nanoseconds quoted by CPU vendors.

No matter how carefully you build the application, you'll be turning interrupts off frequently. Even code that never issues a "disable interrupt" instruction does, indeed, disable them often. For, every time a hardware event issues an interrupt request, the processor itself does an automatic disable, one that stays in effect till you explicitly re-enable them inside of the ISR. Count on skyrocketing latency as a result.

Of course, on many processors we don't so much as turn interrupts off as change priority levels. A 68 K receiving an interrupt on level 5 will prohibit all interrupts at this and lower levels until our code explicitly re-enables them in the ISR. Higher priority devices will still function, but latency for all level 1 to 5 devices is infinity until the code does its thing.

Therefore, in an ISR re-enable interrupts as soon as possible. When reading code one of my "rules of thumb" is that code that does the enable just before the return is probably flawed. Most of us were taught to defer the interrupt enable until the end of the ISR. But that prolongs latency unacceptably. Every other interrupt (at least at or below that priority level) will be shut down until the ISR completes. Better, enter the routine, do all of the nonreentrant things (like handling hardware), and *then* enable interrupts. Run the rest of the ISR, which manages reentrant variables and the like, with interrupts on. You'll reduce latency and increase system performance.

The downside might be a need for more stack space if that same interrupt can re-invoke itself. There's nothing wrong with this in a properly designed and reentrant ISR, but the stack will grow until all pending interrupts get serviced.

The second biggest cause of latency is excessive use of the disable interrupts instruction. Shared resources—global variables, hardware, and the like—will cause erratic crashes when two asynchronous activities try to access them simultaneously. It's up to us to keep the code reentrant by either keeping all such accesses atomic, or by limiting access to a single task at a time. The classic approach is to disable interrupts around such accesses. Though a simple solution, it comes at the cost of increased latency.

9.13   Taking Data

So what is the latency of your system? Do you know? Why not?

It's appalling that so many of us build systems with a "if the stupid thing works at all, ship it" philosophy. It seems to me there are certain critical parameters we must understand in order to properly develop and maintain a product. Like, is there any free ROM space? Is the system 20% loaded . . . or 99%? How bad is the maximum latency?

Latency is pretty easy to measure, sometimes those measurements will yield surprising and scary results.

Perhaps the easiest way to get a feel for interrupt response is to instrument each ISR with an instruction that toggles a parallel output bit *high* when the routine starts. Drive it low just as it exits. Connect this bit to one input of an oscilloscope, tying the other input to the interrupt signal itself.

The amount of information this simple setup gives is breathtaking. Measure time from the assertion of the interrupt until the parallel bit goes high. That's latency, minus a bit for the overhead of managing the instrumentation bit. Twiddle the scope's time base to measure this to any level of precision required.

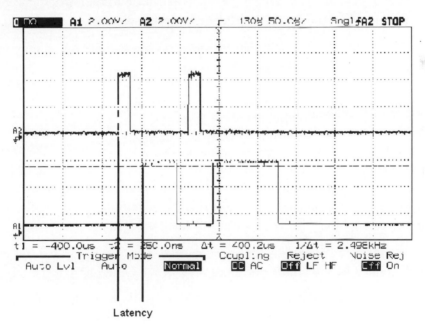

Figure 9.3: The Latency Is the Time from When the Interrupt Signal Appears, Until the ISR Starts

The time the bit stays high is the ISR's total execution time. Tired of guessing how fast your code runs? This is quantitative, cheap, and accurate.

In a real system, interrupts come often. Latency varies depending on what other things are going on. Use a digital scope in storage mode. After the assertion of the interrupt input you'll see a clear space—that's the minimum system latency to this input. Then there will be hash, a blur as the instrumentation bit goes high at different times relative to the interrupt input. These represent variations in latency. When the blur resolves itself into a solid *high*, that's the maximum latency.

All this, for the mere cost of one unused parallel bit.

If you've got a spare timer channel, there's another approach that requires neither extra bits nor a scope. Build an ISR just for measurement purposes that services interrupts from the timer.

On initialization, start the timer counting up, programmed to interrupt when the count overflows. Have it count as fast as possible. Keep the ISR dead simple, with minimal overhead. This is a good thing to write in assembly language to minimize unneeded code. Too many C compilers push *everything* inside interrupt handlers.

The ISR itself reads the timer's count register and sums the number into a long variable, perhaps called `total_time`. Also increment a counter (`iterations`). Clean up and return.

The trick here is that, although the timer reads zero when it tosses out the overflow interrupt, the timer register continues counting even as the CPU is busy getting ready to invoke the ISR. If the system is busy processing another interrupt, or perhaps stuck in an interrupt-disabled state, the counter continues to increment. An infinitely fast CPU with no latency would start the instrumentation ISR with the counter register equal to zero. Real processors with more usual latency issues will find the counter at some positive nonzero value that indicates how long the system was off doing other things.

Therefore, average latency is just the time accumulated into `total_time` (normalized to microseconds) divided by the number of times the ISR ran (`iterations`).

It's easy to extend the idea to give even more information. Possibly the most important thing we can know about our interrupts is the longest latency. Add a few lines of code to compare for and log the maximum time.

Is the method perfect? Of course not. The data is somewhat statistical, so can miss single-point outlying events. Very speedy processors may run so much faster than the timer tick rate that they always log latencies of zero, although this may indicate that for all practical purposes latencies are short enough to not be significant.

The point is that knowledge is power, once we understand the magnitude of latency reasons for missed interrupts become glaringly apparent.

Try running these experiments on purchased software components. One embedded DOS, running on a 100-MHz 486, yielded latencies in the tens of milliseconds!

9.14   Understanding Your C Compiler: How to Minimize Code Size

A C compiler is a basic tool for most embedded systems programmers. It is the tool by which the ideas and algorithms in your application (expressed as C source code) are transformed into machine code executable by your target processor. To a large extent, the C compiler determines how large the executable code for the application will be.

A compiler performs many transformations on a program in order to generate the best possible code. Examples of such transformations are storing values in registers instead of memory, removing code that does nothing useful, reordering computations in a more efficient order, and replacing arithmetic operations by cheaper operations. The C language provides the compiler with significant freedom regarding how to precisely implement each C operation on the target system. This freedom is an important factor in why C can usually be compiled very efficiently, but a programmer needs to be aware of the compiler's freedom in order to write robust code.

To most programmers of embedded systems, the case that a program does not *quite* fit into the available memory is a familiar phenomenon. Recoding parts of an application in assembly language or throwing out functionality may seem to be the only alternatives, while the solution could be as simple as rewriting the C code in a more compiler-friendly manner.

In order to write code that is compiler friendly, you need to have a working understanding of compilers. Some simple changes to a program, like changing the data type of a frequently-accessed variable, can have a big impact on code size while other changes have no effect at all. Having an idea of what a compiler can and cannot do makes such optimization work much easier.

9.15   Modern C Compilers

Assembly programs specify both what, how, and the precise order in which calculations should be carried out. A C program, on the other hand, only specifies the calculations that should be performed. With some restrictions, the order and the technique used to realize the calculations are up to the compiler.

The compiler will look at the code and try to understand what is being calculated. It will then generate the best possible code, given the information it has managed to obtain, locally within a single statement and also across entire functions and sometimes even whole programs.

9.15.1   The Structure of a Compiler

In general, a program is processed in six main steps in a modern compiler (not all compilers follow this blueprint completely, but as a conceptual guide it is sufficient):

- **Parser**—The conversion from C source code to an intermediate language.

- **High-level optimization**—Optimizations on the intermediate code.

- **Code generation**—Generation of target machine code from the intermediate code.

- **Low-level optimization**—Optimizations on the machine code.

- **Assembly**—Generation of an object file that can be linked from the target machine code.

- **Linking**—Linking of all the code for a program into an executable or downloadable file.

The parser parses the C source code, checking the syntax and generating error messages if syntactical errors are found in the source. If no errors are found, the parser then generates intermediate code (an internal representation of the parsed code), and compilation proceeds with the first optimization pass.

The high-level optimizer transforms the code to make it better. The optimizer has a large number of transformations available that can improve the code, and will perform those that it deems relevant to the program at hand. Note that we use the word "transformation" and not "optimization." "Optimization" is a bit of a misnomer. It conveys the intuition that a change always improves a program and that we actually find optimal solutions, while in fact optimal solutions are very expensive or even impossible to find (undecidable, in computer science lingo). To ensure reasonable compilation times and the termination of the compilation process, the compiler has to use heuristic methods ("good guesses"). Transforming a program is a highly nonlinear activity, where different orderings of transformations will yield different results, and some transformations may actually make the code worse. Piling on more "optimizations" will not necessarily yield better code.

When the high-level optimizer is done, the code generator transforms the intermediate code to the target processor instruction set. This stage is performed, piece-by-piece, on the intermediate code from the optimizer, and the compiler will try to do smart things on the level of a single expression or statement, but not across several statements.

The code generator will also have to account for any differences between the C language and the target processor. For example, 32-bit arithmetic will have to be broken down to 8-bit arithmetic for a small-embedded target (like an Intel 8051, Motorola 68HC08, Samsung SAM8, or Microchip PIC).

A very important part of code generation is allocating registers to variables. The goal is to keep as many values as possible in registers, since register-based operations are typically faster and smaller than memory-based operations.

After the code generator is done, another phase of optimization takes place, where transformations are performed on the target code. The low-level optimizer will clean up after the code generator (which sometimes makes suboptimal coding decisions), and perform more transformations. There are many transformations that can only be applied on the target code, and some that are repeated from the high-level phase, but on a lower level. For example, transformations like removing a "clear carry" instruction if we already know that the carry flag is zero are only possible at the target code level, since the flags are not visible before code generation.

After the low-level optimizer is finished, the code is sent to an assembler and output to an object file.

All the object files of a program are then linked to produce a final binary executable ROM image (in some format appropriate for the target). The linker may also perform some optimizations, for example, by discarding unused functions.

Thus, one can see that the seemingly simple task of compiling a C program is actually a rather long and winding road through a highly complex system. Different transformations may interact, and a local improvement may be worse for the whole program. For example, an expression can typically be evaluated more efficiently if given more temporary registers. Taking a local view, it thus seems to be a good idea to provide as many registers as necessary. A global effect, however, is that variables in registers may have to be spilled to memory, which could be more expensive than evaluating the expression with fewer registers.

9.15.2   The Meaning of a Program

Before the compiler can apply transformations to a program, it must analyze the code to determine which transformations are legal and likely to result in improvements. The legality of transformations is determined by the semantics laid down by the C language standard.

The most basic interpretation of a C program is that only statements that have side effects or compute values used for performing side effects are relevant to the meaning of a program. Side effects are any statements that change the global state of the program. Examples that are generally considered to be side effects are writing to a screen, changing a global variable, reading a volatile variable, and calling unknown functions.

The calculations between the side effects are carried out according to the principle of "do what I mean, not what I say." The compiler will try to rewrite each expression into the most efficient form possible, but a rewrite is only possible if the result of the rewritten code is the same as the original expression. The C standard defines what is considered "the same," and sets the limits of allowable optimizations.

9.15.3   Basic Transformations

A modern compiler performs a large number of basic transformations that act locally, like folding constant expressions, replacing expensive operations by cheaper ones ("strength reduction"), removing redundant calculations, and moving invariant calculations outside of loops. The compiler can do most mechanical improvements just as well as a human programmer, but without tiring or making mistakes. The table below shows (in C form for readability) some typical basic transformations performed by a modern C compiler. Note that an important implication of this basic cleanup is that you can write code in a readable way and let the compiler calculate constant expressions and worry about using the most efficient operations.

Before Transformation	After Transformation
```unsigned short int a;``` ```a /= 8;```	```unsigned short int a;``` ```a >>= 3; /* shift replaced divide*/```
```a *= 2;```	```a += a; /* multiply replaced by add */``` ```a <<= 1; /* or a shift */```

```a = b + c * d;```
```e = f + c * d;```
	```temp = c * d; /* common expression */```
	```a = b + temp; /* saves one multiply */```
	```e = f + temp;```

```a = b * 1;```	```a = b; /* x*1 == x */```

```a = 17;```
```b = 56 + (2 * a);```
	```a = 17;```
	```b = 90; /* constant value evaluated */```
	``` /* at compile time */```

```#define BITNO 4``` ```port |= (1 << BITNO);```	```port |= 0x10; /* constant value */```

```if(a > 10)```
```{```
```   b = b * c + k;```
```   if(a < 5)```
```   a+=6;```
```}```
	```if(a > 10)```
	```{```
	``` b = b * c + k;```
	```/* unreachable code removed */```
	```}```

```a = b * c + k;``` ```a = k + 7;```	```/* useless computation removed */``` ```a = k + 7;```

```for(i=0; i<10; i++)```
```{```
```   b = k * c;```
```   p[i] = b;```
``` }```
	```b = k * c; /* constant code moved */```
	``` /* outside the loop */```
	```for(i=0; i<10; i++)```
	```{```
	``` p[i] = b;```
	```}```

All code that is not considered useful—according to the definition in the previous section—is removed. This removal of unreachable or useless computations can cause some unexpected effects. An important example is that empty loops are completely discarded, making "empty delay loops" useless. The code shown below stopped working properly when upgrading to a modern compiler that removed useless computations:

Code that Stopped Working	. . . After Compiler Optimizations
| ```
void delay(unsigned int time)
{
  unsigned int i;
  for (i=0; i<time; i++)
    ;
  return;
}

void InitHW(void)
{
  /* Highly timing-dependent code */
  OUT_SIGNAL (0x20);
  delay(120);
  OUT_SIGNAL (0x21);
  delay(121);
  OUT_SIGNAL (0x19);
}
``` | ```
void delay(unsigned int time)
{
  /* loop removed */
  return;
}


void InitHW(void)
{
  /* Delays do not last long here...*/
  OUT_SIGNAL (0x20);
  delay(120);
  OUT_SIGNAL (0x21);
  delay(121);
  OUT_SIGNAL (0x19);
}
``` |

9.15.4   *Register Allocation*

Processors usually give better performance and require smaller code when calculations are performed using registers instead of memory. Therefore, the compiler will try to assign the variables in a function to registers. A local variable or parameter will not need any RAM allocated at all if the variable can be kept in registers for the duration of the function. If there are more variables than registers available, the compiler needs to decide which of the variables to keep in registers, and which to put in memory. This is the problem of *register allocation*, and it cannot be solved optimally. Instead, heuristic techniques are used. The algorithms used can be quite sensitive, and even small changes to a function may considerably alter the register allocation.

Note that a variable only needs a register when it is being used. If a variable is used only in a small part of a function, it will be register allocated in that part, but it will not exist in the rest of the function. This explains why a debugger sometimes tells you that a variable is "optimized away at this point."

The register allocator is limited by the language rules of C—for example, global variables have to be written back to memory when calling other functions, since they can be accessed by the called function, and all changes to global variables must be visible to all functions. Between function calls, global variables can be kept in registers.

Note that there are times when you *do not* want variables to be register allocated. For example, reading an I/O port or spinning on a lock, you want each read in the source code to be made from memory, since the variable can be changed outside the control of your program. This is where the volatile keyword is to be used. It signals to the compiler that the variable should not ever be allocated in registers, but read from memory (or written) each time it is accessed.

In general, only simple values like integers, floats, and pointers are considered for register allocation. Arrays have to reside in memory since they are designed to be accessed through pointers, and structures are usually too large. In addition, on small processors, large values like 32-bit integers and floats may be hard to allocate to registers, and maybe only 16-bit and 8-bit variables will be given registers.

To help register allocation, you should strive to keep the number of simultaneously live variables low. Also, try to use the smallest possible data types for your variables, as this will reduce the number of required registers on 8-bit and 16-bit processors.

9.15.5   Function Calls

As assembly programmers well know, calling a function written in a high-level language can be rather complicated and costly. The calling function must save global variables back to memory, make sure to move local variables to the registers that survive the call (or save to the stack), and parameters may have to be pushed on the stack. Inside the called function, registers will have to be saved, parameters taken off the stack, and space allocated on the stack for local variables. For large functions with many parameters and variables, the effort required for a call can be quite large.

Modern compilers do their best, however, to reduce the cost of a function call, especially the use of stack space. A number of registers will be designated for parameters, so that short parameter lists will most likely be passed entirely in registers. Likewise, the return value will be put in a register, and local variables will only be put on the stack if they cannot be allocated to registers.

The number of register parameters will vary wildly between different compilers and architecture. In most cases, at least four registers are made available for parameters. Note also that just like for register allocation, only small parameter types will be passed in registers. Arrays are always passed as pointers to the array (C semantics dictate that), and structures are usually copied to the stack and the structure parameter changed to a pointer to a structure. That pointer might be passed in a register, however. To save stack space, it is thus a good idea to always use pointers to structures as parameters and not the structures themselves.

C supports functions with variable numbers of arguments. This is used in standard library functions like `printf()` and `scanf()` to provide a convenient interface. However, the implementation of variable numbers of arguments to a function incurs significant overhead. All arguments have to be put on the stack, since the function must be able to step through the parameter list using pointers to arguments, and the code accessing the arguments is much less efficient than for fixed parameter lists. There is no type-checking on the arguments, which increases the risk of bugs. Variable numbers of arguments should not be used in embedded systems!

9.15.6   *Function Inlining*

It is good programming practice to break out common pieces of computation and accesses to shared data structures into (small) functions. This, however, brings with it the cost of calling a function each time something should be done. In order to mitigate this cost, the compiler transformation of *function inlining* has been developed. Inlining a function means that a copy of the code for the function is placed in the calling function, and the call is removed.

Inlining is a very efficient method to speed up the code, since the function call overhead is avoided but the same computations carried out. Many programmers do this manually by using preprocessor macros for common pieces of code instead of functions, but macros lack the type checking of functions and produce harder-to-find bugs. The executable code will often grow as a result of inlining, since code is being copied into several places.

Inlining may also help shrink the code: for small functions, the code size cost of a function call might be bigger than the code for the function. In this case, inlining a function will actually save code size (as well as speed up the program).

The main problem when inlining for size is to estimate the gains in code size (when optimizing for speed, the gain is almost guaranteed). Since inlining in general increases the code size, the inliner has to be quite conservative. The effect of inlining on code size cannot be exactly determined, since the code of the calling function is disturbed, with nonlinear effects.

To reduce the code size, the ideal would be to inline all calls to a function, which allows us to remove the function from the program altogether. This is only possible if all calls are known, i.e., are placed in the same source file as the function, and the function is marked `static`, so that it cannot be seen from other files. Otherwise, the function will have to be kept (even though it might still be inlined at some calls), and we rely on the linker to remove it if it is not called. Since this decreases the likely gain from inlining, we are less likely to inline such a function.

9.15.7 Low-Level Code Compression

A common transformation on the target code level is to find common sequences of instructions from several functions, and break them out into subroutines. This transformation can be very effective at shrinking the executable code of a program, at the cost of performing more jumps (note that this transformation only introduces machine-level subroutine calls and not full-strength function calls). Experience shows a gain from 10–30% for this transformation.

9.15.8 Linker

The linker should be considered an integral part of the compilation system, since there are some transformations that are performed in the linker. The most basic embedded-systems linker should remove all unused functions and variables from a program, and only include the parts of the standard libraries that are actually used. The granularity at which program parts are discarded varies, from files or library modules down to individual functions or even snippets of code. The smaller the granularity, the better the linker. Unfortunately, some linkers derived from desktop systems work on a per file basis, and this will give unnecessarily big code.

Some linkers also perform postcompilation transformations on the program. Common transformation is the removal of unnecessary bank and page switches (that cannot be done at compile-time since the exact allocation of variable addresses is unknown at that time) and code compression as discussed above extended to the entire program.

9.15.9 Controlling Compiler Optimization

A compiler can be instructed to compile a program with different goals, usually speed or size. For each setting, a set of transformations has been selected that tend to work towards the goal—maximal speed (minimal execution time) or minimal size. The settings should be considered approximate. To give better control, most compilers also allow individual transformations to be enabled or disabled.

For size optimization, the compiler uses a combination of transformations that tend to generate smaller code, but it might fail in some cases, due to the characteristics of the compiled program. As an example, the fact that function inlining is more aggressive for speed optimization makes some programs smaller on the speed setting than on the size setting. The following example data demonstrates this, the two programs were compiled with the same version of the same compiler, using the same memory and data model settings, but optimizing for speed or size:

Program 1	Speed optimization	1301 bytes
	Size optimization	1493 bytes
Program 2	Speed optimization	20,432 bytes
	Size optimization	16,830 bytes

Program 1 gets slightly smaller with speed optimization, while program 2 is considerably larger, an effect we traced to the fact that function inlining was lucky on program 1. The conclusion is that one should always try to compile a program with different optimization settings and see what happens.

It is often worthwhile to use different compilation settings for different files in a project: put the code that must run very quickly into a separate file and compile that for minimal execution time (maximum speed), and the rest of the code for minimal code size. This will give a small program, which is still fast enough where it matters. Some compilers allow different optimization settings for different functions in the same source file using #pragma directives.

9.15.10   Memory Model

Embedded microcontrollers are usually available in several variants, each with a different amount of program and data memory. For smaller chips, the fact that the amount of memory that can be addressed is limited can be exploited by the compiler to generate smaller code. An 8-bit direct pointer uses less code memory than a 24-bit banked pointer where software has to switch banks before each access. This goes for code as well as data. For example, some Atmel AVR chips have a code area of only 8 kb, which allows a small jump with an offset of $+/-$ 4 kb to reach all code memory, using wraparound to jump from high addresses to low addresses. Taking advantage of this yields smaller and faster code.

The capacity of the target chip can be communicated to the compiler using a memory model option. There are usually several different memory models available, ranging from "small" up to "huge." In general, function calls get more expensive as the amount of code allowed increases, and data accesses and pointers get bigger and more expensive as the amount of accessible data increases. Make sure to use the smallest model that fits your target chip and application—this might give you large savings in code size.

9.16   Tips on Programming

The previous section discussed how a compiler works, and gave some examples of how to code better. In this section, we will look at some more concrete tips on how to write compiler-friendly code.

9.16.1   Use the Right Data Size

The semantics of C state that all calculations should have the same result as if all operands were cast to `int` and the operation performed on `int` (or `unsigned int` or `long int` if values cannot fit in an `int`). If the result is to be stored in a variable of a smaller type like `char`, the result is then (conceptually at least) cast down. On any decent 8-bit micro compiler, this process is short-circuited where appropriate, and the entire expression calculated using `char` or `short`.

Thus, the size of a data item to be processed should be appropriate for the CPU used. If an unnatural size is chosen, the code generated might get much worse. For example, on an 8-bit micro, accessing and calculating 8-bit data is very efficient. Working with 32-bit values will generate much bigger code and run more slowly, and should only be considered when the data being manipulated need all 32 bits. Using big values also increases the demand for registers for register allocation, since a 32-bit value will require four 8-bit registers to be stored.

On a 32-bit processor, working with smaller data might be inefficient, since the registers are 32 bits. The results of calculations will need to be cast down if the storing variable type is smaller than 32 bits, which introduces shift, mask, and sign-extend operations in the code (depending on how smaller types are represented). On such machines, 32-bit integers should be used for as many variables as possible. `chars` and `shorts` should only be used when the precise number of bits are needed (like when doing I/O), or when big types would use too much memory (for example, an array with a large number of elements).

9.16.2 Use the Best Pointer Types

A typical embedded micro has several different pointer types, allowing access to memory in a variety of ways, from small zero-page pointers to software-emulated generic pointers. It is obvious that using smaller pointer types is better than using larger pointer types, since both the data space required to store them and the manipulating code is smaller for smaller pointers.

However, there may be several pointers of the same size but with different properties, for example two banked 24-bit pointers huge and far, with the sole difference that huge allows objects to cross bank boundaries. This difference makes the code to manipulate huge pointers much bigger, since each increment or decrement must check for a bank boundary. Unless you really require very large objects, using the smaller pointer variant will save a lot of code space.

For machines with many disjoint memory spaces (like Microchip PIC and Intel 8051), there might be "generic" pointers that can point to all memory spaces. These pointers might be tempting to use, since they are very convenient, but they carry a cost in that special code is needed at each pointer access to check which memory a pointer points to and performing appropriate actions. Also note that using generic pointers typically brings in some library functions.

In summary: use the smallest pointers you can, and avoid any form of generic pointers unless necessary. Remember to check the compiler default pointer type (used for unqualified pointers, and determined by the data memory model used). In many cases it is a rather large pointer type.

9.16.3 Structures and Padding

A C struct is guaranteed to be laid out in memory with the fields in the order of the declaration. However, on processors with alignment restriction on loads and stores, the compiler will probably insert padding between structure members, in order to align each member efficiently. This will make the struct larger than the sum of the sizes of the types of the members, and could break code written under the assumption that structs are laid out contiguously in memory.

C Declaration	Actual Memory Layout
| ```
struct s {
  uint8_t   a;
  uint32_t  b;
  uint16_t  c;
  int8_t    d;
  int32_t   e;
};
``` | ```
struct s {
  uint8_t   a; /* 1 byte            */
               /* 3 bytes of padding */
  uint32_t  b; /* 4 bytes           */
  uint16_t  c; /* 2 bytes           */
  int8_t    d; /* 1 byte            */
               /* 1 byte padding    */
  int32_t   e; /* 4 bytes           */
};
``` |

Alignment requirements are rare on 8- and 16-bit CPUs, but quite common on 32-bit CPUs. Some CPUs (like the Motorola ColdFire and NEC V850) will generate errors for misaligned loads, while other will only lose performance (Intel x86).

Padding will be inserted at the end of a structure if necessary, to align the size of the structure with the biggest alignment requirement of the machine (typically 4 bytes for a 32-bit machine). This is because every element in an array of structures must start at an aligned boundary in memory. The sizeof() operator will reveal the total size of a struct, including padding at the end. Incrementing a pointer to a structure will move the pointer sizeof() bytes forward in memory, thus reflecting end padding. When a struct contains another struct, the padding of the member structure is maintained.

In some cases, the compiler offers the ability to pack structures in memory (by #pragma, special keywords, or command-line options), removing the padding. This will save data space, but might cost code size, since the code to load misaligned members is potentially much bigger and more complex than the code required to load aligned members.

To make better use of memory, sort the members of the struct in order of decreasing size: 32-bit values first, then 16-bit values, and, finally, 8-bit values. This will make internal padding unnecessary, since each member will be naturally aligned (there will still be padding at end of the struct if the size of the struct is not an even multiple of the machine word size).

Note that the compiler's padding can break code that uses structs to decode information received over a network or to address memory-mapped I/O areas. This is especially dangerous when code is ported from an architecture without alignment requirements to one with them.

9.16.4   Use Function Prototypes

Function prototypes were introduced in ANSI C as a way to improve type checking. The old style of calling functions without first declaring them was considered unsafe, and is also a hindrance to efficient function calls.

If a function is not properly prototyped, the compiler has to fall back on the language rules dictating that all arguments should be promoted to `int` (or `double`, for floating-point arguments). This means that the function call will be much less efficient, since type casts will have to be inserted to convert the arguments. For a desktop machine, the effect is not very noticeable (most things are the size of `int` or `double` already), but for small embedded systems, the effect is potentially great. Problems include ruining register parameter passing (larger values use more registers) and lots of unnecessary type conversion code.

In many cases, the compiler will give you a warning when a function without a prototype is called. Make sure that no such warnings are present when you compile!

The old way to declare a function before calling it (Kernighan & Ritchie or "K&R" style) was to leave the parameter list empty, like "`extern void foo().`" This is not a proper ANSI prototype and will not help code generation. Unfortunately, few compilers warn about this by default.

The register to parameter assignment for a function can always be inferred from the type of the function, i.e., the complete list of parameter types (as given in the prototype). This means that all calls to a function will use the same registers to store parameters, which is necessary in order to generate correct code. The code in a function does not in any way affect the assignment of registers to parameters.

9.16.5   Use Parameters

As discussed above, register allocation has a hard time with global variables. If you want to improve register allocation, use parameters to pass information to a called function and not shared global variables. Parameters will often be allocated to registers both in the calling and called function, leading to very efficient calls.

Note that the calling conventions of some architectures and compilers limit the number of available registers for parameters, which makes it a good idea to keep the number of parameters down for code that needs to be portable and efficient across a wide range of platforms. It might pay off to split a very complex function into several smaller ones, or to reconsider the data being passed into a function.

9.16.6 Do Not Take Addresses

If you take the address of a local variable (the "&var" construction), it is not likely to be allocated to a register, since it has to have an address and, thus, a place in memory (usually on the stack). It also has to be written back to memory before each function call, just like a global variable, since some other function might have gotten hold of the address and is expecting the latest value. Taking the address of a global variable does not hurt as much, since they have to have a memory address anyway.

Thus, you should only take the address of a local variable if you really must (it is very seldom necessary). If the taking of addresses is used to receive return values from called functions [from scanf(), for example], introduce a temporary variable to receive the result, and then copy the value from the temporary to the real variable.[1] This should allow the real variable to be register allocated.

Making a global variable static is a good idea (unless it is referred to in another file), since this allows the compiler to know all places where the address is taken, potentially leading to better code.

An example of when not to use the address-of operator is the following, where the use of addresses to access the high byte of a variable will force the variable to the stack. The good way is to use shifts to access parts of values.

Bad example	Good example

`#define highbyte(x) (*((char *)(&x)+1))`
 	
`short a;`
`char  b = highbyte(a);`
	`#define highbyte(x) ((x>>8)&0xFF)`
	
	`short a;`
	`char  b = highbyte(a);`

9.16.7 Do Not Use Inline Assembly Language

Using inline assembly is a very efficient way of hampering the compiler's optimizer. Since there is a block of code that the compiler knows nothing about, it cannot optimize across that block. In many cases, variables will be forced to memory and most optimizations turned off.

---

[1] Note that in C++, reference parameters ["foo(int &)"] can introduce pointers to variables in a calling function without the syntax of the call showing that the address of a variable is taken.

The output of a function containing inline assembly should be inspected after each compilation run to make sure that the assembly code still works as intended. In addition, the portability of inline assembly is very poor, both across machines (obviously) and across different compilers for the same target.

If you need to use assembler, the best solution is to split it out into assembly source files, or at least into functions containing only inline assembly. Do not mix C code and assembly code in the same function!

9.16.8 Do Not Write Clever Code

Some C programmers believe that writing fewer source code characters and making clever use of C constructions will make the code smaller or faster. The result is code that is harder to read, and that is also harder to compile. Writing things in a straightforward way helps both humans and compilers understand your code, giving you better results. For example, conditional expressions gain from being clearly expressed as conditions.

Consider the two ways to set the lowest bit of variable b if the lower 21 bits of another (32 bit) variable are nonzero as illustrated below. The clever code uses the ! operator in C, which returns zero if the argument is nonzero ("true" in C is any value except zero), and one if the argument is zero.

The straightforward solution is easy to compile into a conditional followed by a set bit instruction, since the bit-setting operation is obvious and the masking is likely to be more efficient than the shift. Ideally, the two solutions should generate the same code. The clever code, however, may result in more code since it performs two ! operations, each of which may be compiled into a conditional.

"Clever" solution	Straightforward solution

`unsigned long int a;`
`unsigned char     b;`
 	
`/* Move bits 0..20 to positions 11..31`
` * If nonzero, first ! gives 0      */`
`b |= !!(a << 11);`
	`unsigned long int a;`
	`unsigned char     b;`
	
	`/* Straight-forward if statement */`
	`if( (a & 0x1FFFFF) != 0)`
	`  b |= 0x01;`

Another example is the use of conditional values in calculations. The "clever" code will result in larger machine code, since the generated code will contain the same test as the straightforward code, and adds a temporary variable to hold the one or zero to add to str.

The straightforward code can use a simple increment operation rather than a full addition, and does not require the generation of any intermediate results.

"Clever" solution	Straightforward solution

```int bar(char *str)```
```{```
  ```/* Calculating with result of */```
  ```/* comparison.   */```
  ```return foo(str+(*str=='+'));```
```}```
	```int bar(char *str)```
	```{```
	  ```if(*str=='+')```
	    ```str++;```
	  ```return foo(str);```
	```}```

Since clever code almost never compiles better than straightforward code, why write clever code? From a maintenance standpoint, writing simpler and more understandable code is definitely the method of choice.

9.16.9   Use Switch for Jump Tables

If you want a jump table, see if you can use a `switch` statement to achieve the same effect. It is quite likely that the compiler will generate better and smaller code for the `switch` rather than a series of indirect function calls through a table. Also, using the `switch` makes the program flow explicit, helping the compiler optimize the surrounding code better. It is very likely that the compiler will generate a jump table, at least for a small dense `switch` (where all or most values are used).

Using a `switch` is also more reliable across machines; the layout that may be optimal on one CPU may not be optimal on another, but the compiler for each will know how to make the best possible jump table for both. The `switch` statement was put into the C language to facilitate multiway jumps: use it!

9.16.10   Investigate Bit Fields Before Using Them

Bit fields offer a very readable way to address small groups of bits as integers, but the bit layout is implementation defined, which creates problems for portable code. The code generated for bit fields will be of very varying quality, since not all compilers consider them very important. Some compilers will generate incredibly poor code since they do not consider them worth optimizing, while others will optimize the operations so that the code is as efficient as manual masking and shifting.

The advice is to test a few bit field variables and check that the bit layout is as expected, and that the operations are efficiently implemented. If several compilers are being used, check that they all have the same bit layout. In general, using explicit masks and shifts will generate more reliable code across more targets and compilers.

9.16.11   Watch Out for Library Functions

As discussed above, the linker has to bring in all library functions used by a program with the program. This is obvious for C standard library functions like `printf()` and `strcat()`, but there are also large parts of the library that are brought in implicitly when certain types of arithmetic are needed, most notably floating point. Due to the way in which C performs implicit type conversions inside expressions, it is quite easy to inadvertently bring in floating point, even if no floating point variables are being used.

For example, the following code will bring in floating point, since the `ImportantRatio` constant is of floating point type—even if its value would be `1.95*20==39`, and all variables are integers:

Example of Accidental Floating Point

`#define ImportantRatio (1.95*Other)`
`… int`
`temp = a * b + CONSTANT * ImportantRatio;`

If a small change to a program causes a big change in program size, look at the library functions included after linking. Especially floating point and 32-bit integer libraries can be insidious, and creep in due to C implicit casts.

Another way to shrink the code of your program is to use limited versions of standard functions. For instance, the standard `printf()` is a very big function. Unless you really need the full functionality, you should use a limited version that only handles basic formatting or ignores floating point. Note that this should be done at link time: the source code is the same, but a simpler version is linked. Because the first argument to `printf()` is a string, and can be provided as a variable, it is not possible for the compiler to automatically figure out which parts of the function your program needs.

9.16.12   Use Extra Hints

Some compilers allow the programmer to specify useful information that the compiler cannot deduce itself to help optimize the code. For example, DSP compilers often allow users to specify that two pointer or array arguments are unaliased, which helps the compiler optimize code accessing these two arrays simultaneously. Other examples are the specification of function as ***pure*** (without side effects) or ***tasks*** (will loop forever, thus no need to save registers on entry). A common example is ***inline***, which might be considered a hint or an order by the compiler. This information is usually introduced using nonportable keywords and should be put in tuned header files (if possible). It might give great benefits in code efficiency, however.

9.17　Final Notes

The above sections have tried to give you an idea of how a modern C compiler works, and to give you some concrete hints on how you can get smaller code by using the compiler wisely. A compiler is a very complex system with highly nonlinear behavior, where a seemingly small change in the source code can have big effects on the assembly code generated.

The basis for the compilation process is that the compiler should be able to understand what your code is supposed to do, in order to perform the operations in the best possible way for a given target. As a general rule, code that is easy to understand for a fellow human programmer—and thus easy to maintain and port—is also easier to compile efficiently.

Note that unless you let your compiler use its higher optimization levels, you have wasted a lot of your investment. What you pay for when you buy a compiler is mostly the work put into developing and tuning optimizations for a certain target, and if you do not use these optimizations, you are not using your compiler to its best effect.

Choose your compiler wisely: different compilers for the same chip can be very different. Some are better at generating fast code, other at generating small code, and some may be no good at all. To evaluate a compiler, the best way is to use a demo version to compile small portions of your own "typical" code. Some chip vendors also provide benchmark tests of various compilers for their chips, usually targeted toward the intended application area for their chips. The compiler vendor's own benchmarks should be taken with some skepticism, it is (almost) always possible to find a program where a certain compiler performs better than the competition.

For more tips on efficient C programming for embedded systems, you should check out the classes presented at embedded systems trade shows (around the world). The web sites of the companies making compilers often contain some technical notes or white papers on particular tricks for their compilers (but make sure to watch out for those that are pure marketing material!). In particular, I would like to recommend the IAR web site section with white papers and articles: http://www.iar.com/Press/Articles.

9.18   Acknowledgments

This text in this chapter was developed with much help from the compiler developers at IAR Systems, and most of the work was performed during my time at IAR. I had some really fun and educating years at IAR.

CHAPTER 10

Development Technologies and Trends

Colin Walls

10.1 How to Choose a CPU for Your System on Chip Design

There are several factors to consider when choosing a CPU for your next system on chip (SoC) design. If you consider that the CPU is to the SoC what an engine is to an automobile, you would not put a Volkswagen engine into a Hummer and expect it to perform. Similarly, a Ferrari engine would also be unsuitable in a Hummer. Although it may deliver similar horsepower to the Hummer engine, it would fail due to a lack of torque. Simple assessments of "horsepower" are just as misleading in CPU selection as they are in the automobile world. There is an optimal solution for the desired functionality. The same holds true for the CPU choice in an SoC. Many times the CPU is chosen based purely on the system architect's knowledge of, and past experience with, a particular device. The decision of which CPU to use should also consider the overall system metrics: complexity of overall design, design reuse, protection, performance, power, size, cost, tools, and middleware availability.

10.1.1 Design Complexity

The design's complexity is critical to the choice of CPU. For example, if the design calls for a single-state machine to be executed with interrupts from a small set of peripherals, then you may be better off with a small CPU and/or microcontroller, such as the 8051 or the Z80. Many systems may fit this category initially. An example might be a pager. The memory footprint is small, the signal is slow, and battery consumption is required to be extremely low.

The algorithms and their interaction will dictate the complexity. They may or may not also dictate the need for an RTOS. Typically, as the application complexity increases, the need for a greater bit-width processor increases.

10.1.2 Design Reuse

Designs are continuing to be reused and are growing in complexity; that pager designed in 2000 may have to be upgraded to play MP3s in 2007. Now the 8-bit CPU may not be enough to keep up with the task at hand. How many interfaces a design contains is a good indicator of the amount of processor power required. In our pager example, initially there were two main interfaces: the user interface and the radio link. For the new design, which adds an MP3 player, we will need to add a memory interface for storing and transferring the data, and an audio interface for playing the data. Now the system complexity is greatly increased from its initial conception, and if we have taken a forward-looking approach to the design, we can reuse much of this earlier work. Make sure that you have room for growth. Today your 8-bit design may be good for the MP3 player, but when the design gets reused and placed in a set-top box application, which has a much higher bandwidth peripheral set, you may need to reengineer the complete solution to migrate to an ARM-, MIPS-, or PowerPC-based architecture to deal with the new constraints.

10.1.3 Memory Architecture and Protection

The system may need to protect itself from outside attack or even from itself. This causes us to look at CPUs that include (or can include) memory management units (MMUs) to address this issue. Virtual memory will allow trusted programs access to the entire system, and untrusted ones to access just the memory they have been allocated. A 3-G cell phone—a phone with Internet connectivity—is a prime example of the need for protection. No longer can you use a CPU that lacks an MMU, since a rogue program will crash your phone. Although an MMU does not eliminate the possibility of system crashes, it reduces the likelihood of hard-to-resolve system failures.

Three main CPU architectures center around 8-, 16-, and 32-bit data registers with 16-, 24-, and 32-bit address buses. The main difference between these CPUs is how much information one particular register can hold and how much it can address directly:

- 8-bit data/16-bit address = (0 . . . 256), with 64-K address space

- 16-bit data/24-bit address = (0 . . . 65,536), with 16-M address space

- 32-bit data/32-bit address = (0 . . . 4 billion), with 4-G address space

Why would an embedded system ever need to access 4 G of address space? The answer is simple: as the system is asked to perform more complex tasks, the size and complexity of the code it runs increases. In the early days, CPM on a Z80 utilized a process of banking memory and page swapping in order to run programs that are more complex on an 8-bit machine. Space of 64 K was not sufficient, and a solution was to make the system more complex by overlaying memory and pages to get more out of the CPU.

It seems like a 24-bit address bus would be adequate for many designs. A couple of factors drive us to a 32-bit address space: protection and pointers. For protection, the CPU with virtual memory can use the entire address range to divide up the physical memory into separate virtual spaces, thus providing protection from bad pointers. And the ability for any register to become a pointer to memory without the need for indexing simplifies the software.

10.1.4 CPU Performance

The performance of the overall system will be greatly impacted by the selection of CPU. Specifically, features like cache, MMU, pipelining, branch prediction, and superscalar architecture all affect the speed of a system. Depending on the needs of the SoC, these features may be necessary to achieve system performance.

10.1.5 Power Consumption

The end use of the SoC will determine how much power your design can consume. If your design is battery operated, the CPU will need to be as power conscious as possible. For instance, some CPUs have the ability to sleep, doze, or snooze. These modes allow the CPU, when idle, to suspend operation and consume less power by shutting down various parts of the CPU. Different CPUs perform the same task with different results.

10.1.6 Costs

The cost of the CPU can be measured in several ways. First the intellectual property (IP) cost, which is the cost to acquire the IP for your SoC and any derivative products. Then there is the system integration cost. Which tools are available for design and implementation of your SoC? Finally, is the CPU variant silicon proven, and is it available on the bus architecture that your SoC is utilizing?

10.1.7 Software Issues

The availability of an RTOS and middleware may dictate your choice as well. For instance, in designing a PDA, you may want the middleware that is available for Linux, but the choice of a virtual operating system will dictate that you migrate away from small non-MMU CPUs.

Is there a graphics system or a file system necessary in the design? If so, then the choice of RTOS will dictate the type of CPU that is needed as well. Many RTOS vendors target specific families, leaving others untouched. Most 8-bit CPUs have simple schedulers that are adequate for small designs that utilize little outsourced code. They are not likely to be adequate for designs that consume any quantity of outsourced code. The outsourcing of the solution will strongly influence the RTOS choice, which, in turn, dictates what types of CPU are possible.

The tools necessary to do the design: are they available for the standard ANSI C/C++ compiler that you may use? How will you debug your design, either in the hardware/software co-simulation environment or on the SoC after it exists? Does a JTAG port exist, and is the CPU using this channel for debug, or is a dedicated serial port necessary? The choice of a higher level language like C++ or code generated from a design in UML may also dictate the need for a higher bus width and clock frequency to deal with the code size and complexity.

10.1.8 Multicore SoCs

The SoC may be better off if partitioned into several processor subsystems that communicate via a loosely connected FIFO or serial channel. Many designs incorporate a DSP (digital signal processor) and a RISC CPU to share the workload and simplify the design of each processor domain. But this further complicates the CPU choice, which may now be multiplied several times over.

10.1.9 Conclusions

Modern SoC design has presented new challenges for the system architect. No longer is the choice of CPU trivial. Utilizing metrics such as the complexity of overall design, design reuse, protection, performance, power, size, cost, tools, and middleware availability can simplify the decision.

10.2 Emerging Technology for Embedded Systems Software Development

It is easy to think of embedded systems development as state of the art and leading edge. However, since microprocessors were first introduced in the early 1970s and the business has been developing more than 30 years—more than a quarter of a century—it is now a mature technology. By "mature," I do not mean "stagnant" or "boring." Embedded systems software development is far from boring. It is hard to identify any other business that is more dynamic, fast moving, and forward looking.

That maturity may be used to real advantage. After 30 years of growing, it is possible to identify a number of clear trends in the evolution of embedded systems development as a whole. Those trends point to the emergence of key technologies, upon which we may confidently focus to address the challenges ahead.

In this chapter, we endeavor to identify some of those trends and single out the technologies that they drive, resulting in an agenda for our attention over the coming months and years.

10.2.1 *Microprocessor Device Technology*

The earliest microprocessors were 4- and 8-bit devices. As fabrication techniques became more sophisticated, integrated 8-bit microcontrollers began to appear and the first 16-bit microprocessors came into use. Once again, silicon technology moved on, and 16-bit microcontrollers were introduced and widely applied, as demand grew for more sophisticated embedded systems. Devices with 32-bit architecture gradually took hold in higher-end applications, and these too were complemented by highly integrated microcontrollers. The first 32-bit devices were all CISC architecture, but increasingly RISC chips are providing even higher performance.

It would be easy to interpret this "potted history" of the embedded microprocessor, as illustrated in Figure 10.1, as a description of a timeline: 8-bit micros were yesterday; 32-bit RISC is today. However, this is not the case. As the more powerful devices have become available and found application, they have not, for the most part, replaced the earlier parts but have augmented the range of options available to the designer. An embedded systems designer has a wider choice of microprocessors than ever before and must make a choice based on functionality, specification, support, availability, and price.

This increasingly wide range of devices has a number of possible impacts on the software designer. Obviously, suitable programming tools must be available to support this array of processors; it is preferable that the tools are consistent from one device to another. More importantly, the necessity of migrating both code and programming expertise from one device to another is becoming commonplace. This need not present major problems. By careful code design, use of off-the-shelf components, and adherence to recognized standards, porting may be quite straightforward.

Figure 10.1: Microprocessor Technology

10.2.2 System Architecture

As microprocessors have evolved, the architecture of the systems in which they are used has progressed as well. The earliest systems were comprised of the CPU and a selection of logic devices. More highly integrated devices reduced the chip count, and higher-performance devices presented many design challenges to the hardware developer. From the software engineer's point of view, nothing really changed. For many years, the same debugging techniques could be employed as the system became more complex: in-circuit emulation, on-chip debug, ROM monitors, and instruction set simulation. This situation began to change.

As embedded systems become more powerful, with ever-increasing levels of demanded functionality, many designers are taking a fresh look at their use of microprocessors and microcontrollers. In many cases, instead of following the obvious path of simply incorporating more powerful devices, an alternate choice is made: the application of multiple processors. This choice may be driven simply by a desire to distribute the processing power (which would be typical in a telephone switch, for example). Alternatively, an additional processor may be added to provide specific functionality (e.g., a DSP—digital signal processor—in a mobile phone).

One of the biggest challenges faced by software developers when confronted with a multiprocessor system is debugging. It is, of course, possible to simply run one debugger for each device. However, that is not really addressing the problem. What is needed is the means to debug the system, the functioning of each processor and the interaction between them needs to be debugged. The requirement is for a debug technology that supports multiple processors in a single debug session, even when a variety of architectures are represented.

10.2.3   Design Composition

In the earliest days of embedded systems, all of the development—both hardware and software design—was typically undertaken by a single engineer. The software element represented a small part of the entire effort: perhaps 5–10%. As illustrated in Figure 10.2, over time, the proportion of the engineering time dedicated to software development increased substantially. By the mid-1980s, this work was done by software specialists and comprised more like 50% of the development effort.

In the last few years, although hardware design has become more complex, the amount of software has grown drastically, now often being 70–80% of the total design effort. The result is that teams of software engineers are involved and new challenges arise. Among these is the availability of hardware to facilitate software testing. Since more software needs to be developed (in a shorter time), an environment for testing is required sooner. Various solutions are available, including native code execution prototyping environments, instruction set simulation, and the use of standard, low-cost, off-the-shelf evaluation boards. In addition, low-cost host-target connection technologies are becoming common, typically using a JTAG interface.

This climate represents an ideal opportunity for hardware and software teams to work together. By using codesign and, in particular, co-verification techniques, software engineers can test on "real" hardware sooner, and the hardware designers are able to prove their designs earlier, with less prototyping cycles.

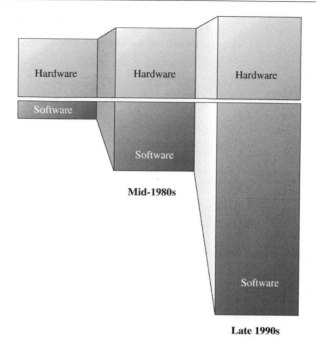

Figure 10.2: Design Composition

10.2.4 Software Content

The proportion of development time dedicated to software has been increasing. Meanwhile, under pressure from worldwide trade and truly global competition, time to market has been decreasing. This has radically influenced the design strategy. The earliest designs were quite simple, being comprised solely of in-house designed applications code. As systems became more complex, a multitasking model was widely adopted for software development, and many developers opted for standard, commercial real-time operating system (RTOS) products.

As shown in Figure 10.3, the proportion of bought-in software, or "intellectual property" (a term borrowed from the hardware design world), has steadily increased, as further standards are adopted.

This trend has a number of implications for the software developer. The integration of standard software components—with the applications code and with one another—are a matter of concern. Debugging in a multitasking context is another issue. The business decision associated with the selection of intellectual property is particularly complex, future (e.g., migration to different processors) as well as immediate requirements must be taken into consideration.

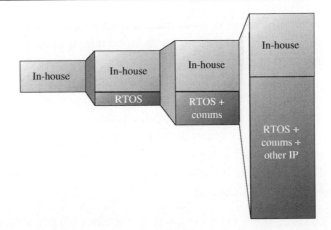

Figure 10.3: Software Content

10.2.5   *Programming Languages*

For the first 4- and 8-bit microprocessors, there was no choice of programming language. Assembler was the only option. Since the applications were relatively simple, this was not a big problem.

As 16-bit technology became viable, the need for a practical high-level language became apparent, and several options emerged. Pascal and C were both in use on the desktop, and these languages were adapted for embedded systems. Intel developed PL/M specifically for this kind of application. Forth was also very popular for certain types of systems. Over time, with the increasing use of 32-bit technology, the two languages that persisted were C and Ada. The latter is prevalent in defense-oriented systems.

It has been known for some years that C++ would start to replace C for embedded software development. Now, between one-quarter and one-third of embedded systems code is written in C++. What was not anticipated a few years ago was the emergence of new languages and approaches, which are set to play a strong role in applications development in the future. The Java language was developed specifically for embedded applications and has found a niche where runtime reconfigurability is demanded. The Unified Modeling Language (UML) has become the most popular choice for a higher-level design methodology.

10.2.6   *Software Team Size and Distribution*

As discussed earlier, the initial embedded system designs were one-man efforts. In due course, specialization resulted in engineers being dedicated to software development.

The next step was the establishment of embedded software development teams. Managing software development is challenging in any context; embedded systems development is no exception and brings its own nuances. Using conventional programming techniques—procedural languages like C and assembler—most members of the team need to have a thorough knowledge of the whole system. As the team grows, this becomes less and less feasible. Typically, specific members of the team have expertise in particular areas. To manage the team effectively, a strategy must be in place that permits the encapsulation of their expertise. It must be possible for the work of an expert to be applied by the nonspecialist in a safe, secure, and straightforward manner. Object-oriented programming techniques find application in this context.

With many very large companies, the software teams are not simply growing; they are becoming distributed. Some members of the team are located at one site, while others are elsewhere. The sites may even be in different countries. This arrangement is common in Europe, where (spoken) language may be a concern. Elsewhere, time zones may be an issue (or an advantage, as a distributed team can work around the clock). This is increasingly the case as emerging technology centers (e.g., in India) are widely utilized. The need for reusable software components becomes even more apparent in this context.

10.2.7   UML and Modeling

The UML has become a key design methodology in recent years, which goes hand in hand with increasing embedded software team size. There are broadly two ways to use a design tool: either as a guide to writing the actual code or as a means of generating the code directly. Code generation is controversial for embedded software, as it may be argued, quite validly, that every system is different and has very specific needs in this respect. This is where xtUML (executable and translatable UML) is attractive because it enables the application and architecture to be clearly separated. This follows the same philosophy as object-oriented programming—leveraging expertise through tools and technology.

10.2.8   Key Technologies

All of these trends, which have become established over 30 years of embedded systems development, point to some key technologies:

Microprocessor technology—Leading to a proliferation in devices that involved the consideration of migration issues and writing portable code. That, in turn, drives a requirement for **compatible tools** and **RTOS products** across microprocessor families.

System architecture—Progressing so that multiprocessor embedded systems are becoming commonplace. This drives a requirement for a **debug technology** that addresses these needs.

Design composition—Changing, with a much greater part of the design effort being expended on software. This drives a requirement for **instruction set simulator technology** and **host-based prototyping** and the application of **on-chip debug facilities** and **hardware/software co-verification**.

Software content—Moving from entirely in-house design to the wide use of intellectual property. This drives a requirement for **standards-based RTOS technology** and appropriate **debug technology**.

Programming language—Narrowing choices somewhat. Although a strong requirement for **C tools** still prevails, compatible **C++ products** are in strong demand.

Software team size and composition—Changing from one engineer (or less) to the employment of large, evenly distributed teams. This drives a requirement for tools to support **object-oriented programming** and **RTOS technology with a familiar or standard API.** There is also an increasing demand for **modeling and design tools**.

10.2.9   Conclusions

Tracking all the emerging technologies, which are driven by the ongoing trends in embedded systems development, is no easy task. Taking any one in isolation is also fruitless because of the many interrelationships. For example, multitasking and multiprocessor debugging go hand in hand; standards-based RTOS technology is a real boon to processor migration; using a design methodology that flows naturally toward an implementation makes complete sense.

10.3   Making Development Tool Choices

This chapter is a review of available tools and techniques for program development in embedded systems, and it discusses the implications of the availability of development tools on selection of a target microprocessor and real-time operating system. This chapter addresses the following questions regarding the selection process:

- What build tools will be needed?

- What features should be sought?

- What about the debugging parameters and options?

- What about tool integration?

10.3.1   The Development Tool Chain

A useful way to view software development tools for this purpose is from the perspective of a tool chain, where each component forms a tight link to the next: from a high-level language, through the assembler, linker, and so on, to one or more debugger variants. (See Figure 10.4.)

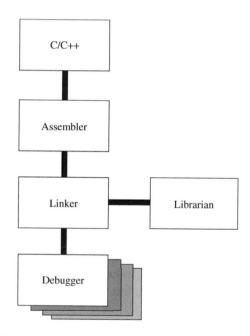

Figure 10.4: The Development Tool Chain

There are two distinct parts to the tool chain:

- The body of the chain consists of the tools that take source code and generate an executable: the *build tools*.

- The base of the chain includes the tools used to execute and verify the executable program: the *debug tools*.

Almost without exception, the use of the build tools (options, controls, formats, and so on) should be quite unaffected by the proposed execution environment and the variant of debug tool employed. For example, it should not be necessary to build using a special library in order to use a particular debug tool. The clear requirement is to test exactly the same code at all stages of the development process.

The options, with respect to the execution environment, offered by debug tools are numerous. These options will be reviewed in turn, but let us first consider the build tools.

10.3.2   Compiler Features

Most commonly, software for embedded systems is written in C. However, C++ is increasing in popularity, as object-oriented design becomes the norm. The main parameters governing the choice of cross-compiler are similar to those applied to native products, but other factors must be taken into consideration, such as:

- **Programming language accepted**—The primary requirement in a compiler is that it accept the programming language in use. For both C and C++, full compliance with the ANSI specifications is essential.

- **Libraries provided**—An ANSI-compliant C compiler need not, according to the specification, include a full runtime library. In reality, such a library is very useful, and its absence would hinder efficient program development. Unfortunately, a number of the standard library functions, as specified by ANSI, are intrinsically nonreentrant, which may be a problem for some embedded system designs. Because particular demands may be placed on library code by an embedded system, access to the library source code is particularly desirable. A common reason for using C++ is to facilitate code reuse and to be able to employ standard class libraries. It is, therefore, reasonable to expect a compiler to be supplied with such a library as standard.

- **Build tools that support an entire microprocessor family**—Typically, an engineer selects a cross-compiler to support development for a specific target microprocessor that will be used for the current project. It is quite likely that future projects will use a different device, but it is commonly another member of the same family. With this in mind, choose build tools that support an entire microprocessor family. This support should, of course, go beyond the generation of code for the "baseline" device. For example, a Freescale 68000 family toolkit should feature specific code-generation capabilities for, say, 68332, 68020, and 68040 and not generate "plain vanilla" 68000 instructions at all times.

- **Manufacturer support**—Beyond the technical requirements of the build tools, it is at least as important to look at the "pedigree" of the build tools: consider the reputation of the company who produces them, their technical support facilities, and the size of the current user base.

10.3.3 Extensions for Embedded Systems

A cross-compiler is intrinsically a more complex tool than its native equivalent. This primarily comes about because very few assumptions about the target environment may be made by the compiler developer. To maintain the appropriate level of flexibility, the compiler manufacturer must implement a number of special features.

In particular, embedded systems usually have complex memory configurations. The simplest have read-only memory (ROM) for code and constant data and random access (read/write) memory (RAM) for variable data. To accommodate this, the minimum required of the compiler is the generation of ROMable code, with the data clearly separated from the code. In most systems, a greater degree of control is needed, and being limited to this simple memory model would be a serious restriction.

A further implication of the memory structure of embedded systems is a clash with a language construct in C. In C, a static variable may be given an initial value. This was intended to avoid the necessity for initialization code for variables whose location in memory could be predicted at compile time. The intention was that such variables would be preset to their starting value in the executable file (memory image) on disk and loaded into memory with the program. For an embedded system, where the program is already in ROM, this mechanism does not work. This situation has three possible outcomes:

- Static variables cannot be initialized.

- Initialized statics can only be used as constants because they must be stored in ROM.

- The build tools must readily accommodate the copying of data from ROM to RAM at startup.

Since the C and C++ languages permit direct access to specific memory addresses, these languages are often useful for embedded system development, particularly for code that is closely associated with the hardware. Naturally, the compiler should not restrict this capability in any way.

As for assembler code, even though nowadays a high-level language is usually chosen for software development, it is inevitable that, at some time, the programmer will write some assembler code. The use of assembler code may be necessary to permit the programmer to extract the last ounce of performance from the target chip, but, more likely, the programmer uses assembler code to access some microprocessor facility that does not map into C (for example, enabling or disabling of interrupts). In the interests of efficiency and code portability, the use of assembler should be minimized, and the facilities for its development should be as flexible as possible. The ability to write a complete assembler module should be augmented by the means to include one or more lines of low-level code among the C language.

10.3.3.1 Impact of Real-Time Systems

The majority of embedded microprocessors are employed in real-time systems, and this puts further demands on the build tools. A real-time system tends to include interrupt service routines (ISRs); it should be possible to code these in C or C++ by adding an additional keyword `interrupt` declaring that a specific function is an ISR. This performs the necessary context saving and restoring and the return from interrupt sequence. The interrupt vector may usually be defined in C using an array of pointers to functions.

Furthermore, in a real-time system, it is common for code to be shared between the mainline program and ISRs or between tasks in a multitasking system. For code to be shared, it must be reentrant. C and C++ intrinsically permit reentrant code to be written, because data may be localized to the instance of the program (since it is stored on the stack or in a register). Although the programmer may compromise this capability (by using static storage, for example), the build tools need to support reentrancy. As mentioned previously, some ANSI-standard library functions are intrinsically nonreentrant, and some care is required in their use.

Since memory may be in short supply in an embedded system, cross-compilers generally offer a high level of control over its usage. A good example is a language extension supporting the `packed` keyword, which permits arrays and structures to be stored more efficiently. Of course, more memory-efficient storage may result in an access time increase, but such trade-offs are typical challenges of real-time system design.

One of the enhancements added to the C language during the ANSI standardization was the `volatile` keyword. Applying this qualifier to a variable declaration advises the compiler that the value of the variable may change in an unexpected manner. The compiler is thus prevented from optimizing access to that variable. Such a situation may arise if two tasks share a variable or if the variable represents a control or data register of an I/O device. It should be noted that the `volatile` keyword is not included in the standard C++ language. Many compilers do, however, include it as a language extension. The usefulness of a C++ cross-compiler without this extension is dubious.

10.3.4 Optimizations

All modern compilers make use of optimization techniques to generate good quality code. These optimization techniques can result in software that rivals handcrafted assembler code for size and efficiency. Optimizations can be local to a function or global across a whole module.

Many optimizations may be applied in a generalized way, regardless of the target. More interesting are those that take specific advantage of the architectural characteristics of a specific microprocessor. These include instruction scheduling, function inlining, and `switch` statement tuning.

Instruction scheduling is a mechanism by which instructions are presented to the microprocessor in a sequence that ensures optimal usage of the CPU. This technique is a common requirement for getting the best out of RISC architectures. However, CISC devices can benefit from such a treatment.

The inlining of functions is the procedure whereby the actual code of a (small) function is included instead of a call. This is useful to maximize execution speed of the compiled code. Some compilers require specific functions to be nominated for this treatment, but automatic selection by the compiler is preferable. The optimization can yield very dramatic improvements in runtime performance.

In C, `switch` statements lend themselves to optimal code generation. Depending upon the values and sequence of the `case` constants, quite different code-generation techniques may be appropriate. Explicit tests, lookup tables, or indexed jump tables are all possible, depending on the number and contiguousness of the `case` constants. It may even be efficient to generate a table with dummy entries if the constants are not quite contiguous. Since the compiler can "rewrite" the code each time it is run, efficiency rather than future flexibility can be the sole priority. This example gives a compiler a distinct advantage over a human assembler code writer.

Manufacturers of development tools for embedded systems have very limited knowledge of the architecture of individual configurations—every system is unique. As a result, fine control over the optimization process is essential. At a minimum, there should be a user-specified bias toward either execution time or memory usage.

10.3.5   Build Tools: Key Issues Recapped

In selecting build tools for embedded system software development, consider these two key issues:

- Do the tools provide extensive accommodation for the special needs of embedded system development?

- Does the compiler perform a high standard of optimization, with extensive user control of the process?

10.3.6   Debugging

Having designed and written a program and succeeded in getting it compiled and built, the programmer's next challenge is to verify the program's operation during execution. This is a challenge for any programming activity and never more than when working on an embedded system. In this context, many external influences on the debugging process and stringent requirements dictate the selection of tools.

10.3.6.1   Debugger Features

Some debugger features are desirable or vital in any context; others are specific to embedded systems work. We will concentrate on the latter.

A key capability of a debugger is the ability to debug fully optimized code. Although this sounds quite straightforward, it is not a facility offered by all debuggers. Often, there is a straight choice: ship optimized code or fully debugged code. It is common to select a microprocessor based on its performance and to rely on the compiler to deliver this performance. This is particularly true of high-performance RISC devices. It is unacceptable to be limited by available debugging technology.

In reality, debugging fully optimized code may be challenging for the programmer. The results of some optimizations (e.g., code motion and register coloring) can make it difficult to follow the execution process. It is, therefore, common to perform initial debugging with optimization "reigned in." However, for the final stages of testing, the debugger should not preclude the use of maximum optimization.

Programmers write software in a high-level language (usually C or C++) primarily in the interests of efficiency. The debugger should pursue this philosophy fully and operate entirely in high-level terms. Code execution should be viewed on a statement-by-statement basis; line-by-line is not good enough. Data should be accessible in appropriate terms. Expression evaluation, structure expansion, and the following of pointers should all be straightforward. On the other hand, low-level access to code and data should also be available, when required.

C++ presents additional requirements: function names should always be shown "unmangled" and constructors and destructors should be visible, for example.

Since the suppliers of tools for embedded systems development cannot predict exactly what a given embedded system is like, they are unable to predict the precise functional requirements of the debugger. In the same way as with the build tools, this problem may be circumvented by providing enough flexibility to the user. For a debugger, this flexibility is manifest in the availability of a powerful scripting language. This might permit I/O device modeling, test automation, and code patching, for example.

The user interface of a debugger is of primary concern, because its design can directly affect the user's productivity. It must be powerful, consistent, and intuitive, which is particularly important when debuggers are to be used in a variety of execution environments. It is clear that if a single debugger family can fulfill all the differing requirements, hours of operator training time can be saved.

10.3.6.2   Development Phases

Before considering how code debugging can be performed, it is useful to review the total development cycle of the embedded system. This process can be divided into five phases, as illustrated in Figure 10.5. At each phase, work on the software may progress, but the scope for progress and the techniques employed change as the system development continues.

In phase 1, although the system hardware is undefined, initial work developing algorithms and trying ideas can proceed. At this stage, it is wise to train the engineers who are going to use the debugger in the use of the software development tools.

At phase 2, since the hardware configuration is known, the engineer performs detailed software design and a large part of the implementation.

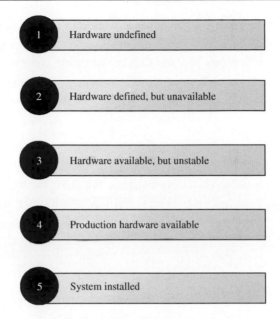

Figure 10.5: Development Phases

In phase 3, although hardware is available, a software engineer may often wish that it was not because the hardware will probably be unstable. However, the engineer can now begin the software/hardware integration.

In phase 4, the availability of stable hardware, maybe in multiple units, permits the engineer to complete final integration and testing.

Some development projects can be completed entirely within the factory, without requiring phase 5. Commonly, however, on-site installation requires final tuning of the software. At some later time, enhancements to the system may necessitate on-site work on the software.

Each development phase calls for a particular type of debugger, as described in the sections that follow.

10.3.6.3   *Native Debugger*

At first sight, a native debugger (i.e., one running on the host computer, executing code in that environment) seems inappropriate for embedded systems development. However, there are two contexts in which such a tool may be useful.

During phase 1 of the project, with no clear idea of the target hardware configuration, a native debugger can provide a useful environment in which to develop ideas and formulate algorithms, particularly for sections of code that are not time critical. This idea can be extended further if a host-based prototyping environment is available. This permits a significant amount of development to proceed on parts of the application that interact with the hardware.

If a native debugger is available, one that has the same (or very similar) user interface to debuggers being used at later stages of the project, the native debugger can offer an ideal training ground, since even if the target hardware is available for training purposes, it may not be the safest place to "play around." The worst that can happen with a native debugger is to crash the computer. The consequences of some embedded systems going out of control may be more dire.

10.3.6.4   Debugger with Simulator

The simulation of the target chip, instruction by instruction, on the host computer provides a very useful environment for software testing at almost any phase of the project. In particular, at phase 2, when the hardware is known but unavailable, a simulator will make rapid progress possible.

A simulator allows very detailed debugging to be performed. Although not running at anything like full speed, the simulator keeps track of execution time and permits accurate timings to be taken. This means the engineer can fine-tune critical code sections early in the development cycle. Since the simulator can effectively add functionality to the microprocessor it is simulating, the execution of the code may be monitored in great detail without any intrusion at all. This facilitates 100% performance analysis and code coverage, which is not possible using other techniques.

Of course, a simulator limited to the simulation of just the core CPU would be of limited utility. The simulator must also address the interrupt and I/O systems.

10.3.6.5   Debugger with ICE Interface

An in-circuit emulator (ICE) for the microprocessor is a very powerful tool for software/hardware integration, particularly when the hardware is exhibiting instability. An ICE enables the software to be run at full speed on the target, while permitting a real-time trace and the specification of complex breakpoint conditions.

Unfortunately, while they were once a ubiquitous tool in any embedded development lab, ICEs are no longer available for most high-end processors. The clock speed of processors made the devices more difficult and expensive to produce, and other technologies have become accepted alternatives.

If one is available, the usefulness of an ICE is influenced critically by the user interface, whose operation in high-level language terms is assumed. However, an interface that is compatible with other debuggers in use during the project is a real bonus. An important parameter in the selection of a debugger is support for industry-standard ICEs.

10.3.6.6   Debugger with Monitor

Once stable and fully-functional hardware is available, the exceptional power of an ICE is less necessary. This is partly because ICEs can be overkill once the hardware is working reliably. Additionally, a cost-effective means of performing on-target debugging for large teams is increasingly required.

This situation led to the development of monitor debuggers where the target hardware is connected to the host computer by a communications link (serial line, Ethernet, and so on) and the target runs a small (<10 K) monitor program that provides a debug environment to the debugger itself, which runs on the host. The result is a low-cost, highly functional debugging solution that enables code to be run at full speed on the target with very little overhead. The ICE may be retained for use in particularly tricky situations.

For a monitor debugger to be viable, the monitor itself must be highly configurable. Standard boards (VME cards and evaluation boards) should be supported "out of the box." Tools and services must be available to facilitate the rapid accommodation of custom hardware.

Although the use of a monitor debugger is most common during phase 4 of a project, it can also be used in phase 5. If the target monitor is included in the shipped software (after all, its memory overhead is likely to be very small), on-site debugging may be possible using just a laptop computer running the debugger.

10.3.6.7   Debugger with Hardware Assist

As the speed and complexity of microprocessors increases, the likely cost (and lack of feasibility) of in-circuit emulators increases. As a result, semiconductor manufacturers are increasingly adding debug facilities to the silicon itself. This may vary from the provision of hardware breakpoints (address/data comparators), which should be supported by a monitor debugger, to a special "debug mode" that requires specific debugger support.

An early example of such a debug mode is background debug mode (BDM), which is featured in Freescale 683xx (CPU32) series devices. Most commonly, devices use a JTAG connection to provide on-chip debug (OCD). Assertion of OCD mode stops the processor and enables a debugger to read and write information to and from the machine registers and memory. To utilize OCD, an appropriate connector must be included on the target board, but this low-cost connector does not represent a significant overhead. Between the host computer and the target board, an OCD adapter is required. Like a monitor, a debugger with OCD (also termed "hardware assist") provides some ICE functionality at a much lower cost. Unlike a monitor, such a technique does not require an additional debug communications port(s) or code on the target.

10.3.6.8   Debugger with RTOS

As embedded applications become more complex, the use of a real-time operating system (RTOS) is increasingly common. Debugging such a system has its own challenges, and they dictate specific requirements in a debugger. Two particular areas of functionality are required in an RTOS debugger:

- Code debugging must be "task aware." Setting a breakpoint on a line of code should result in a break only when the code is being executed by the task being debugged. Code shared between tasks is very common, so this requirement can easily arise. Similarly, data belonging to a specific task instance must be accessible to the engineer.

- Information about the multitasking environment (system data) is required: task status, queues, intertask communications, and so on.

It is clearly desirable that both these requirements are addressed in the same debug tool. If an in-house designed RTOS is used, particular debug challenges arise.

RTOS awareness may be implemented using all of the previously mentioned debug technologies. In particular, OCD and monitor debuggers are most commonly adapted. It is, however, quite possible to enhance simulators or even native debug environments to be RTOS aware.

10.3.7   Debug Tools: Key Issues Recapped

In selecting debug tools for embedded systems software development, there are two key issues:

- Does the debugger permit the use of fully optimized code?

- Do the tools provide support for a wide selection of execution environments used in various phases of the development?

10.3.8   Standards and Development Tool Integration

When selecting development tools, attention to standards is essential. For build and debug tools, it is worth investigating the tools that colleagues and associates are using. Industry standards are likely to enjoy long-term support and "grow" with the target chip. Apart from the development tools themselves, integration with standard version management systems is increasingly a requirement with larger project teams. Similarly, clear links to design techniques must be sought.

Beyond industry standards, attention should be paid to the adherence to "real" standards; that is, those set by international standards bodies. An obvious starting point is the programming language itself. Although the use of pure ANSI C/C++ is desirable, in reality a few specific language extensions are essential to make the language useful and efficient for embedded systems development. Such extensions are provided by suppliers of appropriate compilers (i.e., compilers specifically designed for working with embedded systems), and their use is, of course, very reasonable. A good example of an essential extension to the C language is the keyword `interrupt`, which enables a C function to be declared an interrupt service routine. Then the compiler can take care of the necessary context saving and restoring. However, some nonessential language extensions, provided by a few suppliers, should be avoided to aid code portability between compilers. Similarly, the use of a standard object module format (OMF) for relocatable and absolute binary files may remove the necessity of using build and debug tools from a single source.

In broad terms, choosing tools developed with open interfaces ensures interoperability with other products now and in the future.

10.3.9   Implications of Selections

Although selection of the software development tools is important in itself, it is one of a number of such selections that must be made during the development of an embedded system. Other selections include the target microprocessor, the development host computer, and the RTOS. It is important to appreciate the interaction between these various selection processes, some of which may be less obvious than others.

10.3.9.1 Target Chips

Many reasons can be cited for the selection of a particular microprocessor:

- It has the right range of features.

- The price was right.

- Low power consumption.

- It is fast.

- I have used it before.

- A colleague is using it.

- I liked the salesman.

These reasons are all valid, and a combination of them may be justification for selecting a device. However, another criterion should also be applied:

- A good range of software development tools is available.

Purchasing something from a single, unique source rarely is an acceptable decision. Why should it be the case with software tools? If a microprocessor is supported by a very limited range of tools—perhaps from a single vendor—its use should be called into question.

10.3.9.2 Host Computers

The choice of development platform is largely driven by the local culture. It is likely to be a PC (Windows or Linux) or a UNIX workstation. Software tools vendors offering support on an incredibly wide selection of hosts may be guilty of redefining the word "support." Often, on the less-popular platforms, the product versions on offer are extremely old and have not been maintained.

10.3.9.3 RTOS

The choice of an RTOS (along with the decision to use one or not) is influenced by a number of factors. This topic is worthy of a chapter by itself; however, the availability of development tools is a significant factor, which I address here.

An RTOS with a suitably open architecture makes the most sense. It should accept the output generated by a range of build tools. Suitable debugging tools must also be available.

An option, which is considered under some circumstances, is the use of an in-house developed RTOS. This often represents the worst case in terms of tool availability and compatibility.

10.3.10 Conclusions

The selection of development tools for embedded systems software is not an easy task, with many vendors offering partial or even complete selections of products. A good appreciation of the possibilities and a checklist of questions to pose to vendors are key prerequisites.

10.4 Eclipse—Bringing Embedded Tools Together

Development tools are widely known to be key to the success of microprocessors. Although powerful embedded tools have been developed over the last two decades, little progress has been made in integrating multivendor tools on multiple hosts. Without good integration, communication between tools is restricted, and the full potential of the tools is untapped.

Proprietary IDEs (integrated development environments) limit integration and prevent use of best in class or preferred tools. This inflexibility frustrates developers and curbs productivity. De facto proprietary standards partially address this problem but are restricted to a single host. Thus, embedded developers have long wished for a host-agnostic open IDE that they can enhance with their own or third party tools.

The new Eclipse platform, an open host-independent, industry-standard base, makes this possible. On the desktop, the Eclipse platform is already noted for its excellence and is used in numerous business applications. The benefits seen on the desktop—a common tool interface and integration platform—can be brought to the embedded world.

10.4.1   Eclipse Platform Philosophy

During the Internet boom days, the availability of tools mushroomed for the various Internet business applications. Since these tools were built by diverse organizations, most of them had their own GUI paradigms and rarely worked well with each other. It became apparent that a standard IDE and framework were required. To address this need, IBM started the Eclipse project to build a well-designed tool integration platform so that independently built tools could be part of a single environment. The result was the Eclipse platform.

Originally, IBM released the Eclipse platform into Eclipse Open Source, and later, on February 2, 2004, the Eclipse Foundation reorganized into a not-for-profit corporation. "Eclipse became an independent body that would drive the platform's evolution to benefit the providers of software development offerings and end-users. All technology and source code provided to this fast-growing ecosystem will remain openly available and royalty-free."

Unlike other open source organizations, the Eclipse Foundation is driven by business needs; hence, it is also known as the "directed" open source organization.

A major goal for Eclipse is to provide a well-planned and secure platform for commercial tool vendors. In addition, the Eclipse Foundation constantly works to remove hurdles in licensing the platform for commercial use. Contributed code is thoroughly scrubbed before it is committed; to ensure ease of licensing, there are plans to replace the existing CPL (Common Public License), which is already much simpler than the GPL (General Public License), with a more relaxed EPL (Eclipse Public License).

10.4.2   Platform

The Eclipse design focuses on a new paradigm—an open platform to integrate tools. In the old paradigm, individual tools are integrated, one at a time, either into an IDE or with another tool. This is a workable patch for a small set of proprietary tools but fails to scale in the larger multivendor context.

To address scalability, the Eclipse platform uses the innovative plug-in architecture. The platform, developed from the ground up, comprises well-defined GUI and framework mechanisms that provide a standard interface, facilitate integration, and are extensible. Tool developers, who no longer have to worry about GUI and framework issues, can concentrate on their tool-specific advancements—for example, multicore debug.

Extension points, extensions, and plug-ins form the underlying mechanisms of the plug-in architecture. Plug-ins are the smallest functional entities. Eclipse plug-ins from any source can be plugged into the platform for a single integrated environment. Except for the platform runtime, Eclipse itself is implemented as a set of plug-ins as shown in Figure 10.6.

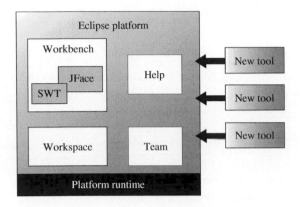

Figure 10.6: Eclipse Platform Architecture

The core of Eclipse is its user interface made up of the Workbench, JFace, and the Standard Widget Toolkit (SWT). The combination of these plug-ins is known as the Rich Client Platform (RCP).

- **SWT and JFace**—SWT and JFace take care of the windowing system in an OS independent way allowing portability across hosts.

- **Workbench**—The Workbench is the Eclipse UI. It is a collection of editors, views, perspectives, and dialogs provided as a common base for tools to use and extend.

- **Workspace**—Resources (projects, folders, and files) reside in the Eclipse Workspace where you can navigate them at will. The manipulation of resources, of course, provokes automatic incremental builds.

- **Team**—The Team plug-in takes care of source control. CVS (Concurrent Versions System) is the default, but other source control systems, including ClearCase, Source Integrity, and Visual SourceSafe, can be plugged in.

- **Help**—The Help plug-in does what the name implies. Integrated tools can extend help for tool-specific needs.

10.4.3   How Eclipse Gets Embedded

As many in the Eclipse community have said, "The Eclipse platform by itself is an IDE for everything and nothing in particular." CDT (C/C++ Development Tools) and JDT (Java Development Tools) are open source incarnations of the Eclipse platform for the desktop C++ and Java developer, respectively. They do not address the complexities of embedded development.

An Eclipse-based embedded IDE is a powerful, self-contained environment for building and debugging embedded systems, integrating program management and debug tools, and allowing users to drop in their favorite Eclipse plug-ins—for example, an editor or source control system.

It is important that such a development strictly adheres to Eclipse principles and is itself implemented as a set of plug-ins. This methodology allows the inheriting of today's key features in Eclipse, but also future ones as they become available. For example, platform runtime changes made in Eclipse 3.0 would be automatically reflected in the embedded IDE.

The key embedded development technologies that need to be made available as Eclipse plug-ins include:

- **Build tools**—To accommodate the great variability between embedded systems, compilers, assemblers, linkers, and so on, build tools tend to be significantly more complex than their native development counterparts are. The return, in terms of improved usability, of their incorporation into an IDE is very significant.

- **Debug**—Software engineers spend more time debugging than anything else, and embedded programmers tend to use a variety of debug tools, which may accommodate different execution environments, RTOS awareness, or multicore debug capabilities.

- **Target connection**—An embedded development environment generally consists of a host computer with one or more target devices connected to it. These targets may be local, or they may be located remotely and reached via a network; they may be "real" targets (i.e., actual boards) or "virtual" (i.e., provided by some kind of simulation or emulation facility). Selecting and configuring the connections to multiple targets is another complex matter that may be readily simplified in this way.

- **Simulation**—Availability (or, rather, nonavailability) of hardware is an increasingly difficult challenge for embedded software developers, as the beginning stages of software development are brought forward to an earlier point in the project cycle.

A number of simulation tools may be employed: native execution, instruction-set simulation, and hardware/software co-verification are all options. These also need to be brought into the IDE.

- **Profiling**—Ensuring that an embedded application functions correctly is the first priority, but since resources are always scarce, profiling tools are employed and need to be contained within the IDE. These tools analyze how resources, time, and memory primarily, are used by the application (and any associated RTOS).

10.4.4   Conclusions

The need for an IDE for embedded software is apparent. The use of Eclipse as the basis for such an environment is clearly a very flexible approach, which is gaining ground across the embedded software development industry and will yield benefits for both suppliers and users alike.

10.5   Embedded Software and UML

10.5.1   Why Model in UML?

Yes—why? For all the usual reasons: to reduce costs, time to market, and unnecessary redevelopment and to increase productivity, quality, and maintainability. How can UML models do all that?

That depends on what a model is, and how it relates to the systems development process. There are at least three meanings of "model," and each meaning has different uses and implications. Let's take a look at each meaning.

One meaning for the word "model" is a "sketch." For example, we might sketch out a hardware configuration on the back of a beer mat, showing a few boxes for processors and lines for communication or adding a few numbers to indicate bandwidth or expected usage. The sketch is not precise or complete, nor is it intended to be. Often, a sketch of this nature is "talked to" by pointing at various boxes to explain what is happening there and how it relates to other elements. The purpose is to communicate a rough idea, or to try one out just to see if it will work. The sketch is neither maintained nor delivered.

A second meaning for "model" is "blueprint"—a classical example is the set of plans for a house. The blueprint lays out what must be done, describing properties needed to build the real thing, as determined by an architect. Because blueprints are intended to be plans for construction, they often map closely to the artifact that is to be built, so for each important element in construction, there is a "symbol." Because software is a complex beast, the set of symbols—the vocabulary—can become quite large, and without standards, chaos can ensue.

Enter the Unified Modeling Language (UML), which is a language that can be used for building software blueprints. (It has other uses too, as we shall see.) The UML is the result of an effort to reduce needless differences between different systems development methods and establish a common vocabulary for software modeling.

Why would you want to use UML? For all the reasons we outlined previously. Thinking about what you intend to build carefully—to the point of defining it exactly so that someone else can build it—will reduce costs and decrease defects, similar to the efficiency of writing a detailed, reviewed shopping list that avoids all the effort involved in returning a wrong item and getting the right one.

However, as anyone who has built a house knows, the blueprint is rarely followed to the letter. Instead, as the builder (in contrast to the architect) constructs the house, the facts on the ground cause some modifications to be made.

The same argument can be applied to models: as we write code, we discover that our design wasn't as clever as we thought. This critique has led to the deprecation of models as "paper mills" that deliver pictures but not working systems. Instead, it has been argued, we should just hack—sorry, write—code because it executes.

Execution is important because it closes the verification gap between a concept on paper and a reality that either works or not. Code either runs right or it doesn't. You can't be certain of that one way or the other with a blueprint, even with the best review team in the world.

The third meaning for "model" then, is an "executable." When we build an executable model, we have described the behavior of our system just as surely as if we had written a program in C. Indeed, when you have a software model that can be compiled and executed, there's no need to distinguish between the model and the "real thing." It is the software.

So does this mean we should "program in UML"? In addition, if so, why should that reduce costs, time to market, and unnecessary redevelopment, as well as increase productivity, quality, and maintainability?

The answer to the first question is "Yes, but at a higher level of abstraction." For example, when you declare an association between two classes, you do not say whether that will be implemented by a pointer, a reference, or a list (just as when you program in C, you don't think about allocating registers). Therefore, while you are "programming," when you build an executable UML model, you don't have to think about a lot of things you normally worry about when programming in a language at a lower level of abstraction.

This approach reduces costs (the first of our reasons for modeling) because the cost of writing a line of code is the same irrespective of language. Studies as far back as 30 years ago showed that, on average, a developer produces 8 to 12 lines of assembly code, or C, or FORTRAN, or whatever per day. These numbers are "fully loaded," meaning that we're taking into account the time we spend in meetings, unjamming the printer, dealing with performance reviews, fighting the configuration management system, running tests, and all that other stuff. Although some programmers are much more productive, their productivity is also the same irrespective of language.

When we program in an executable UML, we write at a higher level of abstraction, thus reducing costs and increasing productivity. One user of executable UML generates 7 to 10 lines of C++ for each line of logic written in UML; the amount of code would be greater if this user's projects were written in C. Because the number of lines of code per day is the same, this translates directly into a decrease in time to market and an increase in productivity. For these reasons, we developed higher-level programming languages as sketched in Figure 10.7.

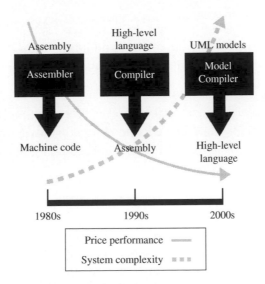

Figure 10.7: The Evolution of Software Development

Using an executable UML also increases quality. Not only is the number of defects reduced, but the errors are found earlier, providing time to react. It is better to know you have a problem when you have six months to go on the project than six weeks! Figure 10.8 shows the effect of applying this methodology.

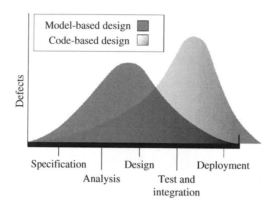

Figure 10.8: Model- versus Code-based Design

Early error identification is achieved by building test cases and running them against the executable model. Because the model is executable, we can provide real values and get real results immediately, using a model executer that interprets the models. You and other experts can see immediately whether the model is doing the right thing. If it is not, you make the change in the models, then and there, and run the tests again.

We must emphasize that model testing occurs early in the life cycle, thus removing downstream defects. In turn, this reduces the effort involved in implementing the wrong thing, just as with that shopping list. The combination of removing defects early and avoiding wasted effort implementing the wrong thing reduces costs and time to market, while increasing productivity and quality.

Models are also more maintainable than code because it is easier to manipulate concepts at a high level of abstraction than a lower one. The careful reader will have noted that we have discussed all of our reasons to model except reduction of unnecessary development, or—putting it in the positive—maximizing reuse. While it is certainly easier to reuse models than code (that higher level of abstraction argument again), the main reason you can reuse models is the same reason you are more likely to reuse a C program than one written in M68000 assembly code—namely, you can port the C program across multiple hardware platforms.

The same concept applies to executable models. When we built an association, we did not specify whether it was implemented as a pointer, a reference, or a list. This allows us to decide later, once we better understand the speed and performance constraints of our system. In other words, executable models confer independence from the software platform, just as writing in C made us independent of the hardware platform. We can then redeploy the executable model onto different software platforms and implementation environments. This is actually something of an understatement. Models can be translated into just about any form, so long as their application behavior, as defined in the executable model, is preserved.

This brings us to a concern. When we moved from assembly code to C, we lost control of, for example, register allocation, which could lead to a reduction in the performance of the system—a killer concern in an embedded system. The keyword here is "control." If the compiler does a good enough job we don't care, but if the compiler doesn't know enough about our environment to make sound decisions, and we have no control over those decisions, we're in trouble.

For this reason, models need not only to be executable, but also to be translatable onto any software platform, and you, the developer, have to have control of how that translation process takes place, reducing performance concerns to zero. After all, if you can write the code, you can also describe how to go from a concept, as expressed in an executable model, to that code. A translatable UML also offers complete control of how that code is produced.

It is for this reason that we support executable and translatable UML, or xtUML, for short. xtUML models are both executable and translatable to any target software platform in an open manner. We do this by using a trick that differentiates blueprint models from executable models: separation of application from architecture.

10.5.2   Separating Application from Architecture

The separation of the application from the architecture differentiates blueprint-type models from executable ones. To understand that, we first need to understand how "blueprint" model-driven developers do their work.

10.5.2.1   Blueprint Development

After some initial requirements work, which can be supported by models such as use cases, the blueprint developer builds an analysis model, in UML, that captures the problem domain under study. This model will use various elements of the UML, but there is no universal agreement as to what those elements should be. As a simple example, the UML allows for attributes to be tagged with a visibility (`public`, `private`, and so on). Should an analysis model include this information? That depends upon taste—some do, some don't. Everyone is agreed that an analysis model should not contain design details, but there is little agreement on what that means exactly.

The next stage is to build a design model that does incorporate all that design detail. The design model is a blueprint that captures the software structure of the intended implementation. The work of transforming the analysis model to a design model exercises embedded systems design expertise. For example, we know, as embedded system designers, that a good way to store fixed-size data elements in a memory-limited environment is to pre-allocate memory—or whatever your expertise tells you to do. This expertise is applied to the analysis model to produce the design blueprint.

The next step is to code it up from the blueprint. Putting aside possible errors in the design, this means filling in code bodies. There are two ways to do that. One is to add the code directly to the model, and have a tool generate code according to the software structure. Another way is to code up the software structure suggested by the blueprint, adding in coding details. The process is sketched in Figure 10.9.

10.5.2.2   What's Wrong with That?

Nothing, if you like doing all that work over and over every time the technology—and therefore the software structure—changes. And if you like reinventing and reapplying the same programming constructs when you add new system functionality.

This approach to software development is rather like using C-like pseudo code to outline your design, then hand coding the assembler. Each time you add new application functionality, you have to decide over again how to pass parameters to a function, how to allocate registers to compute an expression, and so on. Each time you port to a new hardware platform, you have to work out what the assembly code meant (you wouldn't trust the pseudo code, would you?) and rewrite it for a new processor.

At root, you have failed to leverage and capture the embedded systems design expertise represented by going from analysis to design to code. Alternatively, to use the pseudo code analogy, you have failed to leverage and capture the expertise involved in assembly coding.

10.5.2.3   Model Compilers

The solution, of course, is to build a compiler from a more formalized pseudo code (which we may call C) for each of the various processors. Certain parts of the compilers are

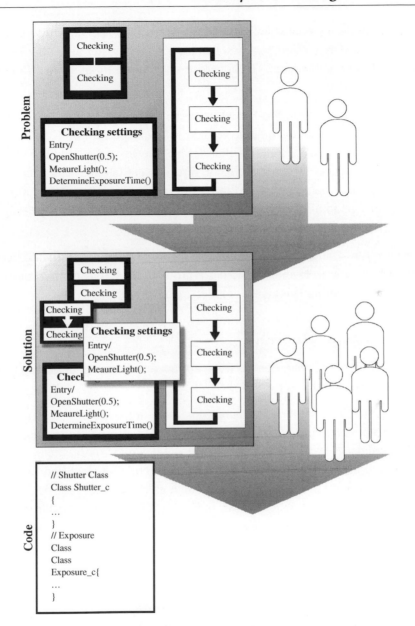

Figure 10.9: The Blueprint Development Process

common, such as building an abstract syntax tree. Others are specialized to the target processor, though they may share common techniques for register allocation, expression ordering, or peephole optimization. The expertise is captured in an artifact (a compiler) that can be reused as required.

The same concept applies to xtUML model compilers. We can build model compilers for each software platform. (Note the adjective: each *software* platform.) A software platform is simply that set of technology that defines the software structure, such as choice of data structure and access to it; concurrency, threads, and tasking; and processor structure and allocation. All these details are filled in by the model compiler, just as a programming language compiler fills in all the details of register allocation, parameter passing, and so on, as determined by the hardware platform.

This approach captures the expertise involved in making embedded software design decisions and allows you to leverage it across a project and across many projects. Model compilers, like programming language compilers, can be bought.

10.5.2.4   Sets, States, and Functions

Figure 10.10 illustrates the separation between application and architecture. The element to focus on is the dotted line that separates the two.

When we build an xtUML model, it is represented in a simple form for translation, as sets of data that are to be manipulated, states the elements of the problem go through, and some functions that execute to access data, synchronize the behavior of the elements, and carry out computation. The UML is just an accessible graphical front end for those simple elements. When you build a "class" in xtUML, such as `CookingStep` in a microwave oven, it represents a set of possible cooking steps you might execute, each with a cooking time and power level. Similarly, when you describe the life cycle of a cooking step using a state chart diagram, it follows a sequence of states as synchronized by other state machines (when you open the microwave door, it had better stop cooking!), external signals (such as a stop button), and timers. In addition, in each state, we execute some functions.

Naturally, it's a bit more complicated than that, but the point is that any xtUML model can be represented in terms of these primitive concepts. And once that's done, we can manipulate those primitive concepts completely independently of the application details.

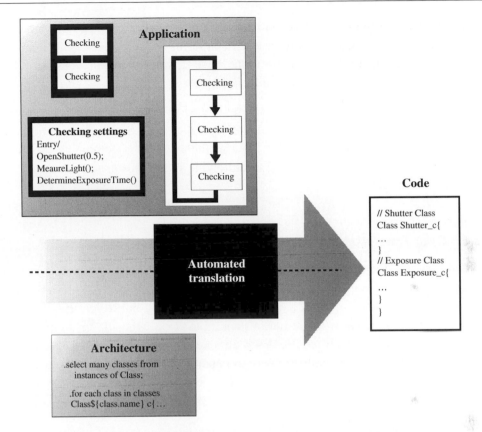

Figure 10.10: Separation of Application and Architecture

10.5.2.5 Rules

The ability to perform this independent manipulation allows us to write rules. One rule might take a "class" represented as a set CookingStep(cookingTime, powerLevel) and produce a C++ class declaration. Crucially, the rule could just as easily produce a struct for a C program, or even a COMMON block in FORTRAN. Similarly, we may define rules that turn states into arrays, lists, switch statements, or follow the state pattern from the Design Patterns community. (This is why I put "class" in quotation marks. A "class" in an executable model represents the set of data that can be transformed into anything that captures that data; in a blueprint-type model, a class is an instruction to construct a class in the software.)

These rules let us separate the application from the architecture. The xtUML model captures the problem domain graphically and represents it in terms of sets, states, and functions. The rules read the application as stored in terms of sets, states, and functions, and turn that into code. This leads to the process shown in Figure 10.10.

The value here is that application models can be reused by applying different sets of rules (a different model compiler) to target a new software platform. Similarly, the model compiler can be reused in any project that requires the same architecture. The applications and the model compilers can each evolve separately, reducing costs, and increasing productivity and reuse.

10.5.2.6   Open Translation

There is one critical difference between today's programming language compilers and model compilers. With a programming language compiler, you have limited control over the output. Sure, you can apply a few flags and switches, but if you truly dislike the generated code for any reason, you're out of luck unless you persuade the vendor to make the changes to the compiler you require. With a model compiler, the translation rules are completely open. If you can see a better way to generate code because of the particular pattern of access to data, say, you can change the rule to generate exactly what you want. This completely removes any concerns about optimization. It is totally under your control.

I should emphasize that you rarely need to change the model compiler, still less write one of your own. But the knowledge that you can change it should increase your confidence in the technology. Another analogy to programming languages: when they were new, people were concerned about the quality of the output and having some control over it. Over time, of course, those concerns have diminished, even in the embedded space.

10.5.3   xtUML Code Generation

We will now take a look at the code that will be produced from xtUML models. Obviously, what we want is executable code. For an example, we will look at the safety-related logic of a simple microwave oven. The oven components are the door, which must be closed while cooking, and the actual cooking element. There will be some code to manage the cooking times and power levels.

Take a look at some representative code for such a microwave oven:

```
struct Oven_s
{
    ArbitraryID_t OvenID;

    /* Association storage */
    Door_s *Door_R1;
    Cooking_Step_s *Cooking_Step_R2;
    Cooking_Step_s *Cooking_Step_R3;
    Magnetron_s *Magnetron_R4;

    /* State machine current state */
    StateNumber_t current_state;
};
```

The C `struct` captures information about the oven, which has an arbitrary ID
(an identifier to distinguish a particular instance) and some pointers that reference its
components. The oven `struct` also has a `current_state` that captures
the—well—current state of the oven.

The "Cooking step" in Figure 10.10 allows the microwave oven to be programmed to cook
in steps, each at a different power level for a certain time. Each step describes cooking
parameters for the oven. Typical uses are to program a cooking step to defrost by pushing
one button (Time 1, say) with low power and a long time, followed by pushing another
button (Time 2, for a second unimaginative name, say) to cook ready to eat at high power
for a shorter time. There are twin steps because there are two buttons.

Here is the code for `Cooking_Step_s`:

```
struct Cooking_Step_s
{
    i_t stepNumber;
    i_t cookingTime;
    i_t powerLevel;
    Timer_s *executionTimer;

    /* Association storage */
    Oven_s *Oven_R2;
    Oven_s *Oven_R3;

    /* State machine current state */
    StateNumber_t current_state;
};
```

This structure includes a step number (used also as an identifier), cooking time, and power level. In the scenario described previously, there could be two instances of this `struct`, say:

Step Number	Cooking Time	Power Level
1	10 mins	20%
2	3 mins	100%

An execution timer is also used to refer to one of several potential timers. This reference is required so the timer can be interrogated, reset, or deleted. In addition, association storage refers back to the oven. There are two associations because we can program two cooking steps.

Again, the cooking step has a current state attribute to capture whether the cooking step is ready (i.e., has been programmed), executing, or complete.

We need to understand the conceptual entities in the problem and how they are described by data. Figure 10.11 shows a so-called "class diagram," which does just that.

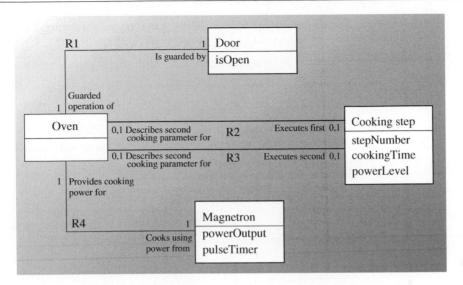

Figure 10.11: Microwave Oven Class Diagram

This diagram declares four conceptual entities (the oven and the cooking step we have already discussed, plus a door interlock and a magnetron tube), with several associations identified with "R numbers." The numbers are simply a way to identify each association uniquely; the "R" comes from the real-world relationship captured by the associations. The associations also have names that capture the real-world relationship. R2, for example, is read: Oven executes first CookingStep and CookingStep describes first cooking parameters for Oven.

The associations also have a multiplicity that indicates how many instances participate in each association. For R2, each oven may have zero or one (0, 1) first cooking steps. And each cooking step may or may not be the first cooking step for this oven (hence the 0, 1 again). The association (R4) between the oven and the magnetron is "1" in both directions because one Oven *houses* one Magnetron, and one Magnetron *is housed in* one Oven. In general, an association can have many instances, as when a dog owner owns (one or more) dogs, which would be written "1 . . . *". A person, on the other hand, owns "0 . . . *" dogs, because you have to own at least one dog to be considered a dog owner, but a person is free to choose not to own any.

Let's compare now the "class diagram" of Figure 10.11 with the declaration of Oven_s. We can see that the name of the struct is the same as the name of the box but with a suffix _s, and the OvenID has a type ArbitraryID with a suffix _t; both these coding conventions remind us of the purpose of the name symbols.

The second section of the code, marked "Association storage," has a pointer of type Door_s, named Door_R1, which implements the R1 association. Remember, the _s indicates the pointer is to a struct capturing information about the door. Similarly, the other pointers implement the other associations, R2, R3, and R4.

The current_state attribute is part of an underlying state machine mechanism; it's a little special and deserves its own type—state_number_t. It happens to be an integer, but because we know something about its likely values, we may choose to implement it as an unsigned char, for instance.

The primary observation to make here is the close correlation between the oven diagram and the corresponding declaration. Note that the declaration of Cooking_Step_s and the Cooking Step "class" have the same close correspondence too.

So, why do we put quotation marks around "class"? We do so because the class oven isn't a class at all. It's a C struct! This goes back to the separation between application and architecture we discussed earlier. The application model describes the fundamentals of the solution, while the architecture defines the mapping to the implementation—just as we showed here in this extended example.

Moreover, the C we showed illustrates the point we made right at the beginning. The models we build are executable and translated into the implementation. This is in contrast to "blueprint-type" models that are intended to direct the implementation. There, a "class" means that we should build a class in the code. Here a "class" simply declares important data and houses behavior as defined by a state machine, the trace of which is the current-state attribute.

10.5.4 Conclusions

Modeling and the use of the UML mean different things to different people; the terminology can be confusing and is often misused or abused. There are various possible goals that come out of the use of modeling, but an approach that reduces rework and leverages the diverse expertise of the embedded development team must be a clear winner.

A Note on Terminology

A software platform is analogous to a hardware platform. It is the set of technologies that we rely on to make our software work: linked lists, middleware, operating systems, IPC mechanisms, and so on. These technologies don't know about an application, although the choice of software platform will have an impact on system performance.

An architecture (more precisely, an application-independent software architecture) is an abstract representation of how to map an application to a specific target platform. (It includes the software platform by reference.)

A model compiler is the physical realization of the architecture; it is a program like a programming language compiler.

These terms have analogies in the programming language compiler world: a compiler embodies choices about the architecture based on, and relying on, the facilities provided by the hardware platform.

10.6   Model-based Systems Development with xtUML

Developers of embedded systems have always faced a number of challenges. From determining the hardware/software partition to meeting performance and cost objectives, the job of building these systems has never been easy. With ever-increasing demand for more functionality packed into smaller spaces consuming less power, building embedded systems definitely is becoming more complex every day. Add to these demands the need to shrink development cycles and reduce the overall cost of the system, and you have the state of the embedded systems industry today.

This brief chapter provides a glimpse at a new way to overcome many of the challenges confronting embedded systems developers today.

10.6.1   Why Is Building an Embedded System so Difficult?

Perhaps because it involves the coordination of at least two very creative and rather complex disciplines: software engineering and logic design. Maybe because it often involves creating something new—something nobody has ever built before. Or it may simply be the multitude of choices to be made: allocation of function among logic and software, which processor to use, how to arrange the bus, which programming language to employ, and whether or not to use an operating system. Most of these choices interact with one another in interesting ways, making the prospect of building a new product more than a little daunting.

10.6.1.1   Every Project Has at Least Two Fundamental Challenges

By their very nature, embedded systems meld software and hardware together to form a coherent solution. The user of the resulting product never fully appreciates the effort that goes into getting the partition between the two right. With current development practices, getting it wrong is very costly at best and disastrous at worst.

Verifying the hardware/software partition requires the ability to test the system, and producing a version of the system that can represent the real thing accurately enough to verify the partition requires a huge intellectual investment. First, a hardware/software interface specification must be painstakingly written (or scratched out on a bar napkin, as the case may be). Then, the logic designers and the software engineers must construct, using implementation languages like VHDL and C, behavioral models of the function to be mapped to hardware and software. In most cases, the software engineers are not building models at all, but instead are writing the application code for the system.

Working at a level of abstraction somewhere between that of C and assembly language, the logic designers toil away to produce behavioral models of the hardware, often writing small test drivers and diagnostics to test their models along the way.

Integration: Where the Fun Begins   After months of effort, the two teams are ready to test their prototype system. Most folks do this in one of two ways: simulation or prototype hardware. With recent advances in FPGA technology, it has become significantly easier to test a logic design on an actual chip rather than resorting to simulation on the development workstation. Regardless of the test bed, the integration effort is always interesting.

Recall the development process started with two separate teams, each with different skills, heading off in parallel. The only thing connecting them is that hardware/software interface specification, written in natural language. You know where this is heading.

Two teams with disparate disciplines working against an ambiguous document to produce a coherent system. Sounds like a line from a cheap novel.

Invariably, the two components do not mesh properly. The reasons are myriad: the logic designers didn't really mean what they said about that register in the document; the software engineers didn't read all of the document, especially that part about waiting a microsecond between whacking those two particular bits; and of course, the most common failure mode of all, logic interface changes that came about during the construction of the behavioral models that (horrors!) didn't make it into the interface specification.

10.6.1.2   So What's a Few Interface Problems Among Friends?

Nothing really. Just time and money. And market share. We've been doing it this way for years. It's nothing a few days (well, weeks) in the lab won't solve. Besides, shooting these bugs is fun, and everyone is always so happy when it finally works. It's a great bonding experience.

Eventually, the teams manage to get the prototype running, at least well enough that they can begin measuring the performance of the system. Keep in mind here that "performance" has a number of connotations in this context: along with the obvious execution time and latency issues, memory usage, gate count, power consumption, and its evil twin, heat dissipation, top the list of performance concerns in many of today's handheld devices.

10.6.1.3   The Suboptimal Partition

There's nothing like a performance bottleneck to throw a bucket of cold water on the bonding rituals of the integration heroes. Unlike interface problems, you don't fix hardware/software partition problems with a few long nights in the lab. No, this is when the engineers head back to their desks to ponder why they didn't pursue that career as a long-haul truck driver.

Partition changes are expensive, and they are difficult to do correctly. Since the system is represented only in terms of implementation languages, knowledge of the partition is distributed throughout the software and the logic design. While it's simple enough to say, "Let's move this function into logic," it's quite another matter to make it happen. Remember, we have months invested in the construction of the prototype system. Making fundamental architectural changes cannot be done in a matter of days, at least not with the traditional approach previously described.

Getting the hardware and software to mesh properly and making the right partition between them are two significant challenges faced by all embedded systems developers. Certainly, many others exist, but we promised a *short* chapter.

10.6.1.4   *The State of the Practice*

So, it takes months of effort to produce a prototype that can be executed, but we need to execute it before we will know whether or not the logic designers and the software engineers agree on the hardware/software interface. We need to run the prototype system before we can measure its performance, but if the performance is unacceptable, we'll spend weeks changing the hardware/software partition. That's the state of the practice.

So, what if we had a way to eliminate completely the hardware/software interface problems that are discovered during the initial integration? What if we also had a way to change the partition between the hardware and software in a matter of hours?

10.6.2   *A Better Solution*

We have just such a solution. First, we build abstract models of the system using an executable and translatable UML (xtUML). These models have sufficient detail that they are executable, so we can test the behavior of the system early and continuously before investing in the construction of an actual implementation in C and VHDL. The models are also translatable; that is, they are completely independent of design and implementation concerns. Testing at this level is therefore concerned only with ensuring that the models accurately represent the application to be constructed.

When the models are complete, we can specify an initial partition between hardware and software. A model compiler then translates the models into logic designs (VHDL, Verilog, SystemC, and so on) and software (C, C++, Java, assembler, and so on) according to the specified partitioning.

10.6.2.1   *Interface Problems?*

The interface between the hardware and software is defined by the model compiler. Because the implementation is generated, there can be no interface mismatches. Because we no longer have two separate teams of people working from a natural language interface specification, the generated implementation is guaranteed to have exactly zero interface problems. (This does have the unfortunate side effect of reducing the number of opportunities for logic designers and software engineers to spend long nights together in the integration lab working around interface problems.)

10.6.2.2   *What about the Partition?*

Because the xtUML models accurately and precisely represent the application, and the implementation is generated, with absolute fidelity, from these models, the partition can easily be changed, and a new implementation can then be generated. This replaces weeks of tedious manual changes to an implementation with a few hours of automatic generation.

With the ability to change the partition and regenerate the implementation, the developers can explore much more of the design space, measuring the performance of various allocation arrangements that would otherwise be prohibitively expensive to produce.

10.6.3   *Experience to Date*

An experimental model compiler has been constructed that generates C++ and VHDL. The generated system was then able to execute within a logic simulation environment. There is still a way to go, but the concept is sound.

10.7   *The Future*

Our vision is to provide a complete system-level development environment that allows embedded developers to construct abstract models of their systems and then automatically translate those models into an optimized implementation that includes software and custom logic components. Of course, without the interface problems and the partition issues, we'll need to find another excuse to spend long nights and weekends in the lab.

Index

An entry followed by (f) indicates a figure or that a figure is cited within a page range.
An entry followed by (t) indicates a table or that a table is cited within a page range.
An entry followed by (f,t) indicates a figure and a table or that both are cited within a page range.

A

Acceptable timeliness, 541
ADC (analog-to-digital) converter, 64–66, 450
Address Resolution Protocol (ARP), 276
AES (Advanced Encryption Standard), 278
Alias of the region 0–512KB, 355
American National Standards Institute (ANSI), 17, 266
American Standard Code for Information Interchange (ASCII), 11–12(t)
Analog-to-digital (ADC) converter, 64–66, 450
ANSI (American National Standards Institute), 17, 266
Application specific integrated circuits (ASICs), 556
Application-independent software architecture, definition of, 739
ARP (Address Resolution Protocol), 276
ARRAY, 26–29
ASCII (American Standard Code for Information Interchange), 11–12(t)
ASCII codes, 11(t)

ASICs (application specific integrated circuits), 556
Asymmetric and symmetric dual-core processors, 505–506
Asymmetric programming model, 507–508(f)
Asynchronous events, 540
Asynchronous hardware/firmware, 661–662
ATAPI/Serial ATA, 448
Atomic operations, 475–476
Atomic variables, 656–658
Audio-video synchronization, 523–525(f)
  Buffer basics, 524(f)
  Conceptual diagram of audio-video synchronization, 525(f)
  MpeG-2 encoded transport stream format, 523(f)
Autobuffer scheme, 512
Auto-vectored interrupt scheme, 94

B

Background debug mode (BDM), 718
Backspace (BS), 12
Base register (BR) fields, 114
Base-ten number system, 2–4
BCD (binary-coded decimal), 10

BDM (background debug mode), 718
Berkeley Sockets interface
  Socket connection, making, 297–298
  Sockets, types of
    Datagram (DGRAM), 295, 296, 296n
    Raw, 295n
    Stream, 295, 296
    TCP socket operation, 297(f)
    UDP socket operation, 298(f)
BF (best fit) algorithm, 218
Big endian, 18
Binary addition, 37–39(t)
Binary division, 42–46
Binary multiplication, 40–42
Binary numbers (base two), 4
Binary semaphores, 206
Binary subtraction, 39–40(t)
Binary-coded decimal (BCD), 10
BIT (binary digit) or Boolean, 14–16
Bit reversal, 461–462(f)
Blocks, 111
BLOCK/WAITING queues, 187, 188

Board I/O driver examples,
    143–168(f)
Ethernet driver, initializing
    CDMA/CD (MAC sublayer)
        reception mode,
        147–150(f)
    CDMA/CD (MAC sublayer)
        transmission mode, 147
    Data encapsulation (Ethernet
        frame), 144–146
    Ethernet frames, 145(f)
    Ethernet topologies, 144(f)
    Flowchart of high-level
        functions of full-duplex
        in transmission mode,
        150(f)
    High-level flowchart of MAC
        layer processing a MAC
        client's request to
        transmit a frame, 148(f)
    High-level flowchart of MAC
        layer processing
        incoming bits from the
        physical layer, 149(f)
    Introduction, 143–144
    Media access management,
        144, 146–147
    Motorola/freescale MPC823
        Ethernet example,
        150–154(f)
    MPC823 Ethernet block
        diagram, 151(f)
    MPC823 Ethernet driver
        pseudo code, 154–159
    NET+ARM Ethernet block
        diagram, 159(f)
    NET+ARM40 pseudo code,
        160–162
    NetSilicon NET+ARM40
        Ethernet example,
        159–162(f)
    OSI model, 144(f)
    Introduction, 143
    RS-232 driver, initializing
        Introduction, 162–163

Motorola/freescale MPC823
    RS-232 example,
    164–166
MPC823 serial driver pseudo
    code, 166–168
OSI model, 162(f)
RS-232 frame diagram, 163(f)
RS-232 hardware diagram,
    163(f)
Board Support Package (BSP),
    169, 237–238
Boolean or BIT (binary digit),
    14–16
BR (base register) fields, 114
Broadcast protocol, 31
BS (backspace), 12
BSP (Board Support Package),
    169, 237–238
Buddy system, 218
Buffer protocol (or circular
    buffer), 34–37(t)
Bus protocol, 134–135
Byte addressability, 459
Byte ordering schemes, 111–112

C

C compilers, modern
    Basic transformations, 680–682
    Compiler optimization,
        controlling, 685–686
    Function calls, 683–684
    Function inlining, 684–685
    Introduction, 677–678
    Linker, 685
    Low-level code compression,
        685
    Memory model, 686–687
    Program, meaning of, 680
    Register allocation, 682–683
    Structure of, 678–679
Cacheability protection look-aside
    buffers (CPLBs),
    486–487
CALL statement, 58
CALL/RETURN, 71–72
CAN (controller area network),
    570

Canonical format indicator (CFI),
    146
Carriage return (CR), 12
CASE statement
    (SWITCH/CASE), 53–54
Catch block, 319
CCD AFE (Charge-coupled device
    analog front-end), 550
CDMA/CD (Half-Duplex Carrier
    Sense Multiple
    Access/Collision Detect)
    protocol, 146
CEC (core event controller),
    452–453
CFI (canonical format indicator),
    146
CHAR data type, 16–17
Charge-coupled device analog
    front-end (CCD AFE), 550
Check sum, 12, 13
Chip select (CS), 113
Circular buffer (or buffer
    protocol), 34–37(t)
Circular buffering, 460–461
CL (collision) bit, 141
Client Server paradigm, 293–294
Clock paging, 228
CM (continuous mode) bit, 141
Code, compiled and run
    With Dynamic C, 258–259
    How Dynamic C builds code,
        258(f)
    The traditional build process,
        257(f)
    In traditional development
        environments, 256–257(f)
Coding guidelines, summary of
    Functions, 644–645
    General programming
        guidelines, 642
    Initialization of variables, 643
    Intrinsics, 645
    Loops, 643–644(f)
    Use libraries, 645–646
    Variable declaration, 642
    Variable declaration (data
        types), 642–643

COFF (common object file
format), 217(f)
Collision (CL) bit, 141
Common object file format
(COFF), 217(f)
Communication Processor Module
(CPM), 151
Communication protocols,
29–37
Event-driven multielement
transfers, 34–37(t)
Event-driven single transfer, 32
Introduction, 29
Simple data broadcast, 29–32
Storage and retrieval pointers,
36(t)
Compile All, 256
Compilers and optimization
Architecture and flow, 621(f)
Compile time options
First level of
optimization–register
level optimizations,
631–632
Highest level of
optimization–file
optimization, 632
Second level of
optimization–local
optimization, 632
Third level of
optimization–global
optimization, 632
Compiler optimizations,
621–631(f)
Controlling, 685–686
General architecture, 621(f)
Information generated, output
file, 633
Instruction selection, 625–627(f)
Saturated add function, code
for, 626(f)
Saturated add using DSP
intrinsics, 626(f)
Introduction, 620
Latency and instruction
scheduling, 627–628

Machine dependent
Implementing special
features, 624
Latency, 624–625
Resources, 625
Machine independent
Alias disambiguation, 624
Branch optimization, 621
Common subexpression
elimination, 622
Constant propagation, 622
Dead code elimination, 622
Dead store elimination,
622–623
Example of code motion
optimization, 622(f)
Global register allocation, 623
Inline expansion of
runtime-support library
functions, 624
Inlining, 623(f)
Inlining replaces function
calls with actual code
that increases
performance but may
increase program size,
623(f)
Loop invariant code motion,
622
Loop unrolling, 622
Strength reduction, 623–624
Register allocation, 628–631(f)
C code snippet with a number
of variables, 629(f)
Processor memory hierarchy,
629(f)
Register spilling caused by a
lack of register
resources, 630(f)
Some loops are too large for
the compiler to pipeline,
631(f)
Complex data types
ARRAY, 26–29
STRUCTURE, 22–24
UNION, 25–26

Concurrent Versions System
(CVS), 723
Conditional statements, 47–54
Connecting ISRs to interrupt table,
99
Constants, 254–255
Context switch, 187
Context Switcher, 76–77
Context switching, 98–99
Continuous mode (CM) bit, 141
Control register, 16
Controller area network (CAN),
570
Cooperative operating system,
80–82
Cooperative scheduling, 194(f)
Core event controller (CEC),
452–453
Core processing, 445–446
Counting semaphores, 208–209
Co-verification
Benefits of, 411
Brian Bailey on communication,
413(f)
Versus co-design, 413–414
Versus co-simulation, 413
Co-verification is about
debugging hardware and
software, 410(f)
Defined, 410–411
Definition of co-verification,
410(f)
Hardware/software
co-verification, history of,
406–409
Commercial co-verification
tools appear, 407–409
Improves communication, 413
Introduction, 405–406
Is it necessary, 414
Methods
Code for simple stub, 416(f)
Hardware stubs, 415–416(f)
Instruction set simulation,
415
Introduction, 414

Co-verification (*Continued*)
   ISS with memory model
      interface, 416(f)
   Microprocessor evaluation
      board, 417–418
   Native compiling software,
      415
   Real-Time Operating System
      (RTOS), 417
   Waveforms, log files, and
      disassembly, 418
  Methods, sample of
   C simulation, 423–426(f)
   Cycle-based instruction set
      simulator connected to
      logic simulation, 423(f)
   Evaluation board with logic
      simulation, 430–431(f)
   Example of implicit access,
      420(f)
   FPGA prototype, 434–435
   gdb connected to the RTL
      code of the
      microprocessor, 428(f)
   Hardware model of the
      microprocessor, 429(f)
   Hardware model with logic
      simulation, 429–430(f)
   Host-code execution with
      logic simulation, 419(f)
   Host-code mode example
      function calls, 420(f)
   Host-code mode with logic
      simulation, 418–420(f)
   In-circuit emulation, 431–434
   Instruction set simulation with
      logic simulation,
      421–423(f)
   Instruction set simulator
      connected to logic
      simulation, 422(f)
   JTAG connection speed
      bridge and emulation
      system, 434(f)
   JTAG connection to an
      emulation system, 432(f)

   JTAG connection with test
      chip and emulation
      system, 432(f)
   Microprocessor evaluation
      board with logic
      simulation, 431(f)
   RTL model of CPU with
      software debugging,
      426–428(f)
   RTOS simulation and
      host-code execution,
      421(f)
   Verilog-to-C translator
      requests, 425(f)
  Metrics
   AHB arbitration and cycle
      accuracy issues,
      438–440(f)
   Correct timing of the bus
      request, 439(f)
   Incorrect timing of the bus
      request, 440(f)
   Modeling summary, 440–441
   Other metrics, 442
   Performance, 435–436
   Software, type of, 441–442
   Synchronization, 441
   Verification accuracy,
      436–438
  Project schedule savings,
      411–412
  Project schedule with
      co-verification, 412(f)
  Project schedule without
      co-verification, 412(f)
  visibility, enable learning by
      providing, 412
CPLBs (cacheability protection
   look-aside buffers), 486–487
CPM (Communication Processor
   Module), 151
CPU, choosing for your system on
   chip design
  Conclusions, 700
  Costs, 699
  CPU performance, 699
  Design complexity, 697

  Design reuse, 698
  Introduction, 697
  Memory architecture and
     protection, 698–699
  Multicore SoCs, 700
  Power consumption, 699
  Software issues, 699–700
CR (carriage return), 12
CRC (cyclic redundancy check),
   12, 13, 146, 366
Critical sections, 205
CS (chip select), 113
CVS (Concurrent Versions
   System), 723
Cyclic redundancy check (CRC),
   12, 13, 146, 366

D

D/A (digital-to analog) converters,
   450
Data cache, 480–481(f)
Data encapsulation, 144
Data Encryption Standard (DES),
   278
Data field, 145
Data movement, physics of
  Exploiting priority and
     arbitration schemes
     between system resources,
     494–495
  Grouping like transfers to
     minimize memory bus
     turnarounds, 488–491
  Introduction, 488
  Keeping SDRAM rows open
     and performing multiple
     passes on data,
     492–493(f)
  Optimizing system clock
     settings/ensuring refresh
     rates are tuned for speed at
     which SDRAM runs,
     493–494(t)
  Taking advantage of code and
     data partitioning in external
     memory, 493(f)

Understanding Core and DMA
   SDRAM accesses, 491–492
Using the optimal refresh rate,
   494(t)
Data Segment, 249–250, 254
Data structures
   Complex data types
      ARRAY, 26–29
      STRUCTURE, 22–24
      UNION, 25–26
   Introduction, 13–14
   Simple data types
      Boolean or BIT (binary digit),
         14–16
      CHAR data type, 16–17
      DOUBLE (double precision
         floating-point), 20–21
      FLOAT (floating point),
         20–21
      INT (integer), 17–19
      LONG (long integer), 19–20
      Pointers, 21–22
      STRUCTURE, 15, 16
Datagram (DGRAM), 295, 296,
   296n, 299n
Deadline, 198
Deadlock, 209
Death traps, 367
Debugging
   Debug tools: key issues
      recapped, 719
   Debugger features, 713–714
   Debugger with hardware assist,
      717–718
   Debugger with ICE interface,
      716–717
   Debugger with simulator, 716
   Development phases,
      714–715(f)
   Monitor debugger, 717
   Native debugger, 715–716
   RTOS debugger, 718
Decision operators, 445–446
Decision tree, 51–52
Decoders, 445
Demand paging, 227
Demilitarized zone (DMZ), 279

DES (Data Encryption Standard),
   278
Design sign-off model (DSM), 440
DEVELOPER, hat, the, 334–336
   Case study—big bang
      integration and no
      regression test suite,
      335–336
   Guidelines, 340–341
   Regression testing—test early
      and test often, 334–335
Development technologies and
   trends
   CPU, choosing for your system
      on chip design
      Conclusions, 700
      Costs, 699
      CPU performance, 699
      Design complexity, 697
      Design reuse, 698
      Introduction, 697
      Memory architecture and
         protection, 698–699
      Multicore SoCs, 700
      Power consumption, 699
      Software issues, 699–700
   Embedded systems software
      development
      Conclusions, 707
      Design composition,
         703–704(f)
      Introduction, 700–701
      Key technologies, 706–707
      Microprocessor device
         technology, 701–702(f)
      Microprocessor technology,
         702(f)
      Programming languages, 705
      Software content, 704, 705(f)
      Software team size and
         distribution, 705–706
      System architecture, 702–703
      ULM and modeling, 706
   The future, 743
Development tool choices, making
   Build tools: key issues recapped,
      713

Compiler features, 709–710
Debug tools: key issues
   recapped, 719
Debugging
   Debugger features, 713–714
   Debugger with hardware
      assist, 717–718
   Debugger with ICE interface,
      716–717
   Debugger with simulator, 716
   Development phases,
      714–715(f)
   Monitor debugger, 717
   Native debugger, 715–716
   RTOS debugger, 718
Development tool chain,
   708–709(f)
Eclipse platform
   Conclusions, 725
   Eclipse platform architecture,
      723(f)
   Embedded development
      technologies, necessary,
      724–725
   Help plug-in, 723
   Introduction, 721
   Philosophy, 722
   Platform, 722–723(f)
   SWT and JFace, 723
   Team plug-in, 723
   User interface, 723
   Workbench, 723
   Workspace, 723
Extensions for embedded
   systems, 710–711
   Assembler code, 711
   Impact of real-time systems,
      711
   Memory configurations,
      710–711
Introduction, 707–708
Optimizations, 712–713
Selections, implication of
   Conclusions, 721
   Host compilers, 720
   Introduction, 719–720

Development tool choices, making
(*Continued*)
  RTOS, 721
  Target chips, 720
  Standards and development tool
      integration, 719
Device drivers
  Architecture-specific, 86, 87(f)
  Board I/O driver examples; *see*
      Board I/O driver examples
  Definition of, 85
  Driver code layers, 89(f)
  Embedded system board
      organization, 86(f)
  Embedded systems model and
      device drivers, 85(f)
  Example: device drivers for
      interrupt handling, 89–110
  Example: memory device
      drivers, 110–134
  Functions, 88
  Generic, 86, 87(f)
  Hardware states, 88
  Interrupt device driver pseudo
      code examples; *see*
      Interrupt device driver
      pseudo code examples
  Interrupt handling, for; *see*
      Interrupt-handling, device
      drivers for (example)
  Interrupt-handling and
      performance, 109–110
  Introduction, 85–89
  Memory device drivers; *see*
      memory device drivers:
      example
  MPC860 architecture specific
      device driver system stack,
      87(f)
  MPC860 hardware block
      diagram, 87(f)
  Onboard bus; *see* Onboard bus
      device drivers: example
  Summary, 168
  von Neumann model,
      85–86

Device drivers for
    interrupt-handling: example;
    see interrupt-handling, device
    drivers for (example)
DFT (Discrete Fourier transform),
    587
DGRAM (Datagram), 295, 296,
    296n, 299n
DHCP (Dynamic Host
    Configuration Protocol), 277,
    278
Digital multimeter (DMM), 337
Digital Signal Processing (DSP);
    see DSP (Digital Signal
    Processing)
Digital still camera (DSC), 445,
    550
Digital-to analog (D/A) converters,
    450
Direct memory access (DMA); *see*
    DMA (direct memory access)
"Direction Control" facility, 489
Discrete Fourier transform (DFT),
    587
Dispatcher, 191
DMA (direct memory access)
  Access data in on-chip memory,
      598(f)
  Code snippet that pends on a
      semaphore for DMA
      completion, 602(f)
  Code snippet that polls for
      DMA completion, 603(f)
  Code to set up and enable A
      DMA operation, 597(f)
  Example, 599–601
  Internal memory, managing,
      603–604(f)
  Internal memory of a DSP must
      be managed by the
      programmer, 604(f)
  Introduction, 595–596(f)
  Pending versus polling,
      601–603(f)
  Simple function consisting of
      five executable lines of
      code, 599(f)

Staging data, 596–599(f)
  Template for using DMA to
      stage data on- and off-chip,
      598(f)
  Using, 596–604(f)
DMA bandwidth, 503
DMZ (demilitarized zone), 279
DNS (Domain Name System), 277
DOUBLE (double precision
    floating-point), 20–21
DO/WHILE statement, 55–56
DSC (digital still camera), 445,
    550
DSM (design sign-off model), 440
DSP (Digital Signal Processing),
    529–654
  Compiler output, helping the
      programmer, 640–641(f)
  Compilers and optimization,
      more on
    Architecture and flow, 621(f)
    Compile time options,
        631–633
    Compiler optimizations,
        621–631(f)
    General architecture of a
        compiler, 621(f)
    Introduction, 620
  DMA (direct memory access);
      see DMA (direct memory
      access)
  DSP acceleration decisions
    Computational complexity,
        563
    Cost, 564–565
    Data locality, 563
    Data parallelism, 563
    Hardware/software mix in an
        embedded system, 564(f)
    Signal processing algorithm
        parallelism, 562–563
  DSP and μP (microprocessor),
      535
  DSP systems are hard real-time
    Based on signal sample, time
        to perform actions before
        next sample arrives, 538

Hard real-time systems,
538–540
Introduction, 538
Efficient execution and the
execution environment
Efficiency overview, 541
Resource management,
541–542
Embedded system life cycle; *see*
Embedded system life
cycle using DSP
Embedded systems and real-time
systems, overview of, 536
Embedded systems development
life cycle using DSP, 554
Example of a DSP-based
"system" for embedded
video applications, 554(f)
Flexibility and power,
summarization of, 534(f)
Hard real-time and soft real-time
systems, 537–541(t)
Hard real-time and soft real-time
systems
Introduction, 537
Real-time and time-shared
systems, differences
between, 537–538(t)
Real-time event
characteristics—real-time
event categories,
540–541
Inline, definition of, 633
Inlining, 639
Intrinsics
C code to perform saturated
add, 635(f)
Definition of, 633
Introduction, 635
Some intrinsics for the
TMS320C55 DSP, 637(f)
TMS320C55 DSP assembly
code for the saturated
add routine, 636(f)
TMS320C55 DSP assembly
code for the saturated
add routine using a

single call to an intrinsic,
636(f), 637(f)
Introduction, 529–536(f,t)
Keywords
Common, 638
Definition of, 633
Loop unrolling
Execution units, filling the,
605–607(f)
Introduction, 604
Loop overhead, reducing,
607–609
Register space, fitting the
loop to, 609–610(f)
Trade-offs, 610(f)
Make the common case fast, 584
Make the common case
fast—DSP algorithms
DSP algorithms share
common characteristics,
587–588(f)
FFT versus DFT for various
sizes of transforms
(logarithmic scale),
587–588(f)
Make the common case
fast—DSP architectures,
584–586(f,t)
DSP algorithms share
common characteristics,
586(t)
DSP Harvard
Architecture–multiple
address and data busses
accessing multiple banks
of memory
simultaneously, 586(f)
Special multiplication
hardware speeds up DSP
processing, 585(f)
Make the common case
fast—DSP compilers
C example and the
corresponding pipelined
assembly language
output, 591(f)

Corresponding pipelined
assembly language
output, 592–593(f)
DSP architectures, 589(f)
FFT versus DFT for various
sizes of transforms
(logarithmic scale),
588–594
Five stage loop with three
iterations, 589(f)
Optimization
DMA (direct memory access),
595–604(f)
Loop unrolling, 604–610(f)
Software pipelining; *see*
Software pipelining
Optimizing DSP software
Definition of, 580–581
General DSP code
optimization process,
583(f)
Process, the, 581–584
Pragmas
Definition of, 633
Example of using pragmas to
improve the efficiency of
the code, 634(f)
Introduction, 634
Profile-based compilation
Achieving full entitlement,
652(f)
Advantages, 646–647(f)
Code optimization process,
summary of, 648–653(f)
Debugging optimized code,
issues with, 647–648
Expanded code optimization
process for DSP,
649(f)
Introduction, 646
Profile-based compilation
shows the various
trade-offs between code
size and performance,
647(f)
Summary, 653

DSP (Digital Signal Processing)
(*Continued*)
Real-time system design,
challenges in
Distributed and
multiprocessor
architectures, 544–545
Embedded systems,
545–553(f)
Failures, recovering from, 544
Introduction, 542
Response time, 543–544
Real-time systems
Definition of, 536, 537(f)
- A real-time system reacts to
inputs from the
environment and
produces outputs that
affect the environment,
537(f)
Real-time systems—soft and
hard, 536–537
Reducing stack access time,
639–640
Modifying the stack pointer to
point to on-chip
memory, 640(f)
Moving the stack pointer
on-chip to increase
performance in a DSA
application, 640(f)
Signal processing algorithms,
531–532
Signal processing, definition of,
530–532
Software pipelining
Enabling, 618–619(f)
Example; simple loop
implemented in C,
612–618(f)
Interrupts and pipelined code,
619
Introduction, 610–612(f)
Summary of coding guidelines
Functions, 644–645
General programming
guidelines, 642

Initialization of variables, 643
Intrinsics, 645
Loops, 643–644(f)
Use libraries, 645–646
Variable declaration, 642
Variable declaration (data
types), 642–643
Summary of signal processing
algorithm, 531(t)
DSP and μP (microprocessor), 535
Dual-core processors, 505–506
Duration, 198
Dynamic addressing, 283, 283n
Dynamic C development system
Debugging, 244–246
Development, 244–245
Dynamic C libraries, brief
introduction to, 246
Dynamic C, memory spaces in
Dynamic C's usage of the
Rabbit 3000 segments,
250(f)
Introduction, 247
Memory usage without separate
I & D space, 248–251(f)
Quick summary, 256
The Rabbit 3000 MMU
segments, 248(f)
Rabbits memory segments,
247–248
Separate instruction and data
memory, 252–255
XMEN, placing functions in,
252
Dynamic C support for networking
protocols, 275–279(f)
The four-layer networking
model and related
protocols, 275(f)
Introduction, 275
Networking protocols, common,
275(f), 276–278(f)
Optional modules for Dynamic
C, 278–279
Dynamic Host Configuration
Protocol (DHCP), 277, 278

E

E (empty) bit, 141
Earliest deadline first (EDF)
algorithms, 540n. 2
Earliest Deadline First (EDF) clock
driven scheduling, 198, 199(f)
EBIDIC (8-bit code), 11
Eclipse platform
Conclusions, 725
Eclipse platform architecture,
723(f)
Embedded development
technologies, necessary,
724–725
Help plug-in, 723
Introduction, 721
Philosophy, 722
Platform, 722–723(f)
SWT and JFace, 723
Team plug-in, 723
User interface, 723
Workbench, 723
Workspace, 723
ECU (embedded control unit), 548
EDF (earliest deadline first)
algorithms, 540n. 2
EDF (Earliest Deadline First)/clock
driven scheduling, 198, 199(f)
Effective Page Number Register
(EPN), 125
Efficient programming,
architectural features for
Bit reversal, 461–462
Byte addressability, 459
Circular buffering, 460–461
Example of single-cycle,
multi-issue instruction,
457(f)
Hardware loop constructs,
457–459
Interlocked instruction pipelines,
462–465(f)
Multiple operations per cycle,
456–457(f)
Electromagnetic interference
(EMI), 13

ELF (executable and linking format), 215, 216(f)
Embedded coding techniques
Asynchronous hardware/firmware, 661–662
Atomic variables, 656–658
C compilers, modern
Basic transformations, 680–682
Compiler optimization, controlling, 685–686
Function calls, 683–684
Function inlining, 684–685
Introduction, 677–678
Linker, 685
Low-level code compression, 685
Memory model, 686–687
Program, meaning of, 680
Register allocation, 682–683
Structure of, 678–679
Final notes, 695–696
Firmware, not hardware, 668–671
Gray code, 669–670
Interrupt latency, 671–674
Metastable state, 667(f)
Metastable states, 666–668(f)
Nonreentrant code, eliminating, 659–661
Options, 664–665
Other RTOSes, 665–666
Programming tips
Addresses, do not take, 691
Bit fields, investigate before using, 693–694
Clever code, do not write, 692–693
Extra hints, use, 694–695
Function prototypes, use, 690
Inline assembly language, do not use, 691–692
Jump tables, use a switch for, 693
Library functions, 694
Parameters, use, 690

Pointer types, use the best, 688
Right size data, use, 687
Structures and padding, 688–689
Variables, 691n
Race conditions, 662–663
Recursive function, 661
Reentrancy, 655–656
Setup and hold times, 667(f)
Taking data, 674–677(f)
Two more rules, 658
Embedded control unit (ECU), 548
Embedded media processing
Asymmetric and symmetric dual-core processors, 505–506
Audio-video synchronization, 523–525(f)
components of
ATAPI/Serial ATA, 448
Classes of peripheral interfaces and representative examples, 446(t)
Components of a typical media processing system, 445(f)
Connectivity, 449–450
Core processing, 445–446
Data movement, 450
External memory interface, 448
Flash storage card interfaces, 449
Host interface, 448
I²C (Inter-IC Bus), 447
IEEE 1394 ("FireWire"), 450
Input/Output subsystems—peripheral interfaces, 446–450
Memory subsystem, 450–451
Parallel Video/Data Port, 450
PCI (Peripheral Component Interconnect), 449–450

Programmable Flags/GPIO (General Purpose Inputs/Outputs), 447–448
Programmable timers, 447
Real-Time Clock (RTC), 447
SPI (Serial Peripheral Interface), 447
Storage, 448–449
Subsystem control—low-speed serial interfaces, 446–448
Synchronous Serial Audio/Data Port (SPORT), 450
UART (Universal Asynchronous Receiver/Transmitter), 446–447
Universal Serial Bus (USB) 2.0, 449
Watchdog timer (WDT), 448
Data types, choosing
Arrays versus pointers, 467
C data types and their mapping to Blackfin registers, 466(t)
Data buffers, 469–470
Division, 467–468
Intrinsics and in-lining, 470–471
Loops, 468–469
Volatile data, 471–472
Efficient programming, architectural features for
Bit reversal, 461–462
Bit reversal in hardware, 462(f)
Byte addressability, 459
Circular buffer in hardware, 460(f)
Circular buffering, 460–461
Example of interlocked pipeline architecture with stalls inserted, 463(f)

Embedded media processing
(*Continued*)
Example of single-cycle,
multi-issue instruction,
457(f)
Hardware loop constructs,
457–459
Interlocked instruction
pipelines, 462–465(f)
Multiple operations per cycle,
456–457(f)
Efficient programming, complier
considerations for
C data types and their
mapping to Blackfin
registers, 466(t)
Data types, choosing,
466–472(t)
Introduction, 465–466
Event generation and
handling
Sample system-to-core
interrupt mapping,
453(f)
System interrupts, 452–455(f)
Frameworks; *see* Frameworks
Master-Slave and Pipelined
model representations,
509(f)
Memory architecture–need for
management
Data cache, 480–481(f)
Data cache and DMA
coherency, 480(f)
Data memory management,
479–481
DMA and cache, system
guidelines for
choosing between,
481–486(f)
Instruction memory
management—to cache
or DMA, 477–479
Memory access trade-offs,
476–477(f)
Memory Management Unit
(MMU), 486–488

Physics of data movement
Core and DMA SDRAM
access, understanding,
491–492
"Direction Control" facility,
489
Grouping like transfers to
minimize memory bus
turnarounds, 488–491
Introduction, 488
Optimizing system clock
settings/ensuring refresh
rates are tuned for speed
at which SDRAM runs,
493–494(t)
Priority and arbitration
schemes, 494–495
SDRAM rows open and
performing multiple
passes on data, keeping,
492–493(f)
Taking advantage of code and
data partitioning in
external memory, 493(f)
Programming methodology,
455–456
Programming models
Asymmetric, 507–508
Asymmetric model, 507(f)
Homogeneous, 508–510
Master-Slave programming
model, 509–510(f)
Programming model
summary, 510(t)
System and core synchronization
Atomic operations, 475–476
Introduction, 472
Load/store synchronization,
472–473
Ordering, 473–475
System resource partitioning and
code optimization, 451
Techniques for, 443–44
The "ultimate" portable device,
444(f)
Embedded networks, 274–275
Embedded operating systems

Introduction, 169–175
General OS model, 170(f)
Interrupt and error detection
management, 170
I/O system management,
171
kernel subsystem
dependencies, 171(f),
172(f)
Layered OS block diagram,
174(f)
Memory management, 170
Microkernel-based OS block
diagram, 174(f)
Monolithic OS, 173(f)
OSes and the embedded
systems model, 169(f)
OSes models, 172–175
Process management, 170
Security system management,
170
I/O and file system
management, 230–232(f)
Memory management
Introduction, 213–214
Kernel memory space,
229–230
User memory space,
214–229(f)
Multitasking and process
management
Intertask communication and
synchronization,
204–213(f)
Introduction, 177–178(f)
Process implementation,
178–191(f)
Process scheduling,
191–204(f)
OS performance guidelines,
235–236
OS standards example: POSIX
(portable operating system
interface), 232–235(f)
OSes and board support
packages (BSPs), 237–238

Process (or task)
  definition of, 175
  Multitasking OSes, 175–176
  OS task, 175(f)
  Tasks and threads, 177(f)
  Unitasking OSes, 175, 177
Summary, 239
Embedded programming concepts
  Communication protocols
    Event-driven multielement
      transfers, 34–37
    Event-driven single transfer,
      32–34
    Introduction, 29
    Simple data broadcast, 29–32
  Data structures
    Complex data types, 22–29
    Introduction, 13–14
    Simple data types, 14–22
  Introduction, 1–2
  Mathematics
    Binary addition and
      subtraction, 37–40
    Binary division, 42–46
    Binary multiplication, 40–42
    Introduction, 37
  Multitasking
    Basic requirements of, four,
      76–78
    Introduction, 74–76
    Operating systems, 78–82
    State machine multitasking,
      82–84
  Numbering systems
    Base-ten number system, 2–4
    Binary numbers (base two), 4
  Numeric comparison
    Conditional statements, 47–54
    Introduction, 46–47
    Loops, 54–57
    Other flow control statements,
      57–59
  Signed binary numbers, 5–13
    Alternate numbering systems,
      9–10(t)
    ASCII, 11–12(t)
    Binary-coded decimal, 10

  Error detection, 12–13
  Fixed-point binary numbers,
    7–8
  Floating-point binary
    numbers, 8–9
  Negative numbers, represent
    in, 5–7
State machines
  Data-indexed, 64–66
  Execution-indexed, 67–72
  Hybrids, 72–74
  Introduction, 59–64
Embedded system design process
  Embedded system design
    process, 399(f)
  Hardware and software
    integration, 401
  Hardware design, 400–401
  Introduction, 399
  Microprocessor selection, 400
  Requirements, 400
  Software design, 401
  System architecture, 400
Embedded system life cycle using
  DSP
  Acceleration decisions,
    562–565(f)
  Architectural block diagram of a
    DSP, 567(f)
  Basics and architecture,
    understanding, 565–568
  Compiler techniques for
    producing high
    performance code, 577(f)
  Determining what to do based
    on available CPU
    bandwidth, 575(f)
  Digital signal processors,
    560–561(f)
  DSP CPU architectural
    highlights, 568(f)
  DSP filtering using a FIR filter,
    566(f)
  DSP processor solutions, 561(f)
  DSP software, 575–577(f)
  DSP solution, definition of,
    555–556(f)

  Example DSP application with
    major building blocks,
    580(f)
  Example DSP reference
    Framework, 579(f)
  Example—performance
    calculation, 572(f)
  FPGA solutions, 559–560
  Frameworks, 578–580(f)
  General signal processing
    solution, 561–562(f)
  General-purpose processor
    solutions, 558(f)
  General-purpose processors, 558
  Hardware components, selection
    of, 556
  Hardware gates, 557
  Input/Output options,
    570–571(f)
  Introduction, 554–555
  Many applications, many
    solutions, 557(f)
  Microcontroller (mC), 559
  Microcontroller solutions,
    559(f)
  A model of a DSP Framework
    for signal processing,
    578(f)
  Models of DSP processing,
    568–570(f)
  performance, calculating,
    571–575(f)
  Performance calculation
    analysis, 574(f)
  Pros and cons of writing DSP
    code in assembly language,
    577(f)
  Reuse opportunities using DSP
    libraries and third parties,
    576(f)
  Selection criteria, 555–556(f)
  Single sample and block
    processing, 568(f)
  Software-programmable,
    557–558
  Typical DSP algorithms,
    565(f)

Embedded systems software
    development
  Conclusions, 707
  Design composition, 703–704(f)
  Introduction, 700–701
  Key technologies, 706–707
  Microprocessor device
      technology, 701–702(f)
  Microprocessor technology,
      702(f)
  Programming languages,
      705
  Software content, 704, 705(f)
  Software team size and
      distribution, 705–706
  System architecture, 702–703
  ULM and modeling, 706
EMI (electromagnetic
    interference), 13
Empty (E) bit, 141
Enabling/disabling interrupts, 99
Encoders, 445
End string, 310
EPN (Effective Page Number
    Register), 125
Error checking field, 146
Error handling and debugging,
    333–398
  Building a great watchdog,
      379–382
  External WDTs, 384–385,
      390(f), 391–392
    TI's TPS3813, 385(f)
    Window timing of Maxim's
        equally cool MAX6323,
        385(f)
  Hardware alternatives
    Flexible and safe, 363–364
    Introduction, 362
    Memory diagnostics, 364–365
    Separating code and data,
        362–363
  Implementing downloadable
      firmware with flash
      memory, 350
  Internal WDTs, 382–384,
      389–390(f)

Microprogrammer
  Advantages of, 351
  A basic microprogrammer,
      354–355
  Definition of, 350–351
  Disadvantages of, 351–352
  Receiving, 352–353
MSP430–a 16 bit processor that
    uses no PCB real estate,
    391(f)
Multibyte writes, 374–377
Nonvolatile memory, 372
Problems (common) and their
    solutions
  Can't generate
      position-independent
      code, 359
  Debugger doesn't like
      self-relocating code,
      356–359
  Debugger doesn't like
      writeable code, 355–356
  No firmware at boot time,
      359–360
  Persistent watchdog time-out,
      360
  Unexpected power
      interruption, 360–361
Pseudocode for handling an
    internal WDT, 390(f)
RAM tests, 367–372
ROM tests, 365–367
Summary and other thoughts,
    395–397
Supervisory circuits, 372–374
Testing, 378
Watchdog for a dual-processor
    system, 392(f)
WDTs, characteristics of great,
    386–389
WTDs for multitasking,
    393–395
The Zen of embedded systems
    development and
    troubleshooting

  Avoid debugging
      altogether—code smart,
      340–341
  Conclusion, 349–350
  The DEVELOPER hat,
      334–336
  The FINDER hat, 337–339
  The FIXER hat, 339
  Guideline # 1: use small
      functions, 340
  Guideline # 2: use pointer
      with great care, 340
  Guideline # 3: comment code
      well, 340
  Guideline # 4: avoid "magic
      numbers", 341
  Introduction, 333–334
  MMUs, 348–349
  Proactive debugging, 341–342
  Seeding memory, 344–345
  Special decoders, 347–348
  Stacks and heaps, 342–344
  Wandering code, 346–347
ESC (escape), 12
Ethernet driver, initializing; see
    Board I/O driver examples
Ethernet Interface, 144
Events, 452
Exceptions, 91
Executable and linking format
    (ELF), 215, 216(f)
Execution time, 236
Exponent, 8
Extended Memory Segment; see
    XPC Segment
External memory interface, 448
External WDTs, 384–385,
    391–392(f)

F
Fairness, 191
Fast Fourier transform (FFT), 569,
    587
FF (first fit) algorithm, 217
FFT (fast Fourier transform), 569,
    587

Field-programmable gate array
(FPGA), 534–535, 556,
559–560
FIFO (first-in-first-out), 228
File system management, 170
File Transfer Protocol (FTP), 276
FINDER hat, the, 337–339
FIR (finite impulse response)
filter, 461, 563, 566(f)
FireWire (IEEE 1394), 450
First-Come-First-Serve
(FCFS)/Run-to-Completion,
193(f)
First-in-first-out (FIFO), 228
Fixed-point binary numbers, 7–8
FIXER hat, the, 339, 340–341
Flash storage card interfaces, 449
FLOAT (floating point), 8, 20–21
Floating-point binary numbers, 8–9
Fork call, 179–180
Fork/exec model, 178, 179–180(f),
181
Format coders, 445
FOR/NEXT statement, 56–57
Fourier transform (FT), 587
FPGAs (field-programmable gate
arrays), 534–535, 556,
559–560
Framework, performance-based
Data movement (L1 to L3
memory), 521(f)
Data movement (L3 to L1
memory), 520(f)
High performance decoder
example, 518–521(f)
Image pipe and compression
example, 516–518
Introduction, 516
Performance-based framework,
516(f)
Tips, 522
Typical video decoder, 519(f)
Frameworks
Algorithm complexity, 528
Application with device drivers,
527(f)

Application with device drivers
and OS, 527(f)
Architecting, strategies for
Autobuffer scheme, 512
Ease of use, 513–515(f)
Edge detection, 514–515(f)
Framework that focuses on
ease of use, 513(f)
Introduction, 510–511
Performance-based,
516–521(f)
"Ping-pong" buffer, 512(f)
Processing data as it enters
the system, 511(f)
Processing data on the fly,
511–513(f)
Audio-video synchronization,
523–525(f)
Buffer basics, 524(f)
Conceptual diagram of
audio-video
synchronization, 525(f)
Defining
Application timeline,
498–499(f)
Block diagram of video
display system, 500(f)
Evaluating bandwidth,
500–505(f)
A line of video with
cache-line misses
overlayed onto it, 505(f)
Minimum timeline examples,
499(f)
Questions to consider,
497–498
Definition of, 496–497
Introduction, 495
Moving an application to an
embedded processor,
496(f)
MpeG-2 encoded transport
stream format, 523(f)
System flow, managing,
525–528(f)
System services, 525–526,
526(f)

Frequency, 198
FT (Fourier transform), 587
FTP (File Transfer Protocol),
276
Function calls, 683–684
Function inlining, 684–685

G

Garbage collector algorithms,
218–221(f)
gdb (GNU debugger), 427–428
General Purpose Inputs/Outputs
(GPIO), 447–448
General-purpose chip-select
machine (GPCM), 113
General-purpose processor (GPP),
556
Generations, 220–221
GNU debugger (gdb), 427–428
GOTO statement, 55, 57–58
GPCM (general-purpose
chip-select machine), 113
GPIO, 570
GPIO (General Purpose
Inputs/Outputs), 447–448,
570
GPP (general-purpose processor),
556
Graph coloring algorithm, 623n
Gray code, 669–670
GUI (graphical user interface),
241

H

Half-Duplex Carrier Sense
Multiple Access/Collision
Detect (CDMA/CD) protocol,
146
"Hard" breakpoints, 265
Hardware alternatives
Introduction, 362
Separating code and data,
362–363
Flexible and safe, 363–364
Hardware gates, 557
Hardware loop buffers, 458–459
Hardware loop constructs, 457–459

Hardware/software co-verification, 399–442
  Co-verification; *see* Co-verification
  Embedded system design process, 399(f)
    Hardware and software integration, 401
    Hardware design, 400–401
    Introduction, 399
    Microprocessor selection, 400
    Requirements, 400
    Software design, 401
    System architecture, 400
  Human interaction, 403–405(f)
    Management structure with combined engineering teams, 404(f)
    Management structure with separate engineering teams, 403(f)
  Validation, 402
  Verification, 401–402
Harmonic analysis, 587
Heaps and stacks, 342–344
Heap segments, 217–226(f)
Heuristics, 588n, 620n
Hexadecimal, 9–10(t)
High-level language (HLL), 455–456
HLL (high-level language), 455–456
Homogeneous programming model, 508–510
  Master-Slave, 509–510(f)
  Pipelined, 509(f), 510
Host interface, 448
Host port interface (HPI), 570
HPI (host port interface), 570
HTTP (Hypertext Transfer Protocol), 277

I

I (Interrupt) bit, 141
IACK (interrupt acknowledgement), 93, 94
I²C (Inter-IC Bus), 447

ICE (In-Circuit Emulator), 242, 716–717
ICF (Internet Connection Firewall), 315
ICMP (Internet Control Message Protocol), 276
IDE (integrated development environment), 242, 244, 721
IEEE 1394 (FireWire), 450
IETF (Internet Engineering Task Force), 292, 292n
IF statement (IF/THEN/ELSE), 48–53
ILP (instruction level parallelism), 562
IMMR (Internal Memory Map Register), 119(f)
In-Circuit Emulator (ICE), 242, 716–717
INLINE statement, 58
Inlining, 639
Input/Output subsystems—peripheral interfaces; *see* Peripheral interfaces
"Instance", 657
Instruction level parallelism (ILP), 562
Instruction set simulator (ISS), 408, 413, 415, 416(f), 421–423(f)
INT (short for integer), 17–19
Integral programmer, 356
Integrated development environment (IDE), 242, 244, 721
Integrated memory managers (MMUs), 86, 348–349, 486–488
Intel's Programmable Interrupt Controller (PICs), 92–93
Inter-IC Bus (I²C), 447
Interlocked instruction pipelines, 462–465(f)
Internal Memory Map Register (IMMR), 119

Internal Request Level (IRQ) pin or port, 91
Internal WDTs, 382–384, 389–390
Internet Connection Firewall (ICF), 315
Internet Control Message Protocol (ICMP), 276
Internet Engineering Task Force (IETF), 292, 292n
Internet Protocol (IP), 276
Inter-process communication (IPC), 204, 419, 421
Interrupt (I) bit, 141
Interrupt acknowledgement (IACK), 93, 94
Interrupt and error detection management, 170
Interrupt device driver pseudo code examples, 99–109
  CICR register, 100(f)
  CIMR register, 103–104(f)
  CIPR register, 101–102(f)
  Initializing CPM for interrupts—4 step process, 100–107
  Interrupt-handling disable on MPC860, 107
  Interrupt-handling enable on MPC860, 107
  Interrupt-handling servicing on MPC860, 108–109
  Interrupt-handling shutdown on MPC860, 107
  Interrupt-handling startup (initialization) MPC860, 100
  Introduction, 99
  SCC priorities, 101(f)
  SIEL register, 106–107(f)
  SIMASK register, 104–105(f)
Interrupt driven, 154, 159, 166
Interrupt handler; *see* Interrupt service routine (ISR)
Interrupt latency, 109–110(f), 671–674
Interrupt latency/response, 212

Interrupt priorities, 94–98
  Dynamic multilevel, 96
  Equal single level, 96
  Introduction, 94–95
  Mitsubishi M37267M8 8-bit TV
    microcontroller interrupt
    table, 98(f)
  Mitsubishi M37267M8 8-bit TV
    microcontroller interrupts,
    98(f)
  Motorola/freescale 68000 IRQs,
    97(f)
  Motorola/Freescale MPC860
    interrupt levels, 96(f)
  Motorola/Freescale MPC860
    interrupt pins and table,
    95(f)
  Motorola/freescale 68K IRQs
    interrupt table, 97(f)
  Static multilevel, 96
Interrupt recovery, 212
Interrupt service routine (ISR),
    78–79, 94, 99, 242, 252,
    474–475
Interrupt services, 99
Interrupt vector, 94
Interrupt vectored scheme, 94
Interrupt-handling, device drivers
    for (example), 89–110(f)
  Context switching, 98–99
  Interrupt device driver pseudo
    code examples; *see*
    Interrupt device driver
    pseudo code examples
  Interrupt latency, 109(f)
  Interrupt priorities
    Dynamic multilevel, 96
    Equal single level, 96
    Introduction, 94–95
    Mitsubishi M37267M8 8-bit
      TV microcontroller
      interrupt table, 98(f)
    Mitsubishi M37267M8 8-bit
      TV microcontroller
      interrupts, 98(f)
    Motorola/freescale 68000
      IRQs, 97(f)

Motorola/Freescale MPC860
    interrupt levels, 96(f)
Motorola/Freescale MPC860
    interrupt pins and table,
    95(f)
Motorola/freescale 68K IRQs
    interrupt table, 97(f)
Static multilevel, 96
Interrupt-handling and
    performance, 109–110(f)
Interrupts
  Definition of, 89–90
  Exceptions or traps, 91
  Interrupt acknowledgement
    (or IACK), 93, 94
  Interrupt vectored scheme, 94
  Interrupt-handling
    mechanisms, 92–93
  Raised by external events,
    91–92
  Software for handling, 90
  types of, 90–92
Interrupt-handling and
    performance, 109–110
Interrupt-Handling Disable, 90
Interrupt-Handling Enable, 90
Interrupt-Handling Servicing, 90
Interrupt-Handling Shutdown, 90
Interrupt-Handling Startup, 90
Interrupts
  Auto-vectored interrupt scheme,
    94
  Controllers, 92–93
  Definition of, 89–90
  Device drivers for handling, 90
  Edge-triggered, 91(f), 92
  Edge-triggered interrupts, 91,
    91(f), 92
  Edge-triggered interrupts
    drawbacks, 92(f)
  Exceptions or traps, 91
  External hardware interrupts, 90,
    91–92
  Internal hardware interrupts,
    90–91
  Interrupt acknowledgement (or
    IACK), 93, 94

Interrupt priorities; *see* Interrupt
    priorities
Interrupt vector, 94
Interrupt vectored scheme, 94
Interrupt-handling mechanisms,
    92–93
IRQ (Internal Request Level)
    pin or port, 91
Level-triggered, 91(f), 92
Level-triggered drawbacks,
    92(f)
Level-triggered interrupts, 91(f),
    92
Level-triggered interrupts
    drawbacks, 92(f)
Mitsubishi M37267M8 circuitry,
    93(f)
Motorola/Freescale MPC860
    interrupt controllers, 93(f)
Raised by external events,
    91–92
Software for handling, 90
Software interrupts,
    90, 91
Types of, 90–92
Intertask communication and
    synchronization,
    204–209(f)
  Example: interrupt handling in
    vxWorks, 212–213(t)
  Example: message passing in
    vxWorks, 210–211
  Interrupt routines in vxWorks,
    213(t)
  Memory sharing, 204(f)
  Message passing, 209–211(f)
  Message queues, 209(f)
  Mutual exclusion techniques,
    205–206
  OS interrupt subroutine, 212
  Processor assisted locks, 205
  Semaphores, 206–209
  Signals and interrupt handling
    (management) at the kernel
    level, 211–213(f)
  vxWorks processor assisted
    locks, 205(f)

Intertask communication and
synchronization (*Continued*)
vxWorks semaphores
Binary semaphores, 206
Counting semaphores,
208–209
Mutual exclusion semaphores,
207–220
vxWorks signaling mechanism,
211(f)
Intertask communications, 77
Intrinsics
C code to perform saturated add,
635(f)
Coding guidelines, 645
DSP, examples of, 625–627(f)
Introduction, 635
Some intrinsics for the
TMS320C55 DSP, 637(f)
TMS320C55 DSP assembly
code for the saturated add
routine, 636(f)
TMS320C55 DSP assembly
code for the saturated add
routine using a single call
to an intrinsic, 636(f),
637(f)
I/O and file system management,
230–232(t)
Buffers, 232
Device driver code, 232
File system algorithms, 230
File system management
mechanisms, 230
Functions of, 231–232
Middleware file system
standards, 231(t)
Primitives used for file
manipulation, 231
IP (Internet Protocol), 276
IP address, setting up
Dynamic addressing, 283, 283n
Static addressing, 283, 283n
Table used to set the
TCPCONFIG macro, 285(t)
IPC (interprocess communication),
204, 419, 421

IRQ (Internal Request Level) pin
or port, 91
Isochronous events 540-541
ISR (interrupt service routine),
78–79, 94, 99, 242, 252,
474–475
ISS (instruction set simulator),
408, 413, 415, 416(f),
421–423(f)

J

Jbed (Java), 184

K

Kb/s (kilobits/second), 446
Kernel, 170
Kernel memory space, 229–230
Kernel mode, 214
Kernel subsystem dependencies,
171(f)
Key technologies
Design composition, 707
Microprocessor technology, 706
Programming language, 707
Software content, 707
Software team size and
distribution, 707
System architecture, 707
Keywords
Common, 638
Definition of, 633
Kilobits/second (kb/s), 446

L

L (last) bit, 141
LANs (local area networks), 274
Last bit (L), 141
Last in, first out (LIFO) queue, 217
Latencies, 109
Layered design, 172, 173, 174(f)
LCALL assembly instructions, 252
Least recently used (LRU), 228
Least recently used (LRU)
algorithm, 475
Least significant bit (LSB), 4
Length/Type field, 145
LF (line feed), 12

LIFO (last in, first out) queue, 217
Link layer selection, 285–286
Linker, 685
Linux OS block diagram, 173(f)
Little endian, 18
Local area networks (LANs), 274
Locking/unlocking of interrupts, 99
LONG (long integer), 19–20
Loop unrolling, 604–610(f)
Execution units, filling the,
605–607(f)
Implementation of a loop
unrolled eight times,
610(f)
Implementation of a loop
unrolled four times,
609(f)
Implementation of a simple
loop, 608(f)
Introduction, 604
Loop overhead, reducing,
607–609
Loop unrolling, 608(f)
A more parallel implementation
of the C loop, 607(f)
Register space, fitting the loop
to, 609–610(f)
Serial assemble language
implementation of C loop,
606(f)
Simple for loop in C, 606(f)
Trade-offs, 610(f)
Loops, 54–57
CALL statement, 58
DO/WHILE statement, 55–56
FOR/NEXT statement, 55,
56–57
GOTO statement, 55,
57–58
INLINE statement, 58
REPEAT/NEXT statement, 56
RETURN statement, 58
STACK, 58
WHILE/DO statement, 55–56
Loops, coding guidelines,
643–644(f)

Low-level code compression, 685

Low-speed serial interfaces
Host interface, 448
I²C (Inter-IC Bus), 447
Programmable Flags/GPIO (General Purpose Inputs/Outputs), 447–448
Programmable timers, 447
Real-Time Clock (RTC), 447
SPI (Serial Peripheral Interface), 447
UART (Universal Asynchronous Receiver/Transmitter), 446–447
Watchdog timer (WDT), 448

LRET assembly instructions, 252
LRU (least recently used), 228
LRU (least recently used) algorithm, 475
LSB (least significant bit), 4

M

MAC (media access control) addresses, 145
MAC (multiple and accumulate), 585
MAC sublayer, 146
Machine independent compilers, 621–624(f)
MANs (metropolitan area networks), 274
Mantissa, 8
Masked, 95
Mathematics
Binary addition, 37–39(t)
Binary division, 42–46
Binary multiplication, 40–42
Binary subtraction, 39–40(t)
Introduction, 37
Maximum receiver buffer length register (MRBLR), 139
Mbps (megabits per second), 446
Mbytes/s (megabytes per second), 446
mC (microcontroller), 556, 558, 559

McBSP (multichannel buffered serial port), 570
MCR (memory command register), 117
MDR (memory data register), 117
Media access control (MAC) addresses, 145
Media access management, 144, 146–147
Media processing frameworks; *see* Frameworks
Megabits per second (Mbps), 446
Megabytes per second (Mbytes/s), 446
MEM, 343–344
Memory alias, 355–356, 358
Memory architecture–need for management
Data memory management
Benchmarks (relative cycles per frame) for G.729a algorithm with cache enabled, 486(f)
Benchmarks (relative cycles per frame) for GSM aMr algorithm with cache enabled, 486(f)
Checklist for choosing between data cache and DMA, 485(f)
Checklist for choosing between instruction cache and DMA, 484(f)
Data cache and DMA coherency, 480(f)
Data cache, what about, 480–481
Instruction cache, data DMA, 481–482
Instruction cache, data DMA/cache, 482
Instruction DMA, data DMA, 483–486(f)
Introduction, 479
Instruction memory management—to cache or DMA, 477–479

Memory access trade-offs, 476–477
Typical memory configuration, 477(f)
Memory bandwidth, 503–504
Memory banks, 112
Memory command register (MCR), 117
Memory data register (MDR), 117
Memory device drivers: example, 110–134(f)
Big endian, 111, 112
Byte ordering schemes, 111–112
Endianess, 112(f)
Introduction, 110–112
Little endian, 111, 112
Memory banks, 112
Memory subsystem startup (initialization) on MPC860: pseudo code example; *see* Memory subsystem startup (initialization) on MPC860
Memory subsystem writing/erasing flash, 132–134
Memory system disable on MPC860, 131–132
Memory system enable on MPC860, 132
Memory diagnostics, 364–365
Memory disambiguation, 618
Memory management
Example: Jbed memory management and segmentation, 224–225
Example: Linux memory management and segmentation, 225–226
Example: vxWorks memory management and segmentation, 222–223
Introduction, 213–214
Kernel memory space, 229–230

Memory management (*Continued*)
  User memory space
    Class executable file format, 216(f)
    ELF executable file format, 216(f)
    Heap segments, 217–226(f)
    Introduction, 214
    Paging and virtual memory, 227–228(f)
    Segmentation, 215–226(f)
    Stack segments, 217
    Virtual memory, 228–229(f)
Memory map, 110–111(f), 214
Memory model, 686–687
Memory periodic timer prescaler register (MPTPR), 117
Memory subsystem startup (initialization) on MPC860, 113–131(f)
  Base and option registers, 116(f)
  IMMR (Internal Memory Map Register), 119(f)
  Initializing the MMU, 120–131(f)
    16 kB effective address format, 123(f)
    2-level translation table for 4 Kb page scheme, 124(f)
    4 kB effective address format, 122–123(f)
    512 kB effective address format, 123(f)
    8 MB effective address format, 123(f)
    L1 descriptor, 125(f)
    L1/L2 configuration, 129(f)
    L2 descriptor, 125(f)
    Level 1 and 2 entries, 122(f)
    MD_CR, 124(f)
    MI_CTR, 124(f)
    Mx-EPN, 126(f)
    Mx-RPN, 126(f)
    Mx-TWC, 126(f)
    Physical memory map, 129(f)

TLB, 121(f)
TLB within VM scheme, 120(f)
Initializing the internal memory map, 119–120
Initializing the memory controller and connected ROM/RAM, 113–119
Introduction, 113
Memory controller pins, 115(f)
MPC860 integrated memory controller, 114(f)
PowerPC connected to DRAM, 116(f)
PowerPC connected to SRAM, 115(f)
Sample memory map, 113(f)
Message passing, 209–211(f)
Metastable states, 666–668
  Metastable state, 667(f)
  Setup and hold times, 667(f)
Metropolitan area networks (MANs), 274
Microcontroller (mC), 556, 558, 559
Microkernel (client-server) design, 172, 173–175(f)
Microprocessors (mP), 558
Microprogrammer, 350–355
  Advantages of, 351
  Basic, 354–355
  Definition of, 350–351
  Disadvantages of, 351–352
  Microprogrammer-based firmware update process, 351
  Receiving, 352–353
MIPS (millions of instructions per second), 456
MMUs (integrated memory managers), 86, 348–349, 486–488
Model compilers, definition of, 739
Model-based systems development with xtUML; *see* xtUML model system development

Modules, 172
Monolithic design, 172–173, 173(f)
Monolithic OS, 173(f)
Monolithic-modularized algorithm, 172
Most significant bit (MSB), 4
mP (microprocessors), 558
MPTPR (memory periodic timer prescaler register), 117
MRBLR (maximum receiver buffer length register), 139
MSB (most significant bit), 4
Multibyte writes, 374–377
Multichannel buffered serial port (McBSP), 570
Multimedia processing, 445–446
Multiple and accumulate (MAC), 585
Multitasking
  Automotive multitasking, 75(f)
  Basic requirements of, four, 76–78
  Communications, 77
  Context switching, 76–77
  Introduction, 74–76
  Managing priorities, 77
  Operating systems
    Cooperative, 80–82
    Preemptive, 78–80
  State machine, 82–84(f)
  Timing control, 77–78
Multitasking and process management, 177–213
  Interleaving tasks, 178(f)
  Intertask communication and synchronization
    Example: interrupt handling in vxWorks, 212–213(t)
    Example: message passing in vxWorks, 210–211
    Example: vxWorks semaphores, 206–210
    Memory sharing, 204(f)
    Mutual exclusion techniques, 205–206

Processor assisted locks, 205
Semaphores, 206
vxWorks processor assisted
    locks, 205(f)
Introduction, 177–178
Process implementation
    Embedded Linux and
        fork/exec task deleted,
        186(f)
    Embedded Linux and states,
        190–191(f)
    Example: creating a task in
        vxWorks, 181–182
    Example: embedded Linux
        and fork/exec,
        184–187(f)
    Example: embedded Linux
        and states, 190–191(f)
    Example: Jbed kernel and
        states, 189–190
    Example: Jbed RTOS and
        task creation, 182–184
    Example: vxWorks Wind
        kernel and states,
        189–190(f)
    Fork/exec model, 178,
        179–180(f), 181
    FORK/EXEC process
        creation, 179(f)
    Spawn model, 178–179,
        180–181(f)
    Spawn process creation,
        180(f)
    State diagram for Jbed
        interrupt tasks, 189(f)
    State diagram for Jbed joined
        tasks, 190(f)
    State diagram for Linux tasks,
        191(f)
    State diagram for oneshot
        tasks, 190(f)
    State diagram for periodic
        tasks, 190(f)
    State diagram for vxWorks
        tasks, 188(f)
    Task hierarchy, 178(f)
    Task state diagram, 187(f)

Task states, 186–187(f)
Task states and queues, 187(f)
vxWorks and spawn tasks
    deleted, 186(f)
vxWorks tasks and queues,
    188(f)
Process scheduling
    Cooperative scheduling,
        194(f)
    Dispatcher, 191
    Example: Jbed and EDF
        scheduling, 201
    Example: TimeSys embedded
        Linux priority-based
        scheduling, 201–204(f)
    Example: vxWorks
        scheduling, 200(f)
    First-Come-First-Serve
        scheduling, 193(f)
    Nonpreemptive scheduling,
        192–194(f)
    Preemptive priority
        scheduling augmented
        with Round-Robin
        scheduling, 200(f)
    Preemptive scheduling,
        194–200(f)
    Priority (preemptive)
        scheduling, 196(f)
    Round-Robin/FIFO
        scheduling, 195(f)
    Scheduler, 191
    Scheduling algorithms,
        191–192
    Shortest process next
        scheduling, 194(f)
    Shortest Process Next
        (SPN)/Run-to-
        completion,
        193–194(f)
    Task structure, 192(f)
Multitasking OSes, 175–176, 177
Mutual exclusion (Mutex for
    short), 205–206
Mutual exclusion semaphores,
    207–220

N
NAK bit, 141
Nanokernels, 174
Networking, 241–233(f)
    Berkeley Sockets interface,
        294–298(f)
    Client server paradigm,
        293–294
    Code, compiled and run
        With Dynamic C, 258–259(f)
        In traditional development
            environments,
            256–257(f)
    Core module's network
        configuration, setting up
        Debugging macros for
            networking, 287–288
        IP address, setting up,
            282–285(t)
        Link layer selection, 285–286
        TCP/IP definitions at compile
            time, 286–287
        TCP/IP definitions at runtime,
            287
    Corporate network, typical,
        279–281(f)
    Dynamic C development
        system, introduction to
        Debugging, 245–246
        Development, 244–245
    Dynamic C libraries, brief
        introduction to, 246
    Dynamic C library functions for
        socket programming,
        important
        Blocking versus nonblocking
            functions, 302
        Connection initialization or
            termination, functions
            used for, 301
        Determine socket status,
            functions used to, 301
        Introduction, 300
        Send or receive data,
            functions used to,
            302

Networking (*Continued*)

Dynamic C, memory spaces in
  Dynamic C's memory usage
    without separate I & D
    space, 248–251(f)
  Quick summary, 256
  Rabbits memory segments,
    247–248(f)
  Separate instruction and data
    memory, 252–255
  XMEN, placing functions in,
    252
Dynamic C support for
  networking protocols
  Introduction, 275(f)
  Networking protocols,
    common, 276–278(f)
  Optional modules for
    Dynamic C, 278–279
Embedded networks, 274–275
Final thought, 331
Home network, typical,
  281–282(f)
Introduction, 241–243
Networked environment for the
  home, 282(f)
Networking utilities, useful (and
  free!)
  Ethereal, 329–330
  Netcat, 330–331
  Online tools, 331
  Ping, 328–329(f)
  Traceroute, 329
Project 1: bringing up a Rabbit
  core module for networking
  Configuration for dynamic
    addressing, 291–292
  Configuration for static
    addressing, 288–291
  Dynamic addressing, special
    case for, 292
Project 2: implementing a
  Rabbit TCP/IP server
  C++ TCP/IP client, working
    with, 308–310(f)

General-Purpose TCP
  utilities, working with,
  306–307(f)
Introduction, 303
Java TCP/IP client, working
  with, 308
Output of the TCP client,
  310(f)
Rabbit server state machine,
  304(f)
Server TCP/IP state machine,
  303–306(f)
Project 3: implementing a
  Rabbit TCP/IP client
Adding a setting for TCP port
  8000 in Windows XP
  Professional, 315(f)
Client code, verifying,
  317–318(f)
Introduction, 311–315(f)
Java TCP/IP server, working
  with, 318–320(f)
Output from Java server,
  320(f)
Port 8000 opened up for TCP
  and UDP, 316(f)
Rabbit TCP client state
  machine, 312(f)
TCP/IP server in C#, working
  with, 320–322
Two-way communication
  between Rabbit client
  and server, 311(f)
Windows XP firewall,
  disabling, 315–316(f)
Project 4: implementing a
  Rabbit UDP server
C++ UDP client, working
  with, 327
Introduction, 322–325(f)
Java UDP client, working
  with, 325–326
Rabbit-based UDP server,
  323(f)
Rabbit core module in a
  corporate network, 280(f)

RCM3200 development system,
  setting up a PC as, 259
RCM3200 Rabbit Core,
  243–244
TCP versus UDP in an
  embedded application,
  using, 298–300
Writing code, time to start
  Breakpoint, adding, 264–265
  Dynamic C and ANSI C,
    differences between,
    266–270
  Dynamic C Help, 263
  Dynamic C is not ANSI C,
    266–270
  Dynamic C memory spaces,
    270–273
  Dynamic C's debugging
    features, 262
  Rabbit program, everyone's
    first, 260–262
  Introduction, 259–260
  Single stepping, 263–264
  Watch expressions, 265–266
Networking utilities, useful (and
  free!)
  Ethereal, 329–330
  Netcat, 330–331
  Ping, 328–329(f)
  Pinging a URL, 329(f)
  Pinging with Windows XP,
    328(f)
  Traceroute, 329, 330(f)
  Tracing the route to a
    destination, 330
NF (next fit) algorithm, 217
NMI (nonmaskable interrupt), 90,
  95, 374, 382, 386, 387, 452
Nonpreemptive scheduling,
  192–194(f)
  Cooperative, 194(f)
  First-Come-First-Serve (FCFS)/
    Run-to-Completion, 193(f)
  Shortest process next (SPN)/
    Run-to-completion,
    193–194

Nonreentrant code, eliminating, 659–661
Nonvolatile memory, 372
"Normal" breakpoints, 265
Numbering systems
  Base-ten number system, 2–4
  Binary numbers (base two), 4
  Scientific notation, 3–4
Numeric comparison
  Conditional statements, 47–54
    CASE statement
      (SWITCH/CASE), 53–54
    IF statement
      (IF/THEN/ELSE), 48–53
  Introduction, 46–47
  Loops, 54–57
    CALL statement, 58
    DO/WHILE statement, 55–56
    FOR/NEXT statement, 55, 56–57
    GOTO statement, 55, 57–58
    INLINE statement, 58
    REPEAT/NEXT statement, 56
    RETURN statement, 58
    STACK, 58
    WHILE/DO statement, 55–56
  Subtraction-based comparisons, 47(t)
  WINDOW COMPARISON, 50

O

OCD (on-chip debug), 718
Octal, 9–10(t)
Offset, 111
Onboard bus device drivers: example, 134–142(f)
  Bus protocol, 134–135
  I^2C bus startup (initialization) on the MPC860, 135–142(f)
    Collision (CL) bit, 141
    Continuous mode (CM) bit, 141
    Empty (E) bit, 141
    I2ADD, 138(f)
    I2BRG, 138(f)

I2C controller, 136(f)
I_2C parameter RAM, 139–140(f)
I2CER, 138(f)
I2CMR, 138(f)
I2MOD, 137(f)
    Interrupt (I) bit, 141
    Last bit (L), 141
    MPC860 port B pins, 137(f)
    MPC860 port B register, 137(f)
    NAK bit, 141
    Overrun (OV) bit, 141
    SDA and SCL pins, 136(f)
    Underrun condition (UN) bit, 141
    Wrap (W) bit, 140–141
  Introduction, 134–135
On-chip debug (OCD), 718
Online tools, 331
Option register (OR), 114
OR (option register), 114
OS performance guidelines, 235–236
OS standards example: POSIX (portable operating system interface), 232–235
  Example: Linux POSIX example, 234
  Example: vxWorks POSIX example, 235
  The Open Group Base Specifications Issue 6, 232
  POSIX functionality, 233–234(t)
OSes and board support packages (BSPs), 237–239
  Advantages and disadvantages of real-time kernels, 238
  BSP within embedded systems model, 237(f)
  Definition of, 237
  Functions of, 237–238
OV (overrun) bit, 141
Overhead, 191
Overlays, 446
Overrun (OV) bit, 141

P

Packet elementary streams (PES), 523
Pad field, 145
Page fault, 227–228
Pages, 214
Paging and virtual memory, 227–228(f)
Parallel Video/Data Port, 450
Parity bit, 12
Parser, 678
PC (program counter), 99
PCB (Process Control Block), 179, 229–230
PCI (Peripheral Component Interconnect), 449–450
Pend, 601
Peripheral interfaces
  Classes of peripheral interfaces and representative examples, 446(t)
  Connectivity
    IEEE 1394 (FireWire), 450
    Network interface, 450
    PCI (Peripheral Component Interconnect), 499
    Universal Serial Bus (USB) 2.0, 499
  Data movement
    Parallel Video/Data Port, 450
    Synchronous Serial Audio/Data Port (SPORT), 450
  Memory subsystem, 450–451
  Subsystem control—low-speed serial interfaces
    ATAPI/Serial ATA, 448
    Connectivity, 449–450
    External memory interface, 448
    Flash storage card interfaces, 449
    Host interface, 448
    Programmable flags/GPIO (General Purpose Inputs/Outputs), 447–448
    Programmable timers, 447

Peripheral interfaces (*Continued*)
Real-time clock (RTC), 447
Storage, 448–449
UART (Universal Asynchronous Receiver/Transmitter), 446–447
Watchdog timers (WDT), 448
PES (packet elementary streams), 523
PICs (Intel's Programmable Interrupt Controller), 92–93
"Ping-pong" buffer, 512(f)
Pipes, 232
Poc-It, 378
Pointers, 21–22
Point-to-Point Protocol (PPP), 278, 285
POSIX (portable operating system interface), 232–235(f)
PPP (Point-to-Point Protocol), 278, 285
PPP Over Ethernet (PPPOE), 285
PPPOE (PPP Over Ethernet), 285
Pragmas
Definition of, 633
Example of using pragmas to improve the efficiency of the code, 634(f)
Introduction, 634
Preamble bytes, 145
Preemptive operating system, 78–80
Preemptive scheduling, 194–200(f)
EDF (Earliest Deadline First)/clock driven scheduling, 198, 199(f)
OSes and deadlines, 199(f)
Preemptive scheduling and Real-Time Operating System (RTOS), 199–200
Priority (preemptive) scheduling, 196(f)
Introduction, 196
Priorities, determining, 197
Priority inversion, 196
Process starvation, 196

Round Robin/FIFO (First In, First Out) scheduling, 194–196(f)
Round-Robin/FIFO scheduling, 195(f)
Priori, 541n
Priority manager, 77
Priority-based interrupt levels, 94–98
Proactive debugging, 341–342
Process Control Block (PCB), 179, 229–230
Process implementation; *see* Multitasking and process management
Process scheduling; *see* Multitasking and process management
Processor bandwidth, 502
Profile-based compilation
Achieving full entitlement, 652(f)
Advantages, 646–647(f)
Code optimization process, summary of, 648–653(f)
Debugging optimized code, issues with, 647–648
Expanded code optimization process for DSP, 649(f)
Introduction, 646
Profile-based compilation shows the various trade-offs between code size and performance, 647(f)
Summary, 653
Program counter (PC), 99
Programmable timers, 447
Programming methodology, 455–456
Programming tips
Addresses, do not take, 691
Bit fields, investigate before using, 693–694
Clever code, do not write, 692–693
Extra hints, use, 694–695
Function prototypes, use, 690

Inline assembly language, do not use, 691–692
Jump tables, use a switch for, 693
Library functions, 694
Parameters, use, 690
Pointer types, use the best, 688
Right size data, use, 687
Structures and padding, 688–689
Variables, 691n
Protocol, definition of, 29, 293n
"Pure"; *see* reentrancy
Pure (without side effects), 695

Q

QF (quick fit) algorithm, 218

R

Race conditions, 204, 662–663
Radio frequency interference (RFI), 13
RAM (random access (read/write) memory), 710
RAM tests, 367–372
Random access (read/write) memory (RAM), 710
RARP (Reverse ARP), 276
Rate monotonic analysis (RMA), 540n. 1
Rate Monotonic Scheduling (RMS), 197
Rbase (receiver buffer descriptor array), 139
RCM3200 development system, setting up a PC as an, 259
RCM3200 Rabbit Core, Introduction to, 243–244
RCP (Rich Client Platform), 723
RDD (receive data decapsulation), 147
Read-only memory (ROM), 710
READY queues, 187, 188
Real Page Number (RPN), 125
Real-Time Clock (RTC), 447
Real-Time Operating System (RTOS), 238, 417, 639

Real-time system design,
    challenges in
Distributed and multiprocessor
    architectures, 544–545
Embedded systems
    Airbag system: possible
        sensors (including crash
        severity and occupant
        detection), 549(f), 550(f)
    Analog information of various
        types is processed by
        embedded system,
        547(f), 551(f)
    Analog I/O, 546
    Application-specific gates,
        547
    Are reactive systems,
        548–553(f)
    Automotive seat occupancy
        detection, 550(f)
    Block diagram of a cell
        phone, 552
    Characteristics of, 548–549
    Emulation and diagnostics,
        547
    Introduction, 545–546
    Memory, 547
    A model of sensors and
        actuators in embedded
        systems, 549(f)
    Processor core, 546
    Sensors and actuators, 546
    Software, 547
    Summary, 553
    Typical embedded system
        components, 546(f)
Failures, recovering from, 544
Introduction, 542
Response time, 543–544
Receive data decapsulation (RDD),
    147
Receive function code register
    (RFCR), 139
Receive media access management
    (RMAM), 146–147
Receiver buffer descriptor array
    (Rbase), 139

Recursive function, 661
Reentrancy, 655–656
Reentrant functions, 656
Register allocation, 682–683
REPEAT/NEXT statement, 56
Response time, 191, 236
Restrict keyword, 616,
    618, 628
Resume a task, 186
RETURN statement, 58
Reverse ARP (RARP), 276
RFCR (receive function code
    register), 139
RFI (radio frequency interference),
    13
Rich Client Platform (RCP), 723
RIF (Routing Information Field),
    146
RMA (rate monotonic analysis),
    540n. 1
RMAM (receive media access
    management), 146–147
RMS (Rate Monotonic
    Scheduling), 197
ROM (read-only memory),
    710
ROM tests, 365–367
ROMable code, 710
Root Code, 249
Root Constants, 249
Root Data, 249–250
Root Memory, 248(f), 249, 251
Root Segment (Base Segment),
    249, 254
Round trip time (RTT), 299
Routing Information Field (RIF),
    146
RPN (Real Page Number), 125
RS-232 driver, initializing; *see*
    Board I/O driver examples
RS-232 Interface, 162–163
RTC (Real-Time Clock), 447
RTOS (Real-Time Operating
    System), 238,
    417, 639
RTT (Round trip time), 299

S
Saturated add, definition of, 638n
Scheduler, 191, 193(f)
Scientific notation, 3–4
SCL (serial clock line), 135
SDA (serial data line), 135
SDRAM (Synchronous DRAM),
    563
Secure Socket Layer/Secure HTTP
    module (SSL/HTTPS),
    278–279
Security system management, 170
Seeding memory, 344–345
Segment number, 111
Segmentation, 215–226(f)
    BF (best fit) algorithm, 218
    bss segment, 215
    Buddy system, 218
    Class (Java Byte Code),
        215–216(f)
    Class executable file format,
        216(f)
    COFF (common object file
        format), 217(f)
    Copying garbage collector
        diagram, 218(f)
    Data segment, 215
    ELF (executable and linking
        format), 215, 216(f)
    FF (first fit) algorithm, 217
    Garbage collector algorithms,
        218–221(f)
    Generational garbage collector
        diagram, 221(f)
    Heap segments, 217–226(f)
    Mark and sweep and mark and
        compact garbage collector
        diagram, 220(f)
    NF (next fit) algorithm, 217
    QF (quick fit) algorithm, 218
    Segment addresses, 215
    Segment number, 215
    Segment offset, 215
    Segments, 215
    Stack segments, 217
    Static segments, 215
    WF(worst fit) algorithm, 218

Segments, 111, 214, 249
Self-relocate to RAM, 356
Semaphore protocol, 33
Serial clock line (SCL), 135
Serial Communication Controller
 in Ethernet mode (MPC823
 User's Manual), 151–154
Serial communication controllers
 (SCCs), 151
Serial data line (SDA), 135
Serial management controllers
 (2000 MPC823 user's
 manual), 164–166
Serial Peripheral Interface (SPI),
 447, 570
Shortest Process Next
 (SPN)/Run-to-completion,
 193–194(f)
SIC (system interrupt controller),
 452
Sign and magnitude, 5–6
SIGNED, 17
Signed binary numbers, 5–13(t)
 Alternate numbering systems,
  9–10
 ASCII (American Standard
  Code for Information
  Interchange), 11–12(T)
 Binary-coded decimal (BCD),
  10
 Error detection, 12–13
 Exponent, 8
 Fixed-point binary numbers, 7–8
 Floating-point binary numbers,
  8–9
 Mantissa, 8
 Sign and magnitude, 5–6
 Two's complement, 5, 6–7
Silicon software, 238
SIMD (single instruction multiple
 data) capability, 563
Simple data types
 Boolean or BIT (binary digit),
  14–16
 CHAR data type, 16–17
 DOUBLE (double precision
  floating-point), 20–21

FLOAT (floating point), 20–21
 INT (integer), 17–19
 LONG (long integer), 19–20
 Pointers, 21–22
 STRUCTURE, 15, 16
Simple Mail Transfer Protocol
 (SMTP), 277
Single instruction multiple data
 (SIMD) capability, 563
SIU (System Interface Unit), 93
SMTP (Simple Mail Transfer
 Protocol), 277
SoC (system-on-a-chip), 406, 697
Software pipelining, 610–612(f),
 618–619(f)
 Enabling, 618–619(f)
 Breaking up larger loops into
  smaller loops for
  efficiency, 619(f)
 Interrupts may be disabled
  during a software
  pipelined section of
  code, 620(f)
 Example; simple loop
  implemented in C
 Assembly language output of
  the simple loop, 613(f)
 Assembly language output of
  the simple loop
  exploiting the parallel
  orthogonal execution
  units of the DSP, 614(f)
 C example and the
  corresponding pipelined
  assembly language
  output, 615(f)
 Compiler-generated pipeline,
  614–616(f)
 Corresponding pipelined
  assembly language
  output, 617(f)
 Minimally parallel
  implementation, 614
 Restrict keyword,
  implementation with,
  616–618(f)

Serial implementation,
 612–613(f)
 Simple loop in C, 613(f)
 A five stage pipe that is
  software pipelined, 611(f)
 Interrupts and pipelined code,
  619
 Standard loop overhead versus
  loop unrolling and software
  pipelining, 612(f)
Software program, definition of,
 739
SOP (sum of products), 566, 585
Spawn model, 178–179, 181–184
Spawn threading, 182
Special decoders, 347–348
Special purpose registers (SPRs),
 119
Spectral analysis, 587
SPI (Serial Peripheral Interface),
 447, 570
SPN (Shortest Process
 Next/Run-to-completion),
 193–194(f)
SPORT (Synchronous Serial
 Audio/Data Port), 450
SPRs (special purpose registers),
 119
SRAM (static RAM), 563
SSL/HTTPS (Secure Socket
 Layer/Secure HTTP) module,
 278–279
STACK, 58
Stack Segment, 217, 249
Stacks and heaps, 342–344
Standard Widget Toolkit (SWT),
 723
State machine multitasking,
 82–84(f)
State machines
 Advantages of, 59–60
 Data-indexed, 64–66
 Definition of, 59
 Execution-indexed, 63, 67–72
 Hybrid, 64, 72–74
 Introduction, 59–64
State variable, 59

Static addressing, 283, 283n
Static RAM (SRAM), 563
STRING, 16
STRUCTURE, 22–24
Sum of products (SOP), 566, 585
Super loop, 525
Supervisory circuits, 372–374
Suspend a task, 186
Swapping, 228
SWITCH/CASE statement, 59
SWT (Standard Widget Toolkit), 723
Symmetric and asymmetric dual-core processors, 505–506
Synchronous DRAM (SDRAM), 563
Synchronous events, 540
Synchronous Serial Audio/Data Port (SPORT), 450
System and core synchronization
  Introduction, 472
    Atomic operations, 475–476
    Load/store synchronization, 472–473
    Ordering, 473–475
System calls, 214
System Interface Unit (SIU), 93
System interrupt controller (SIC), 452
System level interrupts, 452–455(f)
System-on-a-chip (SoC), 406, 697

T

Tablewalk, 121
Tablewalk Control (TWC) register, 125
Tag Control Information, 146
Taking data, 674–677
Task Control Block (TCB), 179, 229–230
Task states, 186–187(f)
Task throughput, 192
Tbase (transmit buffer descriptor array), 139
TCB (Task Control Block), 179, 229–230
TCB (transfer control block), 595

TCP (Transmission Control Protocol), 276, 278
TDE (transmit data encapsulation) component, 146
TFCR (transmit function code register), 139
TFTP (Trivial File Transfer Protocol), 277
The Open Group Base Specifications Issue 6, 232
Threads (lightweight processes), 176
Throughput, 236
TimeSys embedded Linux priority-based scheduling, 201–204(f)
  Architecture independent scheduler module, 202
  Architecture specific scheduler module, 202
  Embedded Linux block diagram, 202(f)
  Scheduling policy module, 201
  System call interface module, 201
  Task structure, 203(f)
TLBs (translation look-aside buffers), 487
TMAM (transmit media access management), 146
Toggle Breakpoint option, 264
Transcoders, 445
Transfer control block (TCB), 595
Translation look-aside buffers (TLBs), 487
Transmission Control Protocol (TCP), 276, 278
Transmit buffer descriptor array (Tbase), 139
Transmit data encapsulation (TDE) component, 146
Transmit function code register (TFCR), 139
Transmit media access management (TMAM), 146
Traps, 91

Trivial File Transfer Protocol (TFTP), 277
Turnaround time, 191
TWC (Tablewalk Control) register, 125
Two's complement, 5, 6–7
Type auto, variables of, 251

U

UART (Universal Asynchronous Receiver/Transmitter), 446–447, 570
UDP (User Datagram Protocol), 276, 277n, 277–278
UML (Unified Modeling Language), 725–739(f)
  Application from architecture, separating
  Blueprint development, 729–730
  Blueprint development process, 731(f)
  Model compilers, 730–732
  Open translation, 734
  Rules, 733–734
  Separation of application and architecture, 733(f)
  Sets, states, and functions, 732
  Conclusions, 738
  Evolution of software development, 727(f)
  Microwave oven class diagram, 737(f)
  Model, meanings for, 725–726
  Model- versus code-based design, 728(f)
  Reasons for modeling in, 725–729
  Terminology, note on, 739
  xtUML code generation, 734–738(f)
  xtUML models, 729
UN (underrun condition) bit, 141
Unified Modeling Language (UML); *see* UML (Unified Modeling Language)

UNION, 16, 25–26
Unitasking OSes, 175, 177
Universal Asynchronous
    Receiver/Transmitter
    (UART), 446–447, 570
Universal Serial Bus (USB), 449,
    570
UNLOCK function, 60
UNSIGNED, 17
UPF (user priority field), 146
UPMs (user-programmable
    machines), 113
USB (Universal Serial Bus), 449,
    570
User Datagram Protocol (UDP),
    276, 277n , 277–278,
User memory space
    Class executable file format,
        216(f)
    ELF executable file format,
        216(f)
    Heap segments, 217–226(f)
    Introduction, 214
    Paging, 227(f)
    Paging and virtual memory,
        227–228(f)
    Segmentation, 215–226(f)
    Stack segments, 217
    Virtual memory, 228–229(f)
User mode, 214
User priority field (UPF), 146
User-programmable machines
    (UPMs), 113

V

Very long instruction word
    (VLIW), 563
VID (VLAN identifier), 146
Virtual addresses, 229

Virtual local-area network
    (VLAN), 146
Virtual memory, 214
VLAN (virtual local-area
    network), 146, 279–281
VLAN identifier (VID), 146
VLIW (very long instruction
    word), 563
Void exit (int status) system call,
    185
von Neumann model, 85–86
vxWorks, creating a task in,
    181–182
vxWorks tasks and queues, 188(f)

W

W (Wrap) bit, 140–141
Wait time, 236
Wandering code, 346–347
WANs (wide area networks), 274
WDTs (Watchdog timers)
    Building a great watchdog,
        379–382
    Characteristics of great WDTs,
        386–389
    Definition of, 448
    For multitasking, 393–395
WF(worst fit) algorithm, 218
WHILE/DO statement, 55–56
Wide area networks (WANs), 274
WINDOW COMPARISON, 50
Windows XP firewall, disabling,
    315–316(f)
Winsock library, 309
Wrap (W) bit, 140–141
Writing code, time to start
    Breakpoint, adding, 264–265
    Dynamic C and ANSI C,
        differences between,
        266–270

Dynamic C Help, 263
Dynamic C is not ANSI C,
    266–270
Dynamic C memory spaces,
    270–273
Dynamic C shortcut keys, 266
Dynamic C's debugging
    features, 262
Rabbit program, everyone's
    first, 260–262
Introduction, 259–260
Single stepping, 263–264
Watch expressions, 265–266

X

XPC Segment, 251
xtUML model system development
    A better solution
        Experience to date, 743
        Interface problems, 742
        Introduction, 742
        Partition, easily changed, 743
    Building, 732
    Difficulties in building, 739–740
        Fundamental challenges,
            740–741
        Integration, 740–741
        Interface problems, cost of,
            741
        State of the practice, the, 741
        Suboptimal partition, the, 741
    Features, 729

Z

Zero Configuration Networking,
    292, 292n
Zero-overhead hardware loops,
    458–459